Hydrogen-Transfer Reactions

Edited by
James T. Hynes,
Judith P. Klinman,
Hans-Heinrich Limbach,
Richard L. Schowen

1807–2007 Knowledge for Generations

Each generation has its unique needs and aspirations. When Charles Wiley first opened his small printing shop in lower Manhattan in 1807, it was a generation of boundless potential searching for an identity. And we were there, helping to define a new American literary tradition. Over half a century later, in the midst of the Second Industrial Revolution, it was a generation focused on building the future. Once again, we were there, supplying the critical scientific, technical, and engineering knowledge that helped frame the world. Throughout the 20th Century, and into the new millennium, nations began to reach out beyond their own borders and a new international community was born. Wiley was there, expanding its operations around the world to enable a global exchange of ideas, opinions, and know-how.

For 200 years, Wiley has been an integral part of each generation's journey, enabling the flow of information and understanding necessary to meet their needs and fulfill their aspirations. Today, bold new technologies are changing the way we live and learn. Wiley will be there, providing you the must-have knowledge you need to imagine new worlds, new possibilities, and new opportunities.

Generations come and go, but you can always count on Wiley to provide you the knowledge you need, when and where you need it!

William J. Pesce
President and Chief Executive Officer

Peter Booth Wiley
Chairman of the Board

Hydrogen-Transfer Reactions

Volume 4

Edited by
James T. Hynes, Judith P. Klinman,
Hans-Heinrich Limbach, Richard L. Schowen

WILEY-VCH Verlag GmbH & Co. KGaA

The Editors

Prof. James T. Hynes
Department of Chemistry and Biochemistry
University of Colorado
Boulder, CO 80309-0215
USA

Prof. Judith P. Klinman
Departments of Chemistry and
Molecular and Cell Biology
University of California
Berkeley, CA 94720-1460
USA

Département de Chimie
Ecole Normale Supérieure
24 rue Lhomond
75231 Paris
France

Prof. Hans-Heinrich Limbach
Institut für Chemie und Biochemie
Freie Universität Berlin
Takustrasse 3
14195 Berlin
Germany

Prof. Richard L. Schowen
Departments of Chemistry, Molecular
Biosciences, and Pharmaceutical Chemistry
University of Kansas
Lawrence, KS 66047
USA

Cover
The cover picture is derived artistically from
the potential-energy profile for the dynamic
equilibrium of water molecules in the hydration
layer of a protein (see A. Douhal's chapter in
volume 1) and the three-dimensional vibrational
wavefunctions for reactants, transition state,
and products in a hydride-transfer reaction
(see the chapter by S.J. Benkovic and S. Hammes-
Schiffer in volume 4).

■ All books published by Wiley-VCH are carefully
produced. Nevertheless, authors, editors, and
publisher do not warrant the information contained
in these books, including this book, to be free of
errors. Readers are advised to keep in mind that
statements, data, illustrations, procedural details or
other items may inadvertently be inaccurate.

Library of Congress Card No.: applied for

British Library Cataloguing-in-Publication Data
A catalogue record for this book is available
from the British Library.

**Bibliographic information published by
Die Deutsche Bibliothek**
Die Deutsche Nationalbibliothek lists this
publication in the Deutsche Nationalbibliografie;
detailed bibliographic data are available in the
Internet at <http://dnb.d-nb.de>.

© 2007 WILEY-VCH Verlag GmbH & Co. KGaA,
Weinheim

All rights reserved (including those of translation
into other languages). No part of this book may be
reproduced in any form – by photoprinting, micro-
film, or any other means – nor transmitted or trans-
lated into a machine language without written permission from the publishers. Registered names, tra-
demarks, etc. used in this book, even when not
specifically marked as such, are not to be considered
unprotected by law.

Typesetting   Kühn & Weyh, Freiburg;
Asco Typesetters, Hongkong
Printing   betz-druck GmbH, Darmstadt
Bookbinding   Litges & Dopf GmbH, Heppenheim
Cover Design   Adam-Design, Weinheim

Printed in the Federal Republic of Germany.
Printed on acid-free paper.

ISBN: 978-3-527-30777-7

Foreword
The Remarkable Phenomena of Hydrogen Transfer
Ahmed H. Zewail*
California Institute of Technology
Pasadena, CA 91125, USA

Life would not exist without the making and breaking of chemical bonds - chemical reactions. Among the most elementary and significant of all reactions is the transfer of a hydrogen atom or a hydrogen ion (proton). Besides being a fundamental process involving the smallest of all atoms, such reactions form the basis of general phenomena in physical, chemical, and biological changes. Thus, there is a wide-ranging scope of studies of hydrogen transfer reactions and their role in determining properties and behaviors across different areas of molecular sciences.

Remarkably, this transfer of a small particle appears deceptively simple, but is in fact complex in its nature. For the most part, the dynamics cannot be described by a classical picture and the process involves more than one nuclear motion. For example, the transfer may occur by tunneling through a reaction barrier and a quantum description is necessary; the hydrogen is not isolated as it is part of a chemical bond and in many cases the nature of the bond, "covalent" and/or "ionic" in Pauling's valence bond description, is difficult to characterize; and the description of atom movement, although involving the local hydrogen bond, must take into account the coupling to other coordinates. In the modern age of quantum chemistry, much has been done to characterize the rate of transfer in different systems and media, and the strength of the bond and degree of charge localization. The intermediate bonding strength, directionality, and specificity are unique features of this bond.

---

* The author is currently the Linus Pauling Chair Professor of chemistry and physics and the Director of the Physical Biology Center for Ultrafast Science & Technology and the National Science Foundation Laboratory for Molecular Sciences at Caltech in Pasadena, California, USA. He was awarded the 1999 Nobel Prize in Chemistry.
Email: zewail@caltech.edu
Fax: 626.792.8456

Hydrogen-Transfer Reactions. Edited by J. T. Hynes, J. P. Klinman, H. H. Limbach, and R. L. Schowen
Copyright © 2007 WILEY-VCH Verlag GmbH & Co. KGaA, Weinheim
ISBN: 978-3-527-30777-7

The supreme example for the unique role in specificity and rates comes from life's genetic information, where the hydrogen bond determines the complementarities of G with C and A with T and the rate of hydrogen transfer controls genetic mutations. Moreover, the not-too-weak, not-too-strong strength of the bond allows for special "mobility" and for the potent hydrophobic/hydrophilic interactions. Life's matrix, liquid water, is one such example. The making and breaking of the hydrogen bond occurs on the picosecond time scale and the process is essential to keeping functional the native structures of DNA and proteins, and their recognition of other molecules, such as drugs. At interfaces, water can form ordered structures and with its amphiphilic character, utilizing either hydrogen or oxygen for bonding, determines many properties at the nanometer scale.

Hydrogen transfer can also be part of biological catalysis. In enzyme reactions, a huge complex structure is involved in bringing this small particle of hydrogen into the right place at the right time so that the reaction can be catalytically enhanced, with rates orders of magnitude larger than those in solution. The molecular theatre for these reactions is that of a very complex energy landscape, but with guided bias for specificity and selectivity in function. Control of reactivity at the active site has now reached the frontier of research in "catalytic antibody", and one of the most significant achievements in chemical synthesis, using heterogeneous catalysis, has been the design of site-selective reaction control.

Both experiments and theory join in the studies of hydrogen transfer reactions. In general, the approach is of two categories. The first involves the study of prototypical but well-defined molecular systems, either under isolated (microscopic) conditions or in complexes or clusters (mesoscopic) with the solvent, in the gas phase or molecular beams. Such studies over the past three decades have provided unprecedented resolution of the elementary processes involved in isolated molecules and en route to the condensed phase. Examples include the discovery of a "magic solvent number" for acid-base reactions, the elucidation of motions involved in double proton transfer, and the dynamics of acid dissociation in finite-sized clusters. For these systems, theory is nearly quantitative, especially as more accurate electronic structure and molecular dynamics computations become available.

The other category of study focuses on the nature of the transfer in the condensed phase and in biological systems. Here, it is not perhaps beneficial to consider every atom of a many-body complex system. Instead, the objective is hopefully to project the key electronic and nuclear forces which are responsible for behavior. With this perspective, approximate, but predictive, theories have a much more valuable outreach in applications than those simulating or computing bonding and motion of all atoms. Computer simulations are important, but for such systems they should be a tool of guidance to formulate a predictive theory. Similarly for experiments, the most significant ones are those that dissect complexity and provide lucid pictures of the key and relevant processes.

Progress has been made in these areas of study, but challenges remain. For example, the problem of vibrational energy redistribution in large molecules, although critical to the description of rates, statistical or not, and to the separation of intra and intermolecular pathways, has not been solved analytically, even in an approximate but predictive formulation. Another problem of significance concerns the issue of the energy landscape of complex reactions, and the question is: what determines specificity and selectivity?

This series edited by prominent players in the field is a testimony to the advances and achievements made over the past several decades. The diversity of topics covered is impressive: from isolated molecular systems, to clusters and confined geometries, and to condensed media; from organics to inorganics; from zeolites to surfaces; and, for biological systems, from proteins (including enzymes) to assemblies exhibiting conduction and other phenomena. The fundamentals are addressed by the most advanced theories of transition state, tunneling, Kramers' friction, Marcus' electron transfer, Grote-Hynes reaction dynamics, and free energy landscapes. Equally covered are state-of-the-art techniques and tools introduced for studies in this field and including ultrafast methods of femtochemistry and femtobiology, Raman and infrared, isotope probes, magnetic resonance, and electronic structure and MD simulations.

These volumes are a valuable addition to a field that continues to impact diverse areas of molecular sciences. The field is rigorous and vigorous as it still challenges the minds of many with the fascination of how the physics of the smallest of all atoms plays in diverse applications, not only in chemistry, but also in life sciences. Our gratitude is to the Editors and Authors for this compilation of articles with new knowledge in a field still pregnant with challenges and opportunities.

Pasadena, California *Ahmed Zewail*
August, 2006

Contents

Foreword *V*

Preface *XXXVII*

Preface to Volumes 1 and 2 *XXXIX*

List of Contributors to Volumes 1 and 2 *XLI*

I	**Physical and Chemical Aspects, Parts I–III**
Part I	**Hydrogen Transfer in Isolated Hydrogen Bonded Molecules, Complexes and Clusters** *1*
1	**Coherent Proton Tunneling in Hydrogen Bonds of Isolated Molecules: Malonaldehyde and Tropolone** *3* *Richard L. Redington*
1.1	Introduction *3*
1.2	Coherent Tunneling Splitting Phenomena in Malonaldehyde *5*
1.3	Coherent Tunneling Phenomena in Tropolone *13*
1.4	Tropolone Derivatives *26*
1.5	Concluding Remarks *27*
	Acknowledgments *28*
	References *29*
2	**Coherent Proton Tunneling in Hydrogen Bonds of Isolated Molecules: Carboxylic Dimers** *33* *Martina Havenith*
2.1	Introduction *33*
2.2	Quantum Tunneling versus Classical Over Barrier Reactions *34*
2.3	Carboxylic Dimers *35*
2.4	Benzoic Acid Dimer *38*
2.4.1	Introduction *38*

2.4.2	Determination of the Structure	38
2.4.3	Barriers and Splittings	39
2.4.4	Infrared Vibrational Spectroscopy	41
2.5	Formic Acid Dimer	42
2.5.1	Introduction	42
2.5.2	Determination of the Structure	42
2.5.3	Tunneling Path	43
2.5.4	Barriers and Tunneling Splittings	44
2.5.5	Infrared Vibrational Spectroscopy	45
2.5.6	Coherent Proton Transfer in Formic Acid Dimer	46
2.6	Conclusion	49
	References	50

3	**Gas Phase Vibrational Spectroscopy of Strong Hydrogen Bonds**	**53**
	Knut R. Asmis, Daniel M. Neumark, and Joel M. Bowman	
3.1	Introduction	53
3.2	Methods	55
3.2.1	Vibrational Spectroscopy of Gas Phase Ions	55
3.2.2	Experimental Setup	56
3.2.3	Potential Energy Surfaces	58
3.2.4	Vibrational Calculations	59
3.3	Selected Systems	60
3.3.1	Bihalide Anions	60
3.3.2	The Protonated Water Dimer $(H_2O\cdots H\cdots OH_2)^+$	65
3.3.2.1	Experiments	65
3.3.2.2	Calculations	70
3.4	Outlook	75
	Acknowledgments	76
	References	77

4	**Laser-driven Ultrafast Hydrogen Transfer Dynamics**	**79**
	Oliver Kühn and Leticia González	
4.1	Introduction	79
4.2	Theory	80
4.3	Laser Control	83
4.3.1	Laser-driven Intramolecular Hydrogen Transfer	83
4.3.2	Laser-driven H-Bond Breaking	90
4.4	Conclusions and Outlook	100
	Acknowledgments	101
	References	101

Part II Hydrogen Transfer in Condensed Phases *105*

5 Proton Transfer from Alkane Radical Cations to Alkanes *107*
Jan Ceulemans

5.1 Introduction *108*
5.2 Electronic Absorption of Alkane Radical Cations *108*
5.3 Paramagnetic Properties of Alkane Radical Cations *109*
5.4 The Brønsted Acidity of Alkane Radical Cations *110*
5.5 The σ-Basicity of Alkanes *112*
5.6 Powder EPR Spectra of Alkyl Radicals *114*
5.7 Symmetric Proton Transfer from Alkane Radical Cations to Alkanes: An Experimental Study in γ-Irradiated n-Alkane Nanoparticles Embedded in a Cryogenic CCl_3F Matrix *117*
5.7.1 Mechanism of the Radiolytic Process *117*
5.7.2 Physical State of Alkane Aggregates in CCl_3F *118*
5.7.3 Evidence for Proton-donor and Proton-acceptor Site Selectivity in the Symmetric Proton Transfer from Alkane Radical Cations to Alkane Molecules *121*
5.7.3.1 Proton-donor Site Selectivity *121*
5.7.3.2 Proton-acceptor Site Selectivity *122*
5.7.4 Comparison with Results on Proton Transfer and "Deprotonation" in Other Systems *124*
5.8 Asymmetric Proton Transfer from Alkane Radical Cations to Alkanes: An Experimental Study in γ-Irradiated Mixed Alkane Crystals *125*
5.8.1 Mechanism of the Radiolytic Process *125*
5.8.2 Evidence for Proton-donor and Proton-acceptor Site Selectivity in the Asymmetric Proton Transfer from Alkane Radical Cations to Alkanes *128*
References *131*

6 Single and Multiple Hydrogen/Deuterium Transfer Reactions in Liquids and Solids *135*
Hans-Heinrich Limbach

6.1 Introduction *136*
6.2 Theoretical *138*
6.2.1 Coherent vs. Incoherent Tunneling *138*
6.2.2 The Bigeleisen Theory *140*
6.2.3 Hydrogen Bond Compression Assisted H-transfer *141*
6.2.4 Reduction of a Two-dimensional to a One-dimensional Tunneling Model *143*
6.2.5 The Bell–Limbach Tunneling Model *146*
6.2.6 Concerted Multiple Hydrogen Transfer *151*
6.2.7 Multiple Stepwise Hydrogen Transfer *152*
6.2.7.1 HH-transfer *153*

6.2.7.2	Degenerate Stepwise HHH-transfer	159
6.2.7.3	Degenerate Stepwise HHHH-transfer	161
6.2.8	Hydrogen Transfers Involving Pre-equilibria	165
6.3	Applications	168
6.3.1	H-transfers Coupled to Minor Heavy Atom Motions	174
6.3.1.1	Symmetric Porphyrins and Porphyrin Analogs	174
6.3.1.2	Unsymmetrically Substituted Porphyrins	181
6.3.1.3	Hydroporphyrins	184
6.3.1.4	Intramolecular Single and Stepwise Double Hydrogen Transfer in H-bonds of Medium Strength	185
6.3.1.5	Dependence on the Environment	187
6.3.1.6	Intermolecular Multiple Hydrogen Transfer in H-bonds of Medium Strength	188
6.3.1.7	Dependence of the Barrier on Molecular Structure	193
6.3.2	H-transfers Coupled to Major Heavy Atom Motions	197
6.3.2.1	H-transfers Coupled to Conformational Changes	197
6.3.2.2	H-transfers Coupled to Conformational Changes and Hydrogen Bond Pre-equilibria	203
6.3.2.3	H-transfers in Complex Systems	212
6.4	Conclusions	216
	Acknowledgments	217
	References	217
7	**Intra- and Intermolecular Proton Transfer and Related Processes in Confined Cyclodextrin Nanostructures**	**223**
	Abderrazzak Douhal	
7.1	Introduction and Concept of Femtochemistry in Nanocavities	223
7.2	Overview of the Photochemistry and Photophysics of Cyclodextrin Complexes	224
7.3	Picosecond Studies of Proton Transfer in Cyclodextrin Complexes	225
7.3.1	1′-Hydroxy,2′-acetonaphthone	225
7.3.2	1-Naphthol and 1-Aminopyrene	228
7.4	Femtosecond Studies of Proton Transfer in Cyclodextrin Complexes	230
7.4.1	Coumarins 460 and 480	230
7.4.2	Bound and Free Water Molecules	231
7.5.3	2-(2′-Hydroxyphenyl)-4-methyloxazole	236
7.5.4	Orange II	239
7.6	Concluding Remarks	240
	Acknowledgment	241
	References	241

8	**Tautomerization in Porphycenes** *245*	
	Jacek Waluk	
8.1	Introduction *245*	
8.2	Tautomerization in the Ground Electronic State *247*	
8.2.1	Structural Data *247*	
8.2.2	NMR Studies of Tautomerism *251*	
8.2.3	Supersonic Jet Studies *253*	
8.2.4	The Nonsymmetric Case: 2,7,12,17-Tetra-n-propyl-9-acetoxyporphycene *256*	
8.2.5	Calculations *258*	
8.3	Tautomerization in the Lowest Excited Singlet State *258*	
8.3.1	Tautomerization as a Tool to Determine Transition Moment Directions in Low Symmetry Molecules *260*	
8.3.2	Determination of Tautomerization Rates from Anisotropy Measurements *262*	
8.4	Tautomerization in the Lowest Excited Triplet State *265*	
8.5	Tautomerization in Single Molecules of Porphycene *266*	
8.6	Summary *267*	
	Acknowledgments *268*	
	References *269*	
9	**Proton Dynamics in Hydrogen-bonded Crystals** *273*	
	Mikhail V. Vener	
9.1	Introduction *273*	
9.2	Tentative Study of Proton Dynamics in Crystals with Quasi-linear H-bonds *274*	
9.2.1	A Model 2D Hamiltonian *275*	
9.2.2	Specific Features of H-bonded Crystals with a Quasi-symmetric O···H···O Fragment *277*	
9.2.3	Proton Transfer Assisted by a Low-frequency Mode Excitation *279*	
9.2.3.1	Crystals with Moderate H-bonds *280*	
9.2.3.2	Crystals with Strong H-bonds *283*	
9.2.3.3	Limitations of the Model 2D Treatment *284*	
9.2.4	Vibrational Spectra of H-bonded Crystals: IR versus INS *285*	
9.3	DFT Calculations with Periodic Boundary Conditions *286*	
9.3.1	Evaluation of the Vibrational Spectra Using Classical MD Simulations *287*	
9.3.2	Effects of Crystalline Environment on Strong H-bonds: the $H_5O_2^+$ Ion *288*	
9.3.2.1	The Structure and Harmonic Frequencies *288*	
9.3.2.2	The PES of the O···H···O Fragment *291*	
9.3.2.3	Anharmonic INS and IR Spectra *293*	

9.4	Conclusions 296	
	Acknowledgments 297	
	References 217	

Part III	**Hydrogen Transfer in Polar Environments** *301*	

10	**Theoretical Aspects of Proton Transfer Reactions in a Polar Environment** *303*	
	Philip M. Kiefer and James T. Hynes	
10.1	Introduction 303	
10.2	Adiabatic Proton Transfer 309	
10.2.1	General Picture 309	
10.2.2	Adiabatic Proton Transfer Free Energy Relationship (FER) 315	
10.2.3	Adiabatic Proton Transfer Kinetic Isotope Effects 320	
10.2.3.1	KIE Arrhenius Behavior 321	
10.2.3.2	KIE Magnitude and Variation with Reaction Asymmetry 321	
10.2.3.3	Swain–Schaad Relationship 323	
10.2.3.4	Further Discussion of Nontunneling Kinetic Isotope Effects 323	
10.2.3.5	Transition State Geometric Structure in the Adiabatic PT Picture 324	
10.2.4	Temperature Solvent Polarity Effects 325	
10.3	Nonadiabatic 'Tunneling' Proton Transfer 326	
10.3.1	General Nonadiabatic Proton Transfer Perspective and Rate Constant 327	
10.3.2	Nonadiabatic Proton Transfer Kinetic Isotope Effects 333	
10.3.2.1	Kinetic Isotope Effect Magnitude and Variation with Reaction Asymmetry 333	
10.3.2.2	Temperature Behavior 337	
10.3.2.3	Swain–Schaad Relationship 340	
10.4	Concluding Remarks 341	
	Acknowledgments 343	
	References 345	

11	**Direct Observation of Nuclear Motion during Ultrafast Intramolecular Proton Transfer** *349*	
	Stefan Lochbrunner, Christian Schriever, and Eberhard Riedle	
11.1	Introduction 349	
11.2	Time-resolved Absorption Measurements 352	
11.3	Spectral Signatures of Ultrafast ESIPT 353	
11.3.1	Characteristic Features of the Transient Absorption 354	
11.3.2	Analysis 356	
11.3.3	Ballistic Wavepacket Motion 357	
11.3.4	Coherently Excited Vibrations in Product Modes 359	
11.4	Reaction Mechanism 362	
11.4.1	Reduction of Donor–Acceptor Distance by Skeletal Motions 362	

11.4.2	Multidimensional ESIPT Model *363*	
11.4.3	Micro-irreversibility *365*	
11.4.4	Topology of the PES and Turns in the Reaction Path *366*	
11.4.5	Comparison with Ground State Hydrogen Transfer Dynamics *368*	
11.4.6	Internal Conversion *368*	
11.5	Reaction Path Specific Wavepacket Dynamics in Double Proton Transfer Molecules *370*	
11.6	Conclusions *372*	
	Acknowledgment *373*	
	References *373*	
12	**Solvent Assisted Photoacidity** *377*	
	Dina Pines and Ehud Pines	
12.1	Introduction *377*	
12.2	Photoacids, Photoacidity and Förster Cycle *378*	
12.2.1	Photoacids and Photobases *378*	
12.2.2	Use of the Förster Cycle to Estimate the Photoacidity of Photoacids *379*	
12.2.3	Direct Methods for Determining the Photoacidity of Photoacids *387*	
12.3	Evidence for the General Validity of the Förster Cycle and the K_a^* Scale *389*	
12.3.1	Evidence for the General Validity of the Förster Cycle Based on Time-resolved and Steady State Measurements of Excited-state Proton Transfer of Photoacids *389*	
12.3.2	Evidence Based on Free Energy Correlations *393*	
12.4	Factors Affecting Photoacidity *397*	
12.4.1	General Considerations *397*	
12.4.2	Comparing the Solvent Effect on the Photoacidities of Neutral and Cationic Photoacids *398*	
12.4.3	The Effect of Substituents on the Photoacidity of Aromatic Alcohols *400*	
12.5	Solvent Assisted Photoacidity: The ^1L_a, ^1L_b Paradigm *404*	
12.6	Summary *410*	
	Acknowledgments *411*	
	References *411*	
13	**Design and Implementation of "Super" Photoacids** *417*	
	Laren M. Tolbert and Kyril M. Solntsev	
13.1	Introduction *417*	
13.2	Excited-state Proton Transfer (ESPT) *420*	
13.2.1	1-Naphthol vs. 2-Naphthol *420*	
13.2.2	"Super" Photoacids *422*	
13.2.3	Fluorinated Phenols *426*	
13.3	Nature of the Solvent *426*	
13.3.1	Hydrogen Bonding and Solvatochromism in Super Photoacids *426*	

13.3.2   Dynamics in Water and Mixed Solvents *427*
13.3.3   Dynamics in Nonaqueous Solvents *428*
13.3.4   ESPT in the Gas Phase *431*
13.3.5   Stereochemistry *433*
13.4   ESPT in Biological Systems *433*
13.4.1   The Green Fluorescent Protein (GFP) or "ESPT in a Box" *435*
13.5   Conclusions *436*
         Acknowledgments *436*
         References *437*

Foreword   *V*

Preface   *XXXVII*

Preface to Volumes 1 and 2   *XXXIX*

List of Contributors to Volumes 1 and 2   *XLI*

I   **Physical and Chemical Aspects, Parts IV–VII**

Part IV   **Hydrogen Transfer in Protic Systems**   *441*

14   **Bimolecular Proton Transfer in Solution**   *443*
         Erik T. J. Nibbering and Ehud Pines

14.1   Intermolecular Proton Transfer in the Liquid Phase *443*
14.2   Photoacids as Ultrafast Optical Triggers for Proton Transfer *445*
14.3   Proton Recombination and Acid–Base Neutralization *448*
14.4   Reaction Dynamics Probing with Vibrational Marker Modes *449*
       Acknowledgment *455*
       References *455*

15   **Coherent Low-frequency Motions in Condensed Phase Hydrogen Bonding and Transfer**   *459*
         Thomas Elsaesser

15.1   Introduction *459*
15.2   Vibrational Excitations of Hydrogen Bonded Systems *460*
15.3   Low-frequency Wavepacket Dynamics of Hydrogen Bonds in the Electronic Ground State *463*
15.3.1 Intramolecular Hydrogen Bonds *463*
15.3.2 Hydrogen Bonded Dimers *466*
15.4   Low-frequency Motions in Excited State Hydrogen Transfer *471*
15.5   Conclusions *475*
        Acknowledgments *476*
        References *476*

16 Proton-Coupled Electron Transfer: Theoretical Formulation and Applications *479*
Sharon Hammes-Schiffer

16.1 Introduction *479*
16.2 Theoretical Formulation for PCET *480*
16.2.1 Fundamental Concepts *480*
16.2.2 Proton Donor–Acceptor Motion *483*
16.2.3 Dynamical Effects *485*
16.2.3.1 Dielectric Continuum Representation of the Environment *486*
16.2.3.2 Molecular Representation of the Environment *490*
16.3 Applications *492*
16.3.1 PCET in Solution *492*
16.3.2 PCET in a Protein *498*
16.4 Conclusions *500*
Acknowledgments *500*
References *501*

17 The Relation between Hydrogen Atom Transfer and Proton-coupled Electron Transfer in Model Systems *503*
Justin M. Hodgkiss, Joel Rosenthal, and Daniel G. Nocera

17.1 Introduction *503*
17.1.1 Formulation of HAT as a PCET Reaction *504*
17.1.2 Scope of Chapter *507*
17.1.2.1 Unidirectional PCET *508*
17.1.2.2 Bidirectional PCET *508*
17.2 Methods of HAT and PCET Study *509*
17.2.1 Free Energy Correlations *510*
17.2.2 Solvent Dependence *511*
17.2.3 Deuterium Kinetic Isotope Effects *511*
17.2.4 Temperature Dependence *512*
17.3 Unidirectional PCET *512*
17.3.1 Type A: Hydrogen Abstraction *512*
17.3.2 Type B: Site Differentiated PCET *523*
17.3.2.1 PCET across Symmetric Hydrogen Bonding Interfaces *523*
17.3.2.2 PCET across Polarized Hydrogen Bonding Interfaces *527*
17.4 Bidirectional PCET *537*
17.4.1 Type C: Non-Specific 3-Point PCET *538*
17.4.2 Type D: Site-Specified 3-Point PCET *543*
17.5 The Different Types of PCET in Biology *548*
17.6 Application of Emerging Ultrafast Spectroscopy to PCET *554*
Acknowledgment *556*
References *556*

Part V Hydrogen Transfer in Organic and Organometallic Reactions 563

18 Formation of Hydrogen-bonded Carbanions as Intermediates in Hydron Transfer between Carbon and Oxygen 565
Heinz F. Koch

18.1 Proton Transfer from Carbon Acids to Methoxide Ion 565
18.2 Proton Transfer from Methanol to Carbanion Intermediates 573
18.3 Proton Transfer Associated with Methoxide Promoted Dehydrohalogenation Reactions 576
18.4 Conclusion 580
References 581

19 Theoretical Simulations of Free Energy Relationships in Proton Transfer 583
Ian H. Williams

19.1 Introduction 583
19.2 Qualitative Models for FERs 584
19.2.1 What is Meant by "Reaction Coordinate"? 588
19.2.2 The Brønsted α as a Measure of TS Structure 589
19.3 FERs from MO Calculations of PESs 590
19.3.1 Energies and Transition States 590
19.4 FERs from VB Studies of Free Energy Changes for PT in Condensed Phases 597
19.5 Concluding Remarks 600
References 600

20 The Extraordinary Dynamic Behavior and Reactivity of Dihydrogen and Hydride in the Coordination Sphere of Transition Metals 603
Gregory J. Kubas

20.1 Introduction 603
20.1.1 Structure, Bonding, and Activation of Dihydrogen Complexes 603
20.1.2 Extraordinary Dynamics of Dihydrogen Complexes 606
20.1.2 Vibrational Motion of Dihydrogen Complexes 608
20.1.3 Elongated Dihydrogen Complexes 609
20.1.4 Cleavage of the H–H Bond in Dihydrogen Complexes 610
20.2 H_2 Rotation in Dihydrogen Complexes 615
20.2.1 Determination of the Barrier to Rotation of Dihydrogen 616
20.3 NMR Studies of H_2 Activation, Dynamics, and Transfer Processes 617
20.3.1 Solution NMR 617
20.3.2 Solid State NMR of H_2 Complexes 621

20.4	Intramolecular Hydrogen Rearrangement and Exchange *623*	
20.4.1	Extremely Facile Hydrogen Transfer in $IrXH_2(H_2)(PR_3)_2$ and Other Systems *627*	
20.4.2	Quasielastic Neutron Scattering Studies of H_2 Exchange with cis-Hydrides *632*	
20.5	Summary *633*	
	Acknowledgments *634*	
	References *634*	
21	**Dihydrogen Transfer and Symmetry: The Role of Symmetry in the Chemistry of Dihydrogen Transfer in the Light of NMR Spectroscopy** *639*	
	Gerd Buntkowsky and Hans-Heinrich Limbach	
21.1	Introduction *639*	
21.2	Tunneling and Chemical Kinetics *641*	
21.2.1	The Role of Symmetry in Chemical Exchange Reactions *641*	
21.2.1.1	Coherent Tunneling *642*	
21.2.1.2	The Density Matrix *648*	
21.2.1.3	The Transition from Coherent to Incoherent Tunneling *649*	
21.2.2	Incoherent Tunneling and the Bell Model *653*	
21.3	Symmetry Effects on NMR Lineshapes of Hydration Reactions *655*	
21.3.1	Analytical Solution for the Lineshape of PHIP Spectra Without Exchange *657*	
21.3.2	Experimental Examples of PHIP Spectra *662*	
21.3.2.1	PHIP under ALTADENA Conditions *662*	
21.3.2.2	PHIP Studies of Stereoselective Reactions *662*	
21.3.2.3	^{13}C-PHIP-NMR *664*	
21.3.3	Effects of Chemical Exchange on the Lineshape of PHIP Spectra *665*	
21.4	Symmetry Effects on NMR Lineshapes of Intramolecular Dihydrogen Exchange Reactions *670*	
21.4.1	Experimental Examples *670*	
21.4.1.1	Slow Tunneling Determined by ^1H Liquid State NMR Spectroscopy *671*	
21.4.1.2	Slow to Intermediate Tunneling Determined by ^2H Solid State NMR *671*	
21.4.1.3	Intermediate to Fast Tunneling Determined by ^2H Solid State NMR *673*	
21.4.1.4	Fast Tunneling Determined by Incoherent Neutron Scattering *675*	
21.4.2	Kinetic Data Obtained from the Experiments *675*	
21.4.2.1	Ru-D_2 Complex *676*	
21.4.2.2	$W(PCy)_3(CO)_3$ (η-H_2) Complex *677*	
21.5	Summary and Conclusion *678*	
	Acknowledgments *679*	
	References *679*	

Part VI Proton Transfer in Solids and Surfaces 683

22 Proton Transfer in Zeolites 685
Joachim Sauer

22.1   Introduction – The Active Sites of Acidic Zeolite Catalysts 685
22.2   Proton Transfer to Substrate Molecules within Zeolite Cavities 686
22.3   Formation of NH_4^+ ions on NH_3 adsorption 688
22.4   Methanol Molecules and Dimers in Zeolites 691
22.5   Water Molecules and Clusters in Zeolites 694
22.6   Proton Jumps in Hydrated and Dry Zeolites 700
22.7   Stability of Carbenium Ions in Zeolites 703
       References 706

23 Proton Conduction in Fuel Cells 709
Klaus-Dieter Kreuer

23.1   Introduction 709
23.2   Proton Conducting Electrolytes and Their Application in Fuel Cells 710
23.3   Long-range Proton Transport of Protonic Charge Carriers in Homogeneous Media 714
23.3.1 Proton Conduction in Aqueous Environments 715
23.3.2 Phosphoric Acid 719
23.3.3 Heterocycles (Imidazole) 720
23.4   Confinement and Interfacial Effects 723
23.4.1 Hydrated Acidic Polymers 723
23.4.2 Adducts of Basic Polymers with Oxo-acids 727
23.4.3 Separated Systems with Covalently Bound Proton Solvents 728
23.5   Concluding Remarks 731
       Acknowledgment 733
       References 733

24 Proton Diffusion in Ice Bilayers 737
Katsutoshi Aoki

24.1   Introduction 737
24.1.1 Phase Diagram and Crystal Structure of Ice 737
24.1.2 Molecular and Protonic Diffusion 739
24.1.3 Protonic Diffusion at High Pressure 740
24.2   Experimental Method 741
24.2.1 Diffusion Equation 741
24.2.2 High Pressure Measurement 742
24.2.3 Infrared Reflection Spectra 743
24.2.4 Thermal Activation of Diffusion Motion 744
24.3   Spectral Analysis of the Diffusion Process 745
24.3.1 Protonic Diffusion 745

24.3.2	Molecular Diffusion *746*	
24.3.3	Pressure Dependence of Protonic Diffusion Coefficient *747*	
24.4	Summary *749*	
	References *749*	

25 Hydrogen Transfer on Metal Surfaces *751*
Klaus Christmann

25.1	Introduction *751*
25.2	The Principles of the Interaction of Hydrogen with Surfaces: Terms and Definitions *755*
25.3	The Transfer of Hydrogen on Metal Surfaces *761*
25.3.1	Hydrogen Surface Diffusion on Homogeneous Metal Surfaces *761*
25.3.2	Hydrogen Surface Diffusion and Transfer on Heterogeneous Metal Surfaces *771*
25.4	Alcohol and Water on Metal Surfaces: Evidence of H Bond Formation and H Transfer *775*
25.4.1	Alcohols on Metal Surfaces *775*
25.4.2	Water on Metal Surfaces *778*
25.5	Conclusion *783*
	Acknowledgments *783*
	References *783*

26 Hydrogen Motion in Metals *787*
Rolf Hempelmann and Alexander Skripov

26.1	Survey *787*
26.2	Experimental Methods *788*
26.2.1	Anelastic Relaxation *788*
26.2.2	Nuclear Magnetic Resonance *790*
26.2.3	Quasielastic Neutron Scattering *792*
26.2.4	Other Methods *795*
26.3	Experimental Results on Diffusion Coefficients *796*
26.4	Experimental Results on Hydrogen Jump Diffusion Mechanisms *801*
26.4.1	Binary Metal–Hydrogen Systems *802*
26.4.2	Hydrides of Alloys and Intermetallic Compounds *804*
26.4.3	Hydrogen in Amorphous Metals *810*
26.5	Quantum Motion of Hydrogen *812*
26.5.1	Hydrogen Tunneling in Nb Doped with Impurities *814*
26.5.2	Hydrogen Tunneling in α-MnH$_x$ *817*
26.5.3	Rapid Low-temperature Hopping of Hydrogen in α-ScH$_x$(D$_x$) and TaV$_2$H$_x$(D$_x$) *821*
26.6	Concluding Remarks *825*
	Acknowledgment *825*
	References *826*

Part VII	Special Features of Hydrogen-Transfer Reactions 831
27	**Variational Transition State Theory in the Treatment of Hydrogen Transfer Reactions** 833
	Donald G. Truhlar and Bruce C. Garrett

27.1	Introduction 833
27.2	Incorporation of Quantum Mechanical Effects in VTST 835
27.2.1	Adiabatic Theory of Reactions 837
27.2.2	Quantum Mechanical Effects on Reaction Coordinate Motion 840
27.3	H-atom Transfer in Bimolecular Gas-phase Reactions 843
27.3.1	$H + H_2$ and $Mu + H_2$ 843
27.3.2	$Cl + HBr$ 849
27.3.3	$Cl + CH_4$ 853
27.4	Intramolecular Hydrogen Transfer in Unimolecular Gas-phase Reactions 857
27.4.1	Intramolecular H-transfer in 1,3-Pentadiene 858
27.4.2	1,2-Hydrogen Migration in Methylchlorocarbene 860
27.5	Liquid-phase and Enzyme-catalyzed Reactions 860
27.5.1	Separable Equilibrium Solvation 862
27.5.2	Equilibrium Solvation Path 864
27.5.3	Nonequilibrium Solvation Path 864
27.5.4	Potential-of-mean-force Method 865
27.5.5	Ensemble-averaged Variational Transition State Theory 865
27.6	Examples of Condensed-phase Reactions 867
27.6.1	H + Methanol 867
27.6.2	Xylose Isomerase 868
27.6.3	Dihydrofolate Reductase 868
27.7	Another Perspective 869
27.8	Concluding Remarks 869
	Acknowledgments 871
	References 871

28	**Quantum Mechanical Tunneling of Hydrogen Atoms in Some Simple Chemical Systems** 875
	K. U. Ingold

28.1	Introduction 875
28.2	Unimolecular Reactions 876
28.2.1	Isomerization of Sterically Hindered Phenyl Radicals 876
28.2.1.1	2,4,6-Tri–*tert*–butylphenyl 876
28.2.1.2	Other Sterically Hindered Phenyl Radicals 881
28.2.2	Inversion of Nonplanar, Cyclic, Carbon-Centered Radicals 883
28.2.2.1	Cyclopropyl and 1-Methylcyclopropyl Radicals 883
28.2.2.2	The Oxiranyl Radical 884
28.2.2.3	The Dioxolanyl Radical 886

28.2.2.4	Summary *887*	
28.3	Bimolecular Reactions *887*	
28.3.1	H-Atom Abstraction by Methyl Radicals in Organic Glasses *887*	
28.3.2	H-Atom Abstraction by Bis(trifluoromethyl) Nitroxide in the Liquid Phase *890*	
	References *892*	

29	**Multiple Proton Transfer: From Stepwise to Concerted** *895*	
	Zorka Smedarchina, Willem Siebrand, and Antonio Fernández-Ramos	
29.1	Introduction *895*	
29.2	Basic Model *897*	
29.3	Approaches to Proton Tunneling Dynamics *904*	
29.4	Tunneling Dynamics for Two Reaction Coordinates *908*	
29.5	Isotope Effects *914*	
29.6	Dimeric Formic Acid and Related Dimers *918*	
29.7	Other Dimeric Systems *922*	
29.8	Intramolecular Double Proton Transfer *926*	
29.9	Proton Conduits *932*	
29.10	Transfer of More Than Two Protons *939*	
29.11	Conclusion *940*	
	Acknowledgment *943*	
	References *943*	

Foreword *V*

Preface *XXXVII*

Preface to Volumes 3 and 4 *XXXIX*

List of Contributors to Volumes 3 and 4 *XLI*

II	**Biological Aspects, Parts I–II**	
Part I	**Models for Biological Hydrogen Transfer** *947*	
1	**Proton Transfer to and from Carbon in Model Reactions** *949*	
	Tina L. Amyes and John P. Richard	
1.1	Introduction *949*	
1.2	Rate and Equilibrium Constants for Carbon Deprotonation in Water *949*	
1.2.1	Rate Constants for Carbanion Formation *951*	
1.2.2	Rate Constants for Carbanion Protonation *953*	
1.2.2.1	Protonation by Hydronium Ion *953*	

1.2.2.2	Protonation by Buffer Acids	954
1.2.2.3	Protonation by Water	955
1.2.3	The Burden Borne by Enzyme Catalysts	955
1.3	Substituent Effects on Equilibrium Constants for Deprotonation of Carbon	957
1.4	Substituent Effects on Rate Constants for Proton Transfer at Carbon	958
1.4.1	The Marcus Equation	958
1.4.2	Marcus Intrinsic Barriers for Proton Transfer at Carbon	960
1.4.2.1	Hydrogen Bonding	960
1.4.2.2	Resonance Effects	961
1.5	Small Molecule Catalysis of Proton Transfer at Carbon	965
1.5.1	General Base Catalysis	966
1.5.2	Electrophilic Catalysis	967
1.6	Comments on Enzymatic Catalysis of Proton Transfer	970
	Acknowledgment	970
	References	971

2   General Acid–Base Catalysis in Model Systems   975
Anthony J. Kirby

2.1	Introduction	975
2.1.1	Kinetics	975
2.1.2	Mechanism	977
2.1.3	Kinetic Equivalence	979
2.2	Structural Requirements and Mechanism	981
2.2.1	General Acid Catalysis	982
2.2.2	Classical General Base Catalysis	983
2.2.3	General Base Catalysis of Cyclization Reactions	984
2.2.3.1	Nucleophilic Substitution	984
2.2.3.2	Ribonuclease Models	985
2.3	Intramolecular Reactions	987
2.3.1	Introduction	987
2.3.2	Efficient Intramolecular General Acid–Base Catalysis	988
2.3.2.1	Aliphatic Systems	991
2.3.3	Intramolecular General Acid Catalysis of Nucleophilic Catalysis	993
2.3.4	Intramolecular General Acid Catalysis of Intramolecular Nucleophilic Catalysis	998
2.3.5	Intramolecular General Base Catalysis	999
2.4	Proton Transfers to and from Carbon	1000
2.4.1	Intramolecular General Acid Catalysis	1002
2.4.2	Intramolecular General Base Catalysis	1004
2.4.3	Simple Enzyme Models	1006
2.5	Hydrogen Bonding, Mechanism and Reactivity	1007
	References	1010

3 Hydrogen Atom Transfer in Model Reactions   *1013*
Christian Schöneich

3.1 Introduction   *1013*
3.2 Oxygen-centered Radicals   *1013*
3.3 Nitrogen-dentered Radicals   *1017*
3.3.1 Generation of Aminyl and Amidyl Radicals   *1017*
3.3.2 Reactions of Aminyl and Amidyl Radicals   *1018*
3.4 Sulfur-centered Radicals   *1019*
3.4.1 Thiols and Thiyl Radicals   *1020*
3.4.1.1 Hydrogen Transfer from Thiols   *1020*
3.4.1.2 Hydrogen Abstraction by Thiyl Radicals   *1023*
3.4.2 Sulfide Radical Cations   *1029*
3.5 Conclusion   *1032*
Acknowledgment   *1032*
References   *1032*

4 Model Studies of Hydride-transfer Reactions   *1037*
Richard L. Schowen

4.1 Introduction   *1037*
4.1.1 Nicotinamide Coenzymes: Basic Features   *1038*
4.1.2 Flavin Coenzymes: Basic Features   *1039*
4.1.3 Quinone Coenzymes: Basic Features   *1039*
4.1.4 Matters Not Treated in This Chapter   *1039*
4.2 The Design of Suitable Model Reactions   *1040*
4.2.1 The Anchor Principle of Jencks   *1042*
4.2.2 The Proximity Effect of Bruice   *1044*
4.2.3 Environmental Considerations   *1045*
4.3 The Role of Model Reactions in Mechanistic Enzymology   *1045*
4.3.1 Kinetic Baselines for Estimations of Enzyme Catalytic Power   *1045*
4.3.2 Mechanistic Baselines and Enzymic Catalysis   *1047*
4.4 Models for Nicotinamide-mediated Hydrogen Transfer   *1048*
4.4.1 Events in the Course of Formal Hydride Transfer   *1048*
4.4.2 Electron-transfer Reactions and H-atom-transfer Reactions   *1049*
4.4.3 Hydride-transfer Mechanisms in Nicotinamide Models   *1052*
4.4.4 Transition-state Structure in Hydride Transfer: The Kreevoy Model   *1054*
4.4.5 Quantum Tunneling in Model Nicotinamide-mediated Hydride Transfer   *1060*
4.4.6 Intramolecular Models for Nicotinamide-mediated Hydride Transfer   *1061*
4.4.7 Summary   *1063*
4.5 Models for Flavin-mediated Hydride Transfer   *1064*
4.5.1 Differences between Flavin Reactions and Nicotinamide Reactions   *1064*

4.5.2	The Hydride-transfer Process in Model Systems	*1065*
4.6	Models for Quinone-mediated Reactions	*1068*
4.7	Summary and Conclusions	*1071*
4.8	Appendix: The Use of Model Reactions to Estimate Enzyme Catalytic Power	*1071*
	References	*1074*

5   Acid–Base Catalysis in Designed Peptides   *1079*
Lars Baltzer

5.1	Designed Polypeptide Catalysts	*1079*
5.1.1	Protein Design	*1080*
5.1.2	Catalyst Design	*1083*
5.1.3	Designed Catalysts	*1085*
5.2	Catalysis of Ester Hydrolysis	*1089*
5.2.1	Design of a Folded Polypeptide Catalyst for Ester Hydrolysis	*1089*
5.2.2	The HisH$^+$-His Pair	*1091*
5.2.3	Reactivity According to the Brönsted Equation	*1093*
5.2.4	Cooperative Nucleophilic and General-acid Catalysis in Ester Hydrolysis	*1094*
5.2.5	Why General-acid Catalysis?	*1095*
5.3	Limits of Activity in Surface Catalysis	*1096*
5.3.1	Optimal Organization of His Residues for Catalysis of Ester Hydrolysis	*1097*
5.3.2	Substrate and Transition State Binding	*1098*
5.3.3	His Catalysis in Re-engineered Proteins	*1099*
5.4	Computational Catalyst Design	*1100*
5.4.1	Ester Hydrolysis	*1101*
5.4.2	Triose Phosphate Isomerase Activity by Design	*1101*
5.5	Enzyme Design	*1102*
	References	*1102*

Part II   General Aspects of Biological Hydrogen Transfer   *1105*

6   Enzymatic Catalysis of Proton Transfer at Carbon Atoms   *1107*
John A. Gerlt

6.1	Introduction	*1107*
6.2	The Kinetic Problems Associated with Proton Abstraction from Carbon	*1108*
6.2.1	Marcus Formalism for Proton Transfer	*1110*
6.2.2	ΔG°, the Thermodynamic Barrier	*1111*
6.2.3	$\Delta G^\ddagger_{int}$, the Intrinsic Kinetic Barrier	*1112*
6.3	Structural Strategies for Reduction of ΔG°	*1114*
6.3.1	Proposals for Understanding the Rates of Proton Transfer	*1114*
6.3.2	Short Strong Hydrogen Bonds	*1115*

6.3.3	Electrostatic Stabilization of Enolate Anion Intermediates	*1115*
6.3.4	Experimental Measure of Differential Hydrogen Bond Strengths	*1116*
6.4	Experimental Paradigms for Enzyme-catalyzed Proton Abstraction from Carbon	*1118*
6.4.1	Triose Phosphate Isomerase	*1118*
6.4.2	Ketosteroid Isomerase	*1125*
6.4.3	Enoyl-CoA Hydratase (Crotonase)	*1127*
6.4.4	Mandelate Racemase and Enolase	*1131*
6.5	Summary	*1134*
	References	*1135*

7 Multiple Hydrogen Transfers in Enzyme Action *1139*
M. Ashley Spies and Michael D. Toney

7.1	Introduction	*1139*
7.2	Cofactor-Dependent with Activated Substrates	*1139*
7.2.1	Alanine Racemase	*1139*
7.2.2	Broad Specificity Amino Acid Racemase	*1151*
7.2.3	Serine Racemase	*1152*
7.2.4	Mandelate Racemase	*1152*
7.2.5	ATP-Dependent Racemases	*1154*
7.2.6	Methylmalonyl-CoA Epimerase	*1156*
7.3	Cofactor-Dependent with Unactivated Substrates	*1157*
7.4	Cofactor-Independent with Activated Substrates	*1157*
7.4.1	Proline Racemase	*1157*
7.4.2	Glutamate Racemase	*1161*
7.4.3	DAP Epimerase	*1162*
7.4.4	Sugar Epimerases	*1165*
7.5	Cofactor-Independent with Unactivated Substrates	*1165*
7.6	Summary	*1166*
	References	*1167*

8 Computer Simulations of Proton Transfer in Proteins and Solutions *1171*
Sonja Braun-Sand, Mats H. M. Olsson, Janez Mavri, and Arieh Warshel

8.1	Introduction	*1171*
8.2	Simulating PT Reactions by the EVB and other QM/MM Methods	*1171*
8.3	Simulating the Fluctuations of the Environment and Nuclear Quantum Mechanical Effects	*1177*
8.4	The EVB as a Basis for LFER of PT Reactions	*1185*
8.5	Demonstrating the Applicability of the Modified Marcus' Equation	*1188*
8.6	General Aspects of Enzymes that Catalyze PT Reactions	*1194*
8.7	Dynamics, Tunneling and Related Nuclear Quantum Mechanical Effects	*1195*

8.8	Concluding Remarks 1198	
	Acknowledgements 1199	
	Abbreviations 1199	
	References 1200	

Foreword V

Preface XXXVII

Preface to Volumes 3 and 4 XXXIX

List of Contributors to Volumes 3 and 4 XLI

II	**Biological Aspects, Parts III–V**	
Part III	**Quantum Tunneling and Protein Dynamics** 1207	
9	**The Quantum Kramers Approach to Enzymatic Hydrogen Transfer – Protein Dynamics as it Couples to Catalysis** 1209	
	Steven D. Schwartz	
9.1	Introduction 1209	
9.2	The Derivation of the Quantum Kramers Method 1210	
9.3	Promoting Vibrations and the Dynamics of Hydrogen Transfer 1213	
9.3.1	Promoting Vibrations and The Symmetry of Coupling 1213	
9.3.2	Promoting Vibrations – Corner Cutting and the Masking of KIEs 1215	
9.4	Hydrogen Transfer and Promoting Vibrations – Alcohol Dehydrogenase 1217	
9.5	Promoting Vibrations and the Kinetic Control of Enzymes – Lactate Dehydrogenase 1223	
9.6	The Quantum Kramers Model and Proton Coupled Electron Transfer 1231	
9.7	Promoting Vibrations and Electronic Polarization 1233	
9.8	Conclusions 1233	
	Acknowledgment 1234	
	References 1234	
10	**Nuclear Tunneling in the Condensed Phase: Hydrogen Transfer in Enzyme Reactions** 1241	
	Michael J. Knapp, Matthew Meyer, and Judith P. Klinman	
10.1	Introduction 1241	
10.2	Enzyme Kinetics: Extracting Chemistry from Complexity 1242	
10.3	Methodology for Detecting Nonclassical H-Transfers 1245	

10.3.1	Bond Stretch KIE Model: Zero-point Energy Effects	*1245*
10.3.1.1	Primary Kinetic Isotope Effects	*1246*
10.3.1.2	Secondary Kinetic Isotope Effects	*1247*
10.3.2	Methods to Measure Kinetic Isotope Effects	*1247*
10.3.2.1	Noncompetitive Kinetic Isotope Effects: k_{cat} or k_{cat}/K_M	*1247*
10.3.2.2	Competitive Kinetic Isotope Effects: k_{cat}/K_M	*1248*
10.3.3	Diagnostics for Nonclassical H-Transfer	*1249*
10.3.3.1	The Magnitude of Primary KIEs: $k_H/k_D > 8$ at Room Temperature	*1249*
10.3.3.2	Discrepant Predictions of Transition-state Structure and Inflated Secondary KIEs	*1251*
10.3.3.3	Exponential Breakdown: Rule of the Geometric Mean and Swain–Schaad Relationships	*1252*
10.3.3.4	Variable Temperature KIEs: $A_H/A_D \gg 1$ or $A_H/A_D \ll 1$	*1254*
10.4	Concepts and Theories Regarding Hydrogen Tunneling	*1256*
10.4.1	Conceptual View of Tunneling	*1256*
10.4.2	Tunnel Corrections to Rates: Static Barriers	*1258*
10.4.3	Fluctuating Barriers: Reproducing Temperature Dependences	*1260*
10.4.4	Overview	*1264*
10.5	Experimental Systems	*1265*
10.5.1	Hydride Transfers	*1265*
10.5.1.1	Alcohol Dehydrogenases	*1265*
10.5.1.2	Glucose Oxidase	*1270*
10.5.2	Amine Oxidases	*1273*
10.5.2.1	Bovine Serum Amine Oxidase	*1273*
10.5.2.2	Monoamine Oxidase B	*1275*
10.5.3	Hydrogen Atom (H•) Transfers	*1276*
10.5.3.1	Soybean Lipoxygense-1	*1276*
10.5.3.2	Peptidylglycine α-Hydroxylating Monooxygenase (PHM) and Dopamine β-Monooxygenase (DβM)	*1279*
10.6	Concluding Comments	*1280*
	References	*1281*
11	**Multiple-isotope Probes of Hydrogen Tunneling**	*1285*
	W. Phillip Huskey	
11.1	Introduction	*1285*
11.2	Background: H/D Isotope Effects as Probes of Tunneling	*1287*
11.2.1	One-frequency Models	*1287*
11.2.2	Temperature Dependence of Isotope Effects	*1289*
11.3	Swain–Schaad Exponents: H/D/T Rate Comparisons	*1290*
11.3.1	Swain–Schaad Limits in the Absence of Tunneling	*1291*
11.3.2	Swain–Schaad Exponents for Tunneling Systems	*1292*
11.3.3	Swain–Schaad Exponents from Computational Studies that Include Tunneling	*1293*

11.3.4	Swain–Schaad Exponents for Secondary Isotope Effects	1294
11.3.5	Effects of Mechanistic Complexity on Swain–Schaad Exponents	1294
11.4	Rule of the Geometric Mean: Isotope Effects on Isotope Effects	1297
11.4.1	RGM Breakdown from Intrinsic Nonadditivity	1298
11.4.2	RGM Breakdown from Isotope-sensitive Effective States	1300
11.4.3	RGM Breakdown as Evidence for Tunneling	1303
11.5	Saunders' Exponents: Mixed Multiple Isotope Probes	1304
11.5.1	Experimental Considerations	1304
11.5.2	Separating Swain–Schaad and RGM Effects	1304
11.5.3	Effects of Mechanistic Complexity on Mixed Isotopic Exponents	1306
11.6	Concluding Remarks	1306
	References	1307

12 Current Issues in Enzymatic Hydrogen Transfer from Carbon: Tunneling and Coupled Motion from Kinetic Isotope Effect Studies 1311
Amnon Kohen

12.1	Introduction	1311
12.1.1	Enzymatic H-transfer – Open Questions	1311
12.1.2	Terminology and Definitions	1312
12.1.2.1	Catalysis	1312
12.1.2.2	Tunneling	1313
12.1.2.3	Dynamics	1313
12.1.2.4	Coupling and Coupled Motion	1314
12.1.2.5	Kinetic Isotope Effects (KIEs)	1315
12.2	The H-transfer Step in Enzyme Catalysis	1316
12.3	Probing H-transfer in Complex Systems	1318
12.3.1	The Swain–Schaad Relationship	1318
12.3.1.1	The Semiclassical Relationship of Reaction Rates of H, D and T	1318
12.3.1.2	Effects of Tunneling and Kinetic Complexity on *EXP*	1319
12.3.2	Primary Swain–Schaad Relationship	1320
12.3.2.1	Intrinsic Primary KIEs	1320
12.3.2.2	Experimental Examples Using Intrinsic Primary KIEs	1322
12.3.3	Secondary Swain–Schaad Relationship	1323
12.3.3.1	Mixed Labeling Experiments as Probes for Tunneling and Primary–Secondary Coupled Motion	1323
12.3.3.2	Upper Semiclassical Limit for Secondary Swain–Schaad Relationship	1324
12.3.3.3	Experimental Examples Using 2° Swain–Schaad Exponents	1325
12.3.4	Temperature Dependence of Primary KIEs	1326
12.3.4.1	Temperature Dependence of Reaction Rates and KIEs	1326
12.3.4.2	KIEs on Arrhenius Activation Factors	1327

12.3.4.3	Experimental Examples Using Isotope Effects on Arrhenius Activation Factors *1328*	
12.4	Theoretical Models for H-transfer and Dynamic Effects in Enzymes *1331*	
12.4.1	Phenomenological "Marcus-like Models" *1332*	
12.4.2	MM/QM Models and Simulations *1334*	
12.5	Concluding Comments *1334*	
	Acknowledgments *1335*	
	References *1335*	

13 **Hydrogen Tunneling in Enzyme-catalyzed Hydrogen Transfer: Aspects from Flavoprotein Catalysed Reactions** *1341*
Jaswir Basran, Parvinder Hothi, Laura Masgrau, Michael J. Sutcliffe, and Nigel S. Scrutton

13.1	Introduction *1341*	
13.2	Stopped-flow Methods to Access the Half-reactions of Flavoenzymes *1343*	
13.3	Interpreting Temperature Dependence of Isotope Effects in Terms of H-Tunneling *1343*	
13.4	H-Tunneling in Morphinone Reductase and Pentaerythritol Tetranitrate Reductase *1347*	
13.4.1	Reductive Half-reaction in MR and PETN Reductase *1348*	
13.4.2	Oxidative Half-reaction in MR *1349*	
13.5	H-Tunneling in Flavoprotein Amine Dehydrogenases: Heterotetrameric Sarcosine Oxidase and Engineering Gated Motion in Trimethylamine Dehydrogenase *1350*	
13.5.1	Heterotetrameric Sarcosine Oxidase *1351*	
13.5.2	Trimethylamine Dehydrogenase *1351*	
13.5.2.1	Mechanism of Substrate Oxidation in Trimethylamine Dehydrogenase *1351*	
13.5.2.2	H-Tunneling in Trimethylamine Dehydrogenase *1353*	
13.6	Concluding Remarks *1356*	
	Acknowledgments *1357*	
	References *1357*	

14 **Hydrogen Exchange Measurements in Proteins** *1361*
Thomas Lee, Carrie H. Croy, Katheryn A. Resing, and Natalie G. Ahn

14.1	Introduction *1361*	
14.1.1	Hydrogen Exchange in Unstructured Peptides *1361*	
14.1.2	Hydrogen Exchange in Native Proteins *1363*	
14.1.3	Hydrogen Exchange and Protein Motions *1364*	
14.2	Methods and Instrumentation *1365*	
14.2.1	Hydrogen Exchange Measured by Nuclear Magnetic Resonance (NMR) Spectroscopy *1365*	

14.2.2	Hydrogen Exchange Measured by Mass Spectrometry *1367*
14.2.3	Hydrogen Exchange Measured by Fourier-transform Infrared (FT-IR) Spectroscopy *1369*
14.3	Applications of Hydrogen Exchange to Study Protein Conformations and Dynamics *1371*
14.3.1	Protein Folding *1371*
14.3.2	Protein–Protein, Protein–DNA Interactions *1374*
14.3.3	Macromolecular Complexes *1378*
14.3.4	Protein–Ligand Interactions *1379*
14.3.5	Allostery *1381*
14.3.6	Protein Dynamics *1382*
14.4	Future Developments *1386*
	References *1387*

15	**Spectroscopic Probes of Hydride Transfer Activation by Enzymes** *1393*
	Robert Callender and Hua Deng
15.1	Introduction *1393*
15.2	Substrate Activation for Hydride Transfer *1395*
15.2.1	Substrate C–O Bond Activation *1395*
15.2.1.1	Hydrogen Bond Formation with the C–O Bond of Pyruvate in LDH *1395*
15.2.1.2	Hydrogen Bond Formation with the C–O Bond of Substrate in LADH *1397*
15.2.2	Substrate C–N Bond Activation *1398*
15.2.2.1	N5 Protonation of 7,8-Dihydrofolate in DHFR *1398*
15.3	NAD(P) Cofactor Activation for Hydride Transfer by Enzymes *1401*
15.3.1	Ring Puckering of Reduced Nicotinamide and Hydride Transfer *1401*
15.3.2	Effects of the Carboxylamide Orientation on the Hydride Transfer *1403*
15.3.3	Spectroscopic Signatures of "Entropic Activation" of Hydride Transfer *1404*
15.3.4	Activation of CH bonds in $NAD(P)^+$ or $NAD(P)H$ *1405*
15.4	Dynamics of Protein Catalysis and Hydride Transfer Activation *1406*
15.4.1	The Approach to the Michaelis Complex: the Binding of Ligands *1407*
15.4.2	Dynamics of Enzymic Bound Substrate–Product Interconversion *1410*
	Acknowledgments *1412*
	Abbreviations *1412*
	References *1412*

| Part IV | Hydrogen Transfer in the Action of Specific Enzyme Systems 1417 |

| 16 | **Hydrogen Transfer in the Action of Thiamin Diphosphate Enzymes** 1419 |

Gerhard Hübner, Ralph Golbik, and Kai Tittmann

16.1	Introduction 1419
16.2	The Mechanism of the C2-H Deprotonation of Thiamin Diphosphate in Enzymes 1421
16.2.1	Deprotonation Rate of the C2-H of Thiamin Diphosphate in Pyruvate Decarboxylase 1422
16.2.2	Deprotonation Rate of the C2-H of Thiamin Diphosphate in Transketolase from *Saccharomyces cerevisiae* 1424
16.2.3	Deprotonation Rate of the C2-H of Thiamin Diphosphate in the Pyruvate Dehydrogenase Multienzyme Complex from *Escherichia coli* 1425
16.2.4	Deprotonation Rate of the C2-H of Thiamin Diphosphate in the Phosphate-dependent Pyruvate Oxidase from *Lactobacillus plantarum* 1425
16.2.5	Suggested Mechanism of the C2-H Deprotonation of Thiamin Diphosphate in Enzymes 1427
16.3	Proton Transfer Reactions during Enzymic Thiamin Diphosphate Catalysis 1428
16.4	Hydride Transfer in Thiamin Diphosphate-dependent Enzymes 1432
	References 1436

| 17 | **Dihydrofolate Reductase: Hydrogen Tunneling and Protein Motion** 1439 |

Stephen J. Benkovic and Sharon Hammes-Schiffer

17.1	Reaction Chemistry and Catalysis 1439
17.1.1	Hydrogen Tunneling 1441
17.1.2	Kinetic Analysis 1443
17.2	Structural Features of DHFR 1443
17.2.1	The Active Site of DHFR 1444
17.2.2	Role of Interloop Interactions in DHFR Catalysis 1446
17.3	Enzyme Motion in DHFR Catalysis 1447
17.4	Conclusions 1452
	References 1452

| 18 | **Proton Transfer During Catalysis by Hydrolases** 1455 |

Ross L. Stein

18.1	Introduction 1455
18.1.1	Classification of Hydrolases 1455
18.1.2	Mechanistic Strategies in Hydrolase Chemistry 1456
18.1.2.1	Heavy Atom Rearrangement and Kinetic Mechanism 1457

18.1.2.2	Proton Bridging and the Stabilization of Chemical Transition States *1458*
18.1.3	Focus and Organization of Chapter *1458*
18.2	Proton Abstraction – Activation of Water or Amino Acid Nucleophiles *1459*
18.2.1	Activation of Nucleophile – First Step of Double Displacement Mechanisms *1459*
18.2.2	Activation of Active-site Water *1462*
18.2.2.1	Double-displacement Mechanisms – Second Step *1462*
18.2.2.2	Single Displacement Mechanisms *1464*
18.3	Proton Donation – Stabilization of Intermediates or Leaving Groups *1466*
18.3.1	Proton Donation to Stabilize Formation of Intermediates *1466*
18.3.2	Proton Donation to Facilitate Leaving Group Departure *1467*
18.3.2.1	Double-displacement Mechanisms *1467*
18.3.2.2	Single-displacement Mechanisms *1468*
18.4	Proton Transfer in Physical Steps of Hydrolase-catalyzed Reactions *1468*
18.4.1	Product Release *1468*
18.4.2	Protein Conformational Changes *1469*
	References *1469*

19	**Hydrogen Atom Transfers in B_{12} Enzymes** *1473*
	Ruma Banerjee, Donald G. Truhlar, Agnieszka Dybala-Defratyka, and Piotr Paneth
19.1	Introduction to B_{12} Enzymes *1473*
19.2	Overall Reaction Mechanisms of Isomerases *1475*
19.3	Isotope Effects in B_{12} Enzymes *1478*
19.4	Theoretical Approaches to Mechanisms of H-transfer in B_{12} Enzymes *1480*
19.5	Free Energy Profile for Cobalt–Carbon Bond Cleavage and H-atom Transfer Steps *1487*
19.6	Model Reactions *1488*
19.7	Summary *1489*
	Acknowledgments *1489*
	References *1489*

Part V	**Proton Conduction in Biology** *1497*

20	**Proton Transfer at the Protein/Water Interface** *1499*
	Menachem Gutman and Esther Nachliel
20.1	Introduction *1499*
20.2	The Membrane/Protein Surface as a Special Environment *1501*
20.2.1	The Effect of Dielectric Boundary *1501*

20.2.2	The Ordering of the Water by the Surface *1501*	
20.2.2.1	The Effect of Water on the Rate of Proton Dissociation *1502*	
20.2.2.2	The Effect of Water Immobilization on the Diffusion of a Proton *1503*	
20.3	The Electrostatic Potential Near the Surface *1504*	
20.4	The Effect of the Geometry on the Bulk-surface Proton Transfer Reaction *1505*	
20.5	Direct Measurements of Proton Transfer at an Interface *1509*	
20.5.1	A Model System: Proton Transfer Between Adjacent Sites on Fluorescein *1509*	
20.5.1.1	The Rate Constants of Proton Transfer Between Nearby Sites *1509*	
20.5.1.2	Proton Transfer Inside the Coulomb Cage *1511*	
20.5.2	Direct Measurements of Proton Transfer Between Bulk and Surface Groups *1514*	
20.6	Proton Transfer at the Surface of a Protein *1517*	
20.7	The Dynamics of Ions at an Interface *1518*	
20.8	Concluding Remarks *1522*	
	Acknowledgments *1522*	
	References *1522*	

Index *1527*

Preface

As one of the simplest of chemical reactions, pervasive on this highly aqueous planet populated by highly aqueous organisms, yet still imperfectly understood, the transfer of hydrogen as a subject of scientific attention seems hardly to require defense. This claim is supported by the readiness with which the editors of this series of four volumes on *Hydrogen-transfer Reactions* accepted the suggestion that they organize a group of their most active and talented colleagues to survey the subject from viewpoints beginning in physics and extending into biology. Furthermore, forty-nine authors and groups of authors acceded, with alacrity and grace, to the request to contribute and have then supplied the articles that make up these volumes.

Our scheme of organization involved an initial division into physical and chemical aspects on the one hand, and biological aspects on the other hand (and one might well have said biochemical and biological aspects). In current science, such a division may provide an element of convenience but no-one would seriously claim the segregation to be either easy or entirely meaningful. We have accordingly felt quite entitled to place a number of articles rather arbitrarily in one or the other category. It is nevertheless our hope that readers may find the division adequate to help in the use of the volumes. It will be apparent that the division of space between the two categories is unequal, the physical and chemical aspects occupying considerably more pages than the biological aspects, but our judgment is that this distribution of space is proper to the subjects treated. For example, many of the treatments of fundamental principles and broadly applicable techniques were classified under physical and chemical aspects. But they have powerful implications for the understanding and use of the matters treated under biological aspects.

Within each of these two broad disciplinary categories, we have organized the subject by beginning with the simple and proceeding toward the complex. Thus the physical and chemical aspects appear as two volumes, volume1 on simple systems and volume 2 on complex systems. Similarly, the biological aspects appear as volume 3 on simple systems and volume 4 on complex systems.

Volume 1 then begins with isolated molecules, complexes, and clusters, then treats condensed-phase molecules, complexes, and crystals, and finally reaches treatments of molecules in polar environments and in electronic excited states. Volume 2 reaches higher levels of complexity in protic systems with bimolecular reactions in solution, coupling of proton transfer to low-frequency motions and proton-coupled electron transfer, then organic and organometallic reactions, and hydrogen-transfer reactions in solids and on surfaces. Thereafter articles on quantum tunneling and appropriate theories of hydrogen transfer complete the treatment of physical and chemical aspects.

Volume 3 begins with simple model (i.e., non-enzymic) reactions for proton-transfer, both to and from carbon and among electronegative atoms, hydrogen-atom transfer, and hydride transfer, as well as the extension to small, synthetic peptides. It is completed by treatments of how enzymes activate C-H bonds, multiple hydrogen transfer reactions in enzymes, and theoretical models. Volume 4 moves then into enzymic reactions and a thorough consideration of quantum tunneling and protein dynamics, one of the most vigorous areas of study in biological hydrogen transfer, then considers several specific enzyme systems of high interest, and is completed by the treatment of proton conduction in biological systems.

While we do not claim any sort of comprehensive coverage of this large subject, we believe the reader will find a representative treatment, written by accomplished and respected experts, of most of the matters currently considered important for an understanding of hydrogen-transfer reactions. I am enormously grateful to James T. (Casey) Hynes and Hans-Heinrich Limbach, who saw to the high quality of the volumes on the physical and chemical aspects, and to Judith Klinman, who gave me a nearly free pass as her co-editor of the volumes on biological aspects. We are all grateful indeed to the authors who contributed their wisdom and eloquence to these volumes. It has been a very great pleasure to be assisted, encouraged, and supported at every turn by the outstanding staff of VCH-Wiley in Weinheim, particularly (in alphabetical order) Ms. Nele Denzau, Dr. Renate Dötzer, Dr. Tim Kersebohm, Dr. Elke Maase, Ms. Claudia Zschernitz, and – of course – Dr. Peter Gölitz.

Lawrence, Kansas, USA, September 2006 *Richard L. Schowen*

Preface to Volumes 3 and 4

These volumes together address the rather enormous subject of hydrogen transfer in biological systems, volume 3 presenting the role of relatively simple systems in the understanding of hydrogen transfer while volume 4 considers complex systems, for the most part enzymes.

Volume 3 contains two parts that treat basic concepts and systems not limited to a single enzyme or class of enzymes in their significance. Part I consists of five chapters on the chemistry of the transfer of hydrogen in biological model systems: as a proton to and from carbon (Amyes and Richard, Ch. 1); as a proton in acid-base catalysis; i.e., largely among electronegative atoms (Kirby, Ch. 2); as a hydrogen atom (Schöneich, Ch. 3); as a hydride ion (Schowen, Ch..4); as a proton in acid-base catalysis in designed peptides (Baltzer, Ch. 5). Part II is composed of three chapters on generally significant features of biological hydrogen-transfer reactions: in enzyme-catalyzed proton transfer from carbon (Gerlt, Ch. 6); in multiple proton transfers in enzymic systems (Spies and Toney, Ch. 7); and in computer simulations of enzymic hydrogen transfer (Braun-Sand, Olsson, Mavri, and Warshel, Ch. 8).

Volume 4, consisting of three parts, then proceeds to studies in enzyme and protein systems that for the most part serve well as paradigms for broader groups in which hydrogen transfer is important. Part III brings together seven chapters on the subject of quantum tunneling in enzymic hydrogen-transfer and its relationship to protein motions. A relative new theoretical approach is described by Schwartz (Ch. 9), leading into a general consideration of the existing evidence and its significance for the tunneling/dynamics nexus (Knapp, Meyer, and Klinman, Ch. 10), and articles by Huskey (Ch. 11) on the importance of multiple-isotope labeling for characterization of tunneling phenomena, by Kohen on kinetic isotope effects (Ch.12) and by Basran, Hothi, Masgrau, Sutcliffe, and Scrutton on the opportunities afforded by flavoprotein systems (Ch. 13). This part is closed by articles on two important experimental approaches, isotope exchange with solvent as a probe of protein motion (Lee, Croy, Resing, and Ahn, Ch. 14) and resonance Raman spectroscopy as a probe of active-site dynamical properties (Callender and Deng, Ch. 15). Part IV brings into focus several central examples of important enzyme classes: thiamin-dependent enzymes (Ch. 16 by Hübner, Golbik, and

Tittmann), dihydrofolate reductase (Ch. 17 by Benkovic and Hammes-Schiffer), hydrolases (Ch. 18 by Stein), and vitamin B_{12} enzymes (Ch. 19 by Banerjee, Truhlar, Dybala-Defratyka, and Paneth). The volume is the closed by a one-chapter Part V on proton conduction in biology, in which Gutman and Nachliel (Ch. 20) treat the subject of proton conductance at protein surfaces and interfacial regions.

JPK acknowledges the support of grant MCB 0446395 from the US National Science Foundation and of grant GM 025765 from the US National Institutes of Health.

Berkeley, California, USA, September 2006 — *Judith P. Klinman*
Lawrence, Kansas, USA, September 2006 — *Richard L. Schowen*

List of Contributors to Volumes 3 and 4

Natalie G. Ahn
Department of Chemistry and
Biochemistry
Howard Hughes Medical Institute
University of Colorado
Boulder, CO 80309-0215
USA

Tina L. Amyes
Department of Chemistry
University at Buffalo
SUNY
Buffalo, NY 14260-3000
USA

Lars Baltzer
Department of Chemistry
Uppsala University
Box 599
75124 Uppsala
Sweden

Ruma Banerjee
Biochemistry Department
University of Nebraska
Lincoln, NE 68588-0664
USA

Jaswir Basran
Department of Biochemistry
University of Leicester
University Road
Leicester LE1 7RH
UK

Stephen J. Benkovic
Department of Chemistry
104 Chemistry Building,
Pennsylvania State University
University Park, PA 16802
USA

Sonja Braun-Sand
University of Southern California
Department of Chemistry
3620 McClintock Avenue, SGM 418
Los Angeles, CA 90089-1062
USA

Robert Callender
Department of Biochemistry
Albert Einstein College of Medicine
1300 Morris Park Avenue
Bronx, NY 10461
USA

Carrie H. Croy
Department of Chemistry and
Biochemistry
Howard Hughes Medical Institute
University of Colorado
Boulder, CO 80309-0215
USA

Hua Deng
Department of Biochemistry
Albert Einstein College of Medicine
1300 Morris Park Avenue
Bronx, NY 10461
USA

Hydrogen-Transfer Reactions. Edited by J. T. Hynes, J. P. Klinman, H.-H. Limbach, and R. L. Schowen
Copyright © 2007 WILEY-VCH Verlag GmbH & Co. KGaA, Weinheim
ISBN: 978-3-527-30777-7

Agnieszka Dybala-Defratyka
Faculty of Chemistry
Technical University of Lodz
90-924 Lodz
Poland

John A. Gerlt
University of Illinois,
Urbana-Champaign
Departments of Biochemistry and
Chemistry
600 South Mathews Avenue
Urbana, IL 61801
USA

Ralph Golbik
Institute of Biochemistry
Martin Luther University
Halle-Wittenberg
Kurt-Mothes-Strasse 3
06120 Halle/Saale
Germany

Menachem Gutman
Laser Laboratory for Fast Reactions
in Biology
Department of Biochemistry
George S. Wise Faculty of Life Sciences
Tel Aviv University
Tel Aviv 69978
Israel

Sharon Hammes-Schiffer
Department of Chemistry
104 Chemistry Building
Pennsylvania State University
University Park, PA 16802
USA

Parvinder Hothi
Faculty of Life Sciences and
Manchester Interdisciplinary Biocentre
University of Manchester
131 Princess Street
Manchester M1 7ND
UK

Gerhard Hübner
Institute of Biochemistry
Martin Luther University
Halle-Wittenberg
Kurt-Mothes-Strasse 3
06120 Halle/Saale
Germany

W. Phillip Huskey
Department of Chemistry
Rutgers University – Newark
73 Warren Street
Newark, NJ 07102
USA

Anthony J. Kirby
University Chemical Laboratory
Cambridge CB2 1EW
UK

Judith P. Klinman
Departments of Chemistry and
Molecular and Cell Biology
University of California
Berkeley, CA 94720-1460
USA

Michael J. Knapp
Department of Chemistry
710 N. Pleasant Street
University of Massachusetts
Amherst, MA 01003-9336
USA

Amnon Kohen
Department of Chemistry
University of Iowa
Iowa City, IA 52242
USA

Thomas Lee
Department of Chemistry and
Biochemistry
Howard Hughes Medical Institute
University of Colorado
Boulder, CO 80309-0215
USA

Laura Masgrau
School of Chemical Engineering and
Analytical Science Manchester
Interdisciplinary Biocentre
University of Manchester
131 Princess Street
Manchester M1 7ND
UK

Janez Mavri
National Institute of Chemistry
P.O.B. 660
Hajarihova 19
SI-1001 Ljubljana
Slovenia

Matthew Meyer
Merced School of Natural Sciences
University of California
P.O. Box 2039
Merced, CA 95344
USA

Esther Nachliel
Laser Laboratory for Fast Reactions
in Biology
Department of Biochemistry
George S. Wise Faculty of Life Sciences
Tel Aviv University
Tel Aviv 69978
Israel

Mats H. M. Olsson
University of Southern California
Department of Chemistry
3620 McClintock Avenue, SGM 418
Los Angeles, CA 90089-1062
USA

Piotr Paneth
Faculty of Chemistry
Technical University of Lodz
90-924 Lodz
Poland

Katheryn A. Resing
Department of Chemistry and
Biochemistry
Howard Hughes Medical Institute
University of Colorado
Boulder, CO 80309-0215
USA

John P. Richard
Department of Chemistry
University at Buffalo
SUNY
Buffalo, NY 14260-3000
USA

Christian Schöneich
Department of Pharmaceutical
Chemistry
University of Kansas
2095 Constant Avenue
Lawrence, KS 66047
USA

Richard L. Schowen
Departments of Chemistry, Molecular
Biosciences, and Pharmaceutical
Chemistry
University of Kansas
Lawrence, KS 66047
USA

Steven D. Schwartz
Departments of Biophysics and
Biochemistry
Seaver Center for Bioinformatics
Albert Einstein College of Medicine
Bronx, New York
USA

Nigel S. Scrutton
Faculty of Life Sciences and
Manchester Interdisciplinary Biocentre
University of Manchester
131 Princess Street
Manchester M1 7ND
UK

Michael Ashley Spies
Department of Biochemistry
University of Illinois
600 South Mathews Avenue
Urbana, IL 61801
USA

Ross L. Stein
Laboratory for Drug Discovery in Neurodegeneration
Harvard Center for Neurodegeneration and Repair
Department of Neurology
Harvard Medical School
65 Landsdowne Street, Fourth Floor
Cambridge, MA 02129
USA

Michael J. Sutcliffe
School of Chemical Engineering and Analytical Science Manchester
Interdisciplinary Biocentre
University of Manchester
131 Princess Street
Manchester M1 7ND
UK

Kai Tittmann
Martin Luther University
Halle-Wittenberg
Institute of Biochemistry
Kurt-Mothes-Strasse 3
06120 Halle/Saale
Germany

Michael D. Toney
Department of Chemistry
University of California, Davis
1-Shields Avenue
Davis, CA 96616
USA

Donald G. Truhlar
Chemistry Department
University of Minnesota
Minneapolis, MN 55455-0431
USA

Arieh Warshel
University of Southern California
Department of Chemistry
3620 McClintock Avenue, SGM 418
Los Angeles, CA 90089-1062
USA

Part III
Quantum Tunneling and Protein Dynamics

This section deals with the timely subject of the role of quantum mechanical tunneling in enzyme reactions and the way in which this tunneling is linked to protein structure and dynamics. The contributions come primarily from experimentalists, though a stimulating theoretical chapter by Schwartz introduces this section. Schwartz has used a Quantum Kramers approach to model H- transfer in the condensed phase, leading to a formalism for motions within the environment that can be coupled to the H-transfer coordinate in either an anti-symmetric or symmetric manner. The former is similar to the λ parameter in Marcus theory, whereas the latter describes the change in the distance between the reactants, referred to as a promoting vibration(s). Promoting vibrations occur on a very fast time scale and are considered to be "directly" coupled to the reaction coordinate (contrasting with statistical views of protein motions that impact the probability of H-transfer in condensed phases). Schwartz has developed an algorithm to define promoting modes within a protein, applying these methods to both alcohol and lactate dehydrogenases. Recent studies suggest that analyses linking protein motions to the chemical step can be performed also for enzyme catalyzed cleavage of heavy atoms (Antoniou et al. (2006) *Chem. Rev.* **106**, 3170–3187). The chapters by Klinman and co-workers, Kohen, and Scrutton and co-workers all address the growing evidence for H-tunneling in enzyme reactions and the facilitating role of the protein environment. All three chapters lay out methodologies available for detecting tunneling, and their strengths and weaknesses. The range of enzyme reactions that have now been implicated to have tunneling components is quite impressive. The importance of full tunneling models, instead of tunneling correction models, is emphasized, as new data emerge that cannot be easily rationalized by tunneling corrections. The chapter by Huskey focuses on the Swain-Schaad relationship as a basis for detecting tunneling. This is a thoughtful treatise, indicating that single site isotope substitutions are not likely to show deviations indicative of tunneling. By contrast, multi-site substitutions have the potential to be very informative in this regard, especially in the case of secondary isotope effect measurements. Knapp et al., Kohen, Scrutton and Huskey all emphasize the critical importance of isolating single rate-limiting hydrogen-transfer steps for the successful diagnosis of tunneling. The measurement of protein dynamics and its link to catalysis, is a challenging area, normally approached by NMR and/or H/D exchange methodologies.

The chapter by Ahn and co-workers provides a thorough summary of the methods available for the measurement of hydrogen exchange in proteins, which include NMR, mass spectrometry and FT-IR. These methods differ in their spatial resolution, the mass spectrometric approach offering a compromise between moderate resolution and general applicability. This chapter also outlines the full range of questions that can be addressed via H/D exchange including an assessment of the link between protein motions and the hydrogen transfer step. In the final chapter of this section, Callender and Deng describe the application of electronic and vibrational spectroscopy to illuminate the impact of the enzyme active site on the properties and activation of the bound substrates. Temperature jump experiments with lactate dehydrogenase indicate the multi-step nature of substrate binding and show how some protein "melting" may be essential during the formation of the catalytic complex.

9
The Quantum Kramers Approach to Enzymatic Hydrogen Transfer – Protein Dynamics as it Couples to Catalysis

Steven D. Schwartz

9.1
Introduction

Though all life forms are dependent on the catalytic effect of enzymes, detailed understanding of the microscopic mechanism of their action has lagged. This is largely due to the great complexity of enzyme catalyzed chemical reactions. Certainly a large portion of the catalytic effect in enzyme catalyzed reactions comes from the lowering of the free energy barrier to reaction. This preferential binding of the enzyme to the transition state is a concept credited to Pauling [1], and is the origin of the extraordinary potency of transition state inhibitors [2]. This viewpoint is, however, a statistical view of the catalytic process, not a dynamic understanding of how atoms or groups of atoms promote the catalytic event in microscopic detail. One would wish for such a detailed understanding. Work in our group over the past few years has focused on providing a formulation which allows such analysis [3]. Chemical reactions involve the making and breaking of chemical bonds, and so are inherently quantum mechanical in nature. At the very least, a quantum mechanical method is needed to generate a potential energy surface for the reaction of interest. In addition, many of the atom transfer reactions in the chemical step of certain enzymes, inherently involve quantum dynamics – that is if one uses classical mechanics to study their dynamics, the wrong answer will be obtained. One would thus like a fully quantum theory for the study of rate processes in enzymes, but the systems are far too complex for exact solution. We developed our approach, the Quantum Kramers theory to study chemical reactions in condensed phases and then applied it to enzymatic reactions. In fact, it is not an understatement to say that our current work detailing the importance of protein dynamics in the catalytic process of enzymes is in fact due to the failure of our simple theory of condensed phase chemical reactions to be applicable to some enzymatic reactions. The correction of the simple theory resulted in the inclusion of basically different chemical physics than that contained in our earlier work. This chapter will outline the development of our theory and then describe applications of the methodology and further work we have undertaken to understand the importance of protein dynamics in enzymatically catalyzed reactions.

Hydrogen-Transfer Reactions. Edited by J. T. Hynes, J. P. Klinman, H.-H. Limbach, and R. L. Schowen
Copyright © 2007 WILEY-VCH Verlag GmbH & Co. KGaA, Weinheim
ISBN: 978-3-527-30777-7

The structure of this chapter is as follows: Section 9.2 describes the theoretical development of the basic Quantum Kramers methodology [4]. In Section 9.3.1 we proceed to understand the nature of the "promoting vibration," why this physical feature is not present in the basic Quantum Kramers methods and how it can be incorporated. Section 9.3.2 describes why the symmetry of the coupling of promoting vibrations results in the phenomenon known as corner cutting and why, in turn, this masks kinetic isotope effects. In Section 9.4 we begin our study of specific enzyme systems, concentrating first on alcohol dehydrogenases. In Section 9.5 we study lactate dehydrogenase and both identify a unique kinetic control mechanism that may be present in two highly similar human isoforms of the enzyme, and apply a new technique known as Transition Path Sampling [5]. Developed by David Chandler and coworkers, this approach allows the study of an enzymatic reaction in microscopic detail. The atomic motions necessary for chemical reaction to occur are specified. Section 9.6 presents a general expansion of the Quantum Kramers approach to the study of coupled electron proton transfer reactions. A brief Section 9.7 provides preliminary results of a coupled protein motion in a reaction not involving hydrogen transfer, but in which protein motion polarizes bonds and allows leaving group departure. Finally Section 9.8 concludes with discussion of future direction for this area of work.

9.2
The Derivation of the Quantum Kramers Method

It is known that for a purely classical system [6], an accurate approximation of the dynamics of a tagged degree of freedom (for example a reaction coordinate) in a condensed phase can be obtained through the use of a generalized Langevin equation. The generalized Langevin equation is given by Newtonian dynamics plus the effects of the environment in the form of a memory friction and a random force [7].

$$m\ddot{s} = -\frac{\partial V(s)}{\partial s} + \int_0^t dt'\gamma(t-t')\dot{s} + F(t) \tag{9.1}$$

Here the first two terms just give $ma =$ Force as mass times the second time derivative of the friction equal to the F as the negative derivative of potential. γ is the memory friction, and $F(t)$ is the random force. Thus the complex dynamics of all degrees of freedom other than the reaction coordinate are included in a statistical treatment, and the reaction coordinate plus environment are modeled as a modified one-dimensional system. What allows realistic simulation of complex systems is that the statistics of the environment can in fact be calculated from a formal prescription. This prescription is given by the Fluctuation–Dissipation theorem, which yields the relation between the friction and the random force. In particular, this theory shows how to calculate the memory friction from a relatively short-time classical simulation of the reaction coordinate. The Quantum Kramers approach, in turn, is dependent on an observation of Zwanzig [8], that if an interaction potential for a condensed phase system satisfies a fairly broad set of mathematical criteria, the dynamics of the reaction coordinate as described by the generalized Langevin equation can be rigorously equated to a microscopic Hamiltonian in which the reaction coordinate is coupled to an infinite set of Harmonic Oscillators via simple bilinear coupling:

$$H = \frac{P_s^2}{2m_s} + V_o + \sum_k \frac{P_k^2}{2m_k} + \frac{1}{2} m_k \omega_k^2 \left(q_k - \frac{c_k s}{m_k \omega_k^2} \right)^2 \quad (9.2)$$

The first two terms in this Hamiltonian represent the kinetic and potential energy of the reaction coordinate, and the last set of terms similarly represent the kinetic and potential energy for an environmental bath. Here s is some coordinate that measure progress of the reaction (for example in alcohol dehydrogenase where the chemical step is transfer of a hydride, s might be chosen to represent the relative position of the hydride from the alcohol to the NAD cofactor.) c_k is the strength of the coupling of the environmental mode to the reaction coordinate, and m_k and ω_k give the effective mass and frequency of the environmental bath mode. A discrete spectral density gives the distribution of bath modes in the harmonic environment:

$$J(\omega) = \frac{\pi}{2} \sum_k \frac{c_k^2}{m_k \omega_k} [\delta(\omega - \omega_k) - \delta(\omega + \omega_k)] \quad (9.3)$$

Here $\delta(\omega - \omega_k)$ is the Dirac delta function, so the spectral density is simply a collection of spikes, located at the frequency positions of the environmental modes, convolved with the strength of the coupling of these modes to the reaction coordinate. Note that this infinite collection of oscillators is purely fictitious – they are chosen to reproduce the overall physical properties of the system, but do not necessarily represent specific physical motions of the atoms in the system. Now it would seem that we have not made a huge amount of progress – we began with a many-dimensional system (classical) and found out that it could be accurately approximated by a one-dimensional system in a frictional environment (the generalized Langevin equation.) We have now recreated a many-dimensional system (the Zwanzig Hamiltonian.) The reason we have done this is two-fold. First, there is no true quantum mechanical analog of friction, and so there really is no way to use the generalized Langevin approach for a quantum system, such as we would like to do for an enzyme. Second, the new quantum Hamiltonian given by Eq. (9.2) is very much simpler than the Hamiltonian for the full enzymatic system. Harmonic oscillators are the one type of problem that can easily be solved in quantum mechanics. Thus, the prescription is, given a potential for a reaction, we model the exact problem using a Zwanzig Hamiltonian, as in Eq. (9.2), with distribution of harmonic modes given by the spectral density in Eq. (9.3), and found through a simple classical computation of the frictional force on the reaction coordinate.

Then using methods to compute quantum dynamics developed in our group [9], quantities such as rates or kinetic isotope effects may be found. These methods are an approximate but accurate way to compute the quantum mechanical evolution of any systems. The details are given in the literature [10], but in short, we write a general Hamiltonian as:

$$\hat{H} = \hat{H}_a + \hat{H}_b + \hat{f}(a,b) \tag{9.4}$$

where "a" and "b" are shorthand for any number of degrees of freedom. $f(a,b)$ is a coupling, usually only a function of coordinates, but this is not required. Our approach rests on the fact that because these three terms are operators, the exact evolution operator may not be expressed as a product:

$$e^{-iHt} \neq e^{-i\hat{H}_a t} e^{-i\hat{H}_b t + f(a,b)} \tag{9.5}$$

but in fact equality may be achieved by application of an infinite order product of nth order commutators:

$$e^{-iHt} = e^{-i\hat{H}_a t} e^{-i\hat{H}_b t + f(a,b)} e^{c_1} e^{c_2} e^{c_3} \cdots \tag{9.6}$$

This is usually referred to as the Zassenhaus expansion or the Baker Campbell Hausdorf theorem [11]. As an aside a symmetrized version of this expansion terminated at the C_1 term results in the Feit and Fleck [12] approximate propagator. We have shown [13] that an infinite order subset of these commutators, may be re-summed exactly as an interaction propagator:

$$U(t)_{resum} = U(t)_{H_a} U(t)_{H_b + f(a,b)} U^{-1}(t)_{H_a + f(a,b)} U(t)_{H_a} \tag{9.7}$$

The first two terms are just the adiabatic approximation, and the second two terms the correction. For example, if we have a fast subsystem labeled by the "coordinate" a, and a slow subsystem labeled by b; then the approximate evolution operator to first order in commutators with respect to the slow subsystem $b([f(a,b), H_b])$, and infinite order in the commutators of the "fast" Hamiltonian with the coupling: $([f(a,b), H_a])$ is given by:

$$e^{-i(H_a + H_b + f(a,b))t/\hbar} \approx e^{-iH_a t/\hbar} e^{-i(H_b + f(a,b))t/\hbar} e^{+i(H_a + f(a,b))t/\hbar} e^{-iH_a t/\hbar} \tag{9.8}$$

The advantage to this formulation is that higher dimensional evolution operators are replaced by a product of lower dimensional evolution operators. This is always a far easier computation. In addition, because products of evolution operators replace the full evolution operator, a variety of mathematical properties are retained, such as unitarity, and thus time reversal symmetry.

What we have produced so far is an approximate Hamiltonian designed to study chemical reactions in complex condensed phases. We also have a mathematical method to evaluate quantum propagation using this Hamiltonian. We as yet have no practical method to compute observables such as rates. The flux correlation function formalism of Miller, Schwartz, and Tromp [14] provides such a method. Combination of the quantum Kramers idea with the re-summed evolution operators results in a largely analytic formulation for the flux autocorrelation function for a chemical reaction in a condensed phase. After a lengthy but not complex computation the quantum Kramers flux autocorrelation function has been shown to be [15]:

$$C_f = C_f^0 B_1 Z_{bath} - \int_0^\infty d\omega \kappa_f^0 J(\omega) B_2 Z_{bath} \qquad (9.9)$$

Here C_f^0 is the gas phase (uncoupled) flux autocorrelation function, Z_{bath} is the bath partition function, $J(\omega)$ is the bath spectral density (computed as described above from a classical molecular dynamics computation), B_1 and B_2 are combinations of trigonometric functions of the frequency ω and the inverse barrier frequency, and finally:

$$\kappa_f^0 = \frac{1}{4m_s^2} |\langle s=0|e^{-iH_s t_c/\hbar}|s=0\rangle|^2 \qquad (9.10)$$

As in other flux correlation function computations, t_c is the complex time $t - \frac{i\hbar\beta}{2}$. Thus, given the Quantum Kramers model for the reaction in the complex system, and the re-summed operator expansion as a practical way to evaluate the necessary evolution operators needed for the flux autocorrelation function, the quantum rate in the complex system is reduced to a simple combination of gas phase correlation functions with simple algebraic functions.

This approach is able to model a variety of condensed phase chemical reactions with essentially experimental accuracy [16]. We did find, however, one specific experimental system for which this methodology was not able to reproduce experimental results, and that is proton transfer in benzoic acid crystals. In developing a physical understanding of this system, we first identified the concept of the promoting vibration.

9.3
Promoting Vibrations and the Dynamics of Hydrogen Transfer

9.3.1
Promoting Vibrations and The Symmetry of Coupling

The Hamiltonian of Eq. (9.2) couples the reaction coordinate to the environmental oscillator degrees of freedom by terms linear in both reaction coordinate and bath degree of freedom. This is derived in Zwanzig's original approach by an expansion of the full potential in bath coordinates to second order. This innocuous approximation in fact conceals a fair amount of missing physics. We have shown [16a] that this collection of bilinearly coupled oscillators is in fact a microscopic version of the popular Marcus theory for charged particle transfer [17]. The bilinear coupling of the bath of oscillators is the simplest form of a class of couplings that may be termed antisymmetric because of the mathematical property of the functional form of the coupling on reflection about the origin. This property has deeper implications than the mathematical nature of the symmetry properties. Antisymmetric couplings, when coupled to a double well-like potential energy profile, are able to instantaneously change the level of well depths, but do nothing to the position of well minima. This modulation in the depth of minima is exactly what the environment is envisaged to do within the Marcus theory paradigm. As we have shown [16], the minima of the total potential in Eq. (9.2) will occur, for a two-dimensional version of this potential, when the q degree of freedom is exactly equal and opposite in sign to $\frac{cs}{m\omega^2}$, and the minimum of the potential energy profile along the reaction coordinate is unaffected by this coupling. Within Marcus' theory, which is a deep tunneling theory, transfer of the charged particle occurs at the value of the bath coordinates that cause the total potential to become symmetrized. Thus, if the bare reaction coordinate potential is symmetric, then the total potential is symmetrized at the position of the "bath plus coupling" minimum. When this configuration is achieved, the particle tunnels, and in fact the activation energy for the reaction is largely the energy to bring the bath into this favorable tunneling configuration.

The question is if such motions and their mathematical representations encompass all important motions in the coupling of dynamic motions to a reaction coordinate. We became aware of an example in which there is another significant contributor to the chemical dynamics – benzoic acid crystals. There is a long history of the study of proton transfer in crystalline benzoic acid [18]. These experiments seemed to yield anomolous results when compared with quantum chemistry computations. That is, computations showed a reasonably high barrier while experiment showed a low activation energy. That is of course normally indicative of a significant contribution to the chemical reaction from quantum mechanical tunneling. In this system, however, kinetic isotope effects were quite modest (close to three) – classical in behavior. It became clear to us that we could not model such behavior using the mathematical formalism we had developed. The reason for this is apparent in Fig. 9.1. Motions of the carboxyl oxygens toward each other in each dimer that forms the crystal of benzoic acid modulate the potential for proton transfer through symmetric motions of the well bottoms toward each other. This environmental modulation both lowers and thins the barrier to proton transfer. This symmetric coupling of motion to the reaction coordinate requires modification of the Hamiltonian in Eq. (9.2):

$$H = \frac{P_s^2}{2m_s} + V_o + \sum_k \frac{P_k^2}{2m_k} + \frac{1}{2} m_k \omega_k^2 \left(q_k - \frac{c_k s}{m_k \omega_k^2} \right)^2$$

$$+ \frac{P_Q^2}{2M} + \frac{1}{2} M\Omega^2 \left(Q - \frac{Cs^2}{M\Omega^2} \right)^2 \quad (9.11)$$

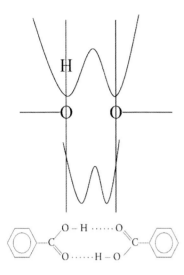

Figure 9.1. A benzoic acid dimer showing how the symmetric motion of the oxygen atoms will affect the potential for hydrogen transfer.

We note that in this case, the oscillator that is symmetrically coupled, represented by the last term in Eq. (9.11), is in fact a physical oscillation of the environment.

9.3.2
Promoting Vibrations – Corner Cutting and the Masking of KIEs

We were able to develop a theory [19] of reactions mathematically represented by the Hamiltonian in Eq. (9.11), and using this method and experimentally available parameters for the benzoic acid proton transfer potential, we were able to reproduce experimental kinetics as long as we included a symmetrically coupled vibration [20]. The results are shown in Table 9.1. The two-dimensional activation energies refer to a two-dimensional system comprised of the reaction coordinate and a symmetrically coupled vibration. The reaction coordinate is also coupled to an infinite environment appropriate for a crystalline phase. Kinetic isotope effects in this system are modest, even though the vast majority of the proton transfer occurs via quantum tunneling. The end result of this study is that symmetrically coupled vibrations can significantly enhance rates of light particle transfer, and also significantly mask kinetic isotope signatures of tunneling. A physical origin for this masking of the kinetic isotope effect may be understood from a comparison of the two-dimensional problem comprised of a reaction coordinate coupled symmetrically and antisymmetrically to a vibration. As Fig. 9.2 shows, antisymmetric coupling causes the minima (the reactants and products) to lie on a line – the minimum energy path, which passes through the transition state. In contradistinction, symmetric coupling causes the reactants and products to be moved from the reac-

Table 9.1. Activation energies for H and D transfer in benzoic acid crystals at $T = 300$ K. Three values are shown: the activation energies calculated using a one- and two-dimensional Kramers problem and the experimental values. The energies are in kcal mol^{-1}.

	E_{1d}	E_{2d}	Experiment
H	3.39	1.51	1.44
D	5.21	3.14	3.01

tion coordinate axis in such a fashion that a straight line connection of reactant and products would pass nowhere near the transition state. This, in turn, results in the gas phase physical chemistry phenomenon known as corner cutting [21]. Physically, the quantity to be minimized along any path from reactant to products is the action. This is an integral of the energy and so, loosely speaking, it is a prod-

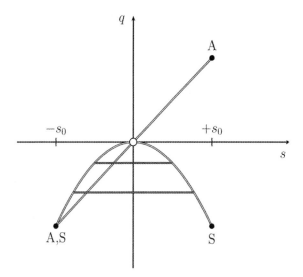

Figure 9.2. The location of stable minima in two-dimensional systems. The figure represents how antisymmetrically and symmetrically coupled vibrations affect the position of stable minima – that is reactant and product. The x-axis, s, represents the reaction coordinate, and q the coupled vibration. The points labeled S and A are the positions of the well minima in the two-dimensional system with symmetric and antisymmetric coupling respectively. An antisymmetrically coupled vibration displaces these minima along a straight line, so that the shortest distance between the reactant and product wells passes through the transition state. In contradistinction, a symmetrically coupled vibration, allows for the possibility of "corner cutting" under the barrier. For example, a proton and a deuteron will follow different paths under the barrier.

uct of distance and depth under the barrier that must be minimized to find an approximation to the tunneling path. The action also includes the mass of the particle being transferred and so, in the symmetric coupling case, a proton will actually follow a very different physical path from reactants to products in a reaction than will a deuteron.

9.4
Hydrogen Transfer and Promoting Vibrations – Alcohol Dehydrogenase

Finding that a promoting vibration, such as that present in benzoic acid crystals, can promote quantum tunneling while inhibiting indicators of tunneling such as kinetic isotope effects we were struck by similar experimental observations in certain enzymes in which the chemical step is thought to involve tunneling. Alcohol dehydrogenase is such an example. Klinman and coworkers have pioneered the study of tunneling in enzymatic reactions. Alcohol dehydrogenases are NAD^+-dependent enzymes that oxidize a wide variety of alcohols to the corresponding aldehydes. After successive binding of the alcohol and cofactor, the first step is generally accepted to be complexation of the alcohol to one of the two bound Zinc ions [22]. This complexation lowers the pK_a of the alcohol proton and causes the formation of the alcoholate. The chemical step is then transfer of a hydride from the alkoxide to the NAD^+ cofactor. They [23] have found a remarkable effect on the kinetics of yeast alcohol dehydrogenase (a mesophile) and a related enzyme from *Bacillus stereothermophilus*, a thermophile. A variety of kinetic studies from this group have found that the mesophile [24] and many related dehydrogenases [25] show signs of significant contributions of quantum tunneling in the rate-determining step of hydride transfer. Their kinetic data seem to show that the thermophilic enzyme actually exhibits less signs of tunneling at lower temperatures. Data of Kohen and Klinman [26] also show, via isotope exchange experiments, that the thermophile is significantly less flexible at mesophilic temperatures, as in the Petsko group's results [27] in studies of 3-isopropylmalate dehydrogenase from the thermophilic bacteria *Thermus thermophilus*.

More detailed studies from the Klinman group analyze changes in dynamics, and seem to localize the largest correlations in changes in tunneling parameters with the substrate binding area of the protein rather than the cofactor side of the protein [28]. As we will discuss in detail below, the promoting vibration seems to originate on the cofactor side of the binding pocket, and so these most recent experimental data are currently difficult to understand. These data have been interpreted in terms of models similar to those we have described above, in which a specific type of protein motion strongly promotes quantum tunneling – thus, at lower temperatures, when the thermophile has this motion significantly reduced, the tunneling component of the reaction is hypothesized to decrease, even though one would normally expect tunneling to increase as temperature decreases.

Hints as to the mechanism causing the odd kinetics are found in mutagenesis experiments. The active site geometry of HLADH is shown in Fig. 9.3. Two specific

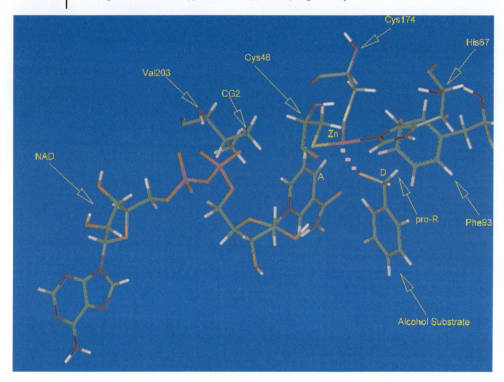

Figure 9.3. A schematic of the active site of horse liver alcohol dehydrogenase. The bound substrate (in this case benzyl alcohol) and the cofactor NAD are shown together with several residues in the active site.

mutations have been identified, Val203 → Ala and Phe93 → Trp, which significantly affect enzyme kinetics. Both residues are located at the active site – the valine impinges directly on the face of the NAD^+ cofactor distal to the substrate alcohol. Modification of this residue to the smaller alanine significantly lowers both the catalytic efficiency of the enzyme, as compared to the wild type, and also significantly lowers indicators of hydrogen tunneling [29]. Phe-93 is a residue in the alcohol-binding pocket. Replacement with the larger tryptophan makes it harder for the substrate to bind, but does not lower the indicators of tunneling [30]. Bruice's recent molecular dynamics calculations [31] produce results consonant with the concept that mutation of the Valine changes protein dynamics, and it is this alteration, missing in the mutation at position 93, which in turn changes tunneling dynamics. (We note that recent experimental results from Klinman's group [32] on mesophilic HLADH do not exhibit a decrease in tunneling as the temperature is raised. This indicates that the mesophile has a basically different coupling of the protein motion to reaction than the thermophile.)

The low level of primary kinetic isotope effect in the benzoic acid crystal when tunneling is the dominant transfer mechanism suggested a similarity between the proton transfer mechanism in the organic acid crystal and that of hydrogen transfer in some enzymatic reactions. We note that there have been previous attempts to understand the anomalously low primary kinetic isotope effects in alcohol dehydrogenases in the presence of a large body of experimental evidence that quantum tunneling is involved in the hydride transfer. Of note, coupled motions of nearby atoms in enzymatic reactions have been shown to result in such anomalous kinetic isotope effects in numerical experiments [33], but these studies were classical kinetics with semiclassical tunneling added (the Bell correction [34]) and they could not be used to account for enzymatic reactions in a deep tunneling regime.

With the suggestion of tunneling with a low kinetic isotope effect, we wish to investigate the dynamics of the enzyme to search for the possible presence of a promoting vibration. The quantity that naturally describes the way in which an environment interacts with a reaction coordinate in a complex condensed phase is the spectral density. In Eq. (9.3), the spectral density can be seen to give a distribution of the frequencies of the bilinearly coupled modes, convolved with the strength of their coupling to the reaction coordinate. The concept of the spectral density is, however, quite general and the spectral density may be measured or computed for realistic systems in which the coupling of the modes may well not be bilinear [35]. We have also shown [36] that the spectral density can be evaluated along a reaction coordinate. One only obtains a constant value for the spectral density when the coupling between the reaction coordinate and the environment is in fact bilinear. We have shown that a promoting vibration is created as a result of a symmetric coupling of a vibrational mode to the reaction coordinate. Analytic calculations demonstrated that such a mode should be manifest by a strong peak in the spectral density when it was evaluated at positions removed from the exact transition state position, in particular in the reactant or product wells. In cases in which there is no promoting vibration, while the spectral density may well change shape as a function of reaction coordinate position, there will be no formation of such strong peaks. Numerical experiments completed in our group have shown a delta function at the frequency position of the promoting vibration as the analytic theory predicted when we study a model problem in which a vibration is coupled symmetrically [37].

Our analysis began with the 2.1 Å crystal structure of Plapp and coworkers [38]. This crystal structure contains both NAD^+ and 2,3,4,5,6-pentafluorobenzyl alcohol complexed with the native horse liver enzyme (metal ions and both the substrate and cofactor.) The fluorinated alcohol does not react and go onto products because of the strong electron withdrawing tendencies of the fluorines on the phenyl ring, and so it is hypothesized that the crystal structure corresponds to a stable approximation of the Michaelis complex. We then replaced the fluorinated alcohol with the unfluorinated compound to obtain the reactive species as in Ref. [31]. This structure was used as input for the CHARMM program [39]. Both crystallographic waters [38] (there are 12 buried waters in each subunit) and environmental waters were included via the TIP3P potential [40]. The substrates were created from the MSI/CHARMM parameters. The NAD cofactor was modeled using the force field of Mackerell et al. [41]. The lengths of all bonds to hydrogen atoms were held fixed using the SHAKE algorithm. A time step of 1 fs was employed. The initial structure was minimized using a steepest descent algorithm for 1000 steps followed by an adapted basis Newton–Raphson minimization of 8000 steps. The dynamics protocol was heating for 5 ps followed by equilibration for 8 ps followed finally by data collection for the next 50 ps. Using CHARMM, we computed the force autocorrelation function on the reacting particle. The force is calculated in CHARMM as a derivative of the velocity. This is a numerical procedure, which can of course introduce error. We have recently found that spectral densities may also be calculated from the velocity autocorrelation function directly, and these spectral densities exhibit exactly the same diagnostics for the presence of a promoting vibration, as do those calculated from the force. In addition, the Fourier transform of the force autocorrelation function can be shown to be related to the Fourier transform of the velocity autocorrelation function times the square of the frequency. This square of the frequency tends to accentuate high frequencies. In a simple liquid this is not a problem because there are essentially no high frequency modes. In a bonded system, such as an enzyme, many high frequency modes remain manifest in autocorrelation functions, and it is advantageous to employ spectral densities calculated from Fourier transforms of the velocity function.

Application of this methodology to this model of horse liver alcohol dehydrogenase yields the results shown in Fig. 9.4. In fact we do see strong numerical evidence for the presence of a promoting vibration – intense peaks in the spectral density for the reaction coordinate are greatly reduced at a point between the reactant and product wells. This is defined as a point of minimal coupling. As we have described, the restraint on the hydride does not impact the spectral density computation. This computation measures the forces on the reaction coordinate, not those

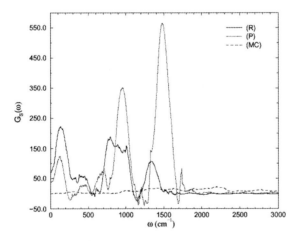

Figure 9.4. The spectral density of the hydride in the reactant well, product well, and at a point of minimal coupling for HLADH.

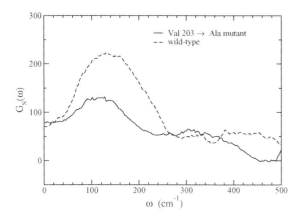

Figure 9.5. The spectral density for the hydride in the reactant well for both wild type HLADH, and one in which we mutate (in the computer) Val203 → Ala. The smaller size of the alanine results in a much smaller effective force on the reaction coordinate.

of the reaction coordinate itself [37]. We are also able to rationalize mutational experimental data. Figure 9.5 shows the results of a mutation of Val203 to a smaller Ala. We note that the intensity of the peak in the spectral density is reduced, indicative of a smaller force on the reaction coordinate. Recall that it is this mutant in which indicators of tunneling decrease. In Fig. 9.6 we show analogous results for a mutation of Phe93 to Trp. This mutation shows no experimental effect on tunneling (though it does affect the rate by lowering binding of substrate,) and in fact the two spectral densities are quite similar.

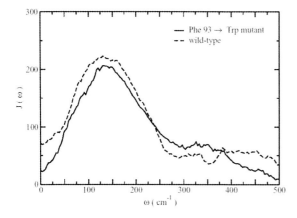

Figure 9.6. The spectral density for the hydride in the reactant well for both wild type HLADH, and one in which we mutate (in the computer) Phe93 → Trp.

These computational experiments were undertaken under the guidance of a large body of experimental literature on this enzyme. In some sense all we have done is rationalize the known experimental results. It is desirable to have a method to identify residues likely to be involved in the creation of a promoting vibration without prior experimental guidance. We have developed such an algorithm [42], and here we sketch the approach and the results found for the HLADH system. The method depends on computing the projection of motions of the center of mass of individual residues along the reaction coordinate axis (by this we mean the donor acceptor axis – we do not mean to imply that the actual set of atomic motions needed for the reaction are identified). A correlation function of this quantity with the donor acceptor motion is found. When Fourier transformed, strong peaks at the frequency location of an identified promoting vibration are indicative of the involvement of a residue in creation of the promoting vibration. The reader is referred to the reference above for mathematical and implementation details.

Eight residues are found to be strongly correlated in their motion to that of the donor and acceptor. They are shown in Fig. 9.7. Some residues identified with this algorithm agree directly with experimental evidence. For example, Val203 has been identified by both Klinman [43], and Plapp [44] as being a residue that on mutagenesis changes kinetic parameters and signatures of tunneling. In addition, Val292 [45], has been found by Plapp et al. to be similarly implicated in tunneling for the hydride reaction coordinate. Phe93 is found by Klinman to not change the indicators of tunneling [46], and we find no evidence for coupling of the dynamics of this residue to the reaction coordinate. There are, however, some potential discrepancies – Plapp finds that Thr178 affects the kinetics. Our algorithm found no evidence of dynamic coupling of this residue to the reaction coordinate. It is possible that there is no contradiction here – clearly static effects such as binding geometries can alter kinetics – they just do it in a different way than the dynamic coupling of residues to the reaction coordinate. It is also important to note that these results fit with the general observations of the Bruice group who find general anticorrelated motions in the protein [47]. They see that one side of the protein generally seems to move towards the other side of the protein. Our results seem to find the dominant motion on the side of the cofactor, but in any case there is clearly a motion of the center of mass of this side of the protein towards the substrate binding side. Viewed from the center of mass of the entire protein, this would be seen as such an anticorrelated motion.

An important question to ask is the extent to which the protein dynamics is actually involved in the catalytic process. If it only produces a tiny fraction of the catalytic effect, then it is of little interest. This is hard to measure experimentally, and difficult to predict accurately from theory. Because alcohol dehydrogenase is a highly studied enzyme, there are some experimental results which seem to indicate that the promoting vibration is a significant contributor to the catalytic effect. The mutagenesis experiments show that 2 residues alone, Val203 and Val292, when mutated to smaller residues contribute at least 3 orders of magnitude of catalytic effect individually. As stated, one cannot "turn off" the promoting vibration. One can lower its effect by mutating the large residues which impact the NAD cofactor to smaller ones. In any case, it is clear that while the promoting vibration is

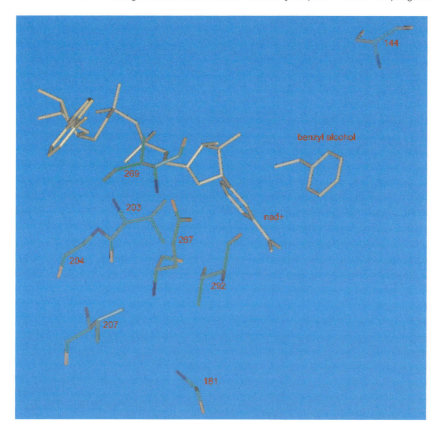

Figure 9.7. Residues found computationally to be important in the creation of a protein promoting vibration in HLADH.

not the only source of the catalytic effect, it is a major contributor. In the last section of this chapter where we examine a very different type of enzyme, purine nucleoside phosphorylase, and a very different type of promoting vibration, we will quantify how protein motion specifically lowers barrier height.

9.5
Promoting Vibrations and the Kinetic Control of Enzymes – Lactate Dehydrogenase

Lactate dehydrogenase (LDH) catalyzes the interconversion of the hydroxy-acid lactate and the keto-acid pyruvate with the coenzyme nicotinamide adenine dinucleotide [48]. This enzyme plays a fundamental role in respiration, and multiple isozymes have evolved to enable efficient production of substrate appropriate for the microenvironment [49]. Two main subunits, referred to as heart and muscle (skeletal), are combined in the functional enzyme as a tetramer to accommodate aerobic and anaerobic environments. Subunit combinations range from pure heart (H_4) to pure muscle (M_4). The reaction catalyzed involves the transfer of a proton between an active site histidine (playing the role of the metal ion in alcohol dehydrogenase) and the C2 bound substrate oxygen, as well as hydride transfer between C4N of the cofactor, NAD(H), and C2 of the substrate. Remarkably, the domain structure, subunit association, and amino acid content of the human isozyme active sites are comparable. In fact the active sites have complete residue identity, with the overall subunits only differing by about 20%. What is astounding is that the kinetic properties of the two isozymes are quite different. The heart isoforms favors the production of pyruvate over lactate as the heart predominantly employs aerobic respiration. In contradistinction, muscles are quite comfortable under periods of stress undergoing anaerobic respiration, and so the muscle isoform favors lactate production. The question that remains is how two proteins that are so strikingly similar in composition can possibly have such different kinetic behavior. We were recently able to propose a solution based on variations in protein dynamics [50]. In addition we will describe a very recent application of the method known as Transition Path Sampling [5] to the actual reactive event in one isoform which shows in microscopic detail how the protein backbone is involved in promoting catalysis.

The first step in the theoretical study of this problem is a molecular dynamics computation on the human proteins. Our methodology is described in detail elsewhere [51], but, in brief: the starting point for computations were crystal structures solved by Read et al. [52] for homo-tetrameric human heart, h-H_4LDH, and muscle, h-M_4LDH, isozymes in a ternary complex with NADH and oxamate at 2.1 Å and 2.3 Å resolution respectively. Numerical analysis of molecular dynamics computations followed our previously published approach [53].

While the chemical step of lactate dehydrogenase and alcohol dehydrogenase is quite similar – transfer of a hydride from or to an NAD cofactor, there have been no mutagenesis studies around the active site to implicate protein dynamics in the preferential formation of one product or another. The first step in the analysis is to search for the presence of a protein promoting vibration. A Fourier transform of the correlation function of the donor acceptor velocity in the two isoforms shows the relative motion that may be imposed on the reaction coordinate. The absence of strong peaks in a similar Fourier transform for the reaction coordinate at a point of minimal coupling – the putative transition state demonstrates, as we have shown, the presence of a symmetrically coupled protein promoting vibration. Such computations are shown in Fig. 9.8 as an example for the heart isoform. A similar set of data obtains for the muscle isoforms. Such figures demonstrate convincing numerical evidence that there is in fact a protein promoting vibration present in both isoforms of this enzyme. A strange result is found, however, when we examine the relative intensity of the peaks in the reaction coordinate figures for the 2 isoforms – Fig. 9.8 and 9.9. It seems that the strength of coupling of the promoting vibration in the heart isoform is larger when pyruvate is bound and, in the muscle isoform, the signal is more intense when lactate is bound. This would seem to favor the production of the opposite chemical species than that which is

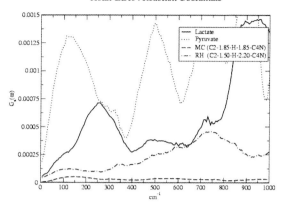

Figure 9.8. The spectral density $G_s(\omega)$ for the reaction coordinate in the wild type human heart lactate dehydrogenase isoform. The solid line represents the configuration where lactate and NAD$^+$ are bound, the dotted line is when pyruvate and NADH are bound, the dashed line is the minimal coupling (MC) simulation with lactate and NAD$^+$ bound and the (hydride–C2) and (hydride–C4N) distances restrained, and the dot-dash line is exemplary of the restrained hydride (RH) simulations to search for the point of minimal coupling. Distances are in Å and defined in the form (C2-Å-Hydride-Å-C4N). The power spectrum is reported in CHARMM units.

required for each tissue. The explanation is found in Fig. 9.10(a) and (b). These figures show time series of the donor–acceptor distance in both the heart and muscle isoforms respectively. Note for example that in the heart isoform the distance between the donor and acceptor when lactate is bound is on average 0.6 Å less than when pyruvate is bound. In contradistinction, in the muscle isoform, the donor acceptor distance is 0.6 Å less when pyruvate is bound. We recall that the in-

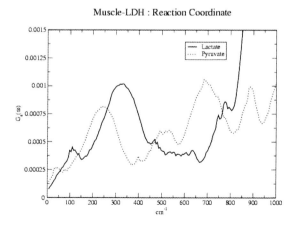

Figure 9.9. Similar to Fig. 9.8, but for the human muscle isoforms.

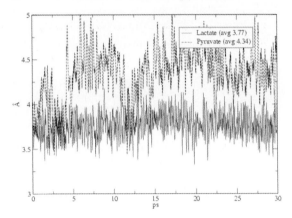

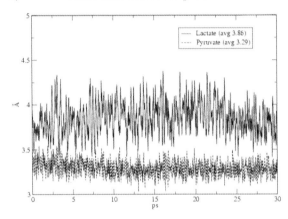

Figure 9.10. (a) Donor–acceptor distance for the wild type human heart lactate dehydrogenase isoform; this is the distance between the C2 carbon of substrate and carbon C4N of the nicotinamide ring of the cofactor. The solid line represents the configuration where lactate and NAD$^+$ are bound, and the dashed line is when pyruvate and NADH are bound. (b) Donor–acceptor distance for the wild type human muscle lactate dehydrogenase isoform; this is the distance between the C2 carbon of the substrate and carbon C4N of the nicotinamide ring of the cofactor. The solid line represents the configuration where lactate and NAD$^+$ are bound, and the dashed line is when pyruvate and NADH are bound.

tensity of the promoting vibration is in fact the product of the strength of the coupling times the distance from the point of minimal coupling – the putative transition state. Thus the argument for these isoforms is that, for example, in the heart isoform when lactate is bound there is rapid conversion to pyruvate followed by relaxation of the protein structure by 0.6 Å. This could be a mechanism for "locking in" the formed pyruvate. For this mechanism to be viable, there either needs to be significant quantum tunneling in the hydride transfer step or significant dissipation to the protein medium as the hydride transfers. If there is tunneling, then clearly the longer distance for the "less preferred" substrate will significantly favor the other substrate. If there is no tunneling, but rather activated transfer across the barrier, the frictional dissipation could lower the probability of transfer across a longer distance.

The foregoing analysis shows through classical mechanics that there is coupling of protein motion to progress along the reaction coordinate. It does not, however, actually chart the course of the chemical reaction as it is proceeding in the enzyme. The potential on which the analysis was done was a simple molecular mechanics potential, and this can certainly never make and break chemical bonds. In order to do this, some form of quantum chemistry must be invoked. One would like to develop a method to use a quantum chemically generated potential energy surface and follow the entire enzyme as the substrates moves from reactants to products. The difficulty with this goal is that there is not a single way for reactants to go to products. The simple one-dimensional view of a reaction coordinate with a free energy barrier to reaction is far too simplistic – the actual surface is highly complex with thousands of "hills and valleys." A single reactive event will not elucidate the nature of the reaction. The difficulty is reaction is an extremely rare event – we note that most enzymes turn over about every millisecond while the actual passage from reactants to products takes picoseconds. Thus one cannot hope to generate a large set of reactive trajectories by simply running a large number of trajectories. This problem has been solved by transition path s ampling developed by Chandler and coworkers [5]. The method has been applied to small systems [54], and we here report on an unpublished first application of TPS to an enzymatically catalyzed chemical reaction. We again focus on LDH, in particular the heart isozyme.

In order to follow the reaction, one must identify "order parameters", simple numerical features which identify whether the reaction is in the reactants or products region. We defined the pyruvate region to include all configurations where the bond length of the reactive proton and the reactive nitrogen of the active site histidine (NE2) was 1.3 Å or shorter *and* the bond length of the reactive hydride and the reactive carbon (NC4) of the NADH coenzyme was 1.3 Å or shorter. The lactate region was defined to include all configurations where the bond length of the reactive proton and the reactive substrate oxygen (O) was 1.3 Å or shorter *and* the bond length of the reactive hydride and reactive substrate carbon (C2) was 1.3 Å or shorter. The transition region was then comprised of all configurations where neither of the above combined bond lengths were satisfied. Order parameters are simply guideline to differentiate regions; 1.3 Å was found to be a viable discriminator. Quantum mechanical/molecular mechanical calculations were performed on a Silicon Graphics workstation using the CHARMM/MOPAC [55] interface with the CHARMM27, all hydrogen force field, and the AM1 semi-empirical method. The CHARMM27 force field includes specific parameters for $NAD^+/NADH$. Oxamate (NH_2COCOO), an inhibitor of LDH, is an isosteric, isoelectronic mimic of pyruvate with similar binding kinetics. Changes to the PDB file included substitution of the oxamate nitrogen with carbon to create pyruvate and replacement of the active site neutral histidine with a protonated histidine to establish appropriate starting conditions with pyruvate and NADH in the active site. A total of 39 atoms were treated with the AM1 potential; 17 or 16 atoms of the NADH or NAD^+ nicotinamide ring, 13 or 12 atoms of the protonated or neutral histidine imidazole ring, and 9 or 11 atoms of the substrate pyruvate or lactate, respectively. The generalized hybrid orbital (GHO) [56] approach was used to treat the two covalent bonds which divide the quantum mechanical and molecular mechanical regions. The two GHO boundary atoms are the histidine Cα atom and the NC1' carbon atom of the NAD^+/NADH adenine dinucleotide structure which covalently bonds to the nicotinamide ring. Protein structure (PSF) and coordinate (CRD) input files were created with CHARMM. Crystallographic waters were treated as TIP3P [57] residues. A single subunit of the enzyme was used in all QM/MM calculations. A TPS interface to CHARMm was created and a transition path ensemble was generated.

In order to start an initial reactive trajectory is needed. This was generated by placing the hydride and proton at the midpoint of their respective donor–acceptor axis. The velocities of all atoms in the protein substrate complex obtained from a 300 K equilibration run were then used with the above coordinates to initiate simulations both forward and backward in time. Recall, for time reversible deterministic dynamics, inverting the sign of each xyz momentum and then integrating forward in time is equivalent to simulating a trajectory backwards in time. The hydride initial velocity was slightly altered to move along the donor–acceptor axis. This produced a 100 fs trajectory that began in the reactant well and finished in the product well. We initially attempted to generate a reactive trajectory using high temperature simulations, but were not successful.

To demonstrate the power of TPS sampling, we first investigated what appeared to be a paradox in the computational literature on LDH. Some studies have found the hydride and proton transfer to be concerted while others found it to be sequential [58]. Our transition path sampling study showed that all paths (with different orders in the case of sequential) are possible. Subtle changes in enzyme motions shift the transfer order and timing. Figure 9.11 shows 100 fs of three reactive trajectories. For each graph, the x-axis is the distance in Å of the proton from the NE2 atom of the active site histidine while the y-axis is the distance in Å of the hydride from the NC4 atom of the coenzyme. The reaction direction, reading from left to right, is pyruvate to lactate (true for all figures unless mentioned otherwise). Part (a) shows all three trajectories plotted together for comparison. Parts (b)–(d) are the three trajectories plotted individually, where each dot represents a 1 fs time step. The time step in which the hydride–cofactor bond breaks, the proton–histidine bond breaks, the hydride–substrate bond forms and the proton–substrate bond forms are color labeled. Using the order parameter values discussed earlier, 1.3 Å for each bond distance involved with transferring atoms, the transfer order can be discerned. In the first trajectory, (trajectory 1 of (b)) the hydride–cofactor bond initially breaks, then the proton–histidine bond, and then the proton–substrate bond forms before the hydride bond. Trajectory 2 is very similar except the hydride–substrate bond forms before the proton–substrate bond. These two

9.5 Promoting Vibrations and the Kinetic Control of Enzymes – Lactate Dehydrogenase

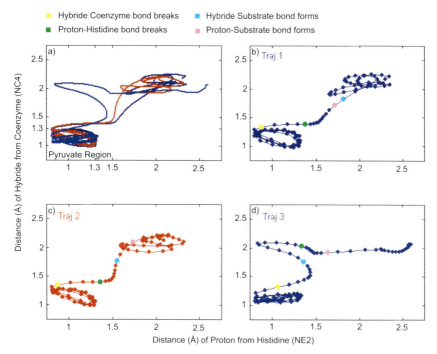

Figure 9.11. Three reactive trajectories demonstrate unique pathways from reactants to products for the lactate dehydrogenase enzymatic reaction. The distance of the proton from the NE2 reactive atom of the active site histidine versus the distance of the hydride from the NC4 reactive atom of the coenzyme nicotinamide is plotted. (b) Reactive trajectory 1 where the hydride–coenzyme bond breaks first, then the proton–histidine bond, with the proton–substrate bond forming before the hydride–substrate bond. (c) Reactive trajectory 2 where again the hydride–coenzyme bond breaks first, but now the hydride–substrate bond forms before the proton–substrate bond. (b) and (c) are examples of a concerted reaction. In (d) the hydride–coenzyme bond breaks and the hydride-substrate bond forms before the proton–histidine bond breaks. (d) is an example of a step-wise, sequential transfer. The enzyme lactate dehydrogenase was previously thought to be limited to one or the other mechanism. Application of the transition path sampling algorithm has demonstrated that either mechanism is possible. Each dot represents a 1 fs time step. Color coding indicates when the hydride coenzyme bond breaks, the proton–histidine bond breaks, the hydride–substrate bond forms, and the proton–substrate bond forms for the pyruvate to lactate reaction direction.

trajectories exemplify the notion of a concerted reaction, where the bond making and bond breaking events occur more or less simultaneously. For Trajectory 3, in (d), the hydride bond breaks and forms with the substrate before the proton–histidine bond even breaks. This trajectory exemplifies a stepwise or sequential mechanism of transfer. Apparently LDH has the capability of interconverting pyruvate and lactate by either mechanism proposed in the literature.

Figure 9.12 plots four distances over a 7.3 ps reactive trajectory (extension of Tra-

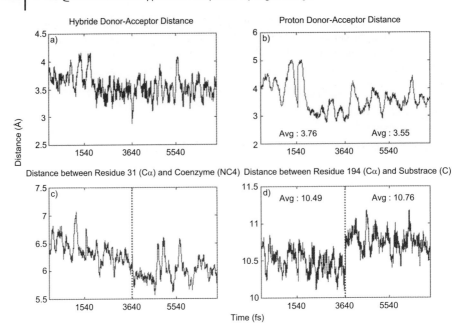

Figure 9.12. Enzyme-wide changes during transition from reactant to product. Exemplary critical distance shifts during the interconversion of pyruvate and lactate by lactate dehydrogenase. Reading the graph from left to right, the pyruvate to lactate reaction direction, the reaction occurs slightly after the minimum hydride donor–acceptor distance. Distances are plotted every 1 fs over a 7.3 ps sampling period. (a) Hydride donor–acceptor distance, (b) the proton donor–acceptor distance, (c) the distance of residue Valine 31, located behind the coenzyme nicotinamide ring, from the active site; (d) the distance of residue Aspartate 194, located behind the substrate, from the active site. Although subtle, these enzyme shifts must occur to have a complete reaction.

jectory 1 forward and backward in time). Plots of specific amino acid distances from the active site, residue Valine 31 and residue Aspartate 194, demonstrate a general pattern of the protein structure proximal to the reaction event. Valine 31, whose distance is plotted in (c), is the first amino acid located directly behind the nicotinamide ring of NAD(H) while Aspartate 194 of (d) is located behind the substrate. The atom transfers begin slightly after the minimum hydride donor–acceptor distance. Valine 31 has been studied and associated with protein promoting vibrations (PPVs) in LDH, as well as, analogously in ADH. Valine 31 is thought to push the coenzyme nicotinamide ring closer to the substrate carbon involved with hydride transfer. It is popular to envision breathing motions in enzymes that drive the reacting species together. However, (c) and (d) show an alternative picture, at least on the picosecond time scale. For the pyruvate to lactate reaction direction all residues located behind the coenzyme compress towards the active site, while all residues located behind the substrate relax away from the active site. The motion is thus akin to a compressional wave, since the compression causes the relaxation. The compression causes the hydride donor–acceptor distance to come sharply closer right before the atom transfers. In fact we have been able to show that nonreactive trajectories can be transformed into reactive ones by imposing this compressional motion on the dynamics. This demonstrates that the motion is both necessary and sufficient for the reaction to occur.

9.6 The Quantum Kramers Model and Proton Coupled Electron Transfer

The foregoing discussion shows how the Quantum Kramers method may be augmented to study complex systems, such as enzymes in which there is a promoting vibration which modulates chemical passage over a barrier. One class of enzymatic reactions for which this model as described will not work is proton coupled electron transfer reactions (PCET). These reactions are clearly of basic biological interest, and have attracted significant biochemical study recently. Two such enzymes are lipoxygenase, which has been studied by Klinman's group [59], and trimethylamine dehydrogenase studied by Scrutton and coworkers [60]. An interesting difference between these reactions and those of alcohol dehydrogenase is the presence of rather large kinetic isotope effects. The amine dehydrogenases exhibit KIEs in the range 15 to 25 while soybean lipoxygenase has one of almost 100. The involvement of protein dynamics is suggested by unexpected temperature dependence of the KIEs.

There have been previous model studies of these systems [61]. These studies, while including the effects of environment, did not address the question of the effect of a promoting vibration. These reactions are inherently electronically nonadiabatic, while the formulation we have thus far presented included evolution only on a single Born–Oppenheimer potential energy surface. We have developed a model system to allow the extension of the Quantum Kramers methodology to such systems, and we now describe that model.

The starting point for the study is a simple model of the coupled process. This model is found from a generalization of the Hamiltonian in Eq. (9.2) to include the modulation of hydrogen transfer potential as a result of electron transfer.

$$H = \frac{P_s^2}{2m_s} + V_D|D\rangle\langle D| + V_A|A\rangle\langle A| + \sum_k \frac{P_k^2}{2m_k} + \frac{1}{2} m_k \omega_k^2 \left(q_k - \frac{c_k s}{m_k \omega_k^2} \right)^2$$

$$+ \frac{P_Q^2}{2M} + \frac{1}{2} M \Omega^2 Q^2 + c_Q (s^2 - s_o^2) Q + \frac{\Delta}{2} \sigma_z + V_c(s) \sigma_x \qquad (9.12)$$

The modifications to the Hamiltonian used previously are:

1. There are two different bare potentials for hydrogen transfer V_D and V_A – the potentials are chosen by the state of the electron given by the projection operators $|D\rangle\langle D|$ and $|A\rangle\langle A|$.

2. The states $|A\rangle$ and $|D\rangle$ are the states of an electron degree of freedom approximated by a 2 state spin system. The rate of electron transfer from one state to the other (from the electron donor to acceptor) is given by V_c. While there are methods available to measure this parameter experimentally, it has not been done for any of the reactions of interest here. We set it to 0.45 eV and 4.5 eV to get a sense of range of effect.

We note that the reaction coordinate is coupled to an infinite bath of harmonic oscillators, which represent the bulk protein, and to a protein promoting vibration. For each mathematical implementation, we here choose the zero of promoting vibration coupling to be in the well rather than at the barrier top, but this is arbitrary. We point out that we can tune this model to allow for both sequential and concerted hydrogen-electron transfer. Sequential transfer is found with a very high transfer rate, and concerted with a lower one.

Our initial results are presented in Table 9.2. Introduction of the electronic degree of freedom significantly raises the KIE. In fact the results for 11 kcal mol^{-1} are similar to those found by Scrutton et al. for amine dehydrogenase. Thus, the stronger the electronic coupling (given by V_c), the lower the enhancement of the KIE. This is understood by the result that very strong electron coupling yields results asymptotic to sequential transfer – in other words, hydrogen transfer in the presence of a promoting vibration in the electron acceptor state alone. This clearly rationalizes the high KIE found in the coupled electron–hydrogen systems with the possibility of the presence of a promoting vibration. In addition, the natural log of the KIEs versus $1/T$ over the biochemically accessible range is essentially temperature independent – again in direct agreement with the amine dehydrogen-

Table 9.2. Kinetic isotope effects from exact quantum rate computations on the model of Eq. (9.12). In one case there is no protein promoting vibration, in the second case there is a promoting vibration coupled with a strength similar to that in our previous model studies. In each case there are two levels of electron coupling – essentially the rate of electron transfer between the two states. Moderate electron transfer enhances the kinetic isotope effect while strong electron coupling enhances it less. We have found that high coupling is asymptotic to sequential transfer.

	No promoting vibration		Moderate promoting vibration	
	Kinetic isotope effect			
Barrier height (kcal mol^{-1})	$V_c = 0.45$ eV	$V_c = 4.5$ eV	$V_c = 0.45$ eV	$V_c = 4.5$ eV
11	56	65	38	24
30	1942	1256	1218	685

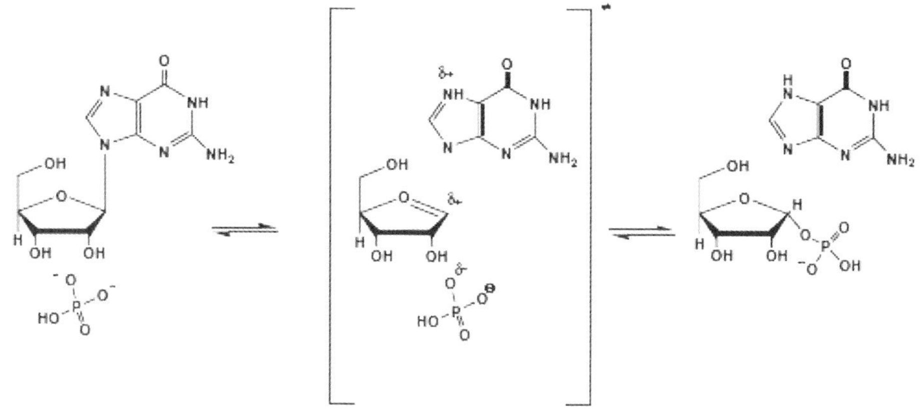

Figure 9.13. hPNP-catalyzed phosphorolysis of the purine nucleoside. The guanine leaving group and phosphate nucleophile are well separated from the oxacarbenium ion, defining a very dissociative TS.

ase results. There is, however a clear activation energy found from plots of k versus $1/T$ – the primary evidence suggested by the Scrutton group for significant "extreme tunneling" in amine dehydrogenase.

9.7
Promoting Vibrations and Electronic Polarization

Before concluding, it is worth mentioning the recent discovery of a very different type of promoting vibration in an enzyme. We have also studied the enzyme purine nucleoside phosphorylase [62]. This enzyme catalyzes the reaction shown in Fig. 9.13. Crystal structures of strongly bound transition state mimics and later transition state analysis [63] showed the oxygen atoms aligned in a closely packed stack. We hypothesized that this stack polarizes the ribosidic bond, and allows leaving group expulsion. Further, it seemed reasonable to search for a protein vibration that would compress this stack, and further destabilize the bond to the leaving group. Using the methods we have described, we did find that the protein imposes a very different vibration than might be found in gas or solution phase substrates. We were also able to show via QM/MM calculations that this vibration in fact results in an average lowering of the chemical barrier to reaction of about 7 kcal mol^{-1} – significant to the overall mechanism [64].

9.8
Conclusions

This chapter has focused on the technology known as the Quantum Kramers methodology, and how it can be used to increase our knowledge of enzymatic catalysis. In fact, a central discovery from the application of this method, the promoting vibration, has come from recognition of when the method does not work. The simple formulation represented by the Hamiltonian of Eq. (9.2) cannot reproduce experimental data in some systems such as benzoic acid crystals and enzymes such as alcohol dehydrogenase. The augmentation of the model with a symmetrically coupled vibration not only permits reproduction of experimentally reasonable results, but, more importantly, also gives physical insight into a portion of the catalytic mechanism of the enzyme. All enzymes incorporate dynamics into their function – for example hinge motions that allow the binding of substrates. Here we refer to something very different.

The fact that enzymes employ dynamics, should in no way be surprising – evolution knows nothing of quantum mechanics, classical mechanics, or vibrationally enhanced tunneling. Rates of reaction are optimized for living systems using all physical and chemical mechanisms available. It is also important to point out that such protein dynamics are far from the only contributor to the catalytic effect. In fact in an enzyme such as alcohol dehydrogenase, transfer of a proton from the alcohol to the coordinated zinc atom is critical to the possibility of the reaction. The specific modulation of the chemical barrier to reaction via backbone protein dynamics is now seen to be part of the chemical armamentarium employed by enzymes to catalyze reactions.

Acknowledgment

The author gratefully acknowledges the support of the Office of Naval Research, The National Science Foundation, and the National Institutes of Health.

References

1 L. PAULING, Nature of forces between large molecules of biological interest, Nature **161**, 707–709 (1948).
2 a) V. L. SCHRAMM, "Enzymatic Transition State Analysis and Transition-State Analogues, Methods Enzymol. **308**, 301–354 (1999); b) R. L. SCHOWEN, Transition States of Biochemical Processes, Plenum Press, New York (1978).
3 D. ANTONIOU, S. D. SCHWARTZ, Internal Enzyme Motions as a Source of Catalytic Activity: Rate Promoting Vibrations and Hydrogen Tunneling, J. Phys. Chem. B, **105**, 5553–5558 (2001).
4 a) S. D. SCHWARTZ, Quantum Activated Rates – An Evolution Operator Approach, J. Chem. Phys., **105**, 6871–6879 (1996); b) S. D. SCHWARTZ, Quantum Reaction in a Condensed Phase – Turnover Behavior from New Adiabatic Factorizations and Corrections, J. Chem. Phys., **107**, 2424–2429 (1997).
5 a) F. S. CSAJKA, D. CHANDLER, Transition pathways in many body systems: application to hydrogen bond breaking in water, J. Chem. Phys., **109**, 1125–1133 (1998); b) P. G. BOLHUIS, C. DELLAGO, D. CHANDLER, Sampling ensembles of deterministic pathways, Faraday Discuss., **110**, 421–436 (1998); c) P. G. BOLHUIS, C. DELLAGO, D. CHANDLER, Reaction coordinates of biomolecular isomerization, Proc. Natl. Acad. Sci. USA, **97**, 5877–5882 (2000).
6 a) J. E. STRAUB, M. BORKOVEC, B. J.

BERNE, Molecular Dynamics Study of an Isomerizing Diatomic in a Lennard-Jones Fluid, *J. Chem. Phys.*, **89**, 4833 (1988); b) B. J. GERTNER, K. R. WILSON, J. T. HYNES, Nonequilibrium Solvation Effects on Reaction Rates for Model SN2 Reactions in Water, *J. Chem. Phys.*, **90**, 3537 (1988).

7 E. CORTES, B. J. WEST, K. LINDENBERG, On the Generalized Langevin Equation: Classical and Quantum Mechanical, *J. Chem. Phys.*, **82**, 2708–2717 (1985).

8 a) R. ZWANZIG, The nonlinear Generalized Langevin Equation, *J. Stat. Phys.*, **9**, 215 (1973); b) R. ZWANZIG, *Nonequilibrium Statistical Mechanics*, Oxford University Press, Oxford (2001).

9 a) S. D. SCHWARTZ, Accurate Quantum Mechanics From High Order Resumed Operator Expansions, *J. Chem. Phys.*, **100**, 8795–8801 (1994); b) S. D. SCHWARTZ, Vibrational Energy Transfer from Resumed Evolution Operators, *J. Chem. Phys.*, **101**, 10436–10441 (1994); c) D. ANTONIOU, S. D. SCHWARTZ, Vibrational Energy Transfer in Linear Hydrocarbon Chains: New Quantum Results, *J. Chem. Phys.*, **103**, 7277–7286 (1995); d) S. D. SCHWARTZ, The Interaction Representation and Non-Adiabatic Corrections to Adiabatic Evolution Operators, *J. Chem. Phys.*, **104**, 1394–1398 (1996); e) D. ANTONIOU, S. D. SCHWARTZ, Nonadiabatic Effects in a Method that Combines Classical and Quantum Mechanics, *J. Chem. Phys.*, **104**, 3526–3530 (1996); f) S. D. SCHWARTZ, The Interaction Representation and Non-Adiabatic Corrections to Adiabatic Evolution Operators II: Nonlinear Quantum Systems, *J. Chem. Phys.*, **104**, 7985–7987 (1996).

10 a) S. D. SCHWARTZ, Accurate Quantum Mechanics From High Order Resummed Operator Expansions, *J. Chem. Phys.*, **100**, 8795–8801 (1994); b) S. D. SCHWARTZ, Vibrational Energy Transfer from Resummed Evolution Operators, *J. Chem. Phys.*, **101**, 10436–10441 (1994); c) D. ANTONIOU, S. D. SCHWARTZ, Vibrational Energy Transfer in Linear Hydrocarbon Chains: New Quantum Results, *J. Chem. Phys.*, **103**, 7277–7286 (1995); d) S. D. SCHWARTZ, The Interaction Representation and Non-Adiabatic Corrections to Adiabatic Evolution Operators, *J. Chem. Phys.*, **104**, 1394–1398 (1996); e) D. ANTONIOU, S. D. SCHWARTZ, Nonadiabatic Effects in a Method that Combines Classical and Quantum Mechanics, *J. Chem. Phys.*, **104**, 3526–3530 (1996); f) S. D. SCHWARTZ, The Interaction Representation and Non-Adiabatic Corrections to Adiabatic Evolution Operators II: Nonlinear Quantum Systems, *J. Chem. Phys.*, **104**, 7985–7987 (1996).

11 W. MAGNUS, On the Exponential Solution of Differential Equations for a Linear Operator, *Commun. Pure Appl. Math.*, **VII**, 649 (1954).

12 M. D. FEIT, J. A. FLECK JR., Solution of the Schrodinger Equation by a Spectral Method II: Vibrational Energy Levels of Triatomic Molecules, *J. Chem. Phys.*, **78**, 301 (1983).

13 S. D. SCHWARTZ, The Interaction Representation and Non-Adiabatic Corrections to Adiabatic Evolution Operators, *J. Chem. Phys.*, **104**, 1394–1398 (1996).

14 Wm. H. MILLER, S. D. SCHWARTZ, and J. W. TROMP, Quantum Mechanical Rate Constants for Bimolecular Reactions, *J. Chem. Phys.*, **79**, 4889–4898 (1983).

15 S. D. SCHWARTZ, Quantum Activated Rates – An Evolution Operator Approach, *J. Chem. Phys.*, **105**, 6871–6879 (1996).

16 a) R. KARMACHARYA, D. ANTONIOU, S. D. SCHWARTZ, Nonequilibrium solvation and the Quantum Kramers problem: proton transfer in aqueous glycine, *J. Phys. Chem. B* (Bill Miller festschrift), **105**, 2563–2567 (2001); b) D. ANTONIOU, S. D. SCHWARTZ, A Molecular Dynamics Quantum Kramers Study of Proton Transfer in Solution, *J. Chem. Phys.*, **110**, 465–472 (1999); c) D. ANTONIOU, S. D. SCHWARTZ, Quantum Proton Transfer with Spatially Dependent Friction:

Phenol-Amine in Methyl Chloride, *J. Chem. Phys.*, **110**, 7359–7364 (1999).

17 a) R. A. MARCUS, Chemical and electrochemical electron transfer theory, *Annu. Rev. Phys. Chem.*, **15**, 155–181 (1964); b) V. BABAMOV, R. A. MARCUS, Dynamics of Hydrogen Atom and Proton Transfer reactions: Symmetric Case, *J. Chem. Phys.* **74**, 1790 (1981).

18 a) K. FUKE, K. KAYA, Dynamics of Double Proton Transfer Reactions in the Excited State Model of Hydrogen Bonded Base Pairs, *J. Phys. Chem.*, **93**, 614 (1989); b) D. F. BROUGHAM, A. J. HORSEWILL, A. IKRAM, R. M. IBBERSON, P. J. McDONALD, M. PINTER-KRAINER, The Correlation Between Hydrogen Bond Tunneling Dynamics and the Structure of Benzoic Acid Dimers, *J. Chem. Phys.*, **105**, 979 (1996); c) B. H. MEIER, F. GRAF, R. R. ERNST, Structure and Dynamics of Intramolecular Hydrogen Bonds in Carboxylic Acid Dimers: A Solid State NMR Study, *J. Chem. Phys.*, **76**, 767 (1982); d) A. STOCKLI, B. H. MEIER, R. KREIS, R. MEYER, R. R. ERNST, Hydrogen bond dynamics in isotopically substituted benzoic acid dimers, *J. Chem. Phys.*, **93**, 1502 (1990); e) M. NEUMANN, D. F. BROUGHAM, C. J. McGLOIN, M. R. JOHNSON, A. J. HORSEWILL, H. P. TROMMSDORFF, Proton Tunneling in Benzoic Acid Crystals at Intermediate Temperatures: Nuclear Magnetic Resonance and Neutron Scattering Studies, *J. Chem. Phys.*, **109**, 7300 (1998).

19 D. ANTONIOU, S. D. SCHWARTZ, Activated Chemistry in the Presence of a Strongly Symmetrically Coupled Vibration, *J. Chem. Phys.*, **108**, 3620–3625 (1998).

20 D. ANTONIOU, S. D. SCHWARTZ, Proton Transfer in Benzoic Acid Crystals: Another Look Using Quantum Operator Theory, *J. Chem. Phys.*, **109**, 2287–2293 (1998).

21 a) V. A. BENDERSKII, S. Yu. GREBENSHCHIKOV, G. V. MIL'NIKOV, Tunneling Splittings in Model 2Dpotentials. II: $V(X, Y) = \lambda(X^2 - X_o^2)^2 - CX^2(Y - Y_o) + \frac{1}{2}\Omega^2(Y - Y_o + CX_o^2/\Omega^2)^2 - C^2X_o^4/2\Omega^2$, *Chem. Phys.* **194**, 1 (1995); b) V. A. BENDERSKII, S. Yu. GREBENSHCHIKOV, G. V. MIL'NIKOV, Tunneling Splittings in Model 2D potentials. III: $V(X, Y) = \lambda(X^2 - X_o^2)^2 - CXY + \frac{1}{2}kY^2 + C^2/2kX^2$ Generalization to N-Dimensional Case, *Chem. Phys.* **198**, 281 (1995); c) V. A. BENDERSKII, V. I. GOLDANSKII, D. E. MAKAROV, Low-temperature Chemical Reactions. Effect of Symmetrically Coupled Vibrations in Collinear Exchange Reactions, *Chem. Phys.* **154**, 407 (1991).

22 P. K. AGARWAL, S. P. WEBB, S. HAMMES-SCHIFFER, Computational studies of the mechanism for proton and hydride transfer in liver alcohol dehydrogenase, *J. Am. Chem. Soc.*, **122**, 4803–4812 (2000).

23 A. KOHEN, R. CANNIO, S. BARTOLUCCI, J. P. KLINMAN, Enzyme dynamics and hydrogen tunneling in a thermophilic alcohol dehydrogenase, *Nature* **399**, 496–499 (1999).

24 Y. CHA, C. J. MURRAY, J. P. KLINMAN, Hydrogen Tunneling in Enzyme Reactions, *Science* **243**, 1325 (1989).

25 a) K. L. GRANT, J. P. KLINMAN, Evidence that both Protium and Deuterium Undergo Significant Tunneling in the Reaction Catalyzed by Bovine Serum Amine Oxidase, *Biochemistry* **28**, 6597 (1989); b) A. KOHEN, J. P. KLINMAN, Enzyme Catalysis: Beyond Classical Paradigms, *Acc. Chem. Res.* **31**, 397 (1998); c) B. J. BAHNSON, J. P. KLINMAN, Hydrogen Tunneling in Enzyme Catalysis, *Methods Enzymol.*, **249**, 373 (1995); d) J. RUCKER, Y. CHA, T. JONSSON, K. L. GRANT, J. P. KLINMAN, Role of Internal Thermodynamics in Determining Hydrogen Tunneling in Enzyme-Catalyzed Hydrogen Transfer Reactions, *Biochemistry*, **31**, 11489 (1992).

26 A. KOHEN, J. P. KLINMAN, Protein Flexibility Correlates with Degree of Hydrogen Tunneling in Thermophilic and Mesophilic Alcohol Dehydrogenases, *J. Am. Chem. Soc.*, **122**, 10738–10739 (2000).

27. P. Zavodsky, J. Kardos, A. Svingor, G. A. Petsko, Adjustment of conformational flexibility is a key event in the thermal adaptation of proteins, *Proc. Natl. Acad. Sci. U.S.A.* **95**, 7406–7411 (1998).
28. Z.-X. Liang, T. Lee, K. A. Resing, N. G. Ahn, J. P. Klinman, Thermal-activated protein mobility and its correlation with catalysis in thermophilic alcohol dehydrogenase, *Proc. Natl. Acad. Sci. U.S.A.*, **101**, 9556–9561 (2004).
29. B. J. Bahnson, T. D. Colby, J. K. Chin, B. M. Goldstein, J. P. Klinman, A link between protein structure and enzyme catalyzed hydrogen tunneling, *Proc. Natl. Acad. Sci. U.S.A.* **94**, 12797–12802 (1997).
30. B. J. Bahnson, D-H. Park, K. Kim, B. V. Plapp, J. P. Klinman, Unmasking of hydrogen tunneling in the horse liver alcohol dehydrogenase reaction by site-directed mutagenesis, *Biochemistry* **32**, 5503–5507 (1993).
31. J. Luo, K. Kahn, T. C. Bruice, The linear dependence of $\log(k_{cat}/K_m)$ for reduction of NAD^+ by $PhCH_2OH$ on the distance between reactants when catalyzed by horse liver alcohol dehydrogenase and 203 single point mutants, *Bioorg. Chem.* **27**, 289–296 (1999).
32. S-C. Tsai, J. P. Klinman, Probes of Hydrogen tunneling with horse liver alcohol dehydrogenase at subzero temperatures, *Biochemistry*, **40**, 2303–2311 (1999).
33. P. Huskey, R. Schowen, Reaction coordinate tunneling in hydride-transfer reactions, *J. Am. Chem. Soc.* **105**, 5704–5706 (1983).
34. R. P. Bell, *The Tunnel Effect in Chemistry*, Chapman & Hall, New York (1980).
35. S. A. Passino, Y. Nagasawa, G. R. Fleming, Three Pulse Stimulated Photon Echo Experiments as a Probe of Polar Solvation Dynamics: Utility of Harmonic Bath Modes, *J. Chem. Phys.* **107**, 6094 (1997).
36. D. Antoniou, S. D. Schwartz, Quantum Proton Transfer with Spatially Dependent Friction: Phenol-Amine in Methyl Chloride, *J. Chem. Phys.* **110**, 7359–7364 (1999).
37. S. Caratzoulas, S. D. Schwartz, A computational method to discover the existence of promoting vibrations for chemical reactions in condensed phases, *J. Chem. Phys.*, **114**, 2910–2918 (2001).
38. S. Ramaswamy, H. Elkund, B. V. Plapp, Structures of horse liver alcohol dehydrogenase complexed with NAD^+ and substituted benzyl alcohols, *Biochemistry*, **33**, 5230–5237 (1994).
39. B. R. Brooks, R. E. Bruccoleri, B. D. Olafson, D. J. States, S. Swaminathan, M. Karplus, CHARMM: A program for macromolecular energy, minimization, and dynamics calculations, *J. Comput. Chem.* **4**, 187–217 (1983).
40. W. L. Jorgensen, J. Chandrasekher, J. D. Madura, R. W. Impey, M. L. Klein, Comparison of .23 Simple Potential Functions for Simulating Liquid Water, *J. Chem. Phys.* **79**, 926 (1983).
41. J. J. Pavelites, J. Gao, P. A. Bash, D. Alexander, J. Mackerell, A molecular mechanics force field for NAD^+ NADH, and the pyrophosphate groups of nucleotides, *J. Comput. Chem.* **18**, 221–239 (1997).
42. J. S. Mincer, S. D. Schwartz, A computational method to identify residues important in creating a protein promoting-vibration in enzymes, *J. Phys. Chem. B*, **107**, 366–371 (2003).
43. B. J. Bahnson, T. D. Colby, J. K. Chin, B. M. Goldstein, J. P. Klinman, A Link between Protein Structure and Enzyme Catalyzed Hydrigen Tunneling, *Proc. Natl. Acad. Sci. U.S.A.*, **94**, 12797–12802 (1997).
44. J. K. Rubach, B. V. Plapp, Amino Acid Residues in the Nicotinamide Binding site Contribute to Catalysis by Horse Liver Alcohol Dehydrogenase, *Biochemistry*, **42**, 2907–2915 (2003).
45. J. K. Rubach, S. Ramaswamy, B. V. Plapp, Contributions of Valine-292 in the Nicotinamide Binding Site of Liver Alcohol Dehydrogenase and Dynamics

to Catalysis, *Biochemistry* **40**, 12686–12694 (2001).

46 a) B. J. Bahnson, D.-H. Park, K. Kim, B. V. Plapp, J. P. Klinman, Unmasking of hydrogen tunneling in the horse liver alcohol dehydrogenase reaction by site-directed mutagenesis, *Biochemistry*, **32**, 5503–5507 (1993); b) J. K. Chin, J. P. Klinman, Probes of a Role for Remote Binding Interactions on Hydrogen Tunneling in the Horse Liver Alcohol Dehydrogenase Reaction, *Biochemistry*, **39**, 1278–1284 (2000).

47 J. Luo, T. C. Bruice, Anticorrelated motions as a driving force in enzyme catalysis: The dehydrogenase reaction, *Proc. Natl. Acad. Sci. U.S.A.*, **101**, 13152–13156 (2004).

48 M. Gulotta, H. Deng, R. B. Dyer, R. H. Callender, Toward an Understanding of the Role of Dynamics on Enzymatic Catalysis in Lactate Dehydrogenase, *Biochemistry*, **41**, 3353–3363 (2002).

49 A. Meister, *Advances in Enzymology and Related Areas of Molecular Biology*, Vol. 37, John Wiley & Sons, New York (1973).

50 J. E. Bassner, S. D. Schwartz, Donor acceptor distance and protein promoting vibration coupling as a mechanism for kinetic control in isozymes of Human Lactate Dehydrogenase, *J. Phys. Chem. B.*, **108**, 444–451 (2004).

51 J. E. Bassner, S. D. Schwartz, How Enzyme Dynamics Helps Catalyze a Chemical Reaction in Atomic Detail: A Transition Path Sampling Study, *J. Am. Chem. Soc.*, submitted.

52 J. A. Read, V. J. Winter, C. M. Eszes, R. B. Sessions, R. L. Brady, Structural basis for altered activity of M- and H-isozyme forms of human lactate dehydrogenase, *Proteins: Struct., Funct., Genet.*, **43**, 175–185 (2001).

53 S. Caratzoulas, J. S. Mincer, S. D. Schwartz, Identification of a protein promoting vibration in the reaction catalyzed by Horse Liver Alcohol Dehydrogenase, *J. Am. Chem. Soc.*, **124** (13), 3270–3276 (2002).

54 C. Dellago, P. G. Bolhuis, D. Chandler, Efficient transition path sampling: application to Lennard-Jones cluster rearrangements, *J. Chem. Phys.*, **108**, 9236–9245 (1998).

55 M. J. Field, A. Bash, M. Karplus, *J. Comput. Chem.*, **11**, 700–733 (1990).

56 J. Gao, P. Amara, C. Alhambra, M. J. Field, *J. Phys. Chem.*, **102**, 4714–4721 (1998).

57 W. L. Jorgensen, J. Chandrasekhar, J. D. Madura, *J. Chem. Phys.*, **79**, 926–935 (1983).

58 a) J. Andres, V. Moliner, J. Krechl, E. Silla, *Bioorg. Chem.*, **21**, 260–274 (1993); b) A. J. Turner, V. Moliner, I. H. Williams, *Phys. Chem. Chem. Phys.*, **1**(6), 1323–1331 (1999); c) S. Ranganathan, J. E. Gready, *J. Phys. Chem. B*, **101**, 5614–5618 (1997).

59 a) M. J. Knapp, K. Rickert, J. P. Klinman, Temperature-Dependent Isotope Effects in Soybean Lipoxygenase-1: Correlating Hydrogen Tunneling with Protein Dynamics, *J. Am. Chem. Soc.* **124**, 3865 (2002); b) M. J. Knapp, J. P. Klinman, Environmentally coupled hydrogen tunneling, *Eur. J. Biochem.* **269**, 3113 (2002).

60 a) J. Basran, M. Sutcliffe, N. Scrutton, Enzymatic H-Transfer Requires Vibration-Driven Extreme Tunneling, *Biochemistry*, **38**, 3218 (1999); b) M. J. Sutcliffe, N. S. Scrutton, A new conceptual framework for enzyme catalysis, *Eur. J. Biochem.*, **269**, 3096 (2002).

61 a) R. I. Cukier, Mechanism for Proton-Coupled Electron-Transfer Reactions, *J. Phys. Chem.* **98**, 2377 (1994); b) X. Zhao, R. I. Cukier, Molecular Dynamics and Quantum Chemistry Study of a Proton-Coupled Electron Transfer Reaction, *J. Phys. Chem.* **99**, 945 (1995); c) R. I. Cukier, Proton-Coupled Electron Transfer through an Asymmetric Hydrogen-Bonded Interface, *J. Phys. Chem.* **99**, 16101 (1995); d) R. I. Cukier, Proton-Coupled Electron Transfer Reactions: Evaluation of Rate Constants, *J. Phys. Chem.* **100**, 15428 (1996); e) S. Shin, H. Metiu, Nonadiabatic effects on the charge transfer rate constant: A

numerical study of a simple model system, *J. Chem. Phys.* **102**, 9285 (1995); f) S. HAMMES-SCHIFFER, Theoretical perspectives on proton-coupled electron transfer reactions, *Acc. Chem. Res.* **34**, 273–281 (2001); g) N. IORDANOVA, S. HAMMES-SCHIFFER, Theoretical investigation of large kinetic isotope effects for proton-coupled electron transfer in ruthenium polypyridyl complexes, *J. Am. Chem. Soc.* **124**, 4848–4856 (2002).

62 S. NUNEZ, D. ANTONIOU, V. L. SCHRAMM, S. D. SCHWARTZ, Promoting vibrations in human purine nucleoside phosphorylase: A molecular dynamics and hybrid quantum mechanical/molecular mechanical study, *J. Am. Chem. Soc.*, **126**, 15720–15729 (2004).

63 A. FEDOROV, W. SHI, G. KICSKA, E. FEDOROV, P. C. TYLER, R. H. FURNEAUX, J. C. HANSON, G. J. GAINSFORD, J. Z. LARESE, V. L. SCHRAMM, S. C. ALMO, *Biochemistry*, **40**, 853–860 (2001).

64 S. NUNEZ, D. ANTONIOU, V. L. SCHRAMM, S. D. SCHWARTZ, Electronic promoting motions in human purine nucleoside phosphorylase: a molecular dynamics and hybrid quantum mechanical/molecular mechanical study, *J. Am. Chem. Soc.*, **126**, 15720–15729 (2004).

10
Nuclear Tunneling in the Condensed Phase: Hydrogen Transfer in Enzyme Reactions

Michael J. Knapp, Matthew Meyer, and Judith P. Klinman

10.1
Introduction

Hydrogen transfer is one of the most pervasive and fundamental processes that occur in biological systems. Examples include the prevalent role of acid–base catalysis in enzyme and ribozyme function, the activation of C–H bonds leading to structural transformations among a myriad of carbon-based metabolites, and the transfer of protons across membrane bilayers to generate gradients capable of driving substrate transport and ATP biosynthesis.

Until quite recently, the kinetic and chemical properties of biological hydrogen transfer had been conceptualized in the same context as reactions involving heavier atoms, i.e., within the framework of transition state theory (TST). This treatment led to a generally accepted theory for the origin of rate discrimination among the isotopes of hydrogen (protium, deuterium and tritium), referred to as the kinetic isotope effect (KIE). Although model reactions were observed, from time to time, to display properties of the KIE that deviated significantly from predictions based on TST, these were often relegated to a "corner of oddities and possible artifacts."

Enzyme reactions offer unique advantages over simple model reactions in the study of fundamental chemical properties. In general, they are specific for a given substrate and, more importantly, lead to a single reaction product. Additionally, in reactions at carbons bearing two hydrogens (methylene centers), enzymes discriminate between the pro-R and pro-S hydrogens, allowing a clear distinction between the properties of the bond that is cleaved from the one that is left behind. It is, perhaps then, not very surprising that the growing evidence for quantum mechanical tunneling in H-transfer has come from the characterization of enzyme reactions.

This chapter has been written largely for the reader who has little or no background in enzyme kinetics. We begin with a simple introduction to the nature of kinetic measurements in enzyme reactions, since many confusing statements have appeared in the literature regarding the definition of catalysis and rate limiting steps. This is followed by a description of the methodology that is currently avail-

Hydrogen-Transfer Reactions. Edited by J. T. Hynes, J. P. Klinman, H.-H. Limbach, and R. L. Schowen
Copyright © 2007 WILEY-VCH Verlag GmbH & Co. KGaA, Weinheim
ISBN: 978-3-527-30777-7

able for the detection of nonclassical hydrogen transfers, together with a discussion of some relevant theoretical treatments of H-transfer. In the final section, we focus on selected experimental examples from the Berkeley laboratory that illustrate the differing ways in which tunneling can be demonstrated, together with the implications of tunneling in enzyme catalyzed hydrogen transfers.

10.2
Enzyme Kinetics: Extracting Chemistry from Complexity

The dominant tool to study hydrogen tunneling in an enzyme reaction is the measurement of isotope effects on the chemical step of catalysis via steady-state kinetics experiments. However, steady-state kinetics are often complicated by the contribution of several microscopic steps to the macroscopically observed rates, making it difficult to study the chemical step. The following section introduces basic enzyme kinetics, with a discussion of the macroscopic rate constants k_{cat} and k_{cat}/K_M and their interpretations. More detailed references on this matter are available [1, 2]. The first concern of the experimentalist is to be able to observe the intrinsic rate of chemistry, thereby allowing probes into the mechanism of hydrogen transfer.

A minimal enzymatic reaction, in which the substrate, S, is converted into the product, P, is shown in Eq. (10.1). Under initial-rate conditions ($[P]_0 = 0$), product release is a kinetically irreversible step ($k_{-3}[P]_0 = 0$) as shown.

$$E + S \underset{k_{-1}}{\overset{k_1}{\rightleftharpoons}} ES \underset{k_{-2}}{\overset{k_2}{\rightleftharpoons}} EP \overset{k_3}{\to} E + P \tag{10.1}$$

The velocity of this reaction ($v = d[P]/dt$) is a function of the bimolecular rate of substrate binding (k_1) and the unimolecular rates of chemistry (k_2, k_{-2}) and substrate and product release (k_{-1}, k_3). The steady-state velocity expression under initial rate conditions (Eq. (10.2)) demonstrates how each microscopic rate constant contributes to the macroscopic reaction rate and the dependence of the velocity upon substrate concentration.

$$v = \frac{k_1 k_2 k_3 [E]_T}{\frac{1}{[S]}(k_2 k_3 + k_{-1}k_{-2} + k_{-1}k_3) + k_1 k_2 + k_1 k_{-2} + k_1 k_3} \tag{10.2}$$

One useful limit is the velocity at saturating substrate ($[S] \to \infty$), which, when normalized for the enzyme concentration, gives the macroscopic rate constant k_{cat}. It can be seen (Eq. (10.3)) that k_{cat} is independent of the rate of substrate binding, a situation that exists for more complex mechanisms as well. Consequently, k_{cat} is a unimolecular rate constant obtained in the limit of infinite substrate concentration that reflects the rate of all steps after the formation of the ES complex.

When one of these unimolecular rates is much slower than the others, it is said to be rate-limiting on k_{cat}; these steps can include chemistry or product release.

$$k_{cat} = \frac{v}{[E]_T} = \frac{k_2 k_3}{k_2 + k_{-2} + k_3} \quad (10.3)$$

Another limiting regime is the velocity under conditions of limiting substrate ($[S] \to 0$), which, when normalized for enzyme concentration, gives the macroscopic rate constant k_{cat}/K_M (Eq. (10.4)).

$$k_{cat}/K_M = \frac{v}{[E]_T[S]} = \frac{k_1 k_2 k_3}{k_{-1} k_{-2} + k_{-1} k_3 + k_2 k_3} \quad (10.4)$$

It can be seen that k_{cat}/K_M reflects a bimolecular binding step, and other subsequent steps, including chemistry. If the chemical step is irreversible ($k_{-2} = 0$), k_{cat}/K_M simplifies to Eq. (10.5).

$$k_{cat}/K_M = \frac{v}{[E]_T[S]} = \frac{k_1 k_2}{k_{-1} + k_2} \quad (10.5)$$

This demonstrates that k_{cat}/K_M reflects all steps from substrate binding up to and including the first irreversible step – whether this step is chemistry (Eq. (10.5)) or product release (Eq. (10.4)). When one of these steps is slow (e.g. substrate binding, chemistry, or product release), it is rate-limiting on k_{cat}/K_M. *We emphasize that the chemical step can be experimentally probed, e.g. by isotope effects, through measurements of the macroscopic rate constant k_{cat}/K_M, despite suggestions to the contrary* [3]. In summary, k_{cat} and k_{cat}/K_M have some microscopic steps in common, the details of which depend upon the particular enzyme mechanism. Careful study of each macroscopic rate constant can reveal which microscopic step or steps are rate-limiting under specific conditions.

Standard kinetic tools to determine whether substrate binding, chemistry, or product release is rate-limiting have been developed over the years [4, 5]. The most straightforward way to demonstrate that chemistry is fully rate-limiting in the steady state is to show that the single-turnover rate of reaction is identical to k_{cat}. Other probes rely on perturbing the experimental conditions, such as site specific mutagenesis of the enzyme or alterations in pH, substrate structure, or viscosity, in a fashion that will affect only one microscopic step. An especially powerful kinetic tool uses substrates deuterated at an appropriate position in order to alter the rate of the chemical step to the exclusion of other steps. Observation of an H/D kinetic isotope effect on the macroscopic rate constants can indicate that chemistry is partially rate limiting. If a KIE is observed on k_{cat} but not on k_{cat}/K_M this suggests that chemistry is at least partially rate limiting on k_{cat}, while being not at all rate limiting on k_{cat}/K_M. Multiple probes can often reveal which steps limit the macroscopic rate constants obtained from steady-state kinetic measurements. For example, the use of slow substrates or site-directed mutants of the enzyme can alter relative microscopic rates such that isotope effects become fully expressed on one or both of the steady-state rate constants [5, 6].

Once the appropriate conditions have been found to isolate chemistry, isotopically labeled substrates can then be used to probe further for nuclear tunneling during the chemical step on the enzyme. There are three limiting kinetic relationships that affect which types of isotope effects can be used to study enzyme chemistry. The first case is when k_{cat} is fully limited by chemistry, the second is when k_{cat}/K_M is limited by chemistry, and the third arises when multiple steps are partially rate limiting. It should be noted that circumstances may also arise where *both* k_{cat} and k_{cat}/K_M reflect the chemical step.

The simplest case conceptually is when k_{cat} is fully rate limited by chemistry and, therefore, probes of chemistry (such as isotope effects) are revealed in k_{cat}. This limit is realized in recombinant soybean lipoxygenase-1 (SLO) and its mutants, greatly facilitating the study of C–H cleavage in this enzyme [7–9]. The study of chemistry in this case simply requires that k_{cat} can be faithfully measured as a function of external perturbation, such as temperature, pH, or substrate deuteration. It is essential to show substrate saturation ($[S] \gg K_M$) under all conditions, however, and this requirement can present experimental limitations. The principal probe for tunneling in enzymes in this kinetic case is the magnitude and temperature dependence of noncompetitive kinetic isotope effects, $^Dk_{cat} = k_{cat(H)}/k_{cat(D)}$.

The next simple case is when k_{cat}/K_M is fully rate limited by chemistry. This limit is realized in yeast alcohol dehydrogenase (YADH), providing a steady-state probe of chemistry in this enzyme [10]. The greatest difficulty here is measuring k_{cat}/K_M precisely, as the exact substrate concentration must be known and K_M can vary with each enzyme preparation. The principal probe for tunneling in this kinetic case is the magnitude of isotope effects, e.g. $^D(k_{cat}/K_M) = (k_{cat}/K_M)_H/(k_{cat}/K_M)_D$, which can be measured noncompetitively or competitively. The latter case bypasses the scatter in individual k_{cat}/K_M determinations, such that the relatively small secondary KIEs (section 10.3.1) can be precisely measured; these have proven to be a particularly powerful tool for demonstrating tunneling.

Finally, many enzymes are kinetically complex, and have multiple steps that partially limit both k_{cat} and k_{cat}/K_M. One approach is to use single-turnover studies to obtain the rate of the chemical step and the kinetic isotope effects by this noncompetitive technique. Several examples of single-turnover studies of enzymes that exhibit the characteristic of tunneling are in the literature [11, 12]. Alternatively, tools that allow microscopic rate constants to be calculated from observed rate constants can be applied. This approach has been documented for peptidylglycine-α-hydroxylating monooxygenase [13], and more recently, for dihydrofolate reductase [14].

In conclusion, steady-state kinetics provide macroscopic rate constants describing enzyme catalysis. Through careful analysis of kinetic data, rate limiting steps on k_{cat} and k_{cat}/K_M can be identified, as can optimal conditions to isolate kinetically the chemical step. Following kinetic isolation, the nature of the chemical steps, including tunneling effects, can be studied with fine detail.

10.3
Methodology for Detecting Nonclassical H-Transfers

Isotope effects are used to probe chemical processes, as isotopic substitution generally alters only the mass of the reacting groups without changing the electronic properties of the reactants. In this fashion, isotope effects can be used as subtle probes of mechanism in chemical transformations. This section will discuss how to use isotope effects to probe for tunneling effects on enzymes. The basic criteria for tunneling are experimental isotope effects that have properties that deviate from those predicted within the semi-classical transition state model, which includes only zero-point energy effects (we refer to this as the "bond stretch model"). Many of the available methods evolved within this context, as described by Bell regarding tunneling corrections [15]. While the Bell model is oversimplified and will not apply to numerous enzyme systems, it has pedagogical value in explaining how certain isotope experiments can demonstrate tunneling. In many cases, multiple anomalous KIEs are required before one can really implicate tunneling as being a likely explanation of the observed KIEs. Detailed interpretation of isotope effects requires theoretical models for hydrogen transfer that incorporate quantum effects (see Section 10.4).

10.3.1
Bond Stretch KIE Model: Zero-point Energy Effects

Conventional theories for kinetic isotope effects (KIEs) start with transition-state theory [16]. Reaction rates within transition-state theory are formulated as the product of three terms (Eq. (10.6)),

$$k_{TST} = \kappa \nu K^{\ddagger} \tag{10.6}$$

where $K^{\ddagger}$ is the equilibrium constant between the ground state and the transition state, ν is the frequency of barrier crossing, and κ is the transmission coefficient. It is conventionally assumed that isotopic substitution at hydrogen will not perturb the potential energy surface, leaving κ and ν largely unchanged but, by virtue of the altered vibrational energy levels, affecting $K^{\ddagger}$. Reaction rates are often reported in terms of the empirical Arrhenius expression, (expressed in terms of E_a, the apparent activation energy, and A, the pre-exponential factor[1]); KIEs can, thus, be reported as simple rate ratios (KIE = k_H/k_D), or as parameter ratios (KIE = A_H/A_D or $\Delta E_a = E_a(D) - E_a(H)$). In this semi-classical context, KIEs arise from the difference in vibrational, rotational, and translational degrees of freedom (quantized properties) in the ground state and in the transition state [16–18], with primary and secondary KIEs distinguished by the position of the isotopic comparison.

---

1) According to Eyring theory, the reaction coordinate frequency is treated classically as an equilibrium process defined by $K^{\ddagger}$, which leads to $k_\nu T/h\nu \exp(-\Delta G^{\ddagger}/RT)$ where k_B is Bolzmann's constant, ν is the reaction coordinate frequency, $\Delta G^{\ddagger}$ is the activation free energy and R is the gas constant. In this instance $E_a = \Delta H^{\ddagger} + RT$.

10.3.1.1 Primary Kinetic Isotope Effects

Primary KIEs for H-transfers are observed when the rate of reaction is studied as a function of isotopic labels at the transferred position. The reaction coordinate for H-transfer is composed of many degrees of freedom, but in the simplest limit will be dominated by the loss of the X–H stretching mode, in that the reaction X–H + B → [X–H–B]‡ → X + B–H converts the X–H stretch to a translation at the transition state. The loss of zero-point energy (ZPE) from this stretch will dominate primary KIEs in H-transfer reactions, due to the extremely large difference in ground-state ZPE for X–H vs. X–D.

The resulting maximal primary k_H/k_D ratios for X–H versus X–D are 6.5, 7.0 and 7.9 when X is C, N, or O. These estimates arise from the ZPE differences which are 1.1, 1.2, and 1.3 kcal mol^{-1}, respectively. It can be seen that the largest primary k_H/k_D ratio is around 8 at 25 °C. This assumes that a stretching mode is lost in the transition state; smaller primary k_H/k_D ratios result if a bending mode is lost at the transition state, as the frequency is less for bonding modes leading to a smaller value for $E_a(D) - E_a(H)$. Additionally, a reduced primary k_H/k_D ratio may arise from vibrational energy in modes perpendicular to the reaction coordinate [18].

The temperature dependence of primary k_H/k_D within the bond-stretch model arises from the zero-point energy differences of the X–H and X–D stretch. The magnitude of k_H/k_D is related to the difference in activation energy for X–H and X–D, as per Eq (10.7) [16],

$$\frac{k_H}{k_D} = \frac{A_H}{A_D} \exp\left(\frac{E_a(D) - E_a(H)}{RT}\right) \approx \exp\left(\frac{\Delta ZPE}{RT}\right) \tag{10.7}$$

where, in the simplest case, ΔZPE is the ground-state zero-point energy difference for X–H and X–D ($1/2 h\nu_H - 1/2 h\nu_D$). Note that similar equations can be written for any two isotopes (e.g. H and T or D and T). The bond-stretch model predicts that the KIE originates almost exclusively from vibrational energy effects, and predicts that ΔZPE is an upper limit to the activation energy difference. Although A_H/A_D is expected to be unity, experimental scenarios can be simulated that predict lower and upper limits of 0.7 and 1.2, respectively [19]. Compensatory motions in the transition state can lead to small deviations of A_H/A_D from unity, but it is generally accepted that this ratio will lie between 0.7 and 1.2 [19].

A traditional use of the primary k_H/k_D ratio is to infer the transition-state structure for a reaction. Large primary k_H/k_D ratios are predicted for symmetric transition-states; k_H/k_D ratios decrease for an early or a late transition state, as compensatory transition-state motions increase in these situations [18]. The simple bond-stretch model predicts a direct relationship between the symmetry of the transition-state structure and the magnitude of the observed primary k_H/k_D. As discussed below and in other sections, this view no longer holds in the context of significant hydrogen tunneling.

10.3.1.2 Secondary Kinetic Isotope Effects

Secondary isotope effects can also arise from differences in vibrational frequencies between the ground state and the transition state. The isotope at a secondary position retains its vibrational modes in the transition state, and consequently experiences much smaller frequency changes than do primary positions. As with primary effects, secondary k_H/k_D ratios have been used to infer transition-state geometry [20, 21]. If the transition state resembles the reactants (an early TS), then there will be little change in vibrational frequencies between the ground state and transition state, leading to a small secondary k_H/k_D that approaches unity. Alternatively, if the transition state resembles the products (a late TS), there will be a large change in vibrational frequencies, leading to a relatively large secondary k_H/k_D effect that approaches the equilibrium K_H/K_D. In the absence of tunneling, secondary k_H/k_D ratios are expected to lie between unity and the equilibrium K_H/K_D value.

The difference between primary and secondary kinetic effects can be elucidated by using the oxidation of benzyl alcohol by nicotinamide adenine dinucleotide (NAD$^+$) as an example (Scheme 10.1 (A)). This reaction is catalyzed by alcohol dehydrogenase (ADH), and has been extensively studied [10, 21–27]. In this reaction, the hydrogen at position L^1 is transferred from benzyl alcohol to NAD$^+$, forming benzaldehyde and reduced nicotinamide (NADH), making L^1 the primary position. Conversely, L^2 is retained upon reaction, making this the secondary position.

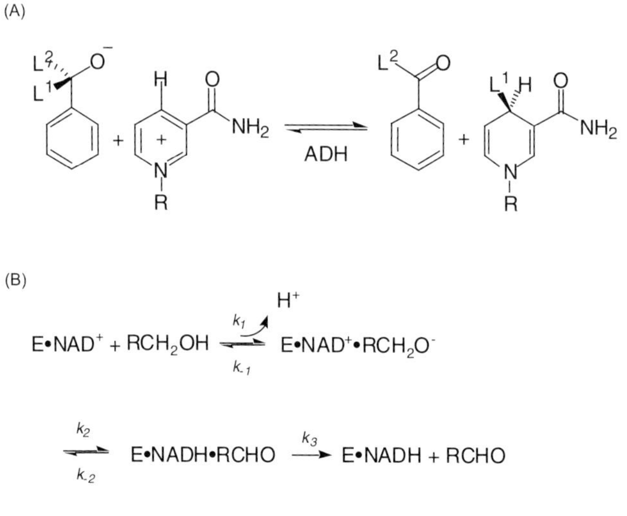

(A)

(B)

$$E \cdot NAD^+ + RCH_2OH \underset{k_{-1}}{\overset{k_1}{\rightleftharpoons}} E \cdot NAD^+ \cdot RCH_2O^- \quad H^+$$

$$\underset{k_{-2}}{\overset{k_2}{\rightleftharpoons}} E \cdot NADH \cdot RCHO \overset{k_3}{\longrightarrow} E \cdot NADH + RCHO$$

Scheme 10.1

10.3.2
Methods to Measure Kinetic Isotope Effects

10.3.2.1 Noncompetitive Kinetic Isotope Effects: k_{cat} or k_{cat}/K_M

Measuring isotope effects on enzyme chemistry requires a careful integration of enzymology and organic chemistry. Enzymology is crucial to ensure that kinetic complexity has been resolved, and that the measured isotope effect is intrinsic to the chemical step, as discussed in Section 10.2. Organic chemistry is crucial to synthesizing substrates that are isotopically labeled in the correct position(s). Clever use of multiple and/or tracer isotopic labels can lead to detailed information on the reaction coordinate via analysis of the KIEs.

The simplest way to measure an isotope effect is the noncompetitive technique, in which the rate (k_H) with fully protiated substrate (^{1}H labeled), is compared to the rate (k_D) at which deuterium labeled substrate (^{2}H labeled) reacts [28]. The label may be in the primary or a secondary position, yielding the primary or secondary KIE, respectively. Steady-state noncompetitive measurements yield the isotope effect on the rate constants k_{cat} or k_{cat}/K_M, but suffer from the requirement of both high substrate purity and isotopic enrichment, and from a large uncertainty in the KIE (ca. 5–10%) due to propagated errors. Single-turnover experiments can yield noncompetitive KIEs on the chemical step, but also generally have large uncertainties. Nevertheless, noncompetitive measurements are the only way to obtain KIEs on k_{cat}, which for certain enzymes may be the sole kinetic parameter that reflects the chemical step(s).

10.3.2.2 Competitive Kinetic Isotope Effects: k_{cat}/K_M

In the competitive technique, the enzyme reacts with a mixture of labeled and unlabeled substrate, yielding isotope effects on k_{cat}/K_M [29]. Competitive measurements, while limited to k_{cat}/K_M isotope effects, are substantially more precise than noncompetitive measurements. In addition, they allow the use of tracer-level radioactive labels, permitting tritium isotope effects at the primary and secondary positions (k_H/k_T or k_D/k_T) to be determined. General methods for determining competitive isotope effects have been published [17b]. One drawback is that multiple isotopic labels must often be used, leading to extensive synthetic efforts.

Radioactive isotopes are commonly used for competitive KIE measurements in a double-label experiment, yielding k_H/k_T or k_D/k_T ratios on k_{cat}/K_M. This technique typically utilizes tracer-level radioactivity in the position of interest (primary or secondary) to monitor the transfer of radioactivity from reactant to product, and requires a remote label (e.g. ^{14}C) in order to measure the conversion of unlabelled substrate to product. As an example, [ring-^{14}C(U)]benzyl alcohol and [1-^{3}H]benzyl alcohol (Scheme 10.2) can be used to simultaneously measure the primary and α-secondary k_H/k_T effects in the reaction catalyzed by alcohol dehydrogenase (ADH), as the tracer tritium is incorporated randomly into primary and α-secondary positions [6, 10].

In summary, competitive measurements yield the kinetic isotope effect on k_{cat}/K_M, and often rely upon tracer-level radioactivity, though recent developments also allow these values to be obtained using natural abundance NMR techniques [123]. Noncompetitive measurements can reveal the kinetic isotope effects on k_{cat} or k_{cat}/K_M, but suffer from larger propagated errors.

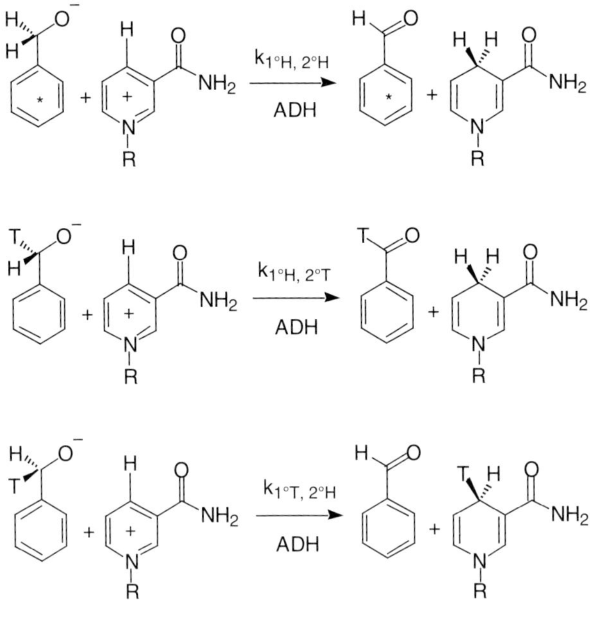

Scheme 10.2. The star (∗) represents C-14.

10.3.3
Diagnostics for Nonclassical H-Transfer

The bond-stretch model provides an upper limit for kinetic isotope effects that arise solely from ground state zero-point energy effects. Observations that deviate from this model imply a nonclassical effect. Provided that potential artifacts are controlled, the observation of KIEs that disobey the bond-stretch predictions calls into question the basic theory.

Theories for hydrogen transfer that treat hydrogen as a quantum mechanical particle have been presented [30–35]; however, most of these models are not fully developed with respect to KIE predictions. These models do agree with some of the basic conclusions taken from the bond-stretch model and the Bell correction, specifically that marked deviations of KIEs from predictions of the bond-stretch model occur when the quantum nature of hydrogen is pronounced. The basic criteria used to evaluate how closely a particular reaction obeys the bond-stretch KIE model and, by extension, a classical reaction model is presented below. In the subsequent sections of theory (Section 10.4) and experimental systems (Section 10.5), more detailed examples of nonclassical KIEs are presented.

10.3.3.1 The Magnitude of Primary KIEs: $k_H/k_D > 8$ at Room Temperature
The magnitude of primary k_H/k_D ratios is a crude yardstick for diagnosing nonclassical hydrogen transfer. These ratios are easily obtained from competitive or

noncompetitive measurements, with the principal difficulty being kinetic complexity (see Section 10.2). Measured k_H/k_D ratios that exceed the limit predicted by the bond-stretch model (ca. 8 at 25 °C), as in several enzymatic reactions [36–40], suggest tunneling [15]. Reactions exhibiting normal k_H/k_D ratios may still have appreciable tunneling components, as coupled motion between primary and α-secondary positions will reduce the size of k_H/k_D [25], as can the contribution of heavy atom motion in controlling tunneling [30–35]. Therefore, hydrogen tunneling is likely to be far more prevalent than is commonly thought on the basis of reported k_H/k_D ratios, and is merely awaiting more refined models.

The potential for quantum behavior by hydrogen isotopes has been recognized from the very earliest models, and was incorporated by Bell as a correction to transition-state theory [15]. In brief, light particles (H, D, and T) can tunnel through narrow portions of the reaction barrier, in particular near the very top (Fig. 10.1). This means that particles can react without attaining enough thermal energy to populate the transition state. The extent of tunneling behavior depends on the mass of the particle and the width of the barrier. The net effect is that lighter isotopes can tunnel lower down on the potential surface than heavier isotopes, leading to k_H/k_D or k_H/k_T ratios that exceed the bond-stretch predictions.

An analysis of a tunneling correction from the experimental k_H/k_D relies upon a calculation of the transition-state structure to obtain the bond-stretch k_H/k_D, and the tunneling effect. Once the transition-state structure is calculated, the tunneling probability is primarily a function of the imaginary frequency ($v^{\ddagger}$) for the reaction coordinate. The truncated Bell correction is shown in Eq. (10.8). This correction

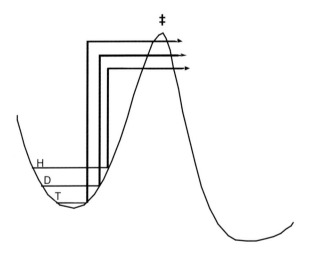

Figure 10.1. A cartoon of the Bell tunneling model, emphasizing that tunneling is more pronounced for lighter particles (H > D > T). Reactants have a probability of forming products, even when their energy is less than that of the transition state ($\ddagger$), via tunneling.

has been applied with notable success by Limbach et al. [41], and has often been used to account approximately for tunneling effects [25, 36]; a more complete description of the Bell model is given in Section 10.4.

$$Q_L = \frac{h\nu^{\ddagger}/2k_B T}{\sin(h\nu^{\ddagger}/2k_B T)} \tag{10.8}$$

Q_L is the ratio of rate that occurs by including tunneling to the rate that would have occurred solely by thermal activation, $\nu^{\ddagger}$ was defined above, and the other symbols have their usual meanings. It should be noted that, as a correction to transition-state theory, large Q_L values (exceeding 2) are almost certainly physically unreasonable, and simply indicate that tunneling is very important to a reaction. This correction works fairly well for reactions that do not deviate appreciably from classical predictions. Explaining primary k_H/k_D ratios greater than 10 usually requires that Q_H and Q_D exceed 2, such that this basic approach becomes dubious.

Many examples of primary k_H/k_D or k_H/k_T ratios are known that exceed the maximally predicted values of the bond-stretch model, some by a very large margin [37–40]. As compensatory motions in the transition state will only reduce the k_H/k_D ratio, these large KIEs suggest nuclear tunneling during the H-transfer. Very large k_H/k_D ratios are generally impossible to interpret within the Bell correction and require a full quantum model to explain their reactions. For example, large k_H/k_D ratios in the reactions catalyzed by the enzymes, methylamine dehydrogenase [42] and lipoxygenase [43], have recently been interpreted within environmentally-mediated tunneling models (cf. Sections 10.4 and 10.5).

Provided that the intrinsic KIE on a single step is observed, a single temperature primary k_H/k_D ratio can suggest tunneling. *It is important to control for artifacts that can inflate the KIE, such as multiplicative isotope effects (two concerted bond cleavages that exhibit ordinary KIEs), kinetic branching, or magnetic isotope effects.* In general, corroborating data are needed to demonstrate conclusively that H-tunneling is important to the reaction.

10.3.3.2 Discrepant Predictions of Transition-state Structure and Inflated Secondary KIEs

The magnitudes of primary and secondary KIEs have been used traditionally to infer transition-state structure [21]. Discrepancies between the transition-state structure predicted by these, as well as by other mechanistic probes, can suggest nonclassical behavior. This has been found in the alcohol dehydrogenase (ADH) reaction, where different transition-state structures were obtained from structure–reactivity correlations [44, 45] and the α-secondary KIEs [26, 27]. This discrepancy was later shown to be the result of hydrogen tunneling [10]. In the reactions catalyzed by dopamine β-monooxygenase (DβM) and peptidyl glycine α-hydroxylating monooxygenase (PHM), the magnitude of the primary KIE implied a symmetric transition state whereas α-secondary KIEs implied a product-like transition-state structure [46–48]. Once again, tunneling has been invoked to explain the observed discrepancies [13, 48].

In fact, some of the earliest suggestions for tunneling came from the observation of an α-secondary k_H/k_D ratio that exceeded the maximally predicted value (secondary $^Dk > {}^DK_{eq}$) [26, 49]. For the sake of the present discussion, discrepant predictors of transition-state structure serve to demonstrate that factors other than changes in vibrational modes are needed to account for the observed KIEs. Two factors that can inflate α-secondary KIEs are coupled motion in the transition state (a classical effect), and tunneling (a quantum mechanical effect), with an estimate of these separate effects requiring computational studies.

The classical effect of coupled motion is to reduce the primary KIE by coupling the primary H translation with the α-secondary H bending modes, which has the effect of 'leaking' some of the primary KIE into the α-secondary KIE. This leads to an enhanced α-secondary KIE, as this position acquires some of the characteristics of the primary position. The extent to which coupled motion can inflate the α-secondary KIEs in the absence of tunneling has been discussed, and serves as a tunneling discriminator [25, 50–52].

It is the combined effect of coupled motion and tunneling that leads to the largest anomalies [25, 52]. In addition to the classical effects of coupled motion, tunneling further increases the α-secondary KIE while significantly increasing the primary KIE. In ADH, as well as several other enzymatic and chemical examples, both tunneling and coupled motion have been invoked to reproduce the experimentally observed primary and α-secondary KIEs [6, 10, 25, 26, 49, 53–55].

Multiple-position KIEs, when combined with computational modeling, can provide enough information to successfully model a reaction coordinate and demonstrate hydrogen tunneling, even when the primary k_H/k_D ratio is not enormous. More complete experimental evidence for tunneling can be obtained by demonstrating a breakdown in the exponential relationships (e.g. k_H/k_T versus k_D/k_T) or by variable-temperature KIE measurements.

10.3.3.3 Exponential Breakdown: Rule of the Geometric Mean and Swain–Schaad Relationships

The bond-stretch model of KIEs results in predictable relationships between k_H, k_D, and k_T due to the ZPEs of X–H, X–D, X–T, as first noted by Swain et al. in 1958 [56]. These Swain–Schaad relationships are historically expressed with X–H as the reference state, $k_H/k_T = (k_H/k_D)^{1.44}$. Using X–T as the reference state leads to a similar relationship in which the exponent, S, is 3.26: $k_H/k_T = (k_D/k_T)^{3.26}$, and facilitates experimental determinations of exponential breakdown [10, 50]. In mixed-label experiments, the rule of the geometric mean (RGM) is an additional factor, causing R to be included in the observed exponent (see Eq. (10.10) below). The experimental KIE exponent, RS is evaluated by Eq. (10.9) as a composite of the Swain–Schaad (S) and RGM (R) exponents. RS is a good indicator of tunneling when it exceeds 3.3 by a large margin, with an extreme semi-classical upper-bound of ca. 5 [57].

$$\left(\frac{k_H}{k_T}\right)_H = \left[\left(\frac{k_D}{k_T}\right)_D\right]^{RS} \tag{10.9}$$

Breakdown in the rule of the geometric mean (RGM) can contribute to inflation of the *RS* exponent. The RGM states that isotope effects are insensitive to remote labels [58]. For example, the magnitude of the secondary k_H/k_T ratio will be independent of the primary label (H or D) so long as the isotope effects arise solely from vibrational modes, which can be expressed as the following exponential relationship:

$$\left(\frac{k_{2°H}}{k_{2°T}}\right)_{1°H} = \left[\left(\frac{k_{2°H}}{k_{2°T}}\right)_{1°D}\right]^R \quad (10.10)$$

where *R* is close to 1 [52]. The combined exponential relationship (*RS*) for secondary KIEs will breakdown if there is mechanical coupling between the primary and secondary positions when tunneling is important.

Measuring exponential KIE relationships generally relies upon the use of tracer-labeled isotopically substituted substrates and, consequently, must be done by competitive methods. Two experiments must be performed, one of which measures k_H/k_T competitively, the other of which measures k_D/k_T competitively. It is then a simple matter to obtain the exponential relationship for both primary and secondary positions, where $\ln(k_H/k_T)_H / \ln(k_D/k_T)_D = RS$.

The primary exponential relationship comes from a comparison of the primary $(k_H/k_T)_{2°H}$ to primary $(k_D/k_T)_{2°D}$ isotope effects. It has been shown that primary exponents are not susceptible to large Swain–Schaad deviations, even in the event of fairly extensive tunneling [59]; furthermore, primary exponents are not susceptible to large RGM deviations [52], and consequently, the composite exponent *RS* should remain close to 3.3 in the absence of kinetic complexity. This is a useful control, as the magnitude of the primary exponent is reduced when chemistry is only partially rate limiting (i.e., it can be use to establish that H-transfer has been kinetically isolated) [60].

For α-secondary KIEs, the exponents turn out to be highly susceptible to RGM deviations when tunneling is important, and can inflate *RS* from 3.3 significantly. Huskey showed that the dominant contributor to such exponential deviations is RGM breakdown, and that this effect was only pronounced in the event of tunneling [51]. The mechanism for large RGM deviations in secondary exponential relationships can be described within the Bell model for tunnel corrections. Since the tunnel correction depends principally upon the mass of the transferred hydrogen, it will be different for a primary H, D, and T. When the primary label is heavy (e.g. primary D) the degree of tunneling is small, causing the perturbation of the α-secondary isotope effect (e.g. secondary k_D/k_T) to also be small. When the primary label is light (e.g. primary H) tunneling is more extensive, and coupled motion can inflate the α-secondary isotope effect (e.g. secondary k_H/k_T) significantly. In this manner, the secondary KIE depends upon the primary label. The *RS* exponential relationship has been successfully used to demonstrate tunneling in several dehydrogenase enzymes and related systems [10, 23, 24, 53, 55, 61]. This approach is quite powerful, and provides an elegant experimental demonstration of a breakdown in the semi-classical KIE model. Its limitation, however, is the requirement

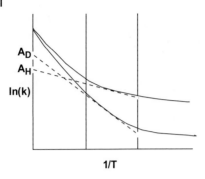

Figure 10.2. Temperature dependence of the rate of hydrogen transfer (k_L), where $L_1 = H$, $L_2 = D$. Experimentally accessible temperatures are indicated by the vertical solid lines, and extrapolations to obtain A_H and A_D are indicated by dashed lines.

for coupled motion between the primary and α-secondary hydrogen positions, as well as the condition that the competitively measured k_{cat}/K_M KIE be fully limited by chemistry.

10.3.3.4 Variable Temperature KIEs: $A_H/A_D \gg 1$ or $A_H/A_D \ll 1$

Variable-temperature kinetic isotope effects are the most widely recognized diagnostic tools for nonclassical H-transfers [62]. The bond-stretch model of isotope effects predicts Arrhenius prefactor isotope effects that are very close to unity, $A_H/A_D \approx 1$, with the limits on A_H/A_D between 1.2 and 0.7 for relatively unusual force constants and mass effects [19]. KIEs on the Arrhenius prefactor that deviate from these limits indicate nonclassical H-transfer, provided that kinetic complexity has been resolved. Arrhenius prefactor ratios that are significantly less than unity ($A_H/A_D < 0.7$) have been recognized as signatures of tunneling for quite some time, with the simple argument that the lighter isotope tunnels more than the heavier isotope. This leads to greater overall curvature in the Arrhenius plot for the lighter isotope. As shown in Fig. 10.2, kinetic measurements are generally restricted to a limited temperature range, such that extrapolation of tangents to the H and D lines within this range leads to an apparent "crossing" and values of $A_H/A_D < 0.7$.

There are a growing number of A_H/A_D ratios that are much greater than unity (Table 10.1) that cannot be readily explained within a bond-stretch or tunnel-correction view. One ad hoc explanation is that both isotopes tunnel appreciably, leading to the observed behavior. An alternative idea, that we advocate, is that the large majority of H-transfer reactions can be viewed as dominantly nonclassical, and should be treated within a quantum mechanical model for hydrogen transfer. In the nonadiabatic limit for hydrogen transfer, this leads to rate equations that are dominated by three contributors: environmental reorganization energy, Franck–Condon factors for hydrogen tunneling, and dynamic modulation of the tunneling

Table 10.1. Examples of enzymatic reactions where $A_H/A_D \gg 1$.

Enzyme	k_H/k_D	A_H/A_D
SLO[a]	81	18
HtADH[b]	3.2	2.2
PHM[c]	10	5.9
MADH[d]	17	13
TMADH[e]	4.6	7.8
SADH[f]	7.3	5.8
AcCoA Desat[g]	23	2.2
DHFR[h]	3.5	4.0

[a] Soybean lipoxygenase, Ref [43]. [b] High temperature alcohol dehydrogenase, Ref [24]. [c] Peptidylglycine α-hydroxylating monooxygenase, Ref [13]. [d] Methylamine dehydrogenase, Ref [42]. [e] Trimethylamine dehydrogenase, Ref [11]. [f] Sarcosine dehydrogenase, Ref [12]. [g] Acyl CoA desaturase, Ref [40]. [h] Dihydrofolate reductase, Ref [14].

barrier [31–34]. These models, which are discussed in more depth in Section 10.4, provide for a range of A_H/A_D ratios that can either exceed or be much less than unity.

The principal difficulty with the use of variable-temperature k_H/k_D measurements is the relatively small range of accessible temperatures (0–50 °C) for most enzymes, though a few enzymes from extremophiles are active over a wider range [24]. This leads to a few limitations worth noting. One is that kinetic complexity, in which steps other than chemistry are partially rate limiting, can have varied effects on A_H/A_D ratios [61]. This is particularly troublesome when the amount of kinetic complexity varies across a temperature range. Another limitation is that propagation of experimental error into the A_H/A_D ratios, particularly from noncompetitive measurements, can make it difficult to diagnose tunneling [11, 13, 37, 42, 43, 63, 64]. Given that the slopes of Arrhenius plots can generally be determined with greater precision than the intercepts, it may be preferable to compare differences in energies or enthalpies of activation, $\Delta E_a = E_a(D) - E_a(H)$ or $\Delta \Delta H^\ddagger = \Delta H(D) - \Delta H(H)$, since changes in ΔE_a correlate with changes in A_H/A_D.

Competitive KIEs can reduce the uncertainty in prefactor isotope effects, and have been used to demonstrate tunneling in several enzymes [24, 36, 65]. As discussed above, the competitive, double-label, technique for measuring KIEs is inherently more precise than noncompetitive techniques, and can reduce the experimental uncertainty in the KIE on the Arrhenius prefactor and energy of activation. The use of tritium also provides multiple ratios A_H/A_T and A_D/A_T, which are helpful in resolving kinetic complexity [61]. It has been noted that a change in the rate-limiting step over the temperature range can lead to anomalous Arrhenius prefactor ratios [65], which could give a false indication of tunneling. However, the reactions of X–D and X–T are slower than the reaction of X–H, making the A_D/A_T ratio relatively unaffected by kinetic complexity and, consequently, a useful diagnostic for tunneling. In recent years, studies of the temperature dependence of isotope effects on k_{cat}/K_M that permit the calculation of the intrinsic primary isotope effect (the isotope effect on a single step that is free of potential complications arising from kinetic complexity) at each temperature have been carried out. In both instances the magnitude of A_H/A_D was found to lie very significantly above unity [13, 14].

10.4
Concepts and Theories Regarding Hydrogen Tunneling

Underlying the development of semi-classical KIEs [17] (referred to as the bond-stretch model in Section 10.3) has been the assumption that the motion along the reaction coordinate itself behaves classically. In truth, the most accurate description of molecular events would be completely quantum mechanical. However, methods and computational power are only now evolving to the point where accurate quantum mechanical rates can be computed for reactions of interest in the condensed phase. The following sections discuss theoretical approaches that have been applied by this laboratory to explain our experimentally observed data. The past five years have been a period during which our conceptual understanding of tunneling in enzymatic systems has evolved away from one in which tunneling effects are treated as a small correction to the semi-classical KIE to one in which quantum mechanical effects are dominant and require full-tunneling models. Under these conditions, the dynamical behavior of heavy atoms surrounding the transferred hydrogen determine the magnitudes and the temperature dependence of KIEs. This section outlines the concept of tunneling and the evolution of the conceptual view of tunneling phenomena for systems studied within the Berkeley laboratory.

10.4.1
Conceptual View of Tunneling

The wave particle duality of matter, first proposed in 1923 by de Broglie, is intimately associated with the concept of tunneling through a classical energy barrier [66]. Spatial delocalization is one major consequence of wave-like behavior. The de Broglie wavelength, Eq. (10.11), is a means of demonstrating how mass affects the quantum nature of objects where h is Planck's constant and p is momentum (equal to mass times velocity). Conceptually, the small wavelength (λ) limit corresponds to more classical or particle-like behavior, while the large wavelength limit corresponds to more quantum or wave-like behavior. The comparison of de Broglie wavelengths associated with several free particles of equivalent kinetic energy (ca. 5 kcal mol^{-1}) shows how this parameter changes substantially among the isotopes of

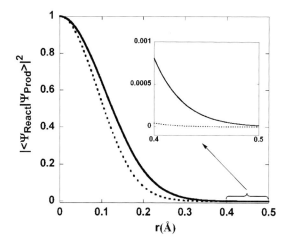

Figure 10.3. Plot of the squared overlap of ground state harmonic oscillator wavefunctions representative of C–H and C–D stretches (3000 cm^{-1} and 2121 cm^{-1}, respectively) separated by r in Å. Inset shows that overlap is greatly attenuated at larger separations and that the ratio of overlap for the light isotopomer to that for the heavy isotopomer increases markedly with increasing distances. C–H (—); C–D (- - -).

hydrogen: $\lambda = 0.63, 0.45$, and 0.36 Å for H, D, and T, respectively, in comparison to $\lambda = 27$ Å for the free electron as a frame of reference.

$$\lambda = h/p \tag{10.11}$$

Tunneling probability is proportional to the overlap between the hydrogen donor and acceptor wavefunctions, which can be understood from the wave picture of matter. Figure 10.3 illustrates the degree to which mass and distance can affect wavefunction overlap between a hydrogen donor and acceptor. Distance is most critical for determining the overlap of less dispersed wavefunctions, such that the overlap between two C–D stretching vibrational wavefunctions is more dependent on distance than the corresponding overlap between the two C–H stretching vibrational wavefunctions. This concept is central to understanding the role of dynamics in catalysis (cf. Eqs. (10.10)–(10.20) and Section 10.4.3).

Another useful viewpoint for introducing tunneling is the classically forbidden transmission of a free particle through a barrier. According to this physical picture introduced in Bell's excellent treatise [15], a potential energy barrier is bombarded from the left by a stream of free particles (A). The barrier reflects some of the particles to the left (C) and allows some of the particles to either penetrate or go over the barrier (B) (depending on the energy of the incident particle). Mathematically, the proportion of incident particles that penetrate or surmount the barrier as a function of energy (E) is expressed as the permeability of the barrier, $G(E)$ in Eq. (10.12), where $|A|^2$ and $|B|^2$ are the fluxes for the incident and transmitted beams, respectively.

$$G(E) = |B|^2/|A|^2 \tag{10.12}$$

The fundamental way in which a quantum physical picture differs from the classical picture is that classical permeability is Boolean in nature: either a particle has an energy equal to or greater than V, the barrier height, and is able to surmount the barrier, or the particle has an energy less than V and is reflected. By contrast, the quantum permeability $[G(E)]$ is a smooth function of E for both light and heavy particles, though it varies more sharply for heavier particles making their behavior more classical. Analogous to the delocalization picture, the barrier penetration picture illustrates that tunneling is more favorable for particles of light mass.

10.4.2
Tunnel Corrections to Rates: Static Barriers

The most utilized treatment of tunneling thus far has been the one developed by Bell [67]. The rate for passage through a barrier, J in Eq. (10.13a), is an integration of the product of the probability $[P(E)]$ for attaining energy E, Eq. (10.13b), and the probability $[G(E)]$ of a particle of that energy being transmitted from the reactant state to the product state (multiplied by the incident particle flux from the reactant side, J_0).

$$J = J_0 \int_0^\infty P(E)G(E)\,dE \tag{10.13a}$$

where $\beta = 1/k_B T$ and k_B is Boltzmann's constant. The tunnel correction is derived from the

$$P(E)\,dE = \frac{e^{-\beta E}\,dE}{\int_0^\infty e^{-\beta E}\,dE} \tag{10.13b}$$

ratio of the quantum rate (Eq. (10.14a)) to the rate predicted from classical mechanics (Eq. (10.14b)). $G(E)$ can be obtained in an analytic form for a parabolic and other types of barrier.

Appendix C of Ref. [15] gives two methods for the development of the tunnel correction for a parabolic barrier; although these expressions contain some typographical errors, they have been corrected by Northrop [68]. The full expression for the Bell tunnel correction, is given in Eq. (10.15a). The two parameters for input into Eq. (10.15a) are the barrier height, V, and the imaginary frequency, $v^{\neq}$, which defines the curvature at the top of the parabolic barrier. These are entered into the equation in their reduced forms which give their magnitude relative to thermal energy (Eqs. (10.15b) and (10.15c)), where N_A is Avogadro's number.

$$J_q = \beta J_0 \int_0^\infty G(E)e^{-\beta E}\,dE \tag{10.14a}$$

$$J_c = \beta J_0 \int_V^\infty e^{-\beta E}\, dE = J_0 e^{-\beta V} \tag{10.14b}$$

$$Q_t = \frac{u_t/2}{\sin(u_t/2)} + \sum_{n=1}^\infty (-1)^{n+1} \frac{\exp\left(\dfrac{u_t - 2n\pi}{u_t}\alpha\right)}{u_t - 2n\pi} \tag{10.15a}$$

$$u_t = \beta h \nu^{\neq} \tag{10.15b}$$

$$\alpha = \beta V / N_A \tag{10.15c}$$

As a matter of convenience, most practitioners use the truncated Bell correction, Eq. (10.8) in Section 10.3, which is the first term of Eq. (10.15a). However, this term inflates rapidly as a function of the reduced imaginary frequency. Inserting a reasonable barrier height of 20 kcal mol^{-1} and a barrier frequency of 800i cm^{-1} into Eq. (10.15a) yields the following parameters: $k_H/k_D = 12.6$; $E_a(D) - E_a(H) = 1.86$ kcal mol^{-1} and $A_H/A_D = 0.55$; these quantities differ substantially from the semi-classical values of $k_H/k_D \cong 7$, $E_a(D) - E_a(H) \cong 1.1$ kcal mol^{-1} and $A_H/A_D \cong 1$. An additional indicator of tunneling that derives from a Bell treatment is inflation of the Swain–Schaad exponent. As discussed in Section 10.3, this can be further altered by breakdowns in the rule of geometric mean, leading to values for RS, Eq. (10.9), that show extensive deviations from semi-classical behavior. Importantly, as with the truncated Bell correction, Eq. (10.15a) has limited applicability as tunneling becomes appreciable. For example, increasing the value of $\nu^{\ddagger}$ to 1000i cm^{-1} (298 K), produces $Q_t > 2$, calling into question the physical relevance of its function as a "correction" as the contribution to the rate by tunneling exceeds 100%.

A number of modern and quite sophisticated treatments of H-transfer in the condensed phase are not that dissimilar from the approach taken by Bell, in that they formulate the rate constant for H-transfer as a semi-classical term multiplied by a tunneling correction factor, e.g. Eqs. (10.16a) and (10.16b).

$$k_{obs} = Q k_{sc} \tag{10.16a}$$

$$= \gamma(T) \nu e^{-\Delta G^{\ddagger}/RT} \tag{10.16b}$$

According to Eq. (10.16b), the tunneling "correction" appears in the prefactor terms $\gamma(T)$ that has been written as the product of $r(T)$, the dynamical re-crossing of the barrier and $\kappa(T)$ the actual tunneling correction [69]. As pointed out in Ref. [69], $r(T)$ is expected to decrease the rate somewhat while $\kappa(T)$ enhances the rate. We have already noted the difficulty in relying on the use of corrections to absolute rate theory to explain the anomalies in H-transfer when reactions deviate very significantly from semi-classical behavior. One recurring type of behavior that eludes treatment via tunnel corrections is the repeated observation of large isotope effects that show very small temperature dependences, leading to values for $A_H/A_D \gg 1$. The growing list of examples of the latter behavior in enzymatic hydrogen transfer reactions has been summarized in Table 10.1.

10.4.3
Fluctuating Barriers: Reproducing Temperature Dependences

Recognizing the potential limitations of using a static barrier model to reproduce tunneling effects on KIEs, Bruno and Bialek attempted to model the effects of a full tunneling model with a fluctuating reaction barrier [32]. This attempt came in response to experimental data published by this laboratory on isotope effects in the oxidation of benzylamine catalyzed by bovine serum amine oxidase (BSAO) [36]. While Grant and Klinman attributed their measurements to a tunnel correction through a static barrier, Bruno and Bialek recognized two potentially conflicting observations: primary H/T and D/T isotope effects that seemed large enough (35 and 3, respectively) to be due to tunneling from ground-state vibrational levels, together with highly temperature dependent KIEs [70]. One expects ground-state tunneling through a static barrier, Eq. (10.17a), to be temperature independent [32, 70]. According to Eq. (10.17a), Δ_0^2 is the square of the coupling between hydrogen donor and acceptor wavefunctions referenced to a given donor–acceptor distance, and S is the tunneling action, Eq. (10.17b), with limits of integration corresponding to classical turning points on the barrier through which the system tunnels. A fluctuating barrier originating from an environmental vibration, as invoked by Bruno and Bialek, can give rise to the observed temperature dependencies through its impact on the tunneling action, S.

In order to understand the rate and KIE behavior that originate from a fluctuational barrier model, it is necessary to understand the physical origins of such a model. Similar in form to the development of the Bell correction, the rate of tunneling results from the multiplication of two probability distributions which depend on the available thermal energy. One probability distribution describes the deformability of the barrier, Eq. (10.18), while the other factor describes the probability of tunneling through the barrier at a given energy, Eq. (10.17a).

$$P_{tunnel} \propto \Delta_0^2 \exp(-2S/\hbar) \tag{10.17a}$$

$$S_{WKB} = \int_a^b \sqrt{2m[V(x) - E]}\, dx \tag{10.17b}$$

$$P_{deformation} \propto \exp\left[-\frac{\beta}{2}\kappa(l - l_{eq})^2\right] \tag{10.18}$$

The tunneling action is influenced by the energy difference between the barrier ($V(x)$) and the energy of the tunneling particle (E) and the distance traversed under the barrier by the tunneling particle (x). In the fluctuating barrier model, the rate is the result of a compromise between the amount of energy needed to reduce the distance between the hydrogen donor and acceptor and the rate enhancement due to deformation of the barrier (Fig. 10.4). In general, hydrogen tunneling is expected to be facilitated by attenuating the barrier in either a lateral or vertical direction (cf. Fig. 10.5). In their treatment, Bruno and Bialek made two simplifying assumptions: first, that there was no need for a vertical deformation and second,

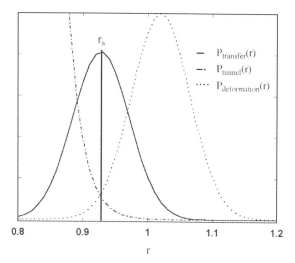

Figure 10.4. Plot showing the peaked nature of the transfer rate as a function of distance in the fluctuational barrier model. The compromise between the probability of attaining a favorable configuration for tunneling ($P_{Deformation}$) and the probability of tunneling for compressed barriers (P_{Tunnel}) is reached for a very small number of configurations centered at r_S.

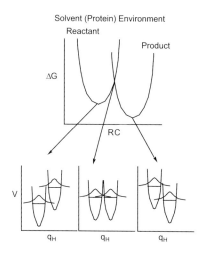
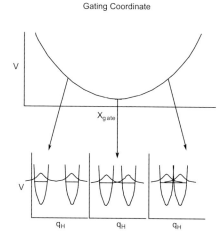

Figure 10.5. The protein environment can influence hydrogenic wavefunction overlap via asymmetrically coupled modes which bring hydrogenic wavefunctions into energetic coincidence [vertical perturbation, shown on left-hand side of figure]. These modes give rise to terms analogous to the Marcus theory of electron transfer [122]. Additionally, symmetrically coupled modes (gating modes) spatially modulate the hydrogenic wavefunction overlap (horizontal perturbation, shown on right-hand side of figure). The separation of the hydrogenic wells when the gating coordinate is at its minimum energy is r_0.

that the energy required to alter the distance between the proton donor and acceptor could be modeled as a classical harmonic oscillator model.

Ultimately, Bruno and Bialek achieved an equation for the KIE with only two adjustable parameters that could accommodate the data for the oxidation of benzylamine by BSAO collected by Grant and Klinman. However, given the modest temperature range and error inherent in the experimental studies, it was not possible to determine whether the observed numbers possessed the curvature in the Arrhenius KIE plot that was predicted from the model. Ultimately, the predominant value of the work of Bruno and Bialek was that it could explain, via a fluctuating barrier model, data that had previously been treated using a tunnel correction through a static barrier. It is notable that earlier, others had observed that "forcing vibrations" which modulate the hydrogen transfer barrier shape can substantially increase hydrogen transfer rates [71].

A more detailed 'gating' model has recently been presented by Kuznetsov and Ulstrup [33]. Once again, the physical picture underlying their model is that modulation of distance between hydrogen donor and acceptor can have enhancing effects on the quantum mechanical rate of H-transfer (Fig. 10.5). This phenomenon has been noted by many authors concerning hydrogen transfer in condensed phase systems [30, 31, 71–76]. In the gating model, a gating coordinate is represented by a classical harmonic oscillator whose temperature dependent motion modulates the distance between donor and acceptor, with the isotope dependence of gating arising from the well-known fact that tunneling is more distance dependent for heavier nuclei [77].

The gating model, described by Kuznetsov and Ulstrup [33] and adapted by Knapp et al. [43], is similar in concept to the model of Bruno and Bialek in two ways: (i) It is a full-tunneling model. (ii) The rate is expressed as the product of the probability of attaining a configuration and the probability of hydrogen transfer at that configuration (Eqs. (10.19)–(10.20)).

$$k_{tun} = \sum_v P_v \sum_w k_{vw} = \sum_v \frac{\exp(-\beta E_v) \sum_w k_{vw}}{\sum_v \exp(-\beta E_v)} \qquad (10.19a)$$

$$k_{vw} = \kappa_{vw} \frac{\omega_{eff}}{2\pi} \exp(-\Delta G^\ddagger_{vw}/RT) \qquad (10.19b)$$

$$\Delta G^\ddagger_{vw} = (\lambda + \Delta G^\circ + E_{vib})^2/4\lambda \qquad (10.19c)$$

$$\kappa_{vw} = |V_{el}|^2 \sqrt{4\pi^3/\lambda RT\hbar^2 \omega^2_{eff}} \; (\text{F.C. term})_{vw} \qquad (10.19d)$$

$$(\text{F.C. term})_{v,w} = \int_{r_1}^{r_0} [e^{(-m_H \omega_H r^2_H)}/2\hbar \; e^{(-E_x/k_\beta T)}] \, dX \qquad (10.19e)$$

$$k_{tun} = \sum_v P_v \sum_w \frac{|V_{el}|^2}{2} \sqrt{\frac{4\pi^3}{\lambda RT\hbar^2}} \exp\left(\frac{-(\Delta G^\circ + E_{vib} + \lambda)^2}{4\lambda RT}\right) (\text{F.C. term})_{v,w} \qquad (10.20)$$

Equation (10.19a) shows that the overall tunneling rate is computed for transfer from each donor hydrogen vibrational level (v) to each acceptor hydrogen vibrational level (w), where k_{vw} is a level-specific rate and P_v is the Boltzmann population level v. This rate is thus summed over all acceptor vibrational levels and Boltzmann weighted for each donor vibrational level. This level-specific rate, Eq. (10.19b), is the product of an exponential term which reflects the probability of attaining a solvent or environmental configuration at which the donor and acceptor vibrational states are isoenergetic and a term, κ_{vw}, which reflects the probability of transmission of the hydrogen atom from donor to acceptor; ω_{eff} is the characteristic average frequency of the environmental modes that are treated classsically. Note that, κ_{vw}, Eq. (10.19d), contains the electronic coupling term $|V_{el}|^2$, under the assumption that the reaction is electronically diabatic, and the Franck–Condon term (originally developed in an ungated context by Ulstrup and Jortner) [78]. The expression in Eq. (10.19e) integrates the probability of the wavefunction overlap ($e^{-m_H \omega_H r^2_H / 2\hbar}$), where m_H, ω_H and r_H are the mass, frequency, and distance, respectively, travelled by the tunneling particle over a range of distances that begin at r_0 and move to the closest possible approach between H-donor and acceptor, r_1. For simplification, this expression is restricted to tunneling from an initial ground state to a final ground state vibrational level. Expressions for the F.C. terms that include excited reactant and product levels can be found in the appendix of ref. [43]. The fluctuating barrier is described by $E_x = 1/2 m_x \omega_x^2 r_x^2$, where m_x, ω_x and r_x represent the mass, frequency and distance traversed by the heavy atoms that control the distance between the H-donor and acceptor (such that $\Delta r = r_0 - r_x$). According to this model, the configuration at which tunneling occurs is comprised of three coordinate systems: (i) the environmental or solvent coordinate parametrized by the reaction driving force ($\Delta G°$) and environmental reorganization (λ); (ii) the gating coordinate parametrized by the gating frequency, ω_x and the reduced mass of the heavy atom motion, m_x; and (iii) the hydrogen transfer coordinate parametrized by the transfer distance (Δr) (Fig. 10.5).

The enzyme, soybean lipoxygenase-1 (SLO), offers an excellent system in which to illustrate the power of Eq. (10.20) in reproducing experimental data. Several characteristics of SLO are impossible to interpret through a Bell-like tunneling correction. The KIE on k_{cat}, $^D k_{\text{cat}} = 81 \pm 5$, is nearly temperature independent, ($E_a(D) - E_a(H) = 0.9 \pm 0.2$ kcal mol^{-1}), and leads to an isotope effect on the Arrhenius pre-factor of $A_H/A_D = 18 \pm 5$ [43]. In the case of several point mutants, $^D k_{\text{cat}}$ is found to be almost unchanged while ΔE_a becomes inflated and A_H/A_D is decreased to below unity.

Using Eq. (10.20), the experimental observables, $[E_a(D) - E_a(H)]$, A_H/A_D and KIE (30 °C) could be reproduced for both WT-SLO and its three active site mutants. Details of this process are included in Knapp et al. [43]. These results show that a stiff gating frequency ($\omega_x \cong 400$ cm^{-1}, $> kT$) leads to a $^D k_{\text{cat}}$ that is nearly temperature independent, as seen for WT-SLO, while the increased temperature dependences seen for L546A/L754A and I553A can be reproduced by reducing the magnitude of the gating frequencies to 165 cm^{-1} and 89 cm^{-1}, respectively.

To summarize, the gating coordinate modulates the height and width of the barrier which separates the reactant and product hydrogenic wells. This type of motion has been referred to as symmetrical coupling to the reaction coordinate, which is rate-enhancing or rate-promoting and affects the system directly along the tunneling coordinate [30, 73, 74]. This occurs because of the extreme distance dependence of tunneling: as the donor and acceptor are forced together, the transfer probability increases extremely rapidly (Fig. 10.4). Antisymmetrically coupled motions modulate the relative energies of the reactant (hydrogen attached to donor) and product (hydrogen attached to acceptor), and arise because of the reorientation that must occur to yield the requisite equivalent energies for a tunneling event. Because of this requirement, antisymmetric modes which can be parametrized into the reorganization energy, λ, have been referred to as "demoting modes" [73, 76]. As a result of the distinctions that have been made between antisymmetrically and symmetrically coupled environmental modes, Knapp et al. referred to the gating coordinate as an example of "active" dynamics; whereas, "passive" dynamics results from the collection of antisymmetrically coupled environmental coordinates [43]. In reality, both types of motions may arise from statistical sampling of a very large number of protein configurations.

Hammes-Schiffer and coworkers have extended the work of Knapp et al. using a multi-state valence bond model, representing the solvent as a dielectric continuum and treating the transferred hydrogen quantum mechanically [80]. An important difference in their approach is that they introduce quantum effects into the oscillating environmental barrier and allow the shape of the tunneling barrier to change as a function of the proton donor–acceptor gating mode. Modeling the data for WT-SLO-1, they reach similar conclusions to Knapp et al. [43]. If they allow the initial donor–acceptor distance to "relax" to a longer value (that is close to the van der Waals radii), they find that the frequency of the gating mode must decrease to allow the donor and acceptor atoms to approach one another [80].

10.4.4
Overview

As our understanding of the tunneling phenomenon evolves, enzyme-catalyzed reactions provide some of the most versatile systems for exploring the experimental parameters and theoretical models associated with different degrees and types of tunneling. In this laboratory we began by considering tunneling as a minor component of reaction rates and, furthermore, thought of reaction barriers as static. Systems like yeast alcohol dehydrogenase led to an understanding of the types of experimental parameters which were indicative of tunneling. Much of this behavior could be explained initially using the Bell correction and the assumption of a static barrier. However, as new data began to emerge, conventional theories were challenged. Beginning with bovine serum amine oxidase, the idea of a fluctuational barrier was introduced to reconcile the large primary KIE with a large temperature dependence in the KIE. Most recently, WT-SLO and its corresponding mutants have provided data that directly implicate a role for "active" dynamics in enzymatic processes. We are now at a critical juncture where our view of hydrogen transfer in the condensed phase has been transformed from a tunnel correction to a semi-classical barrier to a fully quantum mechanical, dynamically enhanced reaction coordinate for hydrogen. *This impacts our view not only of the origin of kinetic hydrogen isotope effects but also of all probes of hydrogen transfer that derive from the traditional approach of transition state theory.*

10.5
Experimental Systems

10.5.1
Hydride Transfers

10.5.1.1 Alcohol Dehydrogenases

Alcohol dehydrogenase (ADH) oxidizes primary alcohols to their corresponding aldehydes via a hydride transfer to the cofactor nicotinamide adenine dinucleotide (NAD^+) (Scheme 10.1). The kinetic mechanism of ADH is well understood, the enzyme has a wide substrate tolerance and, furthermore, ADH has been cloned from several organisms. Cha and Klinman experimentally demonstrated hydrogen tunneling in ADH from yeast (YADH) as the first clear cut example of hydrogen tunneling in an enzyme [10], making ADH one of the bedrocks of enzymatic hydrogen tunneling research.

The kinetic mechanism of ADH is shown in Scheme 10.1 (B) under conditions of saturating NAD^+ and the steady state. Saturating NAD^+ converts all free enzyme into the E•NAD^+ form, which reversibly binds the alcohol to form the enzyme–substrate complex E•NAD^+•RCH_2O^- where alcohol is indicated as the deprotonated alkoxide. The reaction can be driven irreversibly forward by chemically scavenging free aldehyde, making k_3 irreversible. The degree of rate limitation by chemistry depends on the source of the enzyme and the substrate used. In general, k_{cat} is less controlled by hydride transfer and frequently reflects the rate of cofactor dissociation from enzyme. By contrast, through the use of aromatic alcohols as substrates with YADH [10], and with mutagenesis of the ADH from horse liver (HLADH) [6], chemistry can be made to be rate limiting on k_{cat}/K_M.

As defined above, the macroscopic rate constant k_{cat}/K_M reflects all steps from alcohol binding up to and including the first irreversible step, and is probed by competitive isotope effect measurements (cf. Eq. (10.4) in Section 10.2). As pointed out by Northrop, the observed isotope effect on k_{cat}/K_M can be formulated in terms of the intrinsic isotope effect on the chemical step $((k_H/k_T)_{int})$, and the commitment to catalysis C_H [60]. C_H accounts for the relative importance of chemical and nonchemical steps that contribute to k_{cat}/K_M when H is the isotope under study; a similar expression results for D/T isotope effects involving C_D. Under all conditions, $C_D \leq C_H/(k_H/k_D)_{int}$, making D/T KIEs less susceptible to kinetic complexity than H/T KIEs.

$$\frac{(k_{cat}/K_M)_H}{(k_{cat}/K_M)_T} = \frac{(k_H/k_T)_{int} + C_H}{1 + C_H} \tag{10.21}$$

Table 10.2. Kinetic isotope effect data for various ADH enzymes[a].

Enzyme	1° k_D/k_T	α-2° k_H/k_T	α-2° k_D/k_T	2° RS[b,c]
ht-ADH (60 °C)[d]	1.626 ± 0.03	1.23 ± 0.015	1.0158 ± 0.006	13.2 ± 5.03
YADH WT (p-H)[e]	1.73 ± 0.02	1.35 ± 0.015	1.03 ± 0.006	10.2 ± 2.4
YADH WT (p-Cl)[f]	1.59 ± 0.03	1.34 ± 0.01	1.03 ± 0.01	9.9 ± 4.2
HLADH L57F[h]	1.827 ± 0.01	1.318 ± 0.007	1.033 ± 0.004	8.5 ± 1
ht-ADH (20 °C)[d]	1.64 ± 0.013	1.257 ± 0.013	1.028 ± 0.009	8.28 ± 2.6
HLADH F93W[h]	1.858 ± 0.01	1.333 ± 0.004	1.048 ± 0.004	6.13 ± 0.5
HLADH V203A[g]	1.88 ± 0.02	1.316 ± 0.006	1.058 ± 0.004	4.9 ± 0.3
HLADH L57V[h]	1.902 ± 0.021	1.332 ± 0.003	1.065 ± 0.011	4.55 ± 0.75
HLADH V203L[g]	1.89 ± 0.01	1.38 ± 0.005	1.074 ± 0.004	4.5 ± 0.2
HLADH WT[h]	1.894 ± 0.013	1.335 ± 0.003	1.073 ± 0.008	4.1 ± 0.44
HLADH ESE[h]	1.872 ± 0.006	1.332 ± 0.004	1.075 ± 0.003	3.96 ± 0.16
HLADH V203A:F93W[g]	1.91 ± 0.02	1.325 ± 0.004	1.075 ± 0.004	3.9 ± 0.2
HLADH V203G[g]	1.89 ± 0.01	1.358 ± 0.007	1.097 ± 0.007	3.3 ± 0.2
YADH WT (p-MeO)[f]	1.94 ± 0.06	1.34 ± 0.04	1.12 ± 0.02	2.78 ± 0.82

[a] Reported values ± the standard error of the mean. [b] Exponent relating $k_D/k_{T(obs)}$ and $k_H/k_{T(obs)}$: $[k_D/k_{T(obs)}]^{RS} = k_H/k_{T(obs)}$. [c] The error was calculated as follows: error = $\exp[\{\partial \ln(k_H/k_T)/\ln(k_H/k_T)\}^2 + \{\partial \ln(k_D/k_T)/\ln(k_D/k_T)\}^2]^{1/2}$. [d] Ref. [24]. [e] Ref. [10]. [f] Ref. [83]. [g] Ref. [23]. [h] Ref. [6].

Observed isotope effects will approach intrinsic values when the commitment to catalysis is small ($C = 0$). Prior to tunneling analyses, earlier single-turnover experiments [81] and steady-state studies [82] had indicated conditions under which C_H would be small or zero.

Cha et al. provided the first experimental proof of hydrogen tunneling on an enzyme by reporting an elevated RS exponent for benzyl alcohol oxidation by yeast ADH (YADH) [10]. Isotope effects for benzyl alcohol oxidation were determined by the mixed-label tracer method, in which the primary and α-secondary positions of benzyl alcohol are either H or D, with stereochemically random, trace-level T incorporation. In this fashion, the observed ratios between the α-secondary $(k_H/k_T)_{1°H}$ and $(k_D/k_T)_{1°D}$ KIEs are susceptible to both Swain–Schaad and RGM deviations and, thus, are sensitive probes for tunneling (see Section 10.3.3.3). The observed α-secondary RS exponent, $k_H/k_T = (k_D/k_T)^{10.2}$ at 25 °C, greatly exceeded the semi-classical value of 3.3, (Table 10.2).

The results of Cha et al. [10] verified an earlier interpretation of elevated α-secondary KIEs in dehydrogenase reactions as arising from a large tunnel-correction to a semi-classical reaction coordinate [25]. Subsequent force-field calculations on the alcohol oxidation catalyzed by YADH were consistent with the view of a semi-classical reaction coordinate with a significant tunneling correction to the rates and KIEs [52]. As noted by Huskey, the only way to reproduce a large RS exponent is to include a significant tunnel correction [51].

Substituted benzyl alcohols were used to demonstrate that hydrogen tunneling in YADH does not require an isoenergetic reaction [82]. The internal thermodynamics of hydrogen transfer were varied over a 1.9 kcal mol^{-1} range by the use of three *para*-substituted benzyl alcohol substrates. For these substrates, k_{cat}, as well as k_{cat}/K_M, is largely limited by hydride transfer; the finding that k_{cat} at 25 °C varied by less than a factor of two for these substrates indicated a very small change in the rate of hydride transfer [83]. Mixed-label isotope effects on k_{cat}/K_M revealed that the RS exponent for *p*-chlorobenzyl alcohol was very similar to that for benzyl alcohol, while the RS exponent for the reaction with *p*-methoxybenzyl alcohol was below the semi-classical value of 3.3 (Table 10.2). These observations led to the conclusion that tunneling is operative for both the *p*-chloro and the parent benzyl alcohol, despite the fact that the reaction driving force differs by 1.4 kcal mol^{-1}, and that the *p*-methoxy substrate may be kinetically complex.

Hydrogen tunneling in HLADH was examined using active site mutants [23, 84]. While wild-type HLADH is partly limited by binding of aromatic substrates, site-directed mutations in the substrate pocket altered the rate of substrate binding/product release leading to an "unmasking" of chemistry. Mutants produce a 100-fold variation in k_{cat}/K_M, which shows a positive correlation with the magnitude of the α-secondary RS exponent, Table 10.3. At that time, the data were interpreted to indicate that tunneling contributes ca. 100-fold to the catalytic efficiency of HLADH as the RS value increases from 3.3 to 9. A simple energetic argument implies that a reduction in the classical barrier height of 4.5 kcal mol^{-1} would be required to produce an equivalent rate enhancement, making tunneling a significant contributor to catalysis.

The X-ray crystal structures of two HLADH mutants revealed a correlation between the hydride transfer distance and the RS exponent [23]. The high-tunneling F93W mutant was compared with the low-tunneling V203A mutant, each of which was crystallized with the nonreactive substrate-analog trifluoroethanol. It was observed that the distance between the hydride donor and acceptor (C-1 of alcohol

Table 10.3. Rates and α-2° RS exponent for HLADH and mutants, pH 7.00, 25 °C[a].

LADH mutant	k_{cat}/K_M (mM^{-1} s^{-1})[b]	RS[c]
L57F	8.6	8.5 ± 1
F93W	4.7	6.13 ± 0.5
L57V	3.5	4.55 ± 0.75
ESE	3.3	3.96 ± 0.16
V203L	1	4.5 ± 0.2
V203A	0.2	4.9 ± 0.3
V203A:F93W	0.13	3.9 ± 0.2
V203G	0.071	3.3 ± 0.2

[a] Refs. [22, 23]. [b] Errors on k_{cat}/K_M are less than 10% of value.
[c] Exponent relating $k_D/k_{T(obs)}$ and $k_H/k_{T(obs)}$: $[k_D/k_{T(obs)}]^{RS} = k_H/k_{T(obs)}$.

and C-4 of NAD$^+$, respectively) was smaller in the F93W mutant (3.2 Å) than in the V203A mutant (4.0 Å) [23]. This implied that the barrier width will play a role in hydride tunneling and, therefore, must be considered in understanding the factors that impact catalysis.

Hydrogen tunneling was demonstrated over a wide temperature range for the alcohol oxidation catalyzed by a thermophilic ADH (ht-ADH) from *Bacillus stearothermophilus* [24]. For this ADH, both k_{cat} and k_{cat}/K_M appear to be primarily controlled by hydride transfer over a large temperature range. Mixed label substrates were used to measure primary and α-secondary KIEs between 5 °C and 65 °C. As with YADH and HLADH, the α-secondary KIEs were shown to have RS exponents elevated from the semi-classical value of 3.3 (cf. Table 10.2). The RS exponent varied from ca. 5 to almost 20 as the temperature was increased from 5 °C to 65 °C, suggesting that the extent of tunneling increased as the temperature increased. *This contrasts with a simple tunnel correction, which would predict that the extent of tunneling should decrease at elevated temperature.*

Thermophilic proteins are thought to undergo a thermally driven phase transition, from a flexible phase near physiological temperatures, to a more rigid phase at reduced temperature. ht-ADH was shown to exhibit a change in catalytic behavior at 30 °C: above this temperature, the activation energy (E_a) for k_{cat} was moderate ($E_a = 14.0$ kcal mol^{-1}), while below this temperature E_a increased by nearly a factor of two ($E_a = 23.0$ kcal mol^{-1}) [24]. Support for a mechanistic phase transition that is due to altered protein flexibility was obtained from FT-IR H/D amide exchange experiments, which showed a greatly reduced global exchange rate below the 30 °C transition [85]. In more recent studies [86] mass spectrometry was used to quantitate H/D exchange, allowing a spatial resolution of the structural changes that accompany the increase in tunneling above 3 °C. Out of a total of 21 peptides analyzed, 5 showed changes in H/D exchange at the same temperature as the change observed in E_a cited above. All 5 peptides map to the substrate binding domain and 4 out of 5 have loop regions that are positioned to interact with substrate directly. These studies implicate a direct link between changes in local protein flexibility/dynamics and changes in H-transfer efficiency [86].

The primary KIEs on k_{cat} also indicated a transition at 30 °C, below which the primary k_H/k_D ratio is very temperature dependent, extrapolating to $A_H/A_D \ll 1$ [24]. This inverse Arrhenius prefactor ratio is predicted within the Bell tunnel correction for a moderate extent of tunneling, and is consistent with an elevated α-secondary RS exponent. Above 30 °C, the primary k_H/k_D ratio is nearly independent of temperature, resulting in an isotope effect on the prefactor of $A_H/A_D = 2$ [24]. A tunnel correction would also predict such an elevated Arrhenius prefactors ratio when both H and D react almost exclusively by tunneling; however this condition requires a very small activation energy for k_{cat}, while a value of $E_a = 14$ kcal mol^{-1} is observed [24].

The data from ht-ADH raise provocative questions regarding hydride transfer processes. In particular, it would appear that a model that goes beyond a simple tunnel correction is needed to explain the composite data for ht-ADH. One possible explanation is that, at elevated temperatures, hydride transfer is a full tunnel-

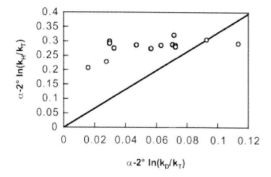

Figure 10.6. Graphical representation of the exponential relationship between $\ln(k_H/k_T)$ and $\ln(k_D/k_T)$ for α-secondary KIEs in alcohol dehydrogenases (YADH, LADH and ht-ADH), open circles. The exponential Swain–Schaad relationship is shown as a line of slope = 3.3.

ing process that is driven by thermal fluctuations of the protein; the latter would give rise to the observed E_a. As the temperature is reduced below 30 °C and the protein stiffens, a new process dominates with the properties of a tunneling correction, leading to the much larger observed E_a value. Given the change in A_H/A_D from >1 (above 30°C) to <1 (below 30°C), remniscent of the trends in A_H/A_D on mutagenesis of soybean lipoxygenase [43], an alternate explanation is full tunneling at all temperatures with varying contributions of the gating mode above and below 30°C (Figure 10.5).

The α-secondary KIE data for all of the ADHs from this laboratory have been summarized (Table 10.2), encompassing ADH from three sources, two alternate substrates with YADH, and two distinct temperature regimes for ht-ADH [6, 10, 23, 24]. A wide range of *RS* values is represented, suggesting, at first glance, that the extent of tunneling varies appreciably over this series. A graphical representation of the exponential relationship between $\ln(k_H/k_T)$ and $\ln(k_D/k_T)$ for the α-secondary KIEs is shown in Fig. 10.6. The semi-classical model predicts that *RS* is equal to 3.3, indicated as a solid line of slope 3.3. Data points above this line indicate an elevated *RS* exponent, which is a signature of tunneling. Most of the ADH data clearly deviate from the semi-classical prediction. Inspection of Table 10.2 reveals that the magnitude of the α-secondary k_D/k_T decreases as the *RS* exponent increases. This phenomenon, illustrated in Fig. 10.7, appears paradoxical in that more extensive tunneling for H than D might be expected to elevate the α-secondary k_H/k_T while leaving the α-secondary k_D/k_T relatively unchanged.

When viewing the ADH reactions through a tunnel-correction model, the k_D/k_T measurements are expected to be a much closer monitor of the semi-classical hydride-transfer coordinate [50]. Thus, it was possible that the small α-secondary k_D/k_T KIEs near the top of Table 10.2 simply reflected an early transition-state structure. This interpretation would require that the slowest LADH mutants (e.g. V203G) have the latest transition states, while the faster LADH mutants (e.g.

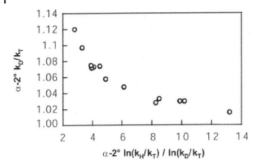

Figure 10.7. Comparison of the α-secondary k_D/k_T values and RS for benzyl alcohol oxidation by alcohol dehydrogenases (YADH, LADH and ht-ADH).

L57F) have the earliest transition states. It is notable that the equilibrium limit for the α-secondary k_D/k_T ratio is 1.09 ± 0.02 [10], approximately what is observed for those entries near the bottom of Table 10.2. While this would be consistent with the Hammond postulate [87] which predicts that earlier transition-state structures should correlate with faster reaction rates [88], a concomitant change in the primary k_D/k_T would also be predicted across this series, assuming relatively little tunneling for the D-transfer. In particular, the primary KIE should be maximal for a symmetric transition-state structure, and decrease for either an early or late structure. This trend is not observed; rather, the primary KIE shows a general increase toward the bottom of Table 10.2.

An alternate view would be that the primary and secondary KIEs simply reflect the properties of tunneling along the reaction coordinate, with tunneling occuring far below the semi-classical transition state, a situation often referred to as large curvature tunneling [89, 90]. We note that a computational study of the ADH reaction could reproduce the large RS exponents with a Bell tunneling correction, but could not reproduce the small α-secondary k_D/k_T ratios unless many vibrational modes were coupled into the hydride transfer coordinate [52]. This calculation showed that the α-secondary k_D/k_T KIEs indicated the extent of tunneling, rather than the structure of the transition state under conditions of pronounced coupled motion. Furthermore, coupled motion can increase or decrease the α-secondary k_D/k_T KIEs [52], making a simple explanation for the trends in Table 10.2 elusive. Krishtalik [77] has, in fact, proposed that the RS exponents may *decrease* under the conditions of a full tunneling model. Clearly, more work is needed before we have a complete theoretical framework for the deviant secondary isotope effects seen in hydride transfer reactions (see note added in proof).

10.5.1.2 Glucose Oxidase

Glucose oxidase (GO) catalyzes the oxidation of sugars to their corresponding lactone products, coupled to the reduction of O_2 to H_2O_2, Scheme 10.3. The reaction

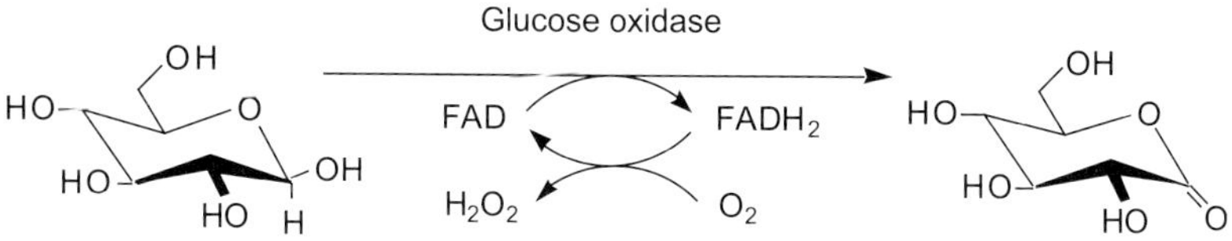

Scheme 10.3

occurs via a ping-pong mechanism that leads to a cycling of the bound flavin between a reduced and oxidized form, Fl_{red} and Fl_{ox}, respectively. Isotopes have been used to probe both the reductive and oxidative half reactions, using hydrogen isotopes in the former [91, 92] and oxygen isotope effects in the latter case [93–96]. Recent studies of the reductive half reaction using enzyme that has been substituted with flavins of modified redox potentials indicate a structure reactivity correlation in support of hydrogen movement from substrate to flavin as a hydride ion [97].

Hydrogen isotope measurements have been carried out primarily for the reaction of the 2′-deoxy-form of sugar rather than glucose itself, as hydride transfer is largely rate determining for 2′-deoxyglucose. Very interestingly, expression of the GO gene in a yeast host gives rise to proteins of varying degrees of glycosylation that yield final molecular masses for the protein rising from 136 kDa (deglycosylated enzyme) to 320 kDa (extensively glycosylated enzyme). Initial studies using the 136, 155 and 205 kDa forms of GO at pH 9 showed similar rates and isotope effects, but different temperature dependences for the isotope effects [91].

It is important to point out that, unlike other enzyme systems where very significant deviations from classical behavior have been observed, the size of the KIEs and their temperature dependences are all seen to hover around the classical values in GO. In a very careful study, Seymour examined the impact of surface modification on the parameters describing H-transfer in GO. The impact of the surface modifier was assessed, by comparing H-transfer in proteins modified with polyethylene glycol to those modified by glycosylation [92].

Although both primary H/T and D/T isotope effects were measured as a function of temperature, the focus of the interpretation was on primary D/T isotope effects, since the smaller commitment for D- than H-transfer leads to a more complete contribution of the chemical step to the measured parameters (see e.g. Section 10.21). The results obtained from this study [92], as well as the earlier work [91], indicate a striking trend in which the value of A_D/A_T moves from near unity to below unity as the surface of the protein is modified (either by glycosylation or addition of polyethylene glycol). A similar type of pattern has been seen in other enzyme systems, such as the thermophilic ADH [24] (Section 10.5.1.1) and soybean lipoxygenase (SLO) [43] (see Section 10.5.3.1 below) where modification of reaction conditions away from either optimal temperature (ht-ADH) or optimal protein packing (via

Table 10.4. Enzymatic examples where perturbations lead to a decrease in A_1/A_2.

Enzyme	A_1/A_2	Ref.
Glucose Oxidase		92
WT:	~1	
surface modified:	<1	
Thermophilic Alcohol Dehydrogenase		24
above 30 °C:	>1	
below 30 °C:	<1	
Soybean Lipoxygenase		43
WT:	≫1	
mutants:	~1 or <1	

site specific mutagenesis in SLO) causes the value of A_H/A_D to fall from greater than unity to less than unity (Table 10.4). *In the context of a full tunneling model (cf. Eq. (10.24)), values for A_1/A_2 less than one reflect a greater need for distance sampling (gating) to achieve the optimal distances between reactant and product that support efficient tunneling.* One curious feature of the data for GO is that surface modification does not, in contrast to an earlier report [91], give rise to greater values for the experimental enthalpies of activation [98]. In the full tunneling model with gating, Eq. (10.20), and in the absence of compensating changes, the value of $\Delta H^\ddagger$ is expected to increase as gating plays a greater role in achieving the optimal configuration(s) for catalysis. As an alternative to a full tunneling model, the hydride transfer catalyzed by GO has also been considered in the context of a tunneling-correction model. In this case, the increased stiffness of protein resulting from surface modification could reduce the distance sampling between reactants, resulting in a higher overall barrier and increased tunneling relative to WT-GO. In this instance the inverse value for A_D/A_T would reflect the greater ease for the lighter isotope to move under the reaction barrier. Within this model, the very small changes in $\Delta H^\ddagger$ among the enzyme forms could reflect the interplay between increases in barrier height, which would increase $\Delta H^\ddagger$, and increased tunneling which would decrease $\Delta H^\ddagger$.

The above discussion of GO illustrates the complexity of this H-transfer systems and the ambiguity that can arise in trying to distinguish between systems in which a tunneling correction [15] is appropriate and systems in which a full tunneling model [33] must be invoked. As will be discussed in Section 5.3.1, there are some enzyme systems where a tunneling correction will simply not apply, providing a contextual understanding for tunneling models that can then be applied to systems with less dramatic deviations from classical behavior.

10.5.2
Amine Oxidases

10.5.2.1 Bovine Serum Amine Oxidase

Bovine serum amine oxidase (BSAO) is a copper-containing amine oxidase which utilizes a covalently bound 2,4,5-trihydroxyphenylalanine quinone (TPQ) cofactor in the two-electron oxidation of a broad spectrum of primary amines [99]. The oxidation is thought to proceed via the formation of an iminium complex between the oxidized form of the cofactor and the primary amine (1 in Scheme 10.4). The substrate imine undergoes deprotonation to form product imine, which, after hydrolysis, releases aldehyde product and reduced cofactor [100]. Proton transfer is either partially or largely rate-limiting for the oxidation of benzylamines, as evidenced by a large deuterium isotope effect at the methylene adjacent to the amino group [36, 101, 102].

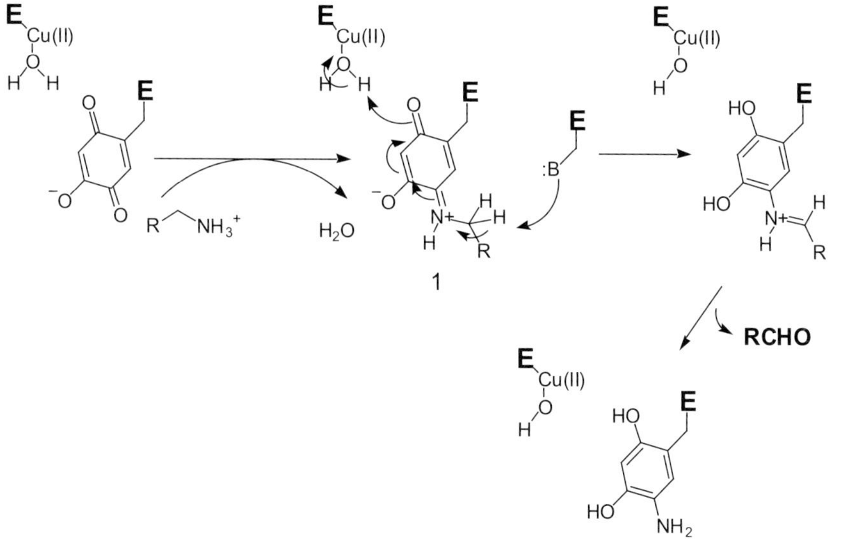

Scheme 10.4

Competitive kinetic isotope effects (H/T and D/T) for the BSAO-catalyzed oxidation of benzylamine were determined using remote labeling methods. The values for primary H/T isotope effects on k_{cat}/K_M average 35.2 ± 0.8 for 6 separate experiments composed of approximately 8 time points each. Correspondingly, the primary D/T isotope effects average to 3.07 ± 0.07 [36]. These values exceed semi-classical expectations of 27 and 2.7 for H/T and D/T primary kinetic isotope effects on elementary reaction steps, respectively [15], over the temperature range 0–45 °C. Arrhenius parameters also exhibit signatures of tunneling from both H/T and

D/T isotope effects. The predicted lower limits for Arrhenius pre-factors (A_L/A_T) are 0.6 and 0.9 for L = H and L = D, respectively [19], in contrast to the measured values for this system of 0.12 and 0.51. The temperature dependence of the isotope effects far exceeds that anticipated by zero-point energy differences in the C–H stretch: for the complete disappearance of the C–H stretching zero-point energy with no compensation from transverse bending modes at the transition state, one would expect ΔE_a values of 2.0 kcal mol^{-1} for the H/T isotope effect and 0.6 kcal mol^{-1} for D/T isotope effects in contrast to observed values of 3.4 and 1.1 kcal mol^{-1}, respectively.

What was most striking in the BSAO system was the evidence for deuterium as well as protium tunneling. First, the value of the D/T isotope effect on k_{cat}/K_M at 25 °C is greater than the maximum expected value arising from zero-point energy differences. Second, the isotope effect on the Arrhenius pre-factor is well below that which is expected from semi-classical predictions. Finally, the differences in enthalpy of activation for D vs. T in the primary position well exceed differences predicted from zero point energies.

While it is conceivable that one could attribute some of the differences in the Arrhenius parameters to kinetic complexity, it is unlikely that kinetic complexity would be present to any significant degree with either deuterium or tritium in the primary position given the significant values of the H/D and H/T primary KIEs. The enthalpy of activation for the oxidation of benzylamine has been computed from the temperature dependence for the BSAO-catalyzed consumption of 1,1-d_2-benzylamine and the difference in the enthalpy of activation, $E_a(D) - E_a(H)$. The value obtained is 13 kcal mol^{-1}, close to the value of 14 kcal mol^{-1} obtained directly from the temperature dependence for the corresponding oxidation of benzylamine [102]. This close correspondence indicates that little of the observed temperature dependence in the H/T isotope effect can be due to the contribution of other partially rate-limiting steps. It should be emphasized that the invocation of kinetic complexity would mean that the absolute magnitude of the intrinsic KIE on the isotopically-sensitive step would be larger than the observed values for the primary competitive kinetic isotope effects, which would further implicate a tunneling mechanism in the BSAO system.

Although it seems evident that tunneling is operative in the proton abstraction step of BSAO-catalyzed oxidations, results obtained regarding competitive secondary isotope effects were initially puzzling in the light of those obtained for the ADH-catalyzed oxidation of alcohols, where secondary kinetic isotope effects have been one of the primary determinants of tunneling behavior. As noted in Section 5.1.1, RS exponents of 10.2 were measured for the α-secondary isotope effect for YADH. In the BSAO-catalyzed oxidation of benzylamines, the Swain–Schaad exponents for both primary and secondary KIEs are near the anticipated 3.26 for relating H/T and D/T isotope effects. Over the temperature range explored, the Swain–Schaad exponent was perhaps slightly reduced from the semi-classical value for the primary position and slightly elevated for the secondary position. These findings were taken to mean that there is much less coupling between the motion of the secondary position and the transferred hydrogen in BSAO than in ADH [36].

From a historical perspective, the BSAO system has played an important role, inspiring a new way of approaching the concept of hydrogen tunneling in enzymes. Bruno and Bialek determined that ground-state tunneling would give rise to H/T isotope effects that far exceed the measured values [32], and introduced a model in which the enzyme environment modulates the barrier for proton transfer. Using a rectangular barrier whose width is altered by thermally-induced protein motions, Bruno and Bialek were able to fit both D/T and H/T kinetic isotope effect data with the same pair of adjustable parameters, changing only the mass of the tunneling particle (see Section 10.4). While their choice of barrier shape was not physically realistic, it illustrated how protein-mediated motions can alter the contribution of tunneling to the reactive flux in enzyme-catalyzed hydrogen/hydride/proton transfers.

10.5.2.2 Monoamine Oxidase B

Monoamine oxidase B (MAO-B) is a membrane-bound flavoprotein that catalyzes the two-electron oxidation of primary, secondary, and tertiary amines with a preference for primary amines. Extensive studies have addressed the issue of whether the hydrogen transfer occurs via a step-wise mechanism (i.e. electron transfer followed by H^+) or a concerted process (H•) (e.g. Refs. 103, 104).

Regardless of the precise nature of the mechanism, the amine must lose an α-hydrogen atom in the course of its oxidation. Jonsson et al. have determined primary and secondary H/T and D/T kinetic isotope effects for the MAO-B catalyzed oxidation of p-methyoxybenzylamine [65]. Initially, these isotope effect studies were performed at pH 7.5; however, it was found that the observed competitive H/T KIE was larger at 25.0 °C than at 2.0 °C. Because the KIE was substantially larger at the higher temperature, it could be ascertained that kinetic complexity was contributing markedly to the kinetic isotope effects. Kinetic complexity was corroborated by the value of the Swain–Schaad exponents for the primary and secondary KIEs. Primary KIEs rarely exhibit abnormal Swain–Schaad exponents due to tunneling, such that a value far below 3.26 relating H/T to D/T isotope effects is indicative of kinetic complexity. The primary Swain–Schaad exponent at pH 7.5 and 2.0 °C is 2.57 ± 0.05, and the exponent for the secondary KIE under the same conditions is 1.20 ± 0.15. At pH 6.1, however, Swain–Schaad exponents for the primary KIEs at all temperatures studied (10.0–43.0 °C) fell very near to the semi-classical expectation of 3.26. The exponents for the secondary KIEs were still somewhat low, however, averaging 2.36 ± 0.13 over the temperature range studied.

It is noteworthy that the Swain–Schaad exponents are temperature independent for both primary and secondary isotope effects at pH 6.1. Two scenarios can be considered. The first is that there is a significant commitment to catalysis which is obscuring the full value of the isotope effect on k_{cat}/K_M. It is anticipated that, because the primary and secondary exponents are temperature independent, this commitment would be temperature independent. Jonsson et al. have used Northrop's expression [60] for correcting observed isotope effects based on the assumption of a temperature-independent commitment (Eq. (10.22)). A commitment of 0.6 for the oxidation of benzylamine brings the secondary exponent to about 3.3, the semi-classical prediction. The commitment for deuterium abstraction is related to the commitment for protium abstraction by the intrinsic deuterium isotope effect, which is about 10 for MAO-B. Using these values, the Arrhenius parameters do not change appreciably from those computed from fits of observed isotope effects. The isotope effect on the Arrhenius pre-factors are still indicative of tunneling with the inclusion of this temperature-independent commitment (C).

$$(k_H/k_T)_{\text{intrinsic}} = [(k_H/k_T)_{\text{observed}}(1 + C)] - C \tag{10.22}$$

The second scenario which might give rise to the observed temperature dependences is a temperature-dependent commitment factor which is exactly compensated by the portion of the reaction which is achieved via tunneling. While it seems highly unlikely that such a scenario could fortuitously exist, the motivation for such a consideration is that the degree to which tunneling participates in reactive flux inversely correlates with temperature. This phenomenon has been clearly demonstrated in the crystalline solid phase and in solid rare gas matrices [105]. Thus, it seems conceivable that, while a larger degree of tunneling at low temperatures could raise the Swain–Schaad exponent, a larger commitment at low temperatures could reduce the exponent, leading to the observed temperature invariant values.

10.5.3
Hydrogen Atom (H•) Transfers

10.5.3.1 Soybean Lipoxygense-1

Some of the most striking evidence for hydrogen tunneling under ambient conditions comes from the kinetic studies of soybean lipoxygenase-1 (SLO) [38, 43, 79, 106]. The $^D k_{\text{cat}} \geq 80$ for WT-SLO at room temperature [38] makes typical views of hydrogen transfer, including tunneling corrections, of dubious relevance. SLO catalyzes the production of fatty acid hydroperoxides at 1,4-pentadienyl positions, and the product 13-(S)-HPOD is formed from the physiological substrate linoleic acid (LA) (Scheme 10.5). SLO is a substrate-activating dioxygenase, activating LA by homolytic C–H cleavage, and is the simplest example of metalloenzyme-catalyzed H• abstraction, in that a stable form of the cofactor rather than a metastable species cleaves the C–H bond. Studies of SLO are useful in illuminating apolar H• transfers, which may have implications for a wide variety of metalloenzyme C–H cleavage reactions.

SLO follows an ordered, bi-uni mechanism, in which linoleic acid (LA) binds and reacts prior to O_2 encounter [8], which has permitted a variety of steady-state and single-turnover studies into chemistry on SLO. The kinetic mechanism can be divided into a reductive half-reaction, described by the rate constant $k_{\text{cat}}/K_M(\text{LA})$, and an oxidative half-reaction described by the rate constant $k_{\text{cat}}/K_M(O_2)$. On the reductive half-reaction, SLO binds LA (k_1), then the Fe^{3+}–OH cofactor abstracts the pro-S hydrogen from C-11 of LA (k_2), forming a substrate-derived radical inter-

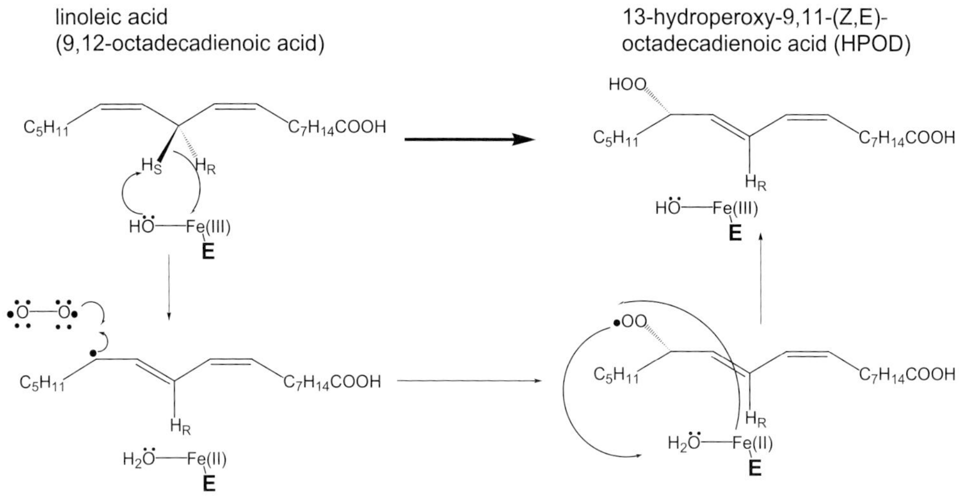

Scheme 10.5

linoleic acid
(9,12-octadecadienoic acid) → 13-hydroperoxy-9,11-(Z,E)-octadecadienoic acid (HPOD)

mediate and $Fe^{2+}–OH_2$ (Scheme 10.5) [8]. H• abstraction is kinetically irreversible [8], making this the final step that appears on $k_{cat}/K_M(LA)$ and $k_{cat}/K_M(LA)$ partially rate limited by both substrate binding and chemistry. Molecular oxygen rapidly reacts with this radical in the oxidative half-reaction, eventually forming 13-(S)-HPOD and regenerating free enzyme at a rate that appears to be limited by local structural features of the protein [107, 108].

Much of the work substantiating hydrogen tunneling in this reaction has relied on steady-state kinetics, in which $^D k_{cat}$ is determined. The magnitude of the KIE was corroborated by viscosity effects, solvent isotope effects, and single-turnover studies [7–9]. One notable study confirmed the magnitude of the KIE as ca. 80 at room temperature, while excluding magnetic effects as the origin of this KIE [38b]. Potential complications in assigning $^D k_{cat}$ to a single chemical step, for example due to a branched reaction mechanism, were also ruled out [7]. All the data indicate that the chemical step (H• abstraction) is fully rate limiting on k_{cat}, in WT-SLO, and that the steady-state KIE ($^D k_{cat}$) represents an intrinsic value.

The bulk of the KIE studies with SLO have used substrate that is dideuterated at the C-11 position. In order to differentiate the impact of deuteration on the cleaved [11,S] vs. the noncleaved position [11,R], stereospecifically deuterated [11,S-^{2}H] – linoleic acid was synthesized. Analysis of enzyme kinetics using either WT- or mutant forms of the enzyme initially yielded anomalously large secondary KIEs. This was subsequently shown, by mass spectrometric analysis of the enzymatic product, to be due to a loss of stereochemistry at C-11, arising, in fact, from the enormous KIE at the primary position. Through a combination of kinetic and mass spectrometric analyses, it could be shown that the bulk of the observed KIE with dideuterated substrate arises from the primary position, $^D[k_{intrinsic}(1°)] = 75$ and $^D[k_{intrinsic}(2°)] = 1.1$ [79].

Though the large KIE is striking, several other enzymatic and nonenzymatic reactions exhibit KIEs of greater than 50 at or near room temperature. Methane monooxygenase exhibits an H/D KIE of between 50 and 100 in the cleavage of the C–H bonds of methane [39] and acetonitrile [109]. A number of polypyridyl ruthenium-oxo compounds of the general formula [LRu^{4+}=O]$^{2+}$ exhibit large KIEs in H• and H$^+$ transfer reactions [110–113]. An example of a KIE of ca. 50 has been reported for H transfer catalyzed by a Cu^{2+}-phenoxyl radical complex [114].

The variable-temperature kinetic data for SLO are unusual, in that the temperature dependence of the rate is very small ($E_a = 2.1$ kcal mol^{-1}) [43]. An Eyring treatment of the variable temperature kinetic data for WT-SLO suggests that the barrier to reaction is mainly entropic, as the enthalpic barrier ($\Delta H^\ddagger = 1.5$ kcal mol^{-1}) is much less than the entropic barrier ($-T\Delta S^\ddagger = 12.8$ kcal mol^{-1}). Such a treatment becomes meaningless when the KIEs are considered, as the isotope effect is predominantly on the entropic term rather than the enthalpic term. In particular, the bond-stretch theory of KIEs predicts that the KIE results entirely from the enthalpic term [16]. It is clear that the bond-stretch model fails to account for the rate and KIE data. Furthermore, a tunnel correction cannot simultaneously reproduce the magnitude and temperature dependence of this KIE, as such corrections result exclusively in inverse Arrhenius prefactors ($A_H/A_D \ll 1$). As discussed in Section 10.4, WT-SLO exhibits a large Arrhenius prefactor KIE ($A_H/A_D = 18$), while mutations near the active site reduce the temperature-dependence of the KIE in a regular manner, leading to an inverse Arrhenius prefactor KIE ($A_H/A_D = 0.12$ for I553A).

That the KIE in WT-SLO and its mutants remains large ($k_H/k_D > 80$ at 30 °C) forces the use of a model in which hydrogen transfer always occurs by tunneling [115]. The variable temperature-KIE behavior observed as a function of mutational position indicates that the active site of SLO plays an important role in optimal positioning of the substrate for H• tunneling. Altering three of the bulky hydrophobic residues at or near the active site residues (Leu^{546}, Leu^{754}, Ile^{553}) makes this positioning sub-optimal, and introduces the requirement for a fluctuating barrier in effecting catalysis. The KIE arises from the differential tunneling probabilities of H and D at the reactive configuration, while the environmental vibration (gating) modulates the width of the tunneling barrier, and leads to the various temperature dependent KIEs. The net rate is always a compromise between an increased tunneling probability at short transfer distances and the energetic cost of decreasing the tunneling barrier (Fig. 10.4). These features directly link enzyme fluctuations to the hydrogen transfer reaction coordinate, making a quantum view of H-transfer necessarily a dynamic view of catalysis.

Explicit tunneling effects are required to accommodate the kinetics of SLO, and may be equally important in many other hydrogen atom transfer reactions. Many H• transfer reactions are characterized by very large inherent chemical barriers, such that movement through, rather than over, the barrier may dominate the reaction pathway as the lowest energy path for conversion of reactants to products.

10.5.3.2 Peptidylglycine α-Hydroxylating Monooxygenase (PHM) and Dopamine β-Monooxygenase (DβM)

DβM and PHM belong to a small class of copper-containing oxygenases that catalyze the formation of norepinephrine and the precursor to C-terminally amidated peptide hormones, respectively (Scheme 10.6A and B) [116]. PHM is often found covalently linked to a second protein domain that is responsible for the hydrolytic breakdown of the immediate hydroxylated intermediate (Scheme 10.6B) to the final peptide product and glyoxylate [117]. Although DβM is significantly larger than PHM, the two proteins share a core of conserved residues that contributes the ligands to the two copper centers per polypeptide chain.

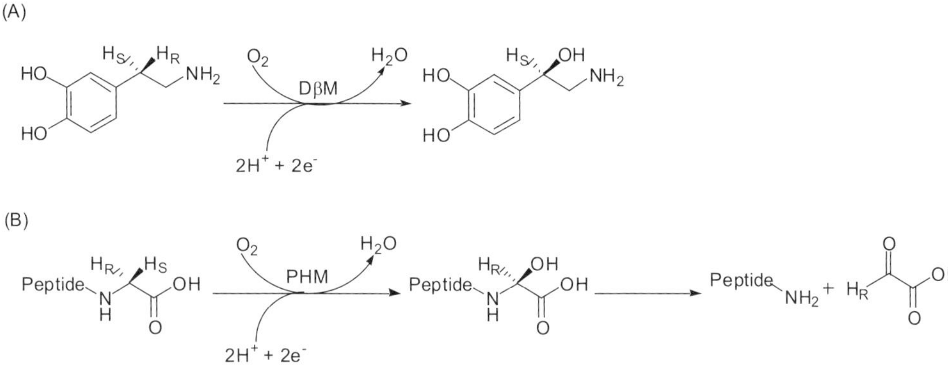

Scheme 10.6

The nature of H-transfer in these reactions is intrinsically linked to the identity of the oxygen species capable of C–H activation. The mechanism of PHM and DβM has been subjected to extensive investigations, including an X-ray structure for PHM that shows the two coppers per subunit located at a distance of ca. 10 Å across a solvent interface [118]. This structure has raised questions regarding how electrons are transferred between the metal centers and how the chemically difficult O_2 and C–H activation reactions can occur at a solvent interface. Recently, Evans et al. have put forth a mechanism that is able to accommodate the very large array of experimental data for both PHM and DβM [119].

As described in Ref. [119], the O_2 and C–H activation chemistry occur at a single copper site (Cu_B), with the donation of the second electron required to complete the reaction sequence (from Cu_A) occurring subsequent to H• transfer from substrate to activated O_2. Although the reactive oxygen species is written as $Cu(II)–(O_2^{\bullet-})$, the failure to observe any uncoupling of O_2 from C–H activation suggests that the amount of this species accumulates to only a very small extent at the enzyme active site [119]. As discussed below, the mechanism of hydrogen transfer displays the properties indicative of H-tunneling, implicating a role for environmental reorganization terms, both to generate the reactive metal superoxospecies and to facilitate a quantum mechanical hydrogen atom transfer from donor to acceptor atoms.

Early experiments with DβM, that provided intrinsic primary and secondary KIEs for the C–H cleavage step, showed a puzzling discrepancy in the context of the bond-stretch model for hydrogen transfer. This discrepancy was that the primary KIE implied a symmetrical transition state while the secondary KIE indicated that the transition state was very product like [120]. The failure of multiple probes of transition state structure to converge on a consistent pattern of behavior is one of the criteria available for the diagnosis of tunneling (cf. Section 10.3.3.2).

In a more recent set of experiments, Francisco et al. turned to the magnitude of the intrinsic KIE in the PHM reaction as a function of temperature [13]. Observing that hydrogen transfer was only partially rate limiting for k_{cat}/K_M, substrate samples were prepared that allowed precise competitive measurement of both H/D and H/T isotope effects between 5 and 55 °C. Subsequent analysis of the breakdown from the Swain–Schaad relationship of the experimental KIEs, according to the methods of Northrop [80], led to the magnitude of the intrinsic primary KIE at each temperature. Although breakdowns from the Swain–Schaad relationship can also be seen under conditions of tunneling, these breakdowns seem restricted to secondary KIEs and, in particular, when comparisons are made between D/T and H/T KIEs (cf. Section 10.3.3.3). The plot of the intrinsic KIEs for PHM as a function of temperature yielded isotopic Arrhenius parameters of $A_H/A_D = 5.9$ (3.2) and $E_a(D) - E_a(H) = 0.37$ (0.33) kcal mol^{-1}.

These data are reminiscent of the data for SLO (Section 5.3.1), with the exception of the smaller magnitude of the KIEs in the case of PHM. Using the expression for the KIE derived by Kuznetsov and Ulstrup and modified by Knapp et al. [43], successful modeling of the temperature dependent KIE data for PHM data has led to a value for the gating term that controls the distance between donor and acceptor atoms that is significantly smaller than that for SLO ($\omega_x = 400$ cm^{-1} for SLO versus 45 cm^{-1} for PHM) [119]. These differences must reflect the very large differences in active site structures, with the SLO active site being deeply buried while the PHM active site is solvent exposed. The results suggest that the SLO site is stiffer and more optimized for H-tunneling, while that of PHM requires more extensive environmental reorganization to achieve the geometries that permit optimal wavefunction overlap between donor and acceptor. This type of analysis opens a new "window" into the properties of enzyme active sites and the nature of the differences from one system to another. *One important aspect of the work on PHM is the demonstration of relatively small KIEs in the context of environmentally modulated H-tunneling reactions.*

10.6
Concluding Comments

The last 15 years has seen a transformation in our understanding of the nature of H-transfer in the condensed phase. Much of this change is due to studies of enzyme catalyzed reactions, though the literature of "anomalous" KIEs in small molecule reactions has received increasing consideration as a result of the findings in enzymatic systems. There is no question that the available data force us to move beyond the simple bond-stretch picture for hydrogen transfer, with the best available explanation for the repeated deviations from semi-classical behavior being the quantum mechanical properties of hydrogen. Theoretical advances have appeared hand in hand with experimental findings, beginning with the treatises by Bell that discuss deviations from semi-classical behavior in the context of a tunneling correction. While this approximation was sufficient for many years, it is no longer able to explain the large body of data that has emerged for numerous enzyme systems. A more satisfactory picture has begun to emerge in which the H particle is treated fully quantum mechanically, with the barriers to reaction and the accompanying energies of activation reflecting the environmental reorganizations that must accompany hydrogen tunneling. This theoretical context is much closer to Marcus theory for electron transfer, with the additional caveat of the importance of distance sampling (gating) between hydrogen donor and acceptor atoms. In the last several years, with the growing recognition that both small molecule and enzyme hydrogen transfer reactions involve tunneling, investigators have begun to examine the degree to which enzymes may enhance tunneling in relation to the solution reactions. This issue has been discussed in a recent review [121], which points out the different kinds of experimental approaches for analysis of the extent of rate acceleration provided by tunneling in enzymes, together with the likelihood that the degree to which an enzyme uses tunneling for its rate acceleration will depend on the nature of the reaction undergoing catalysis.

Note added in proof: A model for the RS deviations in the ADH reactions (Table 10.2) has been recently presented: Klinman, J. P., *Phil. Trans. R. Soc. B* **2006**, *361*, 1323–1331; Nagel, Z., Klinman, J. P., *Chem. Rev.* **2006**, *106*, 3095–3118.

References

1 SEGEL, I. H., *Enzyme Kinetics*, Wiley Classics Edition, 1993 edn. John Wiley & Sons, New York, 1975.
2 CLELAND, W. W., Steady State Kinetics, in *The Enzymes*, 3rd edn., BOYER, P. (Ed.), Academic Press, New York, 1970, Vol. 2, pp. 1–65.
3 VILLA, J., WARSHEL, A., *J. Phys. Chem. B* **2001**, *105*, 7887–7907.
4 FERSHT, A., *Enzyme Structure and Mechanism*, 2nd edn., W. H. Freeman, New York, 1985.
5 Some selected references: (a) SU, Q., KLINMAN, J. P., in *1998 Steenbock Symposium on Enzyme Mechanism*, pp. 20–31; (b) BLACKLOW, S. C., RAINES, R. T., LIM, W. A., ZAMORE, P. D., KNOWLES, J. R., *Biochemistry* **1988**, *27*, 1158–1167; (c) HARDY, L. W., KIRSCH, J. F., *Biochemistry* **1984**, *23*, 1275–1282.
6 BAHNSON, B. J., PARK, D. H., KIM, K., PLAPP, B. V., KLINMAN, J. P., *Biochemistry* **1993**, *32*, 5503–5507.
7 GLICKMAN, M. H., KLINMAN, J. P., *Biochemistry* **1995**, *34*, 14077–14092.
8 GLICKMAN, M. H., KLINMAN, J. P. *Biochemistry* **1996**, *35*, 12882–12892.
9 JONSSON, T., GLICKMAN, M. H., SUN, S. J., KLINMAN, J. P., *J. Am. Chem. Soc.* **1996**, *118*, 10319–10320.
10 CHA, Y., MURRAY, C. J., KLINMAN, J. P., *Science* **1989**, *243*, 1325–1330.

11 Basran, J., Sutcliffe, M. J., Scrutton, N. S., *J. Biol. Chem.* **2001**, *276*, 24581–24587.

12 Harris, R. J., Meskys, R., Sutcliffe, M. J., Scrutton, N. S., *Biochemistry* **2000**, *39*, 1189–1198.

13 Francisco, W. A., Knapp, M. J., Blackburn, N. J., Klinman, J. P., *J. Am. Chem. Soc.* **2002**, *124*, 8194–8195.

14 Sikorski, R. S., Wang, L., Markham, K. S., Rajagopalan, P. T. R., Benkovic, S. J., Kohen, A., *J. Am. Chem. Soc.* **2004**, *126*, 4778–4779.

15 Bell, R. P., *The Tunnel Effect in Chemistry*, Chapman and Hall, New York, 1980.

16 Sühnel, J., Schowen, R. L., Theoretical Basis for Isotope Effects, in *Enzyme Mechanism from Isotope Effects*, 1st edn., Cook, P. F. (Ed.), CRC Press, Boca Raton, FL, 1991, Vol. 1, pp. 3–35.

17 (a) Melander, L., Saunders, W. H., *Reaction Rates of Isotopic Molecules*, R. E. Krieger Publishing, Malabar, FL, 1987; (b) Bigeleisen, J., Goepert-Mayer, M., *J. Phys. Chem.* **1947**, *15*, 261.

18 Westheimer, F. H. *Chem. Rev.* **1961**, *61*, 265.

19 Schneider, M. E., Stern, H. J., *J. Am. Chem. Soc.* **1972**, *94*, 1517–1522.

20 Schowen, R. L., in *Progress in Physical Organic Chemistry*, Streitweiser, A., Jr., Taft, R. W., Jr. (Eds.), Wiley-Interscience, New York, 1972, Vol. 9, p. 275.

21 Klinman, J. P., *Adv. Enzymol. Relat. Areas Mol. Biol.* **1978**, *46*, 415–94.

22 Chin, J. K., Klinman, J. P., *Biochemistry* **2000**, *39*, 1278–1284.

23 Bahnson, B. J., Colby, T. D., Chin, J. K., Goldstein, B. M., Klinman, J. P. *Proc. Natl. Acad. Sci. USA* **1997**, *94*, 12797–12802.

24 Kohen, A., Cannio, R., Bartolucci, S., Klinman, J. P., *Nature* **1999**, *399*, 496–499.

25 Huskey, W. P., Schowen, R. L., *J. Am. Chem. Soc.* **1983**, *105*, 5704–5706.

26 Cook, P. F., Oppenheimer, N. J., Cleland, W. W., *Biochemistry* **1981**, *20*, 1817.

27 Welsh, K. M., Creighton, D. J., Klinman, J. P., *Biochemistry* **1980**, *19*, 2005–2016.

28 Parkin, D. L., Theoretical Basis for Isotope Effects, in *Enzyme Mechanism from Isotope Effects*, 1st edn., Cook, P. F. (Ed.), CRC Press, Boca Raton, FL, 1991, Vol. 1, pp. 269–290.

29 Cleland, W. W., in *Isotope Effects in Chemistry and Biology*, Kohen A., Limbach, H. H. (Eds.), Taylor and Francis – CRC Press, Boca Raton, FL, 2006, pp. 915–929.

30 Antoniou, D., Schwartz, S. D., *J. Phys. Chem. B* **2001**, *105*, 5553–5558.

31 Borgis, D. C., Lee, S. Y., Hynes, J. T. *Chem. Phys. Lett.* **1989**, *162*, 19–26.

32 Bruno, W. J., Bialek, W., *Biophys. J.* **1992**, *63*, 689–699.

33 Kuznetsov, A. M., Ulstrup, J., *Can. J. Chem.* **1999**, *77*, 1085–1096.

34 Levich, V. G., Dogonadze, R. R., German, E. D., Kuznetsov, A. M., Kharkats, Y. I., *Electrochim. Acta* **1970**, *15*, 353–367.

35 Decornez, H., Hammes-Schiffer, S., *J. Phys. Chem. A.* **2000**, *104*, 9370–9348.

36 Grant, K. L., Klinman, J. P., *Biochemistry* **1989**, *28*, 6597–6605.

37 Chowdhury, S., Banerjee, R., *J. Am. Chem. Soc.* **2000**, *122*, 5417–5418.

38 (a) Glickman, M. H., Wiseman, J. S., Klinman, J. P., *J. Am. Chem. Soc.* **1994**, *116*, 793–794; (b) Hwang, C. C., Grissom, C. M., *J. Am. Chem. Soc.* **1994**, *116*, 795–796.

39 Nesheim, J. C., Lipscomb, J. D., *Biochemistry* **1996**, *35*, 10240–10247.

40 Abad, J. L., Camps, F., Fabrias, G., *Angew. Chem. Int. Ed.* **2000**, *39*, 3279–3281, 3157.

41 Meschede, L., Limbach, H. H., *J. Phys. Chem.* **1991**, *95*, 10267–10280.

42 Basran, J., Sutcliffe, M. J., Scrutton, N. S., *Biochemistry* **1999**, *38*, 3218–3222.

43 Knapp, M. J., Rickert, K., Klinman, J. P., *J. Am. Chem. Soc.* **2002**, *124*, 3865–3874.

44 Klinman, J. P., *J. Biol. Chem.* **1972**, *247*, 7977–7987.

45 Klinman, J. P., *Biochemistry* **1976**, *15*, 2018–2026.

46 Miller, S. M., Klinman, J. P., *Biochemistry* **1983**, *22*, 3091–3096.
47 Miller, S. M., Klinman, J. P., *Biochemistry* **1985**, *24*, 2114–2127.
48 Francisco, W. A., Merkler, D. J., Blackburn, N. J., Klinman, J. P., *Biochemistry* **1998**, *37*, 8244–8252.
49 Kurz, L. C., Frieden, C., *J. Am. Chem. Soc.* **1980**, *102*, 4198.
50 Saunders, W. H., Jr., *J. Am. Chem. Soc.* **1985**, *107*, 164–169.
51 Huskey, W. P., *Phys. Org. Chem.* **1991**, *4*, 361–366.
52 Rucker, J., Klinman, J. P., *J. Am. Chem. Soc.* **1999**, *121*, 1997–2006.
53 Amin, M., Price, R. C., Saunders, W. H., *J. Am. Chem. Soc.* **1990**, *112*, 4467–4471.
54 Alston, W. C., Kanska, M., Murray, C. J., *Biochemistry* **1996**, *35*, 12873–12881.
55 Lin, S., Saunders, W. H., *J. Am. Chem. Soc.* **1994**, *116*, 6107–6110.
56 Swain, C. G., Stivers, E. C., Reuwer, J. F., Schaad, L. J., *J. Am. Chem. Soc.* **1958**, *80*, 5885–5893.
57 Kohen, A., Jensen, J. H., *J. Am. Chem. Soc.* **2002**, *124*, 3858–3864.
58 Bigeleisen, J., *J. Chem. Phys.* **1955**, *23*, 2264–2267.
59 Grant, K. L., Klinman, J. P., *Bioorg. Chem.* **1992**, *20*, 1–7.
60 Northrop, D. B., in *Isotope Effects in Enzyme-Catalyzed Reactions*, Cleland, W. W., O'Leary, M. H., Northrop, D. B. (Eds.), University Park Press, Baltimore, MD, 1997, pp. 122–152.
61 Bahnson, B. J., Klinman, J. P., *Methods Enzymol.* **1995**, *249*, 373–397.
62 Kwart, H., *Acc. Chem. Res.* **1982**, *15*, 401–408. Kwart's attribution of temperature independent kinetic isotope effects to a bent transition state was subsequently corrected.
63 Whittaker, M. M., Ballou, D. P., Whittaker, J. W., *Biochemistry* **1998**, *37*, 8426–8436.
64 Harris, R. J., Meskys, R., Sutcliffe, M. J., Scrutton, N. S., *Biochemistry* **2000**, *39*, 1189–1198.
65 Jonsson, T., Edmondson, D. E., Klinman, J. P., *Biochemistry* **1994**, *33*, 14871–14878.
66 de Broglie, L., *Nature* **1923**, *112*, 540.
67 Bell, R. P., *Trans. Faraday Soc.* **1959**, *55*, 1.
68 Northrop, D. B., *J. Am. Chem. Soc.* **1999**, *121*, 3521.
69 Garcia-Viloca, M., Gao, J., Karplus, M., Truhlar, D. G., *Science* **2004**, 303–186–195.
70 Merzbacher, E., in *Quantum Mechanics*, 2nd edn., John Wiley & Sons, New York, 1970, Ch. 18.
71 Borgis, D., Hynes, J. T., in *The Enzyme Catalysis Process*; Cooper, A, Houben, J. L., Chien, L. C. (Eds.), Plenum, New York, 1989, p. 293.
72 Borgis, D., Hynes, J. T., *Chem. Phys.* **1993**, *170*, 315–346.
73 Antoniou, D., Schwartz, S. D., *J. Chem. Phys.* **1998**, *109*, 2287–2293.
74 Caratzoulas, S., Schwartz, S. D., *J. Chem. Phys.* **2001**, *114*, 2910–2918.
75 Caratzoulas, S., Mincer, J. S., Schwartz, S. D., *J. Am. Chem. Soc.* **2002**, *124*, 3270–3276.
76 Cui, Q., Karplus, M., *J. Phys. Chem. B* **2002**, *106*, 7927–7947.
77 Krishtalik, L. I., *Biochim. Biophys. Acta* **2000**, *1458*, 6–27.
78 Ulstrup, J., Jortner, J., *J. Chem. Phys.* **1975**, *63*, 4358–4368.
79 Rickert, K. W., Klinman, J. P., *Biochemistry* **1999**, *38*, 12218–12228.
80 Hatcher, E., Soudacker, A. V., Hammes-Schiffer, S., *J. Am. Chem. Soc.* **2004**, *126*, 5763–5775.
81 Sekhar, V. C., Plapp, B. V., *Biochemistry* **1988**, *27*, 5082–5088.
82 Klinman, J. P., *Biochemistry* **1976**, 2018–2026.
83 Rucker, J., Cha, Y., Jonsson, T., Grant, K. L., Klinman, J. P., *Biochemistry* **1992**, *31*, 11489–11499.
84 Chin, J. K., Klinman, J. P., *Biochemistry* **2000**, *39*, 1278–1284.
85 Kohen, A., Klinman, J. P., *J. Am. Chem. Soc.* **2000**, *122*, 10738–10739.
86 Liang, Z-X., Lee, T., Resing, K. A., Ahn, N. G., Klinman, J. P., *Proc. Natl. Acad. Sci. USA* **2004**, *101*, 9556–9561.
87 Hammond, G. S., *J. Am. Chem. Soc.* **1955**, *77*, 334.
88 Lowry, T. H., Richardson, K. S., in

Mechanism and Theory in Organic Chemistry, 3rd edn., Harper and Row, New York, 1987.

89 KIM, Y., KREENOY, M. M., *J. Am. Chem. Soc.* **1992**, *114*, 7116–7123.

90 TRUHLAR, D. G., ISAACSON, A. D., GARRETT, B. C., in *Theory of Chemical Reaction Dynamics*, BAER, M. (Ed.), CRC Press, Boca Raton, FL, 1985, Vol. 4, pp. 65–137.

91 KOHEN, A., JONSSON, T., KLINMAN, J. P., *Biochemistry* **1997**, *36*, 2603–2611.

92 SEYMOUR, S., KLINMAN, J. P., *Biochemistry* **2002**, *41*, 8747–8758.

93 SU, Q., KLINMAN, J. P., *Biochemistry* **1999**, *38*, 8572–8581.

94 ROTH, J. P., KLINMAN, J. P., *Proc. Natl. Acad. Sci. USA* **2003**, *100*, 62–67.

95 ROTH, J. P., WINCEK, R., NODET, G., EDMONDSON, D. E., MCINTIRE, W. S., and KLINMAN, J. P. *J. Am. Chem. Soc.*, **2004**, *126*, 15120–15131.

96 ROTH, J. P., KLINMAN, J. P., in *Isotope Effects in Chemistry and Biology*, A. KOHEN, H. L. LIMBACH (Eds.), Taylor and Francis – CRC Press, Boca Raton, FL, 2006, pp. 645–669.

97 BRINKLEY, D. W., ROTH, J. P., *J. Am. Chem. Soc.*, **2005**, *127*, 15720–15721.

98 SEYMOUR S. L., The Influence of Enzyme Surface and Environmental Modification on Catalysis Examined with Probes for Hydrogen Tunneling, Ph.D. Thesis, University of California, Berkeley, CA, 2001.

99 JANES, S. M., MU, D., WEMMER, D., SMITH, D., KAUR, A. J., MALTBY, D., BURLINGAME, A. L., KLINMAN, J. P., *Science* **1990**, *248*, 981.

100 KLINMAN, J. P., *J. Biol. Chem.* **1996**, *271*, 27189.

101 PALCIC, M. M., KLINMAN, J. P., *Biochemistry* **1983**, *22*, 5924.

102 HARTMANN, C. KLINMAN, J. P., *Biochemistry* **1991**, *30*, 4605.

103 SILVERMAN, R. B., *Acc. Chem. Res.* **1995**, *28*, 335–342.

104 MILLER, J. R., EDMONDSON, D. E., GRISSOM, C. B., *J. Am. Chem. Soc.* **1995**, *117*, 7830–7831.

105 BENDERSKII, V. A., MAKAROV, D. E., WIGHT, C. A., *Adv. Chem. Phys.* **1994**, *88*, 1.

106 LEWIS, E. R., JOHANSEN, E., HOLMAN, T. R. *J. Am. Chem. Soc.* **1999**, *121*, 1395–1396.

107 KNAPP, M. J., SEEBECK, F. P., KLINMAN, J. P., *J. Am. Chem. Soc.* **2001**, *123*, 2931–2932.

108 KNAPP, M. J., KLINMAN, J. P., *Biochemistry* **2003**, *42*, 11466–11475.

109 AMBUNDO, E. A., FRIESNER, R. A., LIPPARD, S. J., *J. Am. Chem. Soc.* **2002**, *124*, 8770–8771.

110 ROECKER, L., MEYER, T. J., *J. Am. Chem. Soc.* **1987**, *109*, 746–754.

111 BINSTEAD, R. A., MCGUIRE, M. E., DOVLETOGLOU, A., SEOK, W. K., ROEKER, L., MEYER, T. J., *J. Am. Chem. Soc.* **1992**, *114*, 173–186.

112 BINSTEAD, R. A., MEYER, T. J., *J. Am. Chem. Soc.* **1987**, *109*, 3287–3297.

113 BINSTEAD, R. A., MOYER, B. A., SAMUELS, G. J., MEYER, T. J., *J. Am. Chem. Soc.* **1981**, *103*, 2897–2899.

114 CHAUDHURI, P., HESS, M., MULLER, J., HILDEBRANCE, K., BILL, E., WEYHERMULLER, T., WIEGHARDT, J., *J. Am. Chem. Soc.* **1999**, *121*, 9599–9610.

115 KNAPP, M. J., KLINMAN, J. P., *Eur. J. Biochem.* **2002**, *269*, 3113–3121.

116 STEWART, L. C., KLINMAN, J. P., *Annu. Rev. Biochem.* **1988**, *57*, 551–592.

117 EIPPER, B. A., STOFFERS, D. A., MAINS, R. E., *Annu. Rev. Neurosci.* **1992**, *15*, 57–85.

118 PRIGGE, S. T., KOLHEKAR, A. S., EIPPER, B. A., MAINS, R. E., AMZEL, L. M., *Science* **1999**, *278*, 1300–1305.

119 EVANS, J. P., AHN, K., KLINMAN, J. P., *J. Biol. Chem.* **2003**, *278*, 49691–49698.

120 MILLER, S. M., KLINMAN, J. P., *Biochemistry* **1983**, *22*, 3096–3106(b).

121 LIANG, Z.-X., KLINMAN, J. P., *Curr. Opin. Struct. Biol.*, **2004**, *14*, 648–655.

122 MARCUS, R. A., SUTIN, N., *Biochim. Biophys. Acta.* **1985**, *11*, 265–322.

123 SINGLETON, D. A., THOMAS, A. A., *J. Am. Chem. Soc.* **1995**, *117*, 9357.

11
Multiple-isotope Probes of Hydrogen Tunneling

W. Phillip Huskey

11.1
Introduction

Forty years ago, Bell and Goodall [1] and Lewis and Funderburk [2, 3] carried out experimental studies of proton transfers that demonstrated the importance of hydrogen tunneling in chemical reactions. Chemists had recognized for years that tunneling should contribute significantly to observed rates of reactions involving hydrogen transfers, but convincing experimental verification was not available [4–6]. Bell, Goodall, Lewis, and Funderburk found that hydrogen (H/D) isotope effects on proton transfers from 2-nitropropane to substituted pyridines (Fig. 11.1) were too large to be consistent with any reasonable explanations that did not invoke hydrogen tunneling. These isotope effects varied from 9.8 to 24.3 for a series of substituted pyridine bases at room temperature, with the largest isotope effects for proton transfers to the bulkiest bases. In several cases, the researchers measured the temperature dependence of the isotope effects and found that extrapolations to infinite temperature were also consistent with expectations for tunneling. However, it was the fact that the Lewis and Funderburk kinetic isotope effects were too large to be explained without invoking tunneling that proved to be particularly convincing. Not all tunneling systems are expected to show such large isotope effects. Additional experiments, typically involving multiple-isotope probes, are needed to identify reactions with significant tunneling, and to characterize the nature of the tunneling process.

Researchers are now addressing questions about the nature of tunneling and its possible role in catalysis that require comparisons of experimental isotope effects, including multiple-isotope probes, with the results of exacting theoretical and computational efforts [7–14]. This chapter first recounts the basis for the use of simple kinetic isotope effects in studies of tunneling, followed by expectations for H/D/T isotope effect comparisons (Swain–Schaad exponents) and isotope effect on isotope effect experiments (tests of the Rule of the Geometric Mean). The last section reviews the background for experiments to determine mixed isotopic exponents that combine the effects of H/D/T comparisons with isotope effects on isotope effects. Notably, much of this chapter concerns non-tunneling mechanistic effects on

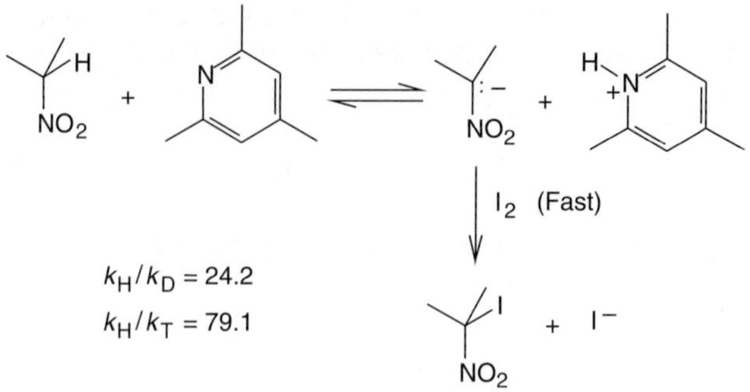

11 Multiple-isotope Probes of Hydrogen Tunneling

$k_H/k_D = 24.2$
$k_H/k_T = 79.1$

Figure 11.1. Isotope effects on a proton transfer from carbon (24.9 °C) measured by Lewis and Funderburk [3] and Lewis and Robinson [29].

multiple-isotope experiments. Appreciating the possible non-tunneling contributions to experimental results is often essential when efforts are made to identify specific tunneling effects.

The isotope effects of particular interest in this chapter are those that arise from substitution of protium, deuterium, or tritium at one or two hydrogenic sites as described in Fig. 11.2. With this many isotopic labels, a few comments about nota-

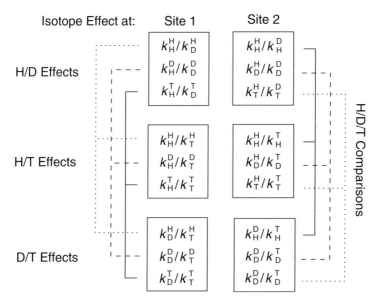

Figure 11.2. Possible hydrogen isotope effects in a study with two isotopic sites. Lines show H/D/T isotope effect comparisons. Ratios of pairs of isotope effects within one of the six boxes constitute isotope effects on isotope effects (tests of the Rule of the Geometric Mean). The subscript shows isotopes at site 1, and the superscript shows isotopes at site 2.

tion are needed. For descriptions of isotope effects and relationships between isotope effects, the notation expressed in Eq. (11.1) below will be used. For any parameter, x, the subscripts i (and sometimes j) refer to isotopes at one site in a reacting system, while the superscripts, k (and sometimes l) refer to isotopes at a possible second site.

$$x_{ij}^{kl} \quad k_H^D/k_T^D \quad k_H/k_T = (k_H/k_D)^{r_{HD}} \tag{11.1}$$

In cases where one site is thought to give rise to a primary isotope effect – one in which bonds are being made or broken at the isotopic element – the subscripts will be used. In cases where secondary isotope effects (by definition, those that are not primary) need to be distinguished, the superscripts will be used. The middle part of Eq. (11.1) illustrates one use of the notation, as a primary H/T isotope effect measured when D occupies a secondary site. As a final example, the right-hand part of Eq. (11.1) shows a Swain–Schaad exponent r defined for a single site of isotopic substitution. The HD subscript serves to connect the exponent to the H/D isotope effect it acts on to produce the H/T isotope effect. In this example, the superscripts were omitted because there is either no second site of isotopic substitution or, in all cases, the isotope is protium.

11.2
Background: H/D Isotope Effects as Probes of Tunneling

Descriptions of the influence of tunneling on isotope effects often use conventional transition state theory (TST) as a convenient reference for discussion. Conventional TST is amenable to a straightforward theory of kinetic isotope effects, and tunneling contributions can be included or discarded. Conventional TST also serves as the starting point for numerous sophisticated modern theories of reaction rates [15], and it can be reduced to very simple terms, as is next described for isotope effects on hydrogen transfers.

11.2.1
One-frequency Models

The useful benchmarks for the Swain–Schaad exponents described in Section 11.3 are obtained from simple one-frequency models for hydrogen transfer reactions. The theoretical basis for the approach comes from the Bigeleisen–Wolfsberg [16] formalism for isotope effects on gas-phase reactions according to transition state theory in the limit of the harmonic approximation for vibrations, and with classical treatments for translations and vibrations. Using reasonable empirical force fields for model reactant-state and transition-state structures, the resulting vibrational frequencies yield isotope effects through the Bigeleisen equation. At this level of theory, primary kinetic hydrogen isotope effects are very well approximated by reducing the problem to very simple one-frequency models [17–20] that can be applied to the reaction shown in Eq. (11.2).

$$\text{C–H} + \text{C} \rightarrow [\text{C--H--C}]^\ddagger \rightarrow \text{C} + \text{H–C} \qquad (11.2)$$

The one-frequency model represented by Eqs. (11.3)–(11.8) shows single isotopic frequency expressions for the MMI (mass/moment of inertia), ZPE (vibrational zero-point energy), and EXC (excited vibrations) terms of the usual Bigeleisen equation [21]. The extra term *tun* is the truncated Bell tunnel correction [22], used here to provide a simple way to express a tunneling effect in terms of a reaction-coordinate frequency, $v^\ddagger$.

$$u_i = v_i hc/k_B T \qquad (11.3)$$

$$mmi = \frac{u_H^\ddagger/u_D^\ddagger}{u_H^R/u_D^R} \qquad (11.4)$$

$$zpe = \exp([u_H^R - u_D^R]/2) \qquad (11.5)$$

$$exc = \frac{1 - \exp(-u_H^R)}{1 - \exp(-u_D^R)} \qquad (11.6)$$

$$tun = \frac{u_H^\ddagger \sin(u_D^\ddagger/2)}{u_D^\ddagger \sin(u_H^\ddagger/2)} \qquad (11.7)$$

$$k_H/k_D = mmi \times zpe \times exc \times tun \qquad (11.8)$$

The utility of the one-frequency model for hydrogen transfer reactions is demonstrated in Fig. 11.3, showing isotope effects calculated down to 250 K (at lower temperatures, the truncated Bell tunnel correction is likely to be particularly unrealistic). The symbols show isotope effects calculated using a more elaborate vibrational model with a 9-atom transition state characterized by 20 real frequencies and one imaginary reaction-coordinate frequency. One set of symbols shows results for a model in which the reaction-coordinate frequency is large enough to make the tunneling term significant. The other set of points shows isotope effects from a model with a reaction-coordinate frequency so small that the tunneling term is not significant. The curves drawn though the two sets of points are least-squares fits to the one-frequency model of Eqs. (11.3)–(11.8). The fact that the one-frequency model is a good approximation for the more complete vibrational model is undoubtedly why the Swain–Schaad exponent of 1.44 (described in Section 11.3) has proved to be a useful benchmark for H/D/T isotope effect studies in many hydrogen transfer reactions.

At 298 K, the one-frequency model without tunneling predicts an isotope effect of 6.4, which is considerably lower than the k_H/k_D of 24.2 observed by Lewis and Funderburk [3]. Observed isotope effects that are larger than expected based on

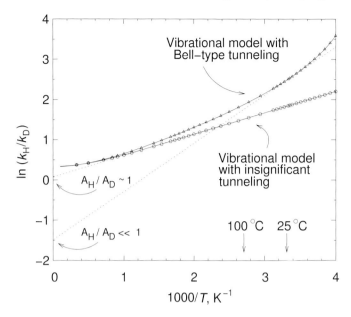

Figure 11.3. Temperature dependence of the primary hydrogen isotope effect calculated using a 9-atom vibrational model for hydrogen transfer (model HHIE3 [37] with simple stretch–stretch coupling to generate a reaction-coordinate frequency). Triangles mark calculated results for models with a reaction-coordinate frequency for H transfer of 984i cm^{-1} including the truncated Bell tunnel correction [6, 22]. The circles show results for models with a sufficiently low reaction-coordinate frequency (90i cm^{-1}) to make the tunnel correction insignificant. The solid curves are least-squares fits to the one-frequency model described in the text represented by Eqs. (11.3)–(11.8). The least-squares estimates of fitted parameters in cm^{-1} were for the triangles, $v_H^R = 3090$, $v_D^R = 2343$, $v_H^\ddagger = 983i$, and $v_D^\ddagger = 708i$. For the circle data, the fits gave $v_H^R = 3124$, $v_D^R = 2380$, and the ratio of reaction coordinate frequencies $v_H^\ddagger/v_D^\ddagger = 1.40$. The dashed lines show tangents at 25 °C with intercepts corresponding to $A_H/A_D = 0.226$ (triangles), and 1.078 (circles).

non-tunneling models still provide the most convincing evidence for tunneling in reacting chemical systems, although other explanations may be important in special cases [23, 24].

11.2.2
Temperature Dependence of Isotope Effects

The results from the rudimentary model for isotope effects on a hydrogen transfer reaction shown in Fig. 11.3 are plotted in Arrhenius style as their logarithmic form versus inverse temperature. An analysis based on the temperature dependence of isotope effects is another standard method for detecting tunneling [25, 26], and again, non-tunneling models are helpful as a basis for comparison. For many experiments, particularly in solution, the range of temperatures that can be studied is fairly narrow, so linear extrapolations of Arrhenius plots are used to determine the intercept at infinite temperature, A_H/A_D. Models without tunneling tend to predict A_H/A_D near unity. The low temperature curvature predicted from models that include tunneling is expected to produce, by extrapolation, values of A_H/A_D that are much less than unity. The tangent at 298 K for the tunneling model in Fig. 11.3 gives $A_H/A_D = 0.23$. An Arrhenius analysis of the isotope effects measured for the proton transfer shown in Fig. 11.1 gives $A_H/A_D = 0.15$, providing further evidence for tunneling. Theories of hydrogen transfer that involve little or no overbarrier reaction and proceed entirely by tunneling predict temperature-independent isotope effects [27], with $A_H/A_D \approx k_H/k_D$.

The effects of complex mechanisms with serial or parallel shifts in rate limiting steps must also be considered in the analysis of the temperature dependence of isotope effects [28]. In addition, tedious attention to details of temperature effects on pH, acidity constants, reaction volumes, and substrate or catalyst stabilities may be needed in some cases to avoid problematic interpretations. For these reasons, temperature studies of isotope effects may not be as convincing as the observation of very large isotope effects in providing evidence for tunneling.

For reactions involving hydrogen transfer that may have substantial tunneling contributions to reaction rates, but do not show exceptionally large isotope effects, other experimental measures that are sensitive to tunneling would be useful. Many researchers have considered the comparisons between tritium and deuterium hydrogen isotope effects as possible tunneling criteria. As the next section describes, these comparisons have not proved to be generally useful for detecting tunneling.

11.3
Swain–Schaad Exponents: H/D/T Rate Comparisons

Soon after the discovery of proton transfers from 2-nitropropane showing large isotope effects consistent with tunneling, Lewis and Robinson [29] wondered if they could use the reaction to develop a tunneling criterion based on relationships between H/D and H/T kinetic isotope effects. Their work did not produce experimental support for a new tunneling criterion, and their early computational results, along with those from More O'Ferrall and Kouba [30] and Stern and Weston [31], further diminished enthusiasm for the idea. Over the ensuing years, most results from studies that involve a single site of isotopic substitution have failed to provide a basis for tunneling criteria, although the temperature dependence of the relationship may prove to be useful [32].

The starting point for defining the relationships between H/D/T isotope effects is the set of equations (11.9) shown below, with exponents r_{HD} and r_{DT} as leading parameters for study. The subscript in the notation used here refers to the type of isotope effect, H/D or D/T, used to predict the H/T effect in an exponential relationship. The two exponents are algebraically related for systems with a single site of isotopic substitution [33]. In this section, results that have been published as r_{DT} were converted to r_{HD} using Eq. (11.9) to simplify comparisons.

$$k_H/k_T = (k_H/k_D)^{r_{HD}} \quad k_H/k_T = (k_D/k_T)^{r_{DT}} \quad r_{DT} = r_{HD}/(r_{HD} - 1) \qquad (11.9)$$

11.3.1
Swain–Schaad Limits in the Absence of Tunneling

Early estimates of the exponent r_{HD} were made by Swain et al. [34] and Bigeleisen [35]. Swain et al. [34] showed that the H/T isotope effect could be predicted from the H/D isotope effect using the one-frequency model described above (Eq. (11.2)) assuming that only the zero-point energy term (*zpe*, Eq. (11.5)) contributed to the kinetic isotope effect. If the single frequency of the model is a diatomic-type stretch, the isotopic *zpe* ratios depend only on the reduced masses for the vibration. For reduced masses of 1, 2, and 3 atomic mass units for H, D, and T transfer, the exponent r_{HD} in Eq. (11.9) turns out to be 1.44. Bigeleisen [35] proposed a range of 1.38–1.55 for r_{HD}. To establish these limits, he first calculated isotope effects for the set of hypothetical equilibria that could be envisioned for pairs of reactants and products selected among H_2, HF, HCl, HBr, HI, and H_2O at 298 K, 600 K, and 1000 K. He next estimated the effects of a reaction-coordinate frequency for hydrogen transfer in rate processes without tunneling, and then reasoned that his limits calculated for equilibria should also apply to rate processes. The Bigeleisen limits are probably too broad to be of general use in hydrogen transfers for most polyatomic systems. The model calculations discussed below provide better guides for expectations of primary kinetic isotope effects in non-tunneling reactions. For similar reasons, Melander and Saunders [21] recommended non-tunneling limits of 1.40–1.45 (or $r_{DT} = 3.2$–3.5).

Calculations using conventional TST and the Bigeleisen–Wolfsberg [16] treatment for isotope effects have demonstrated that $r_{HD} = 1.44$ is a useful benchmark for primary hydrogen isotope effects. Using empirical harmonic force fields and various reactant-state and transition-state geometries, More O'Ferrall and Kouba [30] found, for proton-transfer models, that the exponents were within 2% of the 1.44 value, and similar computational approaches gave $r_{HD} = 1.43$–1.45 (343 K) [36] and 1.43–1.45 (298 K) [37] for calculations that included a wide range of transition-state models and reaction-coordinate motions with insignificant tunneling effects. In their study of tunneling and variational effects on hydride-transfer reactions Kim and Kreevoy [33] obtained vibrational frequencies from extended LEPS potential energy surfaces that also allowed them to calculate results for conventional TST, giving $r_{HD} = 1.47$. Similarly, the theoretical treatments used by Cui et al. permitted conventional TST calculations from their QM/MM (DFT/CHARMM) potentials for studies of liver alcohol dehydrogenase (1.42) [38] and triosephosphate isomerase (1.43) [32]. Melander and Saunders [21] recommended that non-tunneling limits of 1.40–1.45 (or $r_{DT} = 3.2$–3.5) appear to be reasonable for many primary hydrogen isotope effects, at temperatures less than about 1000 K. Given that so many treatments give r_{HD} very near 1.44 for primary H/D isotope effects larger than about 2 at moderate temperatures, r_{DH} for conventional TST could perhaps be taken as a measure of the quality of a potential surface near stationary points for primary hydrogen isotope effect studies.

Limits proposed by Kohen and Jensen [39] should not be considered with the single-site Swain–Schaad values discussed in this section. Kohen and Jensen [39] proposed that a value of 4.8 be treated as an upper limit in a non-tunneling system for a secondary isotope effect exponent defined as r_{DD}^{DT} in Section 11.5 describing "mixed-label" experiments. As shown in Section 11.5, it useful to separate these mixed-label exponents into two factors: one that arises from H/D/T substitutions, and one that arises from isotopic substitution that amounts to an isotope effect on an isotope effect. Also note that it is not straightforward to convert a mixed-label exponent based on D/T isotope effects into one based on H/D isotope effects as Eq. (11.9) shows for single-site exponents.

11.3.2
Swain–Schaad Exponents for Tunneling Systems

Early experimental tests in systems thought to involve significant tunneling were carried out by Lewis and Robinson [29]. In their work, they first extended the treatment of Swain et al. [34] by supposing that tunneling components (Q) could be factored out of the kinetic isotope effect, similar to the approach shown in Eq. (11.8). They used Eq. (11.10), using k^* to represent a rate constant without a tunneling correction, to conclude that tunneling should not cause Swain–Schaad exponents to deviate strongly from a value of 1.44, unless the exponent for the tunnel effect, s_{HD}, is very different from 1.44 and most of the kinetic isotope effect arises from Q_H/Q_D.

$$k_H/k_T = (k_H^*/k_D^*)^{1.44}(Q_H/Q_D)^{s_{HD}} \quad r_{HD} = 1.44 + (s_{HD} - 1.44)\frac{\ln(Q_H/Q_D)}{\ln(k_H/k_D)} \quad (11.10)$$

Lewis and Robinson [29] explored the relationship between H/D and H/T kinetic isotope effects predicted by several one-dimensional models for tunneling, and found no conspicuous deviations from the prediction made by Swain et al. [34].

Noting that there were many uncertain factors in their model calculations, they proceeded to make experimental measurements on systems thought to react with significant tunneling, based on the fact that they showed large kinetic isotope effects. For the proton transfer reaction shown in Fig. 11.1, they measured $r_{HD} = 1.42$. Lewis and Robinson concluded that their results along with the results from five additional proton or hydride transfer reactions were consistent with predictions made using $r_{HD} = 1.44$. The one case that deviated significantly in the direction expected for tunneling was the oxidation of leuco crystal violet by chloranil which showed $r_{HD} = 1.31$. The Lewis and Robinson work demonstrated that Swain–Schaad exponents were not sensitive to the extent of hydrogen tunneling in a reaction, but it did not exclude the possibility of large deviations in the exponents for cases of extreme tunneling.

Jones also summarized early work on H/D and H/T isotope effects on eight reactions [40] with H/D isotope effects large enough to suggest rate-limiting hydrogen transfers that may or may not involve significant tunneling. The values for r_{HD} were in the range 1.38–1.50, except for the case of proton transfer from acetone to hydroxide ion, which requires an uncertain correction for secondary isotope effects. He also reported r_{HD} for proton transfers from 2-carbethoxycyclopentanone to deuterium oxide (1.48 ± 0.02), chloroacetate (1.72 ± 0.05), and fluoride (1.32 ± 0.05). At 298 K, the H/D isotope effects were modest at 3.4, 4.1, and 2.6.

In studies of a different class of proton transfers, Limbach and coworkers [41, 42] used lineshape analysis of ^{1}H and ^{3}H NMR spectra of porphyrin and its monoanion conjugate base to obtain H/D/T rate constants for fast intramolecular proton transfers. They fit their temperature dependence of rate constants to a modified one-dimensional Bell tunneling model [43]. The fit of the data gave r_{HD} very near 1.44 over the temperature range of the data where tunneling was thought to be significant, and down to about 240 K. At lower temperatures, the extrapolated fit predicted r_{HD} to rise to a constant value near 1.7, although the Bell model for tunneling may not be as valid at the very low extrapolated temperatures.

11.3.3
Swain–Schaad Exponents from Computational Studies that Include Tunneling

Computational work has generally agreed with the the conclusions reached from experimental studies of tunneling systems. Vibrational analysis calculations using model reactant and transition states for primary isotope effects on hydrogen transfer reactions give Swain–Schaad exponents that are within 3% of the 1.44 value when tunneling effects are treated using the truncated Bell [22] equation [30, 36, 37]. A slightly wider range for r_{HD} was seen in Stern and Weston's [31] study of tunneling barriers. These authors included tunneling through one-dimensional Eckart barriers [44] for a series of model calculations on hydrogen transfer reactions, and showed that r_{HD} is generally within the Bigeleisen limits [35] of 1.33–1.58 discussed above. However, values as low as 1.25 were calculated for models with the highest reaction barriers. Grant and Klinman tested the exponents that could be generated using the full Bell tunnel correction [22] using assumed barrier heights for H, D, and T transfer and found a very wide range of possible values for r_{HD} from 1.2 to 2.0.

Studies involving multidimensional tunneling treatments have similarly not provided support for r_{HD} as a criterion for tunneling. In their work on hydride transfer reactions, Kim and Kreevoy [33] using extended LEPS potential energy surfaces found r_{HD} to be 1.27–1.50 for six model surfaces using conventional TST with the tunnel correction included, and 1.32–1.50 for the same surfaces using variational TST, again with the tunneling treatment included. The calculations reported by Cui et al. [38], also using variational TST and a multidimensional tunneling treatment gave $r_{HD} = 1.50$. In their study of the triosphosphate isomerase reaction, Cui and Karplus [32] found r_{HD} to be 1.47 for two tunneling treatments at 300 K, and like Tautermann et al. [45], concluded that single-site Swain–Schaad exponents are not reliable indicators for tunneling.

11.3.4
Swain–Schaad Exponents for Secondary Isotope Effects

For secondary isotope effects, no simple one-frequency vibrational models have been devised analogous to the Swain et al. [34] treatment for primary isotope effects. Still, some vibrational analysis calculations tend to give r_{HD} close to 1.44.

A few single-site Swain–Schaad experimental results and several computational studies have been reported for secondary isotope effects. One experimental example comes from the triosephosphate isomerase reaction, which shows a large secondary H/T kinetic isotope effect of 1.27 ± 0.03 and $r_{DT} = 4.4 \pm 1.3$ (therefore, $r_{HD} = 1.29 \pm 0.38$). Vibrational analysis calculations for elimination reactions [36] gave $r_{HD} = 1.36$–1.39 for models with and without truncated Bell [22] tunnel corrections in all cases where k_H/k_D was larger than 1.02. Similar calculations for hydride-transfer reactions gave $r_{HD} = 1.42$–1.44 for a range of models with and without tunnel corrections. Conventional TST produced a single-site secondary exponent of 1.46, and 1.31 for variational TST with multidimensional tunneling in the alcohol dehydrogenase study [38], and for the triosephosphate isomerase reaction [32], $r_{HD} = 1.32$ with conventional TST, and 1.85 and 2.27 for variational TST with respective one-dimensional and multidimensional tunneling treatments. The paper by Cui and Karplus [32] includes a very detailed discussion of the secondary Swain–Schaad exponents, and Hirschi and Singleton [46] report a wide range of secondary Swain–Schaad exponents from a large number of electronic-structure calculations.

The fact that secondary kinetic isotope effects tend to be small causes an additional concern because Swain–Schaad exponents will become very large when one of the isotope effects approaches unity. Cautions about small uncertainties in computational results have been given [47], and a similar sensitivity is expected from experimental errors [48]. Large exponents from small isotope effects may also be quite real. If a secondary isotope effect arises from competing factors involving increasing and decreasing sensitivity of isotopic zero-point energies (from weakening some force constants while tightening others), it can be possible to have a D/T isotope effect very near unity to produce a large exponent.

11.3.5
Effects of Mechanistic Complexity on Swain–Schaad Exponents

Multiple rate-limiting steps can give rise to Swain–Schaad exponents that differ significantly from 1.44. Figure 11.4 gives the general form of the observed isotope effect in cases where two rate-limiting steps appear in series with one or more steady-state intermediates separating the steps, and in cases where rate-limiting steps occur in parallel, after a branch point in a mechanistic scheme. To get a sense

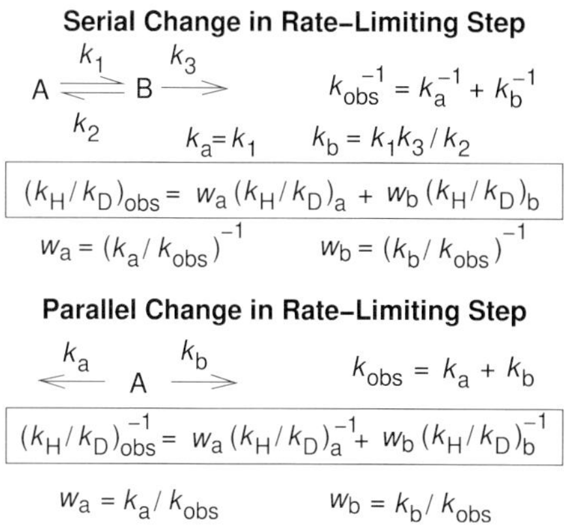

11.3 Swain–Schaad Exponents: H/D/T Rate Comparisons

Serial Change in Rate-Limiting Step

$$A \underset{k_2}{\overset{k_1}{\rightleftharpoons}} B \overset{k_3}{\rightarrow}$$

$$k_{obs}^{-1} = k_a^{-1} + k_b^{-1}$$

$$k_a = k_1 \qquad k_b = k_1 k_3 / k_2$$

$$(k_H/k_D)_{obs} = w_a (k_H/k_D)_a + w_b (k_H/k_D)_b$$

$$w_a = (k_a / k_{obs})^{-1} \qquad w_b = (k_b / k_{obs})^{-1}$$

Parallel Change in Rate-Limiting Step

$$\overset{k_a}{\leftarrow} A \overset{k_b}{\rightarrow} \qquad k_{obs} = k_a + k_b$$

$$(k_H/k_D)_{obs}^{-1} = w_a (k_H/k_D)_a^{-1} + w_b (k_H/k_D)_b^{-1}$$

$$w_a = k_a / k_{obs} \qquad w_b = k_b / k_{obs}$$

Figure 11.4. Observed isotope effects as weighted averages of isotope effects on steps in mechanisms showing either serial or parallel changes in the rate-limiting step.

of the size of the Swain–Schaad deviations, Fig. 11.5 shows predicted exponents that would be observed if only one of two rate-limiting steps has an isotope effect. If that step is arbitrarily identified as the one corresponding to k_a, the equations of interest are shown in Eqs. (11.11) and (11.12).

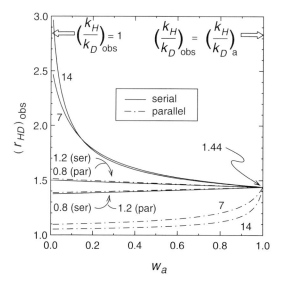

Figure 11.5. Changes in the observed Swain–Schaad exponent as the relative importance of an isotope-dependent step is adjusted by w_a for mechanisms with serial or parallel changes in rate limiting step (see Fig. 11.4). The numbers near the curves show the isotope effect on k_a, which is also the limit of $(k_H/k_D)_{obs}$ when $w_a = 1$. When $w_a = 0$, the isotope-sensitive step no longer limits the reaction rate, so the observed isotope effect becomes 1.

Serial: $(k_H/k_D)_{obs} = w_a(k_H/k_D)_a + 1 - w_a,$

$(k_H/k_T)_{obs} = w_a(k_H/k_D)_a^{1.44} + 1 - w_a$ (11.11)

Parallel: $(k_H/k_D)_{obs} = w_a(k_H/k_D)_a^{-1} + 1 - w_a,$

$(k_H/k_T)_{obs} = w_a(k_H/k_D)_a^{-1.44} + 1 - w_a$ (11.12)

Note that the weighting factors correspond to the common H-isotopomer in the H/D and H/T isotope effects, and expressions for w_a are different for the serial and parallel cases. The equations in Eq. (11.11), developed by Northrop [49, 50], have proved to be useful in obtaining intrinsic kinetic isotope effects in enzymatic reactions with multiple rate-limiting steps that partially mask the full effect of an isotope-sensitive step.

The exponents $(r_{HD})_{obs}$ calculated from Eqs. (11.11) and (11.12) are shown in Fig. 11.5 for various values of $(k_H/k_D)_a$. For mechanisms with multiple rate-limiting steps in series, the observed exponents r_{HD} tend to be larger than 1.44, while a mechanism with parallel rate-limiting steps will give an exponent less than 1.44. The corresponding trends for r_{DT}, according to Eq. (11.9), will be just the opposite: r_{DT} will be less than the benchmark 3.3 value for serial cases and greater than 3.3 for the parallel cases.

The effects of changing rate-limiting steps may appear in temperature studies. With a few assumptions, the Eqs. (11.11) and (11.12) can be modified to allow for a temperature dependent $(k_H/k_D)_a$ and weighting factor, w_a. Equation (11.13) shows w_a expressed using an Arrhenius-type expression for a rate-constant ratio, such that $\alpha = A_a/A_b$ and $\varepsilon = (E_a)_a - (E_a)_b$. The temperature dependence on the isotope effect for k_a assumes the effect arises entirely from zero-point energy terms; a value of ζ can be generated using a specified value for $(k_H/k_D)_a$ at a particular temperature.

$$w_a(\text{serial}) = \frac{1}{1+\alpha e^{-\varepsilon/RT}} \quad w_a(\text{parallel}) = \frac{1}{1+\alpha^{-1}e^{\varepsilon/RT}} \quad (k_H/k_D)_a = e^{\zeta/RT} \quad (11.13)$$

For an assumed H/D isotope effect of 7 for k_H/k_D at 298 K, the effects of temperature on observed values of r_{HD} (upper panels) and the two steps observed H/D isotope effects (lower panels) are shown in Fig. 11.6. The values that define the temperature dependence of the weighting factors for the two steps are also shown on the figure. The range of possible temperature dependences is broad, and more complex temperature effects should be expected if additional rate-limiting steps are included in a mechanism [51]. As has been noted before, serial changes in mechanism only increase r_{HD} from the 1.44 benchmark [52]. Parallel changes tend to decrease the observed exponent. The effects of mechanistic complexity apply potentially to many types of reactions, for both nonenzymic and enzymic systems, and for experiments involving steady-state turnover, rapid-mixing, relaxation kinetics, and other types of rate measurements.

The Swain–Schaad exponents as described here, refer to the use of multiple iso-

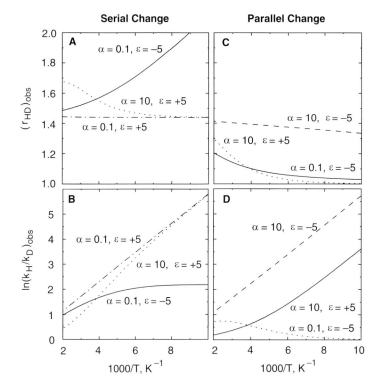

Figure 11.6. Influence of temperature-induced changes in rate-limiting steps on observed Swain–Schaad exponents, $(r_{HD})_{obs}$, and observed H/D isotope effects. Shown above each curve are variables α and ε (kJ mol^{-1}) for the ratio of rate constants k_a/k_b (defined in Fig. 11.4) expressed as $\alpha e^{(-\varepsilon/RT)}$. The curves were generated from a model that had isotope effects arising solely from zero-point energy changes such that $(k_H/k_D)_a = 7$ at 298 K and $(r_{HD})_a = 1.44$. The curve for $\alpha = 10$, $\varepsilon = -5$ is not shown for the serial changes in the rate-limiting step (panels **A** and **B**) because the observed isotope effect is very small. For the same reason, the curve for $\alpha = 0.1$, $\varepsilon = +5$ is not shown for the parallel changes in the rate-limiting step (panels **C** and **D**).

topes at a single site. The remaining sections concern experiments involving a second site of isotopic substitution.

11.4
Rule of the Geometric Mean: Isotope Effects on Isotope Effects

In systems with multiple sites of isotopic substitution, it is possible to determine if isotopic substitution at one site will alter the isotope effect at a second site. One example is the reaction catalyzed by glutamate dehydrogenase shown in Fig. 11.7. Srinivasan and Fisher [53] used stereospecifically labeled NADPH as the hydride

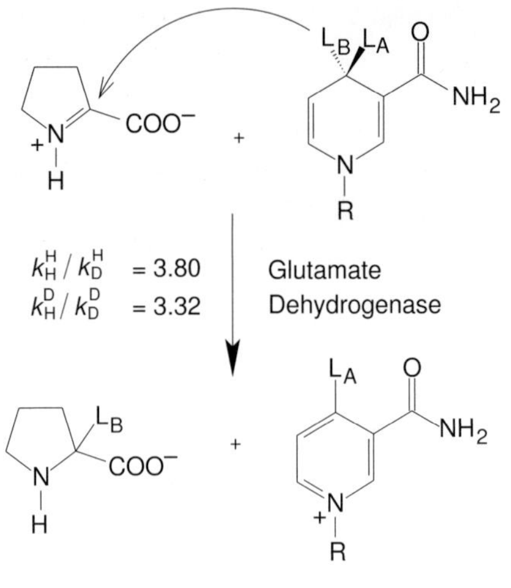

Figure 11.7. Isotope effect on an isotope effect (298.15 K) in Srinivasan and Fisher's study of catalysis by glutamate dehydrogenase [53]. When the isotope at the secondary site A is deuterium (L = H or D), the primary isotope effect at site B is reduced by a factor of 1.14.

$k_H^H/k_D^H = 3.80$
$k_H^D/k_D^D = 3.32$

Glutamate Dehydrogenase

donor in the reaction to measure the H/D primary isotope effect when the non-transferring A-side site contained either protium or deuterium. The ratio of the two values, 1.14, is a significant "isotope effect on an isotope effect", also known as a nonadditive isotopic free energy effect or as a breakdown in the Rule of the Geometric Mean (RGM) [54]. Observations of RGM breakdowns can be indicators of tunneling in reacting systems, and more generally they signify departures from simple two-state models for rates and equilibria. The notation used here to express the RGM effect seen by Srinivasan and Fisher is shown in Eq. (11.14), where the subscripts on g denote the isotopes used in the primary site, and the superscripts denote secondary-site isotopes.

$$g_{HD}^{HD} = \frac{k_H^H/k_D^H}{k_H^D/k_D^D} \qquad (11.14)$$

11.4.1
RGM Breakdown from Intrinsic Nonadditivity

A common use of the RGM is to compare isotope effects among studies with equivalent chemical sites but having different levels of isotopic substitution. The secondary isotope effect arising from H/D substitution at a single hydrogenic site on a methyl group could be compared, for example, with the isotope effect obtained from isotopic substitutions at all three methyl hydrogens [55]. In this case, the RGM predicts that the d_1 effect will be the cube root (the geometric mean) of the d_3 effect. The effects on isotopic free-energy differences are therefore additive, and isotopic substitution at one site does not change the effect of isotopes at other sites. In this way, the RGM predicts no isotope effects on isotope effects.

Experience with model calculations for equilibrium isotope effects and kinetic isotope effects, when using conventional TST, shows that the RGM is valid in the common circumstance in which the effects of coupled vibrational motions cancel between reactant and product states, or between reactant and transition states. The natural coupling expected between the various bends and stretches of the bonds in a methyl group is largely the same in the reactant state and transition state in the acetyl transfer example, so the free-energy effects of multiple isotopic substitutions are strictly additive. In the case of the glutamate dehydrogenase reaction of Fig. 11.7, the RGM would be expected to hold as the effects of coupled L_A–L_B motions cancel between reactant and transition states. Model calculations using conventional TST show that the cancellation of nonadditive effects is nearly exact in spite of the fact that the C–L_B bond is reacting, so the RGM is predicted to hold [56, 57]. Knowing that the experiments show an RGM breakdown is therefore an indication that the simple transition-state picture must be modified to explain this reaction.

In some types of experiments, the effects of vibrational coupling cannot cancel and the RGM breaks down. The case of the isotopic exchange of HOH and DOD to produce HOD is one example [58] of such intrinsic nonadditivity. The reaction is predicted to have an equilibrium constant of 4 on statistical grounds if the RGM holds, while experimental and theoretical results show smaller values of 3.85. The effect of the coupled motions between the two deuterium atoms in DOD cannot cancel in the isotope exchange reaction, and the effect of the extra isotope lowers the zero-point energy by more than expected from the RGM and the singly substituted HOD. In contrast to this specific isotope exchange reaction, most kinetic solvent isotope effect experiments would be expected to show substantial cancellation of nonadditive isotopic effects, and notably, the commonly used theory of reaction rates and equilibria in mixed HOH/DOD solvents relies on the validity of the RGM in these cases [59–62]. A second classic example of intrinsic nonadditivity is the nitrogen isotope effect on the acidity of pyridinium ion in HOH and DOD. Kurz and Nasr [63] found that $K_{a(14)}/K_{a(15)}$ (298 K) was 1.0211 ± 0.0003 in HOH and $1.0250 \pm$ in DOD, and they attributed this isotope effect on an isotope effect to the fact that in the reactant state the nitrogen and hydrogen isotopes are located on the same molecule, while in the product state they are on different molecules. Consequently, the nonadditive, coupled-motion effects of the reactant cannot be expected to be similar in the products.

The RGM is therefore expected to hold for kinetic isotope effects that can be explained with conventional TST involving a single reactant state and a single transition state, with all isotopic sites in roughly similar bonding arrangements in the two states. For equilibrium isotope effects, the RGM should be similarly valid for two-state situations that do not involve the separation of the isotopic sites, as would occur if one isotope is transferred in the reaction.

11.4.2
RGM Breakdown from Isotope-sensitive Effective States

The RGM breakdown reported by Saunders and Cline [64] illustrates a second class of RGM breakdowns for systems with isotope-sensitive effective reactant, product, or transition states. Saunders and Cline used deuterium shifts in carbon-13 NMR signals to measure equilibrium isotope effects on rearrangements of carbocations with different levels of deuterium substitution for protium. The equilibrium constants shown for one set of their results in Fig. 11.8 represent β-deuterium isotope effects for the equilibrium shown, since the reaction with non-deuterated cations must equal unity. The fact that the observed isotope effects are normal (greater than one) can be explained by weaker binding of the isotopes through hyperconjugation with the empty p-orbital of an adjacent sp^2-hybrid carbon. A breakdown in the RGM is apparent as can be see by showing that the cube of the d_1 effect (1.6339) is significantly smaller than the observed d_3 effect (1.7664). Electronic structure calculations were used to show that the RGM breakdown did not arise from intrinsic nonadditivity of isotope effects among single conformers of reactants and products. Instead, Saunders and Cline concluded that the relative populations of isotopically distinguished conformers were shifted by different levels of isotopic substitution. As is shown in Fig. 11.8, the product for the rearrangement with the d_1 isotopomer can adopt conformations that reduce the sensitivity of the equilibrium to isotopic substitution. In this way, the effective product state for the equilibrium is altered by different levels of isotopic substitution, leading to the RGM breakdown.

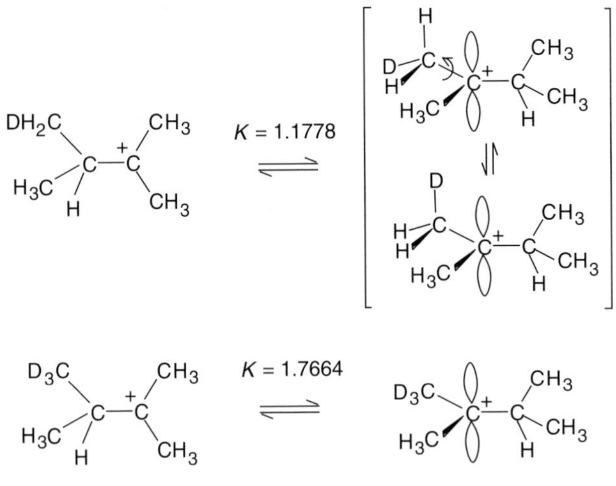

Figure 11.8. Isotope effects (162.5 K) on a carbocation rearrangement studied by Saunders and Cline [64]. The rule of the geometric mean predicts that the three-isotope equilibrium isotope effect should be the cube of the single-isotopic-site effect: $(1.1778)^3 = 1.6338$.

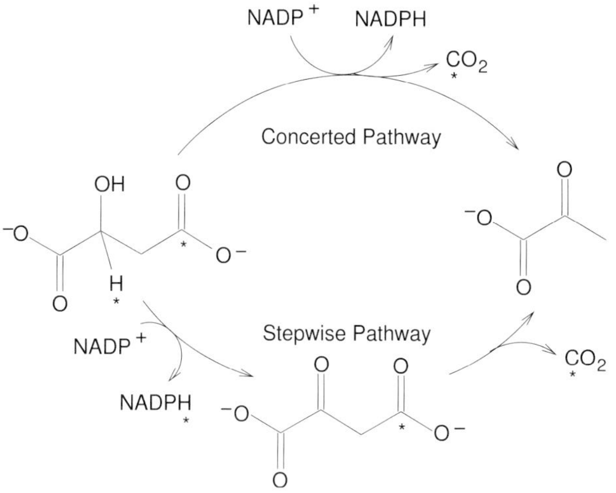

Figure 11.9. Mechanistic pathways for the malic enzyme reaction. Asterisks show the positions of the labels for some of the isotope-effect-on-isotope-effect experiments of Hermes, Cook, Cleland and others [66–70]. The upper pathway shows concerted hydride transfer and decarboxylation; the lower pathway shows the two chemical processes in separate steps. The carbon isotope effects they measured decreased when the transferring hydrogen was deuterium, ruling out the concerted pathway.

Isotope-sensitive effective states can also lead to RGM breakdowns for kinetic isotope effects. In cases involving multiple rate-limiting steps, the effective or *virtual* states [65] that can be deduced from a transition-state analysis may be sensitive to isotopic substitution. The malic enzyme story [66–70] is a comprehensive example of the use of isotope effects on isotope effects to resolve mechanistic details from effective transition states. Two possible routes for the oxidative decarboxylation of the substrate malate are shown in Fig. 11.9, and the two isotope effects of interest are the carbon isotope effect on decarboxylation and the hydrogen isotope effect on hydride transfer. Also of interest is the influence of the two isotope effects on each other – the hydrogen isotope effect on the carbon isotope effect. The distance between the two isotopic sites, along with likely cancellation of coupled-vibration effects, eliminates the possibility of an RGM breakdown from intrinsic nonadditivity. However, there is a chance that an RGM breakdown will be observed if the reaction follows the stepwise path *and* that path happens to have both steps partially limiting the reaction rate. In this case, the effective transition state is a weighted average of the decarboxylation and hydride-transfer transition states, and the relative weighting of the two states becomes sensitive to isotopic substitution. The effective transition state will be different for protium and deuterium transfer. Hermes, Cook, Cleland, and their coworkers [66–70]. found that the $^{12}C/^{13}C$ isotope effect on k_{cat}/K_m at saturating concentrations of $NADP^+$ was 1.0324 ± 0.0003 for protium transfer and 1.0243 ± 0.0004 for deuterium transfer. These results are consistent with a stepwise path in which decarboxylation is less rate-limiting for the deuterated substrate (the hydride transfer step is more rate limiting when deuterium is transfered).

In general, shifting effective states will cause RGM breakdowns according to equations similar to Eqs. (11.15) and (11.16), derived from the equations for serial and parallel changes in the mechanism shown in Fig. 11.4. These equations were simplified from a general set of equations for two isotopic sites and two rate-limiting steps (a and b) by identifying isotopes for one site to be H and D, and using I and J for the isotopes of the other site. A further simplification is the assumption that the H/D isotope effect on step a is unity and the I/J isotope effect on step b is also unity. Intrinsic nonadditivity and tunneling effects are also excluded by using the same $(k_I/k_J)_a$ in both equations.

$$\text{Serial:} \quad g_{IJ}^{HD} = \frac{(k_I^H/k_J^H)_{obs}}{(k_I^D/k_J^D)_{obs}} = \frac{w_{aH}(k_I/k_J)_a + 1 - w_{aH}}{w_{aD}(k_I/k_J)_a + 1 - w_{aD}} \quad (11.15)$$

$$\text{Parallel:} \quad g_{IJ}^{HD} = \frac{(k_I^H/k_J^H)_{obs}}{(k_I^D/k_J^D)_{obs}} = \frac{w_{aD}(k_I/k_J)_a^{-1} + 1 - w_{aD}}{w_{aH}(k_I/k_J)_a^{-1} + 1 - w_{aH}} \quad (11.16)$$

These equations show that if the H/D isotope effect is normal, the isotope effect $(k_I/k_J)_{obs}$ will decrease (move toward unity) for the serial change in the rate-limiting step when H is replaced with D. If the H/D isotope effect is inverse, or the reaction involves a parallel change in mechanism, the RGM breakdown will be reversed and the I/J isotope effect will become larger (move away from unity) when H is replaced with D. There exist a wide range of possible RGM breakdowns from the shifting effective states in complex mechanisms. The full and elaborate analysis of the malic enzyme studies demonstrates the potential for using RGM breakdowns to sort out the details of multiple rate-limiting steps in enzymatic reactions [66–70]. Other approaches to the use of RGM breakdowns to distinguish between mechanistic pathways include studies of proline racemace [71, 72] and studies of multiple intramolecular proton transfers in porphyrins and related systems [43, 73–79], and in hydrogen-bonded dimers [80–82].

Special effects arising from variational TST [15, 83] can also be expected to produce RGM breakdowns in reacting systems with isotopomers having different free-energy bottlenecks. These variational effects are expected to be particularly significant for hydrogen transfer reactions with nearly symmetrical structures at the classical transition-state location at a potential-energy saddle point [14, 84]. The diagram in Fig. 11.10 gives a rough view of how the bottleneck can end up in a different position for hydrogen isotopes. If, near the saddle point, the vibrational potential corresponding to a triatomic symmetric stretch tightens relative to the potential at the saddle point, the resulting difference in the isotopic vibrational energies can be sufficient to shift the bottleneck for the protium isotopomer further from the saddle point than the deuterium isotopomer. In this way, the variational transition state is different for the two isotopomers, and a second isotopic probe would report on the different vibrational properties of the H-transfer and D-transfer bottlenecks. An RGM breakdown is possible since the effective transition state is changed upon isotopic substitution.

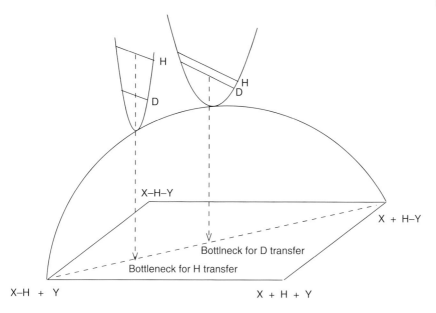

Figure 11.10. Illustration showing how variational transition state theory allows for different bottleneck structures for H and D transfer.

11.4.3
RGM Breakdown as Evidence for Tunneling

The RGM breakdown seen for the glutamate dehydrogenase reaction (Fig. 11.7) was interpreted by Srinivasan and Fisher [53] as evidence for tunneling. Intrinsic nonadditivity had been ruled out based on earlier vibrational analysis calculations for related systems [56, 85].

The same vibrational-analysis studies [56, 85] used truncated Bell tunneling corrections to predict that RGM breakdowns could occur when both isotopic sites in Fig. 11.7 were moving together in the same reaction coordinate. The effective mass for the tunneling motion becomes sensitive to isotopic substitution at both sites giving tunneling terms Eq. (11.7) that are different for the HH, HD, and DD isotopomers. When only one isotopic site has significant reaction-coordinate motion, there is no RGM breakdown. The vibrational analysis calculations predicted that substitution of protium for deuterium could decrease the primary H/D isotope effect by about 10%, in good agreement with $g_{HD}^{HD} = 1.14$ measured by Srinivasan and Fisher [53]. Other early reports of isotope effects on isotope effects in related systems include values of 1.13 and 1.09 for nonenzymic model hydride-transfer reactions [86] and 1.15 for the reaction catalyzed by yeast formate dehydrogenase [87]. The RGM studies in multiple proton transfers also provide valuable insights into tunneling processes [73–76, 80–82, 88].

Cui et al. found an RGM breakdown in their computational study of alcohol dehydrogenase of when tunneling was included, but no breakdown when tunneling was omitted from the calculation. They found for the primary hydrogen isotope effect, $g_{DT}^{HD} = 1.10$ at 300 K and attributed it to the coupled motion of the secondary and primary hydrogen sites along the reaction path.

11.5
Saunders' Exponents: Mixed Multiple Isotope Probes

In 1985, Saunders [36] proposed a new type of isotope-effect experiment to detect tunneling that required using H, D, and T isotopes similar to a Swain–Schaad experiment, but with an additional site of isotopic substitution. The extra isotopic site was shown to be the key reason for the sensitivity to tunneling because it allowed for an RGM breakdown when the two isotopic atoms move together in a tunneling motion [37].

11.5.1
Experimental Considerations

The exponents described by Saunders, sometimes called "mixed isotopic exponents", are shown in Eq. (11.17). The exponent r_{DT}^{DD} describes the relationship between the H/T isotope effect from substitution at site one (determined when protium is at site two), and the site-one D/T isotope effect (determined when deuterium is at site two). If the two sites are distinguished as giving primary and secondary isotope effects, the first exponent in Eq. (11.17) resembles the single-site Swain–Schaad exponent r_{DT} Eq. (11.9) for a primary isotope effect, and the second exponent in Eq. (11.17) resembles a single-site secondary Swain–Schaad exponent. However, the mixed isotopic exponents necessarily involve isotopic substitution at two sites and should not be confused with single-site Swain–Schaad exponents.

$$k_H^H/k_T^H = (k_D^D/k_T^D)^{r_{DT}^{DD}} \quad k_H^H/k_H^T = (k_D^D/k_D^T)^{r_{DD}^{DT}} \tag{11.17}$$

One reason Saunders proposed using the set of isotopomers presented in Eq. (11.17) was for synthetic considerations. For reactants with two hydrogenic sites attached to the same atom, it is typically much simpler to prepare the DD isotopomer than it is to synthesize a compound with high abundance D in an HD compound. The HT and DT isotopomers are generally easier to prepare because the tritium is at tracer levels. Figure 11.11 shows an example [89] of the labeled reactants needed for an experiment.

11.5.2
Separating Swain–Schaad and RGM Effects

The Swain–Schaad and RGM components of mixed isotopic exponents can be readily identified by first defining relevant isotope effects on isotope effects [37].

11.5 Saunders' Exponents: Mixed Multiple Isotope Probes

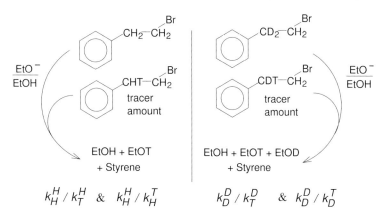

Figure 11.11. Example of competitive experiments used by Saunders et al. [89, 97] to measure isotope effects for mixed-isotope exponent determinations. The primary and secondary isotope effects were determined using the starting tritium activities of the bromide substrate, the product styrene, and the solvent ethanol, all at a known fractional extent of reaction [21]. In double-tracer-label experiments like those used by Cha et al. [90] to study a reaction catalyzed by yeast alcohol dehydrogenase, ^{14}C is included in a remote site (the aromatic ring) to monitor the reaction rates for CD_2 or CH_2 substrates while simultaneously using tritium to monitor reaction rates for the CHT or CDT substrates.

Equation (11.18) shows g_1 as the ratio of primary D/T isotope effects when the secondary site is either H or D, and g_2 as the corresponding H/D isotope effect on a secondary D/T effect. It is also convenient to define the RGM effect in an alternate way using the exponents γ_1 and γ_2, as in Eq. (11.19).

$$g_1 = g_{DT}^{DH} = \frac{k_D^H/k_T^H}{k_D^D/k_T^D} \quad g_2 = g_{DH}^{DT} = \frac{k_H^D/k_H^T}{k_D^D/k_D^T} \tag{11.18}$$

$$\gamma_1 = \gamma_{DT}^{DH} = \frac{\ln(k_D^H/k_T^H)}{\ln(k_D^D/k_T^D)} \quad \gamma_2 = \gamma_{DH}^{DT} = \frac{\ln(k_H^D/k_H^T)}{\ln(k_D^D/k_D^T)} \tag{11.19}$$

With these definitions, a mixed isotopic exponent can be shown to be the product of an RGM exponent and a single-site Swain–Schaad exponent Eq. (11.20). As was noted in a previous section, single-site Swain–Schaad exponents tend to be close to $r_{DT} = 3.3$ (for $r_{DH} = 1.44$, Eq. (11.9)) even for reactions with substantial tunneling. When this is true, a mixed isotopic exponent greater than 3.3 can be explained by γ, the RGM breakdown. And if the RGM breakdown is thought to arise from tunneling, a large mixed isotopic exponent further indicates that the effective tunneling mass is sensitive to both isotopic sites.

$$m_1 = r_{DT}^{DD} = r_{DT}\gamma_1 \quad m_2 = r_{DD}^{DT} = r^{DT}\gamma_2 \tag{11.20}$$

Several experimental studies have shown that the secondary mixed exponents can be especially large. In cases where the single-site Swain–Schaad exponents

(r^{DT}) are expected to be near 3.3, the relationships in Eq. (11.21) demonstrate the extra sensitivity of the secondary mixed isotopic exponents to RGM breakdowns. The RGM effect defined as a ratio of isotope effects tend to be similar in magnitude for g_1 and g_2 (they are identical for g_{HD}^{HD}). In the form of an exponent, however, Eq. (11.21) shows that a large primary k_D/k_T isotope effect will diminish the effect of g_1 on γ_1. A smaller secondary D/T kinetic isotope effect will allow for a greater effect of a similarly valued g_2 on γ_2.

$$\gamma_1 = \left(1 - \frac{\ln(g_1)}{\ln(k_D^H/k_T^H)}\right)^{-1} \quad \gamma_2 = \left(1 - \frac{\ln(g_2)}{\ln(k_H^D/k_H^T)}\right)^{-1} \tag{11.21}$$

The results of Cha et al. [90] illustrate these tendencies. Their isotope effect results give an unexceptional primary mixed exponent, $r_{DT}^{HD} = 3.58 \pm 0.08$, and a much larger secondary mixed exponent, $r_{HD}^{DT} = 10.2 \pm 2.0$. These mixed exponents have been studied recently in several large-scale computational projects [14, 38, 47, 91]. Many other experimental studies [89, 92] involving mixed isotopic exponents are the subject of several reviews [93–96].

11.5.3
Effects of Mechanistic Complexity on Mixed Isotopic Exponents

Efforts to test the influence of mechanistic complexity on the Saunders mixed isotopic exponents would be best approached by first considering the separate treatments presented above for single-site Swain–Schaad exponents and RGM effects according to Eqs. (11.11) and (11.12), (11.15) and (11.16), and (11.21). The net effect on mixed isotopic exponents can then be computed from the product according to Eq. (11.20). The resulting equations needed to explore the influence of either parallel or serial changes in rate-limiting steps will be simpler and the analysis more straightforward than any treatment that does not begin by separating the two components of a mixed isotopic exponent. The combined effects of mechanistic complexity on single-site Swain–Schaad exponents and RGM ratios should provide considerable modeling latitude in explaining the environmental or temperature influences on mixed isotopic exponents.

11.6
Concluding Remarks

Single-site Swain–Schaad exponents have not been found to be useful diagnostics for tunneling. Experimental results in systems thought to involve tunneling typically do not give single-site Swain–Schaad exponents that are outside the range of expected values for non-tunneling reactions, and computational studies have similarly failed to find good evidence for single-site H/D/T tunneling criteria. More work is needed to know if Swain–Schaad criteria could be established for reactions with extreme tunneling but, generally, the effects of H/D/T substitution on reaction rates can arise from a comparatively global influence on vibrational frequencies. Special effects on tunneling coordinates are therefore difficult to distinguish as Lewis and Robinson [29] noted in their discussion of Eq. (11.10) many years ago.

Reactions that involve the motions of two potential sites of isotopic substitution offer better opportunities for studies of tunneling through determinations of isotope effects on isotope effects (RGM tests). The influence of multiple mass sites on a reacting coordinate is somewhat special and can more directly report on tunneling. Mixed isotopic exponents include both Swain–Schaad exponents and RGM tests in a single experiment, and can therefore be expected to provide information related to tunneling if motions of two or more isotopic sites are significant in a tunneling coordinate. The RGM test within the mixed isotopic exponent is unfortunately clouded by the Swain–Schaad component of the experimental probe, particularly for secondary isotope effects. However, for laboratory work, the mixed isotopic probes often prove to give the best precision, and through comparisons with computational results, desired connections can be made between experimental findings and ideas about tunneling, catalysis, and reaction-rate theories.

Finally, the effects of mechanistic complexity must be addressed in any study of tunneling, particularly for enzyme-catalyzed reactions. There is no simple way to avoid the complications from multiple rate-limiting steps – they may appear in rapid-mix experiments, relaxation kinetics, and steady-state turnovers. There is good reason to believe, however, that with sufficient numbers of isotopic probes, many interesting mechanistic details can be resolved.

References

1 R. P. BELL, D. M. GOODALL. *Proc. R. Soc. (London) Ser. A*, **1966**, *294*, 273–297.
2 L. H. FUNDERBURK, E. S. LEWIS. *J. Am. Chem. Soc.*, **1964**, *86*, 2531–2532.
3 E. S. LEWIS, L. H. FUNDERBURK. *J. Am. Chem. Soc.*, **1967**, *89*, 2322–2327.
4 R. P. BELL. *Proc. R. Soc. (London) Ser. A*, **1935**, *148*, 241–250.
5 R. P. BELL, J. A. FENDLEY, J. R. HULETT. *Proc. R. Soc. (London) Ser. A*, **1956**, *235*, 453–468.
6 R. P. BELL. *The Tunnel Effect in Chemistry*. Chapman Hall, London, **1980**.
7 M. GARCIA-VILOCA, J. GAO, M. KARPLUS, D. G. TRUHLAR. *Science*, **2004**, *303*, 186–195.
8 J. GAO, D. G. TRUHLAR. *Annu. Rev. Phys. Chem.*, **2002**, *53*, 467–505.
9 D. G. TRUHLAR, J. GAO, C. ALHAMBRA, M. GARCIA-VILOCA, J. C. CHORCHADO, M. L. SÁNCHEZ, J. VILLÀ. *Acc. Chem. Res.*, **2002**, *35*, 341–349.
10 A. WARSHEL. *Acc. Chem. Res.*, **2002**, *35*, 385–395.
11 J. VILLÀ, A. WARSHEL. *J. Phys. Chem. B*, **2001**, *33*, 7887–7907.
12 S. HAMMES-SCHIFFER. *Biochemistry*, **2002**, *41*, 13335–13343.
13 D. ANTONIOU, S. D. SCHWARTZ. *Proc. Natl. Acad. Sci, USA*, **1997**, *94*, 12360–12365.
14 G. TRESADERN, P. F. FAULDER, M. P. GLEESON, Z. TAI, G. MACKENZIE, N. A. BURTON, I. H. HILLIER. *Theor. Chem. Acc.*, **2003**, *109*, 108–107.
15 D. G. TRUHLAR, B. C. GARRETT, S. J. KLIPPENSTEIN. *J. Phys. Chem.*, **1996**, *100*, 12771–12800.
16 J. BIGELEISEN, M. WOLFSBERG. *Adv. Chem. Phys.*, **1958**, *1*, 15–76.
17 A. STREITWIESER, JR., R. H. JAGOW, R. C. FAHEY, S. SUZUKI. *J. Am. Chem. Soc.*, **1958**, *80*, 2326–2332.

18 L. Melander. *Isotope Effects on Reaction Rates*, Ronald Press Co., New York, **1960**.
19 F. H. Westheimer. *J. Am. Chem. Soc.*, **1961**, *1*, 265–273.
20 J. Bigeleisen. *Pure Appl. Chem.*, **1964**, *8*, 217–213.
21 L. Melander, W. H. Saunders, Jr. *Reaction Rates of Isotopic Molecules*, Wiley, New York, **1980**.
22 R. P. Bell. *Trans. Faraday Soc.*, **1959**, *55*, 1–4.
23 C. B. Grissom, C.-C. Hwang. *J. Am. Chem. Soc.*, **1994**, *116*, 795–796.
24 A. Thibblin, P. Alhberg. *Chem. Soc. Rev.*, **1989**, *18*, 209–224.
25 V. J. Shiner, Jr., M. L. Smith. *J. Am. Chem. Soc.*, **1961**, *83*, 593–598.
26 M. J. Stern, R. E. Weston, Jr. *J. Chem. Phys.*, **1974**, *60*, 2808–2815.
27 J. Basran, M. J. Sutcliffe, N. S. Scrutton. *Biochemistry*, **1999**, *38*, 3218–3222.
28 L. Melander, W. H. Saunders, Jr. *Reaction Rates of Isotopic Molecules*, Wiley, New York, **1980**, Ch. 10: Isotope Effects in Reactions with Complex Mechanisms.
29 E. S. Lewis, J. K. Robinson. *J. Am. Chem. Soc.*, **1968**, *90*, 4337–4344.
30 R. A. More O'Ferrall, J. Kouba. *J. Chem. Soc. B*, **1967**, pp. 985–990.
31 M. J. Stern, R. E. Weston, Jr. *J. Chem. Phys.*, **1974**, *60*, 2815–2821.
32 Q. Cui, M. Karplus. *J. Am. Chem. Soc.*, **2002**, *124*, 3093–3124.
33 Y. Kim, M. M. Kreevoy. *J. Am. Chem. Soc.*, **1992**, *114*, 7116–7123.
34 C. G. Swain, E. C. Stivers, J. F. Reuwer, Jr., L. J. Schaad. *J. Am. Chem. Soc.*, **1958**, *80*, 5885–5893.
35 J. Bigeleisen. *Tritium Phys. Biol. Sci.*, Proc. Symp., Vienna, Austria, **1962**, *1*, 161–168.
36 W. H. Saunders, Jr. *J. Am. Chem. Soc.*, **1985**, *107*, 167–169.
37 W. P. Huskey. *J. Phys. Org. Chem.*, **1991**, *4*, 361–366.
38 Q. Cui, M. Eistner, M. Karplus. *J. Phys. Chem. B*, **2002**, *106*, 2721–2740.
39 A. Kohen, J. H. Jensen. *J. Am. Chem. Soc.*, **2002**, *124*, 3858–3864.
40 J. R. Jones. *Trans. Faraday Soc.*, **1969**, *65*, 2430–2437.
41 J. Braun, H.-H. Limbach, P. G. Williams, H. Morimoto, D. E. Wemmer. *J. Am. Chem. Soc.*, **1996**, *118*, 7231–7232.
42 J. Braun, R. Schwesinger, P. G. Williams, H. Morimoto, D. E. Wemmer, H.-H. Limbach. *J. Am. Chem. Soc.*, **1996**, *118*, 11101–11110.
43 J. Braun, M. Schlabach, B. Wehrle, M. Köcher, E. Vogel, H.-H. Limbach. *J. Am. Chem. Soc.*, **1994**, *116*, 6593–6604.
44 C. Eckart. *Phys. Rev.*, **1930**, *35*, 1303–1309.
45 C. S. Tautermann, M. J. Loferer, A. F. Voegele, K. R. Leidl. *J. Chem. Phys.*, **2004**, *120*, 11650–11657.
46 J. Hirschi, D. A. Singleton. *J. Am. Chem. Soc.*, **2005**, *127*, 3294–3295.
47 C. Alhambra, J. Corchado, M. L. Sánchez, M. Garcia-Viloca, J. Gao, D. G. Truhlar. *J. Phys. Chem. B*, **2001**, *105*, 11326–11340.
48 D. B. Northrop. *Bioorg. Chem.*, **1990**, *18*, 435–439.
49 D. B. Northrop. *Biochemistry*, **1975**, *14*, 2644–2651.
50 D. B. Northrop. in *Istope Effects on Enzyme-Catalyzed Reactions*, W. W. Cleland, M. H. O'Leary, D. B. Northrop (Eds), University Park Press, Baltimore, MD, **1977**, pp. 122–152.
51 I. M. Kovach, J. L. Hogg, T. Raben, K. Halbert, J. Rogers. *J. Am. Chem. Soc.*, **1980**, *102*, 1991–1999.
52 K. L. Grant, J. P. Klinman. *Bioorg. Chem.*, **1992**, *20*, 1–7.
53 R. Srinivasan, H. F. Fisher. *J. Am. Chem. Soc.*, **1985**, *107*, 4301–4305.
54 J. Bigeleisen. *J. Chem. Phys.*, **1955**, *23*, 2264–2267.
55 H. Fujihara, R. L. Schowen. *Bioorg. Chem.*, **1985**, *13*, 57–61.
56 W. P. Huskey, R. L. Schowen. *J. Am. Chem. Soc.*, **1983**, *105*, 5704–5706.
57 M. Amin, R. C. Price, W. H. Saunders, Jr. *J. Am. Chem. Soc.*, **1988**, *110*, 4085–4086.
58 E. K. Thornton, E. R. Thornton. in *Isotope Effects in Chemical Reactios*, C. J. Collins, N. S. Bowman (Eds), Van Nostrand Reinhold Co., New York, **1970**, pp. 213–285.

59 A. J. Kresge. *Pure Appl. Chem.*, **1964**, *8*, 243–258.

60 K. B. Schowen, R. L. Schowen. *Methods Enzymol. C*, **1982**, *87*, 551–606.

61 K. S. Venkatasubban, R. L. Schowen. *Crit. Rev. Biochem.*, **1984**, *17*, 1–44.

62 D. M. Quinn, L. D. Sutton. in *Enzyme Mechanism from Isotope Effects*, P. F. Cook (Ed), CRC Press, Boca Raton, **1991**, pp. 73–126.

63 J. L. Kurz, M. M. Nasr. *J. Phys. Chem.*, **1989**, *93*, 937–942.

64 M. Saunders, G. W. Cline. *J. Am. Chem. Soc.*, **1990**, *112*, 3955–3963.

65 R. L. Schowen. in *Transition States of Biochemical Processes*, R. Gandour, R. L. Schowen (Eds), Plenum Press, New York, **1978**, pp. 77–114.

66 J. D Hermes, C. A. Roeske, M. H. O'Leary, W. W. Cleland. *Biochemistry*, **1982**, *21*, 5106–5114.

67 P. M. Weiss, S. R. Gavva, B. G. Harris, J. L. Urbauer, W. W. Cleland, P. F. Cook. *Biochemistry*, **1991**, *30*, 5755–5763.

68 W. E. Karsten, P. F. Cook. *Biochemistry*, **1994**, *33*, 2096–2103.

69 W. E. Karsten, S. R. Gavva, S.-H. Park, P. F. Cook. *Biochemistry*, **1995**, *34*, 3253–3260.

70 W. A. Edens, J. L. Urbauer, W. W. Cleland. *Biochemistry*, **1997**, *36*, 1141–1147.

71 J. G. Belasco, W. J. Albery, J. T. Knowles. *J. Am. Chem. Soc.*, **1983**, *105*, 2475–2477.

72 J. G. Belasco, W. J. Albery, J. T. Knowles. *J. Am. Chem. Soc.*, **1983**, *105*, 6195.

73 G. Scherer, H.-H. Limbach. *J. Am. Chem. Soc.*, **1989**, *111*, 5946–5947.

74 H. Rumpel, H.-H. Limbach. *J. Am. Chem. Soc.*, **1989**, *111*, 5429–5441.

75 M. Schlabach, G. Scherer, H.-H. Limbach. *J. Am. Chem. Soc.*, **1991**, *113*, 3550–3558.

76 H.-H. Limbach, G. Scherer, M. Maurer, B. Chaudret. *Ber. Bunsenges. Phys. Chem.*, **1992**, *96*, 821–833.

77 M. Schlabach, H.-H. Limbach, E. Bunnenberg, A. Y. L. Shu, B.-R. Tolf, Carl Djerassi. *J. Am. Chem. Soc.*, **1993**, *115*, 4554–4565.

78 H.-H. Limbach, O. Klein, J. M. Lopez Del Amo, J. Elguero. *Z. Phys. Chem.*, **2003**, *217*, 17–49.

79 G. Scherer, H.-H. Limbach. *J. Am. Chem. Soc.*, **1994**, *116*, 1230–1239.

80 D. Gerritzen, H.-H. Limbach. *J. Am. Chem. Soc.*, **1984**, *106*, 869–879.

81 L. Meschede, H.-H. Limbach. *J. Phys. Chem.*, **1991**, *95*, 10267–10280.

82 O. Klein, F. Aguilar-Parrilla, J. M. Lopez, N. Jagerovic, J. Elguero, H.-H. Limbach. *J. Am. Chem. Soc.*, **2004**, *126*, 11718–11732.

83 D.-H. Lu, D. Maurice, D. G. Truhlar. *J. Am. Chem. Soc.*, **1990**, *112*, 6206–6214.

84 B. C. Garrett, D. G. Truhlar, A. F. Wagner, T. H. Dunning, Jr. *J. Chem. Phys.*, **1983**, *78*, 4400–4413.

85 W. H. Saunders, Jr. *J. Am. Chem. Soc.*, **1984**, *106*, 2223–2224.

86 M. G. Ostavić, M. G. Roberts, M. M. Kreevoy. *J. Am. Chem. Soc.*, **1983**, *105*, 7629–7631.

87 J. D. Hermes, S. W. Morrical, M. H. O'Leary, W. W. Cleland. *Biochemistry*, **1984**, *23*, 5479–5488.

88 H.-H. Limbach, J. Hennig, D. Gerritzen, H. Rumpel. *Faraday Discuss. Chem. Soc.*, **1982**, *74*, 229–243.

89 M. Amin, R. C. Price, W. H. Saunders, Jr. *J. Am. Chem. Soc.*, **1990**, *112*, 4467–4471.

90 Y. Cha, C. Murray, J. P. Klinman. *Science*, **1989**, *243*, 1325–1330.

91 C. Alhambra, J. C. Corchado, M. L. Sánchez, J. Gao, D. G. Truhlar. *J. Am. Chem. Soc.*, **2000**, *122*, 8197–8203.

92 W. E. Karsten, C.-C. Hwuang, P. F. Cook. *Biochemistry*, **1999**, *38*, 4398–4402.

93 J. P. Klinman. *Pure Appl. Chem.*, **2003**, *75*, 601–608.

94 A. Kohen. *Progr. React. Kinet. Mech.*, **2003**, *28*, 119–156.

95 A. Kohen, J. P. Klinman. *Chem. Biol.*, **1999**, *6*, R191–198.

96 A. Kohen, J. P. Klinman. *Acc. Chem. Res.*, **1998**, *31*, 397–404.

97 Rm. Subramanian, W. H. Saunders, Jr. *J. Am. Chem. Soc.*, **1984**, *106*, 7887–7890.

12
Current Issues in Enzymatic Hydrogen Transfer from Carbon: Tunneling and Coupled Motion from Kinetic Isotope Effect Studies

Amnon Kohen

12.1
Introduction

12.1.1
Enzymatic H-transfer – Open Questions

Enzymes are the catalysts that direct, control and enhance chemical transformations in biological systems. Enzymes evolved to accomplish two almost contradictory tasks: (i) to catalyze a reaction at a rate most suitable for organism function and (ii) to prevent alternative side-processes that would commonly occur in nonenzymatic reactions. In other words, an enzyme not only catalyzes the reaction of interest, it also inhibits side reactions and the formation of by-products. Commonly, the first effect is denoted as catalysis and the second is denoted as specificity. The rate enhancement is often many orders of magnitude greater than the reaction in solution. *How* enzymes achieve this rate acceleration is a matter of great interest to both chemists and biologists. Different investigators have addressed this question using the tools available to them, sometimes leading to diverse, though not necessarily contradictory, suggestions. The question of whether physical phenomena such as "dynamics" or "quantum mechanical hydrogen tunneling" contribute to enzyme catalysis is one of the "hottest" and most controversial questions in enzymology today. In this chapter I will attempt to present theoretical and experimental approaches to this question. More importantly, I will try to suggest that in many cases the different views result from different definitions rather than from contradictory physical mechanisms.

By way of introduction I will demonstrate this point for a basic, "old", example. The general notion entitled "induced fit" [1, 2] implies that the enzyme's active site acts like a molecular laboratory in which the reactants (substrates) are first recognized during the binding process, and then both the enzyme and the substrate undergo conformational rearrangement. This leads to a substrate conformation that is closer to the desired transition state. This view leads to the generally accepted understanding of the role of transition state stabilization, since the free energy of the whole system is (by definition, as discussed below under Catalysis) lower than that of the slower uncatalyzed reaction. Nevertheless, this has also led to the misconception of "ground state destabilization" [3, 4]. This term was coined since, in a two state system (ground and transition states), either stabilizing the transition state or destabilizing the ground state would have a catalytic effect (lowering the free energy of activation, $\Delta G^{\ddagger}$). Upon binding and rearrangement, the substrate molecule may indeed adopt a conformation that would be less stable in solution, yet the free energy of the whole system ($\Delta G^{\ddagger}$) is commonly lower in its bound-ground-state than in the free state, only the reactant is now in a distorted conformation that would be considered "destabilized" in solution.[1] The reason why this is still a matter of controversy today is, in part, due to the different definitions used by different researchers [5–7].

In this chapter I attempt to distinguish between models that differ due to substantially different physical mechanisms from those that merely use different terminology. Consequently, the following section on terminology and definitions precedes the description of the mechanistic models and experimental examples.

12.1.2
Terminology and Definitions

12.1.2.1 Catalysis

Catalysis or "Catalytic Power" is the ratio between the reaction rate of the catalyzed reaction and that of the uncatalyzed reaction. It is defined as k_{cat}/k_{un} where k_{cat} is the rate of the catalyzed reaction and k_{un} is the rate of the uncatalyzed reaction. By definition, catalysis should be unit-less (a ratio of rate constants), thus care must be practised while determining "Catalytic Power" that k_{cat} and k_{un} have the same units. Alternatively, the second order uncatalyzed reaction's rate ($M^{-1}s^{-1}$ units) can be divided by k_{cat} (s^{-1}) and the ratio then has units of concentration (M). This concentration is called "effective concentration" [2] and could be addressed as the concentration of functional groups or substrates in the enzyme's active site. Since that effective concentration is often in the thousands of M range, it is not a physically meaningful concentration, but rather a manifestation of the role of correct orientation, dynamic, and other catalytic effects induced by the enzyme. A similar approach used the substrate concentration in which the enzymatic and uncatalyzed rates are equal as an indicator for catalytic power [8]. The advantage of the first

---

1) The binding process is a second order event and the relative energies of the free and bound states are concentration dependent. Substrate concentration is reciprocal to the stability of the free state. At a low substrate concentration (k_{cat}/k_M conditions as defined below) the bound state is less stable than the free state but its stabilization or destabilization has no catalytic effect as the barrier for the reaction is now the difference between the free state and the transition state. At high substrate concentration the bound state is more stable than the free state (lower free energy). The bound reactant may be "locally" distorted to a "locally" less stable conformation (one that would be less stable without the enzyme) but the whole complex is more stable than the free enzyme and substrate in solution. Destabilizing the bound state will have a catalytic effect only under these conditions (V_{max}), but most enzymes did not evolve under V_{max} conditions (substrate concentration is rarely higher than k_M).

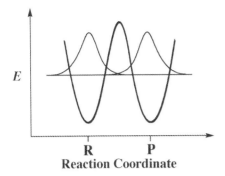

Figure 12.1. An example of ground-state nuclear tunneling. The reactant well (R) is on the left and the product well (P) is on the right. The fine lines represent the probability (nuclear ψ^2) of finding the nuclei in the reactant or the product wells. More overlap between the probability functions of the R and P results in higher tunneling probability.

approach is that the unit-less ratio of rates can be directly converted into the reduction of energy of activation ($\Delta G^{\ddagger}$) induced by the enzyme, which is the most meaningful physical parameter.

Enzyme catalysis can be studied from various points of view: regulation, structural aspects, order of reactant binding and product release, the role of functional residues (e.g., general bases or acids), etc. This chapter presents the use of kinetic isotope effects (KIEs) as tools for studying the physical nature of enzyme catalyzed C–H bond activation.

12.1.2.2 Tunneling

Quantum mechanical tunneling is the phenomenon by which a particle transfers through a reaction barrier by means of its wave-like properties [9]. Figure 12.1 illustrates this phenomenon graphically for a symmetric double well system such as the C–H–C hydrogen transfer. It is important to note that the tunneling probability is affected by both the distance between the R and the P wells and their symmetry. A lighter isotope has a higher tunneling probability than a heavier one, since a heavy isotope has a lower zero point energy and its probability function is more localized in its well. Consequently, kinetic isotope effects (KIEs) are effective tools for studying tunneling. Two practical applications are described in Section 12.3): the Swain–Schaad exponential relationship and the temperature dependence of KIEs.

12.1.2.3 Dynamics

The definition of dynamics and their possible contribution to enzyme catalysis has been a matter of debate in recent years [10–13]. In a couple of recent reviews in *Science*, two groups of prominent researchers appear to disagree on the definition of dynamics [14, 15]. Ref. [15] and several textbooks of physical chemistry prefer the definition that "dynamics is any time dependent process". Any motion in a given system can be considered a dynamic process regardless of whether or not it is in thermal equilibrium with the environment (Boltzman distribution of states). On the other hand, other researchers, e.g., Benkovic, Hammes-Schiffer, Warshel and others, use the term dynamics only in cases of nonequilibrium motions [5, 11, 14, 16]. The reason is probably that motions that are in thermal equilibrium with the environment have no "interesting" dynamic contribution because these are already accounted for in transition state theory. According to these researchers, "dynamics" are only "nonstatistical" modes, namely motion along a coordinate that is not in thermal equilibrium. Direct contribution of such modes to the reaction is depicted as a non-RRKM system. Non-RRKM systems are indeed rare and so far have only been demonstrated in the gas phase [17, 18]. Consequently, such systems will not be discussed in this chapter as I intend to focus on biologically relevant systems in the condensed phase.

12.1.2.4 Coupling and Coupled Motion

These terms are also used extensively, and often loosely, in studies of enzyme catalysis. In mechanistic enzymology the term coupling is commonly used to describe two different phenomena: first, the transferred hydrogen is coupled to another hydrogen bound to the donor or acceptor heavy atom (primary–secondary coupled motion, Section 12.3.3). Second, the tunneling of the transferred hydrogen is coupled to the enzymatic environment (environmentally coupled tunneling, tunneling promoting vibrations, vibrationally enhanced tunneling and other terms are discussed in Section 10.4.1).

A more general definition of physical coupling would be: two coordinates, for which a change in one coordinate affects the potential energy of the other are coupled to each other.[2] This definition is vague because if the two modes of motion are orthogonal, they are not coupled and if they are coupled they do not establish two clearly separable modes. Indeed, if two modes in the same system are fully coherent (the ultimate coupling) they should be redefined as two new modes that result from the mixing of the two original modes. A simple resolution for this enigma is that the energy flow from one mode to the other is slower than the rate of excitation of that mode. For example, the infrared (IR) spectrum of a molecule reflects the different vibrational modes in that molecule. If one irradiates the molecule with a short IR pulse (<100 fs), at the frequency of one of its vibrational modes, then only this mode will be vibrationally excited at first. Within a few ps that vibrational energy will equilibrate among all the normal modes in the system. The energy dissipates due to coupling between the originally excited mode and the other modes in the system. The hydroxyl and the aldehyde of hexan-6-ol-1-al for example, are separated by six bonds and their IR absorbance peaks are separated by 1600 cm^{-1}. Yet irradiation of the OH at 3300 cm^{-1} will quickly lead to more excited carbonyl modes as the vibrational energy dissipates through the molecule via mode–mode coupling.

---

2) From the mathematical point of view, coupling may be defined as mixing between two states (motion along two coordinates). Coupling matrix elements will be proportional to the second derivative of the potential energy with respect to both coordinates.

An example most relevant to the case studies presented in Section 3.3, is the coupling between two hydrogens bound to the same carbon. If the cleaving of one of them is directly affected by changes in the other they are considered coupled. The stretching mode of one (that is converted into translation at the cleavage event) is coupled to vibrational modes of the other, not only in the ground state but also in the TS. Alternatively, one can view these hydrogens as part of the same normal mode throughout. The antisymmetric stretch of the CH_2 system is converted to translation of one of the hydrogens due to greater unharmonicity on one side of that vibration (induced by the environment and the acceptor if present). In such a case, the two hydrogens contribute equally to the same mode in the ground state but have unequal contributions in the transition state. However, both affect the reaction coordinate and the overall effect is commonly denoted as "coupled motion" (see Section 3.3).

12.1.2.5 Kinetic Isotope Effects (KIEs)

The KIE is the ratio of rates between two isotopolog reactants (molecules that only differ in their isotopic composition). For H-transfer reactions, this ratio of rates between light and heavy isotopes is characteristic of the reaction coordinate and the nature of the transition state (TS). The hydrogen KIE is particularly useful since the mass ratio of its isotopes is much larger than that of any other element, resulting in relatively large KIEs [19]. The KIE results from energy of activation differences for the different isotopolog reactants, and much of its magnitude is due to the differences in zero point energy (ZPE) between the ground state and the TS of the reaction [20–22]:

$$k_H/k_D \approx e^{(\Delta G_D^\ddagger - \Delta G_H^\ddagger)/RT} \tag{12.1}$$

where $\Delta G_D^\ddagger - \Delta G_H^\ddagger \approx ZPE_D^\ddagger - ZPE_D^R - ZPE_H^\ddagger + ZPE_H^R$, R is the gas constant and T is the absolute temperature. Two other factors that contribute to the KIE to a lesser extent are the moment of inertia (MMI) and the effect of excitation (EXC). The Bigeleisen equation expresses the combined effects as KIE = ZPE ∗ MMI ∗ EXC [19–21]. Other types of isotope effects, such as magnetic isotope effects are less relevant to enzymology and are not discussed in this chapter.

The primary (1°) KIE is the KIE measured for a bond cleavage or formation that is isotopically substituted on one of the bound atoms. The secondary (2°) KIE is the KIE measured with isotopologs that are labeled on a position other than the one that is being cleaved. 2° KIEs result from a change in bonding force constants and vibrational frequencies during the reaction (e.g., σ (s-sp^3) at the GS to σ (s-sp^2) at the TS). Equilibrium isotope effects (EIEs) are the fractionation of isotopes between stable states, namely, the different ratio between isotopes that are at equilibrium between two systems. 2° KIEs are normally smaller than or equal to the reaction's 2° EIEs. The 2° EIE, results from the change in bond order from reactants to products, while the KIE is only affected by the change from the GS to the TS.

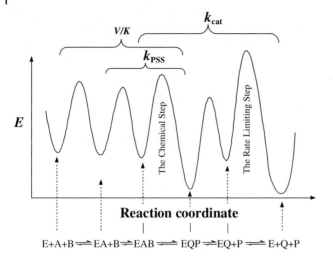

Figure 12.2. A reaction energy profile for enzyme catalyzed $A + B \rightarrow Q + P$ reaction. The chemical step is marked and the rate-limiting step is product Q release. The portions of this reaction profile that are included in three common experimental rate constants (k_{cat}, K_{cat}/K_M, and k_{PSS}, for pre-steady-state rate) are marked.

12.2
The H-transfer Step in Enzyme Catalysis[3]

Many experimental and theoretical studies have attempted to assess the specific contributions of physical phenomena to enzymatic rate enhancement. These contributions include: the relationship between the reactive complex's structure and function; the location and structure of the transition state (TS) along the reaction coordinate; TS stabilization; ground state (GS) destabilization; the significance of quantum mechanical (QM) phenomena; the dynamics of the system; energy distribution through the normal modes of an enzyme; and contributions of entropy and enthalpy. KIEs are one of the most useful parameters that can provide direct information regarding the nature of the H-transfer and its potential surface. However, a major limitation in measuring the KIE on the H-transfer step, or on any other chemical transformation, is that the H-transfer step is rarely the rate limiting step of an enzymatic reaction. Figure 12.2 illustrates a minimal reaction profile for a bi-substrate, bi-product reaction. The kinetic steps included in common rate constants (k_{cat}, k_{cat}/K_M, and pre-steady-state rate k_{PSS}) are marked. It is obvious that none of these kinetic rates represent solely the H-transfer (chemical) step. Conse-

---

3) All the examples and discussions presented below are for C–H bond activation. Some of these discussions and models may not be directly applicable to O–H activation and some other H-transfer phenomena.

quently, conducting measurement that will shed light on that step and will thus be relevant to molecular models and calculations is inherently challenging.

In the investigation of KIEs and their temperature dependence in enzymatic systems (or any other kinetically complex system), explicit care must be exercised with regard to the following issues:

1. Is the KIE measured on a single kinetic step (e.g., internal KIE) or on a kinetically complex rate constant (e.g., k_{cat}/K_M or k_{cat})?
2. Is there an isotope effect on steps other than the one under investigation (e.g., substrate binding)?
3. What is the effect of kinetic steps that are not isotopically sensitive but still mask the isotope effect (e.g., substrate dissociation)?
4. Are the KIEs measured for the same rate constant throughout the whole temperature range (e.g., temperature dependent binding constants may change measurements under substrate saturation to nonsaturated measurements)?
5. Are all the experimental conditions (e.g., pH, ionic strength) consistent at all temperatures (e.g., temperature dependence of the buffer's K_a)?

The first three points are related to 'kinetic complexity', which reflects the fact that the observed KIE (KIE_{obs}) is often smaller than the intrinsic KIE ($KIE_{int.}$). This is due to the ratio between the isotopically sensitive step and the isotopically nonsensitive steps that lead to the decomposition of the same reactive complex. Its mathematical treatment is rigorously described in several published reviews [23, 24] and Chapter 10 (Knapp et al.) in this volume.

$$KIE_{obs} = \frac{KIE_{int} + C_f + C_r \cdot EIE}{1 + C_f + C_r} \quad (12.2)$$

where EIE is the equilibrium isotope effect and C_f and C_r are the forward and reverse commitments to catalysis, respectively. C_f is the ratio between the rate of the isotopically sensitive step forward (e.g., $k_{H\text{-transfer}}$) and the rates of the preceding isotopically nonsensitive steps backward. C_r is the ratio between the rate of the isotopically sensitive step backward and the rates of the succeeding isotopically nonsensitive steps forward. Techniques that allow estimation of the intrinsic effect from the observed one are discussed below (Section 3.2). The importance of the intrinsic KIE is that it imposes a strict constraint on any mechanism, theoretical model, analysis, or simulation addressing the chemical transformation under study. As described in the following section, intrinsic KIEs are unique as they are directly affected by the reaction potential surface and other physical features. In contrast to the KIE_{obs}, KIE_{int} can be compared to theoretical calculations, which commonly only reflect an effect on a single step. The following section presents several approaches towards studies of the H-transfer step in kinetically complex systems.

12.3
Probing H-transfer in Complex Systems

12.3.1
The Swain–Schaad Relationship

12.3.1.1 The Semiclassical Relationship of Reaction Rates of H, D and T

Semiclassical in this context means that some quantum mechanical effects are taken into consideration (e.g., zero-point-energy: ZPE) but others are ignored (e.g., tunneling). When the main origin of KIEs is the differences between the isotopes' ZPEs in the ground state (GS) and transition state (TS), the kinetic relationship between the three isotopes of hydrogen can be predicted [19, 25] and is depicted as the Swain–Schaad relationship. This relationship has wide usage as a mechanistic tool in organic and physical chemistry. The Swain–Schaad exponential relationship (*EXP*, as defined in Eqs. (12.3)–(12.5) is the semiclassical (no tunneling) correlation between the rates of the three isotopes of hydrogen, and was first defined by Swain et al. in 1958 [25]. This relationship can be predicted using the masses (or reduced masses) of the isotopes under examination [19]. Since H has a higher GS ZPE than D and T, it reacts faster than D and T (Fig. 12.3). The Swain–Schaad exponential relationship was originally defined for primary (1°) KIEs [25]:

$$\frac{k_H}{k_T} = \left(\frac{k_H}{k_D}\right)^{EXP} \quad \text{or} \quad EXP = \frac{\ln(k_H/k_T)}{\ln(k_H/k_D)} \quad (12.3)$$

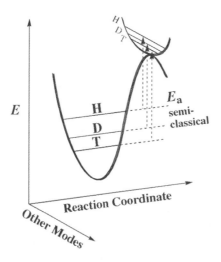

Figure 12.3. Different energies of activation (ΔE_a) for H, D, and T resulting from their different zero-point energies (ZPE) in the ground state (GS) and transition state (TS). The GS-ZPE is constituted by all degrees of freedom but mostly by the C–H stretching frequency, and the TS-ZPE is constituted by all degrees of freedom orthogonal to the reaction coordinate. This type of consideration is depicted as "semiclassical".

where k_i is the reaction rate constant for isotope i. *EXP* can be calculated from [19]:

$$EXP = \frac{\ln(k_H/k_T)}{\ln(k_H/k_D)} = \frac{1/\sqrt{\mu_H} - 1/\sqrt{\mu_T}}{1/\sqrt{\mu_H} - 1/\sqrt{\mu_D}} \quad (12.4)$$

where μ_i is the reduced mass affecting the ZPE of isotope i. The original *EXP* was calculated for H/T vs. H/D KIEs and yielded a value of 1.44 (using atomic masses). Some labeling patterns use T as a frame of reference which is compared to H and D. Their *EXP* follows:

$$EXP = \frac{\ln(k_H/k_T)}{\ln(k_D/k_T)} = \frac{1/\sqrt{\mu_H} - 1/\sqrt{\mu_T}}{1/\sqrt{\mu_D} - 1/\sqrt{\mu_T}} \quad (12.5)$$

Equation (12.5) defines the relationship of H/T to D/T KIEs, for which the semiclassical *EXP* is 3.26 (for atomic masses: $\mu_i = m_i$).

Several investigators examined this relationship under extreme temperatures (20–1000 K), and as a probe for tunneling [26–28]. This isotopic relationship was also used in experimental and theoretical studies to suggest coupled motion between primary and secondary hydrogens for hydride transfer reactions, such as elimination in the gas phase, and in organic solvents [29, 30]. The power of the Swain–Schaad relationship is that it appears independent of the details of the reaction's potential surface and thus can be used to relate unknown KIEs (see Section 12.3.2).

12.3.1.2 Effects of Tunneling and Kinetic Complexity on *EXP*

Two features that might affect the observed Swain–Schaad relationship are kinetic complexity and tunneling. The first would make the fast H-transfer appear slower and would have a smaller such effect on D and T, which are more rate limiting in the first place. As a result $EXP = \ln(H/T)/\ln(D/T)$ will appear smaller and $EXP = \ln(H/T)/\ln(H/D)$ will appear larger than anticipated [26–28]. Tunneling, on the other hand, will make H-transfer even faster relative to D and T than predicted from ZPE considerations. This will have the opposite effect, namely $EXP = \ln(H/T)/\ln(D/T)$ will appear larger and $EXP = \ln(H/T)/\ln(H/D)$ will appear smaller than anticipated. Many researchers have suggested that this phenomenon can serve as an indicator for tunneling [19, 31–34].

Recently, three theoretically calculated phenomenological effects on 1° *EXP* using small molecular systems in the gas phase suggested that tunneling and kinetic complexity may have the same trend. Two of these calculations used a simple, linear double O–H–O transfer in formic acid [35] or a 2-dimensional linear C–H–C transfer (Zorka Smedarchina, personal communication). These calculations assumed parabolic reactant and product ground states, and calculated H-transfer rates while considering the C–H stretch and a symmetric vibration between these two states as two orthogonal modes (full tunneling model). Within this confined model the difference in rates between H and the heavier isotopes was smaller than that predicted from the difference between their ZPEs. Consequently, the Swain–Schaad *EXP* for $\ln(H/T)/\ln(H/D)$ was predicted to be larger than the semiclassical limit (up to 1.58) and the *EXP* for $\ln(H/T)/\ln(D/T)$ was smaller (down to 2.30). Since this trend is the same as that predicted from kinetic complexity, such a prediction could not be used to analyze experimental data. A somewhat more rigorous third study was conducted by Kiffer and Hynes [22, 36] and their conclusion was that, close to ambient temperature, the Swain–Schaad 1° *EXP*, range from 3 to 4. A recent study of 2° *EXP* [37] tested many 2° EIEs and 2° KIEs and concluded that for significant 2° KIEs (>1.1) inflated 2° *EXP* (>4.5) is an indication for tunneling (see more discussion below). In the future, this issue should be further explored to find more realistic models in the condensed phase.

Two common uses of the Swain–Schaad relationship in enzymology are described in the following sections.

12.3.2
Primary Swain–Schaad Relationship

12.3.2.1 Intrinsic Primary KIEs

Both measurements (Eqs. (12.4) and (12.5)) establish the relationship between the three isotopes for a one-step reaction (single barrier). Intrinsically, these relationships are almost independent of the shape of the reaction potential surface [19]. If the chemical step is masked by kinetic complexity (Eq. (12.2)), the observed KIE (KIE_{obs}) will be smaller than the intrinsic one. In the case of Eq. (12.5), this will affect H/T KIE more than D/T KIE and the observed *EXP* will be smaller than the intrinsic one. For the experiment pertinent to Eq. (12.4), such kinetic complexity will have the opposite effect and the observed *EXP* will be larger than the intrinsic one. In the case of H-tunneling, a simple tunneling correction to TST (the Bell correction) [19, 32] predicts that the *EXP* of Eq. (12.4) will not be affected significantly. On the other hand, that of Eq. (12.5) will be inflated (larger than 3.3) for moderate tunneling and will decrease back to values close to 3.3 in the case of extensive tunneling [38]. As mentioned above, some recent calculations of two-dimensional systems have suggested the opposite trend but the relevance of these preliminary studies to H-transfer in the condensed phase is not clear (Ref. [35] and Zorka Smedarchina, personal communication). It is important to note that for 1° KIEs and the resulting 1° Swain–Schaad exponents, no *EXP* values larger than 3.6 are found in the literature. Even in cases in which tunneling was evident from mixed labeling experiments (Section 12.3.3) or from the temperature dependence of KIEs (Section 12.3.4), the 1° *EXP* did not significantly differ from the semiclassically predicted value of 3.3 for $\ln(k_H/k_T)/\ln(k_D/k_T)$ or 1.4 for $\ln(k_H/k_T)/\ln(k_H/k_D)$. This observation suggests that the Swain–Schaad *EXP* can be used to reduce the number of unknowns in the comparison of observed KIEs to an equation with one unknown (an intrinsic KIE).

Northrop [24, 39] developed a simple method for calculating the commitment to catalysis and the intrinsic KIE from the observed KIEs. This method assumes no significant deviation of the intrinsic 1° KIEs from their semiclassically predicted values. By analogy to the method described by Northrop [24, 39], Eq. (12.3) can be written again while subtracting 1 from both sides of the equation:

$$\left(\frac{k_H}{k_T}\right)_{obs} - 1 = \frac{(k_H/k_T)_{int} + C_f + C_r \cdot EIE}{1 + C_f + C_r} - 1$$

$$= \frac{(k_H/k_T)_{int} + C_f + C_r \cdot EIE - 1 - C_f - C_r}{1 + C_f + C_r}$$

$$= \frac{(k_H/k_T)_{int} - 1 + C_r \cdot (EIE - 1)}{1 + C_f + C_r} \tag{12.6}$$

At the limit of the H/T *EIE* goes to 1, which is not a bad assumption for 1° *EIE*, Eq. (12.6) becomes:

$$\left(\frac{k_H}{k_T}\right)_{obs} - 1 = \frac{(k_H/k_T)_{int} - 1}{1 + C_f + C_r} \tag{12.7}$$

Now, Eq. (12.7) for H/T KIE is divided by the same equation for the H/D KIE. Since no isotopic rate constant appears in the denominator, this will cancel leaving the ratio of $KIE_{int} - 1$ on the right-hand side:

$$\frac{(k_H/k_T)_{obs} - 1}{(k_H/k_D)_{obs} - 1} = \frac{(k_H/k_T)_{int} - 1}{(k_H/k_D)_{int} - 1} \tag{12.8}$$

From Eq. (12.5):

$$(k_H/k_D)_{int} = ((k_H/k_T)_{int})^{1/1.44} \tag{12.9}$$

And Eq. (12.8) can be rewritten as:

$$\frac{(k_H/k_T)_{obs} - 1}{(k_H/k_D)_{obs} - 1} = \frac{(k_H/k_T)_{int} - 1}{((k_H/k_T)_{int})^{1/1.44} - 1} \tag{12.10}$$

Even though Eq. (12.10) (and equivalent equations for other KIE experiments) has only one unknown (KIE_{int}), it cannot be solved analytically (due to transcendental functions). After dividing the observed KIEs minus one by each other, a numeric solution can be obtained.[4,5]

---

4) In the original works of Northrop [24, 39], tables for various KIE experiments offer solutions for a wide range of KIEs. Today this can be calculated with most calculators or any computer.
5) In cases where the chemical step is reversible and the assumption of a small 1° EIE is not valid a solution is not possible without measuring the reverse commitment (C_r). In this instance, Cleland [40] has identified a range for the KIE_{int} values between the observed KIE (KIE_{obs}) and the product of EIE and KIE_{obs} for the reverse reaction ($KIE_{obs\text{-rev}} * EIE$).

12.3.2.2 Experimental Examples Using Intrinsic Primary KIEs

Peptidylglycine α-Hydroxylating Monooxygenase (PHM) PHM initiates the oxidative cleavage of C-terminal, glycine-extended peptides by H• abstraction from the α-carbon of glycine. In the PHM reaction, substrate binding and product release contribute significantly to rate limitation under the conditions of steady-state turnover [41]. Francisco et al. [42] studied PHM using H/T and H/D competitive KIEs. The observed KIEs increased with temperature, but the intrinsic KIEs, calculated using the Northrop method, were all close to 10. These intrinsic KIEs exhibited a very small temperature dependence, leading to A_H/A_D of 5.9 ± 3.2. This, together with a large energy of activation ($E_a \sim 13$ kcal mol^{-1}) suggested "environmentally enhanced tunneling" [42]. This model is described in more detail in Chapter 10 Section 10.5.3.2).

Dihydrofolate Reductase (DFHR) DHFR catalyzes the stereospecific reduction of 7,8-dihydrofolate (H_2F) to 5,6,7,8-tetrahydrofolate (H_4F), using nicotinamide adenine dinucleotide phosphate (NADPH) as the hydride donor. Specifically, the *pro-R* hydride is transferred from the C-4 of NADPH to C-6 of H_2F. The complete kinetic scheme for DHFR is complex and the H-transfer step is partly rate determining only at high a pH [43]. Pre-steady state stopped flow measurements resulted in H/D KIE between 2.8 and 3.0 [44]. In a recent study we measured H/T and D/T competitive KIEs and calculated an intrinsic H/D KIE of 3.5 ± 0.2 [45]. This KIE is in excellent agreement with the calculated KIE of 3.4 [46]. This is significant because it demonstrates that the pre-steady state rate is not fully the "the H-transfer rate" (as illustrated in Fig. 12.2). A KIE of 3.5 exposed a commitment of 0.25 on the pre-steady state rate. This commitment suggested that the pre-steady state rate contains an additional step, most likely the reorganization of the nicotinamide ring in and out of the active site [45]. This conclusion is supported by stopped-flow FRET experiments conducted by Benkovic and coworkers with G121V *ec*DHFR [44]. The commitment was temperature dependent and so were the observed KIEs. Nevertheless, the calculated intrinsic KIEs were temperature independent with $A_H/A_T = 7.2 \pm 3.5$, which served as evidence of H-tunneling [45].

Thymidylate Synthase Thymidylate synthase catalyzes the reductive methylation of 2′-deoxyuridine-5′-monophosphate (dUMP) to 2′-deoxythymidine-5′-monophosphate (dTMP). The cofactor N^5,N^{10}-methylene-5,6,7,8-tetrahydrofolate (CH_2H_4folate) serves as a donor of both methylene and hydride [47]. We recently studied the hydride transfer step using competitive H/T and D/T KIEs [48], and the observed KIEs were used to calculate intrinsic KIEs. The observed and intrinsic KIEs were used to calculate the commitment to catalysis (Eq. (12.2)), and it was found that between 20 and 30 °C the hydride transfer is fully rate determining, while at elevated and reduced temperatures the commitment increases. At 20 °C, competitive KIE experiments resulted in 1° H/T KIE on k_{cat}/K_M and D/T KIE on k_{cat}/K_M KIEs of 6.91 ± 0.05 and 1.78 ± 0.02, respectively. The Swain–Schaad exponent for these KIEs is 3.35 ± 0.07, suggesting that the hydride transfer is rate determining overall [24, 48, 49]. H/D KIE on k_{cat} ($^D k = 3.72$) has been measured at 20 °C by Spencer and coworkers [50] under the same experimental conditions as reported here. The Swain–Schaad exponent for the H/T KIEs on k_{cat}/K_M [48] vs. the H/D KIEs on k_{cat} [50] is 1.46 ± 0.4. Taken together, the exponential relationships of k_{cat} and k_{cat}/K_M KIEs suggest no kinetic complexity on either k_{cat} or k_{cat}/K_M, which strongly supports Spencer's suggestion that the hydride transfer step is rate determining at 20 °C.

12.3.3
Secondary Swain–Schaad Relationship

The secondary (2°) Swain–Schaad relationship is calculated from 2° KIEs i.e., not the hydrogen whose bond is being cleaved but its geminal neighbor. In several cases a breakdown of this 2° Swain–Schaad relationship was used as evidence of a tunneling contribution. A number of these reported studies used mixed labeling experiments, as described below. In experiments of this type, the breakdown of the Swain–Schaad relationship indicates both tunneling and coupled motion between the primary and secondary hydrogens [33, 34].

12.3.3.1 Mixed Labeling Experiments as Probes for Tunneling and Primary–Secondary Coupled Motion

Mixed labeling experiments consist of an isotopic labeling pattern that is more complex than that considered in the original Swain–Schaad relationship. Several theoretical studies in the 1980s suggested that mixed labeling experiments would be the most sensitive indicators of H-tunneling [29].

In a mixed labeling experiment, the 1° H/T KIE is measured with H in the 2° position and is denoted as k_{HH}/k_{TH}, where k_{ij} is the rate constant for H-transfer with isotope i in the 1° position and isotope j in the 2° position. The 2° H/T KIE is measured with H at the R position and is denoted as k_{HH}/k_{HT}. The 1° and 2° D/T KIE measurements, on the other hand, are conducted with D in the geminal position, and are denoted as k_{DD}/k_{TD} and k_{DD}/k_{DT}, respectively (Fig. 12.4 and Eq. (12.11)).

$$2° \, ^M EXP = \frac{\ln(k_{HH}/k_{HT})}{\ln(k_{DD}/k_{DT})} \qquad (12.11)$$

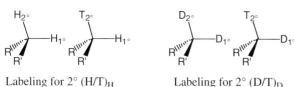

Labeling for 2° (H/T)$_H$ Labeling for 2° (D/T)$_D$

Figure 12.4. The isotopic labeling pattern for a mixed-labeling experiment.

The exponential relationship resulting from such mixed labeling experiments is denoted as $^M EXP$. If the 1° and 2° hydrogens are independent of each other, the isotopic labeling of one should not affect the isotope effect of the other. This is denoted as the rule of geometrical mean (RGM [51]):

$$r = \frac{\ln(k_{Hi}/k_{HT})}{\ln(k_{Di}/k_{DT})} = 1 \tag{12.12}$$

where i is H or D. The RGM predicts that the isotopic label at the geminal position should not affect the $^M EXP$:

$$2° \ EXP = \frac{\ln(k_{HH}/k_{HT})}{\ln(k_{HD}/k_{HT})} = \frac{\ln(k_{HH}/k_{HT})}{\ln(k_{DD}/k_{DT})} = 2° \ ^M EXP \tag{12.13}$$

If the motions of the 1° and 2° hydrogens are coupled along the reaction coordinate a breakdown of the RGM will result in an inflated 2° $^M EXP$. The 1° KIE will have a secondary component, and will be deflated, but since the 2° H/D KIE is very small (~1.2), the expected deflation of the 1° $^M EXP$ is also very small. The 2° KIE on the other hand, will have a primary component and will be significantly inflated. Tunneling of the 1° H will induce a large 2° H/T KIE (k_{HH}/k_{HT}) relative to the more semiclassical 2° D/T KIE (k_{DD}/k_{DT}), due to the reduced effect of D-tunneling in the primary position. In the mixed labeling experiment, when there is coupled motion between the 1° and 2° hydrogens, tunneling along the reaction coordinate will result in the inflation of the 2° $^M EXP$ because H tunneling is more significant than D tunneling.

The $^M EXP$ is a product of the original Swain–Schaad EXP and RGM (r):

$$rEXP = \frac{\ln(k_{HH}/k_{HT})}{\ln(k_{DH}/k_{DT})} \frac{\ln(k_{DH}/k_{DT})}{\ln(k_{DD}/k_{DT})} = \frac{\ln(k_{HH}/k_{HT})}{\ln(k_{DD}/k_{DT})} = ^M EXP \tag{12.14}$$

A mathematically rigorous explanation of the high sensitivity of the mixed labeling experiment to H-tunneling can be found in Refs. [52, 53]. Both Huskey [53] and Saunders [54, 54] have shown independently that exceptionally large values of $^M EXP$ are only computed for 2° KIEs resulting from coupled motion and tunneling. They both concluded that the extra isotopic substitution is an essential feature of the experimental design. The mixed labeling experiment is also presented in similar terms in Chapter 10, Sections 10.3.3.2 and 10.3.3.3, where EXP is denoted S and r is R and $^M EXP$ is RS. Both presentations, are scientifically coherent and thus redundant, but they emphasize different aspects of the issue, so the reader may benefit from this redundancy.

12.3.3.2 Upper Semiclassical Limit for Secondary Swain–Schaad Relationship

For EXP as defined in Eq. (12.5), values smaller than its semiclassical lower limit can be explained by kinetic complexity and values larger than its upper limit serve as evidence of tunneling. Until recently, the upper semiclassical limit used was

3.34 [29, 56]. An upper limit that is more realistic and relevant to the commonly used mixed labeling experiment is calculated below [57]. This limit for *EXP* with no tunneling contribution was calculated using three different approaches: (i) ZPE and reduced mass considerations; (ii) vibrational analysis, and (iii) the effect of kinetic complexity. The results of these calculations suggest that for the mixed labeling method (k_{HH}/k_{HT} vs. k_{DD}/k_{DT}) an experimental 2° $^M EXP$ larger than 4.8 (within statistical experimental error) may serve as a reliable indication of H-tunneling [57]. In the case of experimental 2° $^M EXP$ between 3.3 and 4.8 additional evidence is needed to indicate H-tunneling. Such additional examination consists of simple analytical or numerical solutions such as those described above and, in more detail, in Ref. [57]. In accordance with our conclusion, *ab initio* calculations for many gas phase H-transfer reactions by Hirschi and Singleton [37] demonstrated that for sizable 2° KIEs (e.g., $k_H/k_T > 1.1$), only 2° EXP > 4 may indicate tunneling. Alternatively, a higher level of calculation for a specific enzymatic system could be employed, as discussed in Section 12.4.2. For ADHs, for example, several state of the art theoretical examinations have recently supported the tunneling contribution suggested by the inflated 2° $^M EXPs$ [58–61].

12.3.3.3 Experimental Examples Using 2° Swain–Schaad Exponents

To date, the only experimental examples where a 2° Swain–Schaad relationship resulted in a breakdown of semiclassical models and implicated tunneling and coupled motion were from studies of alcohol dehydrogenases (ADH). Furthermore, all these studies were conducted on the oxidation of the alternative substrate benzyl alcohol to aldehyde. The only attempt so far to conduct similar measurements used a very different system (DHFR). These experiments revealed no deviation from the semiclassical EXP [45]. Until such experiments are extended to other systems or at least extended to the reduction of aldehyde to alcohol for the same system, the generalization of their interpretation should be taken with some discretion. These examples are discussed in great detail in Chapter 10, Section 10.5.1.1, and only a concise summary of two seminal examples is presented below.

Horse Liver Alcohol Dehydrogenase (HLADH) Alcohol dehydrogenases (ADHs) catalyze the reversible oxidation of alcohols to aldehydes with NAD^+ as the oxidative reagent. HLADH has been extensively studied by means of 2° mixed labeling Swain–Schaad relationships [62–64]. Two interesting conclusions of these studies were that (i) For two mutants (F93T and F93T; V203G), a longer donor–acceptor distance (measured by X-ray crystallography) led to a smaller 2° exponent [64]; and (ii) for a series of mutants, a correlation exists between the catalytic efficiency (K_{cat}/K_M) and the 2° exponent [63]. These findings are in accordance with tunneling models in which the barrier width plays a critical role. These models also included a contribution of coupled motion and tunneling to catalysis.

Thermophilic ADH Secondary KIEs were measured using the mixed labeling pattern with thermophilic ADH from *Bacillus stearothermophilus* (*bs*ADH) at temperatures ranging from 5 to 65 °C. At the physiological temperature of this thermophilic ADH (~65 °C) inflated 2° Swain–Schaad exponents (~15) [65] indicated a signature of H-tunneling similar to that of the mesophilic yeast ADH at 25 °C [66]. At temperatures below 30 °C these exponents declined toward the semiclassical region and the enthalpy of activation increased significantly (14.6 to 23.6 kcal mol^{-1} for H-transfer and 15.1 to 31.4 kcal mol^{-1} for D-transfer). This phenomenon was interpreted as indicating a decreased tunneling contribution at reduced temperature due to different environmental sampling at high (physiological) and low temperatures. An alternative interpretation would result from using the environmentally coupled tunneling model. According to that model, at physiological temperature (30–65 °C) the pre-arrangement of the potential surface (the Marcus term) is close to perfect and no gating is needed. At reduced temperature (5–30 °C) the pre-arrangement is not so perfect, leading to gating, which modulates the donor–acceptor distance and results in temperature dependent KIEs. Interestingly, this effect is more pronounced for D-transfer than for H-transfer, possibly due to the higher sensitivity of D-tunneling to the distance between donor and acceptor. These findings were then correlated to the increased rigidity of the enzyme at lower temperatures [67, 68]. These studies suggested that similar enzymes that catalyze the same reaction at very different temperatures evolved to have similar rigidities in their respective physiological conditions and similar tunneling contributions to the H-transfer process. Interestingly, these results suggested possible relationships between protein rigidity and the degree of tunneling. Together with temperature dependence studies that are described below (Section 12.3.3.2), a model was suggested in which the enzyme's fluctuations are coupled to the reaction coordinate [34, 60, 65, 69].

12.3.4
Temperature Dependence of Primary KIEs

12.3.4.1 Temperature Dependence of Reaction Rates and KIEs

Traditional literature treats enzyme catalyzed reactions, including hydrogen transfer, in terms of transition state theory (TST) [4, 34, 70]. TST assumes that the reaction coordinate may be described by a free energy minimum (the reactant well) and a free energy maximum that is the saddle point leading to product. The distribution of states between the ground state (GS, at the minimum) and the transition state (TS, at the top of the barrier) is assumed to be an equilibrium process that follows the Boltzmann distribution. Consequently, the reaction's rate is exponentially dependent on the reciprocal absolute temperature ($1/T$) as reflected by the Arrhenius equation:

$$k = A \cdot e^{-(E_a/RT)} \tag{12.15}$$

where A is the Arrhenius pre-exponential factor, E_a is the activation energy and R is the gas constant. Since the KIE is the ratio of the reactions' rates, its temperature dependence will follow:

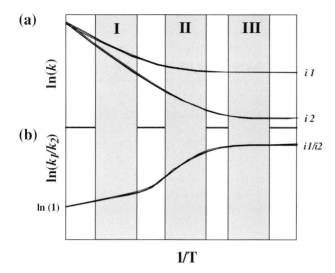

Figure 12.5. An Arrhenius plot of a hydrogen transfer that is consistent with a tunneling correction to transition state theory. (a) Arrhenius plot of a light isotope (i1) and heavy isotope (i2). (b) Arrhenius plot of their KIE (i1/i2). Highlighted are experimental temperature ranges for three regions: I, a system with no tunneling contribution, II, a system with moderate tunneling, and III, a system with extensive tunneling contribution. The dashed lines are the tangents to the plot at each region. This illustration is similar to several schemes we and others have suggested in the past [33, 34, 49, 108].

$$\frac{k_l}{k_h} = \frac{A_l}{A_h} e^{\Delta E_{a(h-l)}/RT} \qquad (12.16)$$

where h and l are the heavy and light isotopes, respectively. This equation is useful as long as the reaction is thermally activated. At low temperatures, the contribution of tunneling becomes significant, as no thermal energy is available for activation. This causes a curvature in the Arrhenius plot as illustrated in Fig. 12.5. Conventional tunneling, through a single, rigid barrier is temperature independent and may affect both the pre-exponential and the exponential factors.

This treatment is confined to tunneling correction to TST and is not valid for any Marcus-like model (e.g., environmentally coupled tunneling). Thus, an Arrhenius plot of KIEs can distinguish between data that might be fitted by a tunneling correction model (e.g., $A_l/A_h > 1$ with $\Delta E_a \sim 0$) and data that can only be fitted by a Marcus-like model (e.g., $A_l/A_h > 1$ with large ΔE_a).

12.3.4.2 KIEs on Arrhenius Activation Factors

Following Eq. (10.16), with no tunneling correction [19, 32] the KIE's temperature dependence will reflect the differences in the energy of activation for the two isotopes, and the KIE on the pre-exponential factors (A_l/A_h) should be close to unity (no-tunneling, region I in Fig. 12.5). Deviation from unity with no tunneling

Table 12.1. Semiclassical limits for the KIE on Arrhenius preexponential factors [19, 32, 74].

	A_H/A_T	A_H/A_D	A_D/A_T
Upper limit	1.7	1.4	1.2
Lower limit	0.3	0.5	0.7

seems to be confined to a limited range, as extensively discussed in the literature [33, 71–74]. These limits follow $\sqrt{\mu_h/\mu_l} > A_l/A_h > \mu_l/\mu_h$ where μ is the reduced mass. These limits for hydrogen KIEs are summarized in Table 12.1.

At a very low temperature, where only tunneling contributes significantly to rates, it is predicted that the KIEs will be very large (over six orders of magnitude [75]) and A_H/A_D will be much larger than unity (extensive tunneling, region III in Fig. 12.5). Between high and low temperature extremes, the Arrhenius plot of the KIEs will be curved, as the light isotope tunnels at a higher temperature than the heavy one. In this region, the Arrhenius slope will be very steep and A_H/A_D will be smaller than unity (moderate tunneling region II in Fig. 12.5). This has been deliberated in several previous reviews [33, 34] and has been used as a probe for tunneling in a wide variety of enzymatic systems (Table 12.2). According to this model, an A_H/A_D smaller than the semiclassical lower limit (Table 12.1) indicates tunneling of only the light isotope ("Moderate Tunneling Region" [34]). Whereas an A_H/A_D larger than unity indicates tunneling of both isotopes ("Extensive Tunneling Region" [34]). Table 12.2 summarizes several reports of H-tunneling based on pre-exponential Arrhenius factors that were outside the semiclassical range (Table 12.1). Several experimental A_H/A_Ds, that do not match the criteria set by the above model are discussed in Section 12.3.3 and alternative models are presented in Section 12.4.

12.3.4.3 Experimental Examples Using Isotope Effects on Arrhenius Activation Factors

Soybean Lipoxygenase-1 (SBL-1) Lipoxygenases catalyze the oxidation of linoleic acid (LA) to 13-(S)-hydroperoxy-9,11-(Z,E)-octadecadienoic acid (13-(S)-HPOD) [76]. This reaction proceeds via an initial, rate-limiting abstraction of the pro-S hydrogen radical from C11 of LA by the Fe^{3+}-OH cofactor, forming a substrate-derived radical intermediate and Fe^{2+}-OH$_2$. Molecular oxygen rapidly reacts with this radical, eventually forming 13-(S)-HPOD and regenerating a resting enzyme. The abstraction of H or D from the pro-S C11 position of LA by the wild type SBL-1 has very large KIEs (~80) and large A_H/A_D (~20) [76–78], which would suggest it fits region III in Fig. 12.5 (extensive tunneling). Yet, its KIEs are "only" around 80, while the above model would predict much larger KIEs [75]. For the wild type

Table 12.2. Enzymatic systems for which tunneling was suggested by the temperature dependences.

Enzyme	k_H/k_D	A_H/A_D	Ref.
Soybean lipoxygenase, wt.	82	18	76
Soybean lipoxygenase, mutants	93–112	4–0.12	76
Methane monooxygenase	50–100		98
Galactose oxidase	16	0.25	99
Methylamine dehydrogenase	17	13	100
Methylamine dehydrogenase (TTQ-dependent)	12.9	9.0	101
Trimethylamine dehydrogenase	4.6	7.8	102
Sarcosine Oxidase	7.3	5.8	103
Methyl Malonyl CoA mutase	36	0.08	104
Acyl CoA desaturase	23	2.2	105
Peptidylglycine α-hydroxylating monooxygenase	10	5.9	42

Enzyme	k_H/k_T	A_H/A_T	Ref.
Bovine serum amine oxidase	35	0.12	106
Monoamine oxidase	22	0.13	107
Thymidylate synthase	7	7	48
Dihydrofolate reductase	6	6	45

SBL-1, the E_a for the H-transfer was small ($\sim$2 kcal mol^{-1}) and the ΔE_a was $\sim$1 kcal mol^{-1}. Several mutants of SBL-1 were also studied and exhibited Arrhenius plots that range between regions II to III in Fig. 12.5 (see Chapter 10).

Thermophilic ADH (ADH-hT) Another example for studies of temperature independent KIEs is taken from our work with thermophilic ADH from *Bacillus stereothermophilus* and is demonstrated in Fig. 12.6 [65, 67, 68]. Under physiological conditions (30–65 °C), this enzyme had A_H/A_T and A_D/A_T larger than the semiclassical limits. However, its KIEs were relatively small ($\sim$3) and the enthalpy of activation for H and D was rather large (14.6 and 15.1 kcal mol^{-1}, respectively). As discussed in Section 12.4.1, this can be explained by a "Marcus-like" model in which the temperature dependences of the reaction rate and of the KIE are separated. Below 30 °C, both isotopes had a much larger energy of activation and large temperature dependence of the KIEs. This result was interpreted as "activity phase transition" due to increased rigidity of this thermophilic enzyme at reduced temperatures [65, 67, 68]. Using Marcus-like models, the low temperature behavior could be rationalized by imperfect pre-organization, more gating, or alternatively, using tunneling correction, the data would fit region II in Fig. 12.5.

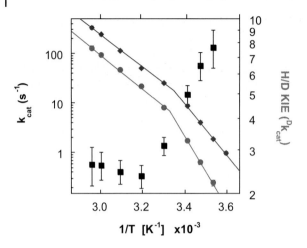

Figure 12.6. A thermophilic ADH (*bs*ADH) with benzyl alcohol (●), [7-^{2}H$_2$] benzyl alcohol (■) and their KIEs (◆). Reproduced using data published in Ref. [65].

Dihydrofolate Reductase (DFHR) Measurements and calculations of intrinsic 1° KIEs with *ec*DHFR are described in Section 12.3.2.2.2 [45]. The commitment was temperature dependent and so were the observed KIEs. Nevertheless, the calculated intrinsic KIEs were temperature independent with $A_H/A_T = 7.2 \pm 3.5$, which served as evidence of H-tunneling [45]. The energy of activation was measured by steady-state [45] and pre-steady state [79] kinetics and was found to be 4 ± 1 kcal mol^{-1}. Taken together with the large A_H/A_T, these findings were in accordance with "Marcus-like" models. Currently, attempts to reproduce these data using QM/MM calculations are being made [80] (see Section 12.4.1). Although this kind of simulation does not use phenomenological models (e.g., Marcus-like models) all the pre-, re-organization and so-called "gating effects" are embedded in the calculations. Such simulation may identify the specific motions that might be coupled to the H-transfer event, and may indicate protein normal modes that are coherent, or that otherwise affect the chemical transformation.

Thymidylate Synthase Measurements and calculations of intrinsic 1° KIEs with *ec*TS are described in Section 12.3.2.2.3 [48]. These intrinsic KIEs were temperature independent with H/T KIEs close to 7 and $A_H/A_T = 6.8 \pm 2.8$. These results served as evidence for QM tunneling, and together with the reaction's energy of activation ($E_a = 4.0 \pm 0.1$ kcal mol^{-1}) suggested a model in which the temperature dependence of the rate results from the reorganization of the system (isotopically insensitive), and an isotopically sensitive H-tunneling step that is temperature independent. In this specific system, since the intrinsic and observed KIES were close (small kinetic complexity), a similar qualitative conclusion would have been reached from the temperature dependence of the observed KIEs.

In addition to the examples mentioned here, several studies by Sutcliffe, Scrutton and coworkers [81–83] have also resulted in temperature independent KIEs, with large A_H/A_D. These works are described in detail in Chapter 13. These systems had enthalpies of activation much larger than the semiclassical, rigid model prediction. As discussed in Section 12.4.1, such findings have led to many theoretical models attempting to explain the experimental results.

It must be emphasized that the semiclassical limits for the energy of activation (the slope of the Arrhenius plot) are not well defined. Consequently, in order to establish that nonclassical features are evident from temperature independent KIEs, the pre-exponential Arrhenius factor must be outside their semiclassical limits. For example, a recent paper misinterpreted "nearly temperature independent" KIEs with A_H/A_D close to unity as "Evidence for environmentally coupled hydrogen tunneling during dihydrofolate reductase catalysis" [84]. Actually, the temperature dependence of the KIEs in that study ($\Delta E_a = 3.0 \pm 0.7$ kcal mol^{-1} above 20 °C) was well within the semiclassical range.

Over the years, TST has been modified and corrected for kinetic effects of tunneling, barrier recrossing and medium viscosity, yet, developing a theory that will explain such a phenomenon is an on-going challenge. The next section describes attempts to lay a general foundation for such a theory.

12.4
Theoretical Models for H-transfer and Dynamic Effects in Enzymes

Most of the studies described above could not be rationalized without invoking contributions from quantum mechanical tunneling and dynamic effects. This conclusion was based on deviations from semiclassical theory that exclude such phenomena. The following section presents attempts to explain those findings using models that were constructed from first principles and that include tunneling and dynamic effects. In the light of the above sections and specifically Section 12.2, it is important to note that all the theoretical treatments presented below assume a single step H-transfer phenomenon. Most of these treatments focus on the transition state of the chemical transformation catalyzed by an enzyme. Since most experimental data represent a more complex system the comparison between the calculations and their experimental counterpart has to be conducted with great care. An additional challenge when comparing theoretical to experimental results is that the experimental data carry an error that can be evaluated by standard statistical methods, while the theoretical results rarely address the accuracy of the calculated values. This having been said, it is well recognized that the only way to interpret the experimental findings on a molecular and energetic level is with a complementary theory, and that a theory that cannot be evaluated by relevant experimental data is rarely meaningful. The two approaches described below present several attempts to explain various experimental findings that could not be rationalized by semiclassical theory or classical phenomenological rate theories (e.g., transition state theory [70]).

12.4.1
Phenomenological "Marcus-like Models"

Models using a uni-dimensional (1D) rigid potential surface that attempted to reproduce temperature independent KIEs required the isotopically sensitive step to have little or no enthalpy of activation (e.g., H-transfer via QM tunneling). In the case of reactions with significant enthalpy of activation (e.g., Section 12.3. 4.3.2) an additional dimension has to be introduced. The temperature dependence of the reaction results from classically activated rearrangements of the potential surface prior to the H-transfer event. Several different models, which were constructed from very different basic principles, are successful because they separate the temperature dependence of the reaction's rate from that of the KIEs [60, 69, 77, 83, 85]. Although these models use different terminology, their common theme involves two requirements for efficient tunnelling: degeneracy of the reactant and product energy levels, and narrow barrier width. These models were developed in an effort to rationalize H-transfer in the condensed phase and particularly in enzymes (e.g., Burgis and Hynes [86, 87], Kuznetzov and Ulstrup [88], Knapp and Klinman (Chapter 10), Benkovic and Bruce [7], Warshel [11, 89], and Schwartz [69]). These models resemble in part the approach of the Marcus theory [90], but with an additional term that accounts for the temperature dependence of the KIEs. Since, in contrast to electron transfer, H-transfer is very sensitive to the donor–acceptor distance, the additional term accounts for fluctuations of that distance (coordinate q in Fig. 12.7). Two common features of these models are the direct effects of the potential surface fluctuations on the reaction rate, and separation of the temperature dependence of the rate and the KIE. An example of a "Marcus-like" model is illustrated in Fig. 12.7.

Environmentally coupled hydrogen tunneling models can accommodate the composite kinetic data for WT-SLO and its mutants [76, 77]. This model is described in more detail in Chapter 10 and was based on the model proposed by Kuznetsov and Ulstrup [88]. In this model, the rate for H• transfer is governed by an isotope-independent term (const.), a Marcus-like term, and a "gating" term (the F.C. Term in Eq. (12.15)). In Eq. (12.15), the Marcus term relates λ, the reorganization energy, to $\Delta G°$, the driving force for the reaction, where R and T are the gas constant and absolute temperature, respectively. This term has a weak isotopic dependence that arises when tunneling takes place from vibrationally excited states. The dominant isotopically sensitive term is the Franck–Condon nuclear overlap along the hydrogen coordinate (F.C. term), which is the weighted hydrogen tunneling probability. This term arises from the overlap between the initial and the final states of the hydrogen's wavefunction and, consequently, depends on the thermal population of each vibration level. The F.C. term is also expected to be affected by the donor–acceptor distance, which is both temperature and isotope dependent. When distance sampling, or gating, occurs, the KIE can become very temperature dependent. The temperature dependence of KIEs arises from the thermal population of excited vibration levels. This model was developed for nonadiabatic radical-transfer (H•) reactions and the full scope of its applications is yet to be explored.

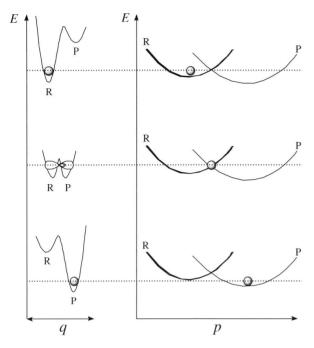

Figure 12.7. Illustration of "Marcus-like" models: energy surface of environmentally coupled hydrogen tunneling. Two orthogonal coordinates are presented: *p*, the environmental energy parabolas for the reactant state (*R*) and the product state (*P*); and *q*, the H-transfer potential surface at each *p* configuration. The gray shapes represent the populated states (e.g., the location of the particle). The original Marcus expression would have fixed *q* distance between donor and acceptor. By adding fluctuations of that distance (see gating by Knapp et al., Chapter 10) a temperature dependence of the KIE can be achieved. For three alternative graphic illustrations of such models see Refs. [33, 34, 109].

$$k = (\text{const.})e^{-(\Delta G^\circ + \lambda)^2/(4\lambda RT)} \cdot (\text{F.C. term}) \qquad (12.15)$$

Do such models suggest that the protein dynamics or the environmental dynamics enhance the reaction rates and maybe contribute to catalysis? The fact that studies of enzymes at this level are so interdisciplinary can result in misunderstandings and disagreements between disciplines. As mentioned under "Definitions", most biochemists consider any motion of the protein or the enzymatic complex to be "dynamics". Most physical chemists on the other hand, will use that term only for motions along the reaction coordinate that are not in thermal (Boltzmann) equilibrium with their environment [14]. By their nomenclature, fluctuations that are in thermal equilibrium (like environmental rearrangement, gating motion, etc.) do not constitute a dynamical effect. For example, in Refs. [33, 34] we suggested models in which dynamic rearrangement of the reaction's potential surface plays a key role in the enhancement of the reaction's rate (illustrated in

Fig. 5 in Ref. [34]). Vilá and Warshel [11], offer a similar graphic presentation (Fig. 2 in Ref. [11]) to argue against such a role. Apparently, differences in terminology may lead to contradictory wording in the conclusions. To date, the temperature dependence of rates and KIEs is still a major challenge for explicit theoretical models or simulations and (to the best of our knowledge) only three studies have reproduced such phenomenon [60, 80, 91].

12.4.2
MM/QM Models and Simulations

Recently, several computational studies and molecular simulations have been conducted in an attempt to reproduce and explain experimental findings such as the breakdown of the Swain–Schaad relationship and the nonclassical temperature dependence of KIEs. These studies employed a molecular mechanics (MM) based simulation of the exterior of a protein with high level *ab initio* calculations along the reaction coordinate and in the vicinity of the reacting atoms. Various methods were used to "buffer" the transition between these two regions. The general name for this kind of calculation is Molecular-Mechanics/Quantum-Mechanics (MM/QM). Most of these studies investigated enzymatic systems for which ample kinetic, structural and other data were available (e.g., TIM [12, 92], carbonic anhydrase [93], ADH [11, 59, 61], LDH [16], methylamine dehydrogenase [94–96], SBL-1 [6, 91] and DHFR [46]). These state of the art simulations were able to reproduce rates and H/D KIEs but had little success in addressing temperature dependences (with the exception of Ref. [91]) and secondary KIEs (with the exception of Refs. [59, 97]). High level calculations of this kind are of critical importance as there is no other way to bring together all the molecular and kinetic data. While direct molecular dynamic simulations are limited to the nanosecond range, free energy perturbation/umbrella sampling calculations [5, 46] allow one to explore the effect of a much larger conformational space by forcing the system to move to different regions upon changing the charge distribution of the reacting system. Currently, the main limitations of QM/MM models are: (i) simulation of temperature dependence is not trivial in most existing methods; (ii) the limited time scale of the simulation; and (iii) the inherent compromise between accuracy (high level of theory) and conformational flexibility (large conformational space). This prohibits the coverage of a more substantial range of motion in the duration of an entire catalytic cycle while investigating quantum mechanical phenomena such as tunneling.

12.5
Concluding Comments

This chapter presents models for H-transfer that are relevant to enzymatic systems, and introduces experimental probes and case studies that have attempted to address this issue in recent years. The chapter focuses on C–H bond activation, which requires significant rate enhancement by the catalyst. Hydrogen kinetic isotope effects, their temperature dependence, and the internal relationships between them, are presented as tools in studying the nature of H-transfer. Studies of H-transfer in enzymatic systems provide a unique opportunity to better understand H-transfer in the condensed phase in general, since the enzyme inhibits most of the side reactions that would take place in solution. Only one stereoselective and specific reaction takes place and can be studied in detail.

Apparently, hydrogen transfer in the condensed phase is a complex phenomenon that includes not only the three atoms obviously involved (donor, hydrogen and acceptor) but many other atoms that constitute the environment of that chemical transformation. Part of this environment is in the immediate vicinity of the reacting atoms. Additionally, in an organized medium such as an enzyme, remote residues may also be coupled to the reaction coordinate. A major contemporary question in enzymology is whether the ability of a protein to serve as the reaction's environment not only electrostatistically stabilizes the reaction's transition state, but also dynamically enhances the reaction's rate. This can occur by statistical re- or pre-arrangement along the reaction path. Such rearrangement can be more efficient in the enzyme than in the uncatalyzed reaction. Several open questions are presented in this chapter, most of which focus on the issue of whether enzymes evolved to better rearrange the reaction's environment, its electrostatics and relevant spins, to enhance the reaction's rate relative to that in solution. Finally, it is suggested that close interaction between theory (calculations and simulations) and experiments is crucial to studying H-transfer in complex systems such as enzymes. The ability of theoreticians and experimentalists to communicate and produce data of relevance and use to each other is most likely to lead to a better understanding of enzyme catalysis and H-transfer phenomena.

Acknowledgments

I thank Judith Klinman for fruitful discussions and the NSF (CHE 01-33117) and the NIH (GM065368) for financial support.

References

1 PAULING, L. (1948) The nature of forces between large molecules of biological interest, *Nature 161*, 707–709.

2 JENCKS, W. P. (1987) *Catalysis in Chemistry and Enzymology*, Dover Publications, New York.

3 SCHOWEN, R. L. (1978) Catalytic power and transition-state stabilization, in *Transition States of Biochemical Processes*, GANDOUR, R. D., SCHOWEN, R. L. (Eds.), Plenum Press, New York.

4 FERSHT, A. (1998) *Structure and Mechanism in Protein Sciences: A Guide to Enzyme Catalysis and Protein Folding*, W. H. Freeman, New York.

5 WARSHEL, A., PARSON, W. W. (2001) Dynamics of biochemical and biophysical reactions: insight from

computer simulations, *Quart. Rev. Biophys.* **34**, 563–679.

6 OLSSON, M. H. M., SIEGBAHN, P. E. M., WARSHEL, A. (2003) Simulating large nuclear quantum mechanical corrections in hydrogen atom transfer reactions in metalloenzymes, *J. Biol. Inorg. Chem.* **9**, 96–99.

7 BRUICE, T. C., BENKOVIC, S. J. (2000) Chemical basis for enzyme catalysis, *Biochemistry* **39**, 6267–6273.

8 NORTHROP, D. B. (1998) On the meaning of Km and Vmax/Km in enzyme kinetics, *J. Chem. Ed.* **75**, 1153–1157.

9 STEINFELD, J. I., FRANCISCO, J. S., HASE, W. L. (1998) *Chemical Kinetics and Dynamics*, 2nd edn., Prentice Hall, Upper Saddle River, N.J.

10 MULHOLLAND, A. J., KARPLUS, M. (1996) Simulations of enzymic reactions, *Biochem. Soc. Trans.* **24**, 247–54.

11 VILLA, J., WARSHEL, A. (2001) Energetics and dynamics of enzymatic reactions, *J. Phys. Chem. B* **105**, 7887–7907.

12 CUI, Q., KARPLUS, M. (2002) Promoting modes and demoting modes in enzyme-catalyzed proton transfer reactions: A study of models and realistic systems, *J. Phys. Chem. B* **106**, 7927–7947.

13 WILSON, E. K. (2000) Enzyme dynamics, *Chem. Eng. News* **78**, July 17, 42–45.

14 BENKOVIC, S. J., HAMMES-SCHIFFER, S. (2003) A perspective on enzyme catalysis, *Science* **301**, 1196–1202.

15 GARCIA-VILOCA, M., GAO, J., KARPLUS, M., TRUHLAR, D. G. (2003) How enzymes work: Analysis by modern rate theory and computer simulations, *Science* **303**, 186–195.

16 HWANG, J. K., CHU, Z. T., YADAV, A., WARSHEL, A. (1991) Simulations of quantum mechanical corrections for rate constants of hydride-transfer reactions in enzymes and solutions, *J. Phys. Chem.* **95**, 8445–8448.

17 OREF. I. (1998) Selective chemistry redux, *Science* **279**, 820–821.

18 DIAU, E. W. G., HEREK, J. L., KIM, Z. H., ZEWAIL, A. H. (1998) Femtosecond activation of reactions and the concept of nonergodic molecules, *Science* **279**, 847–851.

19 MELANDER, L., SAUNDERS, W. H. (1987) *Reaction Rates of Isotopic Molecules*, Krieger, R.E., Malabar, FL.

20 BIGELEISEN, J., MAYER, M. G. (1947) Calculation of equilibrium constants for isotopic exchange reactions, *J. Chem. Phys.* **15**, 261–267.

21 BIGELEISEN, J., WOLFSBERG, M. (1958) Theoretical and experimental aspects of isotope effects in chemical kinetics, *Adv. Chem. Phys.* **1**, 15–76.

22 KIEFER, P. M., HYNES, J. T. (2003) Kinetic isotope effects for adiabatic proton transfer reactions in a polar environment, *J. Phys. Chem. A* **107**, 9022–9039.

23 CLELAND, W. W. (1991) Multiple isotope effects in enzyme-catalyzed reactions, in *Enzyme Mechanism from Isotope Effects*, COOK, P. F. (Ed.), pp. 247–268, CRC Press, Boca Raton, Fl.

24 NORTHROP, D. B. (1991) Intrinsic isotope effects in enzyme catalyzed reactions, in *Enzyme Mechanism from IsotopeEeffects*, COOK, P. F. (Ed.), pp. 181–202, CRC Press, Boca Raton, Fl.

25 SWAIN, C. G., STIVERS, E. C., REUWER, J. F., SCHAAD, L. J. (1958) Use of hydrogen isotope effects to identify the attacking nucleophile in enolization of ketones catalyzed by acetic acid, *J. Am. Chem. Soc.* **80**, 5885–5893.

26 BIGELEISEN, J. (1962) in *International Atomic Energy Agency*, pp. 161, Gerin, P., Vienna.

27 STERN, M. J., VOGEL, P. C. (1971) Relative tritium-deuterium isotope effects in the absence of large tunneling factors, *J. Am. Chem. Soc.* **93**, 4664–4675.

28 STERN, M. J., WESTON, R. E. J. (1974) Phenomenological manifestations of quantum-mechanical tunneling. III. Effect on relative tritium-deuterium kinetic isotope effects, *J. Chem. Phys.* **60**, 2815–2821.

29 SAUNDERS, W. H. (1985) Calculations of isotope effects in elimination reactions. New experimental criteria for tunneling in slow proton transfers, *J. Am. Chem. Soc.* **107**, 164–169.

30 AMIN, M., PRICE, R. C., SAUNDERS, W. H. (1990) Tunneling in Elimination Reactions – Tests of Criteria For Tunneling Predicted by Model Calculations, *J. Am. Chem. Soc.* 112, 4467–4471.

31 BAHNSON, B. J., KLINMAN, J. P. (1995) Hydrogen tunneling in enzyme catalysis, in *Enzyme Kinetics and Mechanism*, pp. 373–397, Academic Press, San Diego.

32 BELL, R. P. (1980) *The Tunnel Effect in Chemistry*, Chapman & Hall, London & New York.

33 KOHEN, A., KLINMAN, J. P. (1998) Enzyme catalysis: beyond classical paradigms, *Acc. Chem. Res.* 31, 397–404.

34 KOHEN, A., KLINMAN, J. P. (1999) Hydrogen tunneling in biology, *Chem. Biol.* 6, R191–198.

35 TAUTERMANN, C. S., LOFERER, M. J., VOEGELE, A. F., LIEDLA, K. R. (2004) Double hydrogen tunneling revisited: The breakdown of experimental tunneling criteria, *J. Chem. Phys.* 120, 11650–11657.

36 KIEFER, P. M., HYNES, J. T. (2006) Interpretation of primary kinetic isotope effects for adiabatic and nonadiabatic proton transfer reactions in a polar environment, in *Isotope Effects in Chemistry and Biology*, KOHEN, A., LIMBACH, H. H. (Eds.), pp. 549–578, Taylor & Francis, LLC CRC, Boca Raton, FL.

37 HIRSCHI, J., SINGLETON, D. A. (2005) The normal range for secondary Swain-Schaad exponents without tunneling or kinetic complexity, *J. Am. Chem. Soc.* 127, 3294–3295.

38 GRANT, K. L., KLINMAN, J. P. (1992) Exponential relationship among multiple hydrogen isotope effects as probes of hydrogen tunneling, *Bioorg. Chem.* 20, 1–7.

39 NORTHROP, D. B. (1977) Determining the absolute magnitude of hydrogen isotope effects, in *Isotope Effects on Enzyme-catalyzed Reactions*, CLELAND, W. W., O'LEARY, M. H., NORTHROP, D. B. (Eds.), pp. 122–152, University Park Press, Baltimore, MD.

40 CLELAND, W. W. (1977) *Adv. Enzymol.* 45, 273–303.

41 FRANCISCO, W. A., MERKLER, D. J., BLACKBURN, N. J., KLINMAN, J. P. (1998) Kinetic mechanism and intrinsic isotope effects for the peptidylglycine α-amidating enzyme reaction, *Biochemistry* 37, 8244–8252.

42 FRANCISCO, W. A., KNAPP, M. J., BLACKBURN, N. J., KLINMAN, J. P. (2002) Hydrogen tunneling in peptidylglycine-hydroxylating monooxygenase, *J. Am. Chem. Soc.* 124, 8194–8195.

43 FIERKE, C. A., JOHNSON, K. A., BENKOVIC, S. J. (1987) Construction and evaluation of the kinetic scheme associated with dihydrofolate reductase from *Escherichia coli*, *Biochemistry* 26, 4085–4092.

44 RAJAGOPALAN, P. T. R., LUTZ, S., BENKOVIC, S. J. (2002) Coupling interactions of distal Residues enhance dihydrofolate reductase catalysis: mutational effects on hydride transfer rates, *Biochemistry* 41, 12618–12628.

45 SIKORSKI, R. S., WANG, L., MARKHAM, K. A., RAJAGOPALAN, P. T. R., BENKOVIC, S. J., KOHEN, A. (2004) Tunneling and coupled motion in the *E. coli* dihydrofolate reductase catalyzed reaction, *J. Am. Chem. Soc.* 126, 4778–4779.

46 AGARWAL, P. K., BILLETER, S. R., HAMMES-SCHIFFER, S. (2002) Nuclear quantum effects and enzyme dynamics in dihydrofolate reductasecatalysis, *J. Phys. Chem. B* 106, 3283–3293.

47 CARRERAS, C. W., SANTI, D. V. (1995) The catalytic mechanism and structure of thymidylate synthase, *Annu. Rev. Biochem.* 64, 721–762.

48 AGRAWAL, N., HONG, B., MIHAI, C., KOHEN, A. (2004) Vibrationally enhanced hydrogen tunneling in the *E. coli* thymidylate synthase catalyzed reaction, *Biochemistry* 43, 1998–2006.

49 KOHEN, A. (2003) Kinetic isotope effects as probes for hydrogen tunneling, coupled motion and dynamics contributions to enzyme catalysis, *Progr. React. Kinet. Mech.* 28, 119–156.

50 SPENCER, H. T., VILLAFRANCA, J. E.,

APPLEMAN, J. R. (1997) Kinetic scheme for thymidylate synthase from Escherichia coli: determination from measurements of ligand binding, primary and secondary isotope effects, and pre-steady-state catalysis, *Biochemistry* 36, 4212–4222.

51 BIGELEISEN, J. (1955) The rule of the geometric mean, *J. Chem. Phys.* 23, 2264–2267.

52 HUSKEY, W. P. (1991) Origin and interpretations of heavy-atom isotope effects, in *Enzyme Mechanism from Isotope Effects*, COOK, P. F. (Ed.), pp. 37–72, CRC Press, Boca Raton, Fl.

53 HUSKEY, W. P. (1991) Origin of apparent Swain-Schaad deviations in criteria for tunneling, *J. Phys. Org. Chem.* 4, 361–366.

54 SAUNDERS, W. H. (1992) The contribution of tunneling to secondary isotope effects, *Croat. Chem. Acta.* 65, 505–515.

55 LIN, S., SAUNDERS, W. H. (1994) Tunneling in elimination reactions – structural effects on the secondary beta-tritium isotope effect, *J. Am. Chem. Soc.* 116, 6107–6110.

56 STREITWIESER, A., JAGOW, R. H., FAHEY, R. C., SUZUKI, F. (1958) Kinetic isotope effects in the acetolyses of deuterated cyclopentyl tosylates, *J. Am. Chem. Soc.* 80, 2326–2332.

57 KOHEN, A., JENSEN, J. H. (2002) Boundary conditions for the Swain-Schaad relationship as a criterion for hydrogen tunneling, *J. Am. Chem. Soc.* 124, 3858–3864.

58 AGARWAL, P. K., WEBB, S. P., HAMMES-SCHIFFER, S. (2000) Computational studies of the mechanism for proton and hydride transfer in liver alcohol dehydrogenase, *J. Am. Chem. Soc.* 122, 4803–4812.

59 ALHAMBRA, C., CORCHADO, J. C., SÁNCHEZ, M. L., GAO, J., TRUHLAR, D. J. (2000) Quantum dynamics of hydride transfer in enzyme catalysis, *J. Am. Chem. Soc.* 122, 8197–8203.

60 ANTONIOU, D., SCHWARTZ, S. D. (2001) Internal enzyme motions as a source of catalytic activity: rate-promoting vibrations and hydrogen tunneling, *J. Phys. Chem. B* 105, 5553–5558.

61 CUI, Q., ELSTNER, M., KARPLUS, M. (2002) A theoretical analysis of the proton and hydride transfer in liver alcohol dehydrogenase (LADH), *J. Phys. Chem. B* 106, 2721–2740.

62 BAHNSON, B. J., PARK, D. H., KIM, K., PLAPP, B. V., KLINMAN, J. P. (1993) Unmasking of hydrogen tunneling in the horse liver alcohol dehydrogenase reaction by site-directed mutagenesis, *Biochemistry* 32, 5503–5507.

63 BAHNSON, B. J., COLBY, T. D., CHIN, J. K., GOLDSTEIN, B. M., KLINMAN, J. P. (1997) A link between protein structure and enzyme catalyzed hydrogen tunneling, *Proc. Nat. Acad. Sci. U.S.A.* 94, 12797–12802.

64 COLBY, T. D., BAHNSON, B. J., CHIN, J. K., KLINMAN, J. P., GOLDSTEIN, B. M. (1998) Active site modifications in a double mutant of liver alcohol dehydrogenase: structural studies of two enzyme-ligand complexes, *Biochemistry* 37, 9295–9304.

65 KOHEN, A., CANNIO, R., BARTOLUCCI, S., KLINMAN, J. P. (1999) Enzyme dynamics and hydrogen tunneling in a thermophilic alcohol dehydrogenase, *Nature* 399, 496–499.

66 CHA, Y., MURRAY, C. J., KLINMAN, J. P. (1989) Hydrogen tunneling in enzyme reactions, *Science* 243, 1325–1330.

67 KOHEN, A., KLINMAN, J. P. (2000) Protein flexibility correlate with degree of hydrogen tunneling in thermophilic and mesophilic alcohol dehydrogenase, *J. Am. Chem. Soc.* 122, 10738–10739.

68 LIANG, Z. X., LEE, T., RESING, K. A., AHN, N. G., KLINMAN, J. P. (2004) Thermal activated protein mobility and its correlation with catalysis in thermophilic alcohol dehydrogenase, *Proc. Nat. Acad. Sci. U.S.A.* 101, 9556–9561.

69 ANTONIOU, D., CARATZOULAS, S., KALYANARAMAN, C., MINCER, J. S., SCHWARTZ, S. D. (2002) Barrier passage and protein dynamics in enzymatically catalyzed reactions, *Eur. J. Biochem.* 269, 3103–3112 and many cited therein.

70 Kraut, J. (1988) How do enzymes work?, *Science* 242, 533–540.
71 Stern, M. J., Schneider, M. E., Vogel, P. C. (1971) Low-magnitude, pure-primary, hydrogen kinetic isotope effects, *J. Chem. Phys.* 55, 4286–4289.
72 Vogel, P. C., Stern, M. J. (1971) Temperature dependences of kinetic isotope effects, *J. Chem. Phys.* 52, 779–796.
73 Schneider, M. E., Stern, M. J. (1972) Arrhenius preexponential factors for primary hydrogen kinetic isotope effects, *J. Am. Chem. Soc.* 94, 1517–1522.
74 Stern, M. J., Weston, R. E. J. (1974) Phenomenological manifestations of quantum-mechanical tunneling. II. Effect on Arrhenius pre-exponential factors for primary hydrogen kinetic isotope effects, *J. Chem. Phys.* 60, 2808–2814.
75 Moiseyev, N., Rucker, J., Glickman, M. H. (1997) Reduction of ferric iron could drive hydrogen tunneling in lipoxygenase catalysis: Implications for enzymatic and chemical mechanisms, *J. Am. Chem. Soc.* 119, 3853–3860.
76 Knapp, M. J., Rickert, K., Klinman, J. P. (2002) Temperature-dependent isotope effects in soybean lipoxygenase-1: Correlating hydrogen tunneling with protein dynamics, *J. Am. Chem. Soc.* 124, 3865–3874.
77 Knapp, M. J., Klinman, J. P. (2002) Environmentally coupled hydrogen tunneling. Linking catalysis to dynamics, *Eur. J. Biochem.* 269, 3113–3121.
78 Glickman, M. H., Wiseman, J. S., Klinman, J. P. (1994) Extremely large isotope effects in the soybean lipoxygenase-linoleic acid reaction, *J. Am. Chem. Soc.* 116, 793–794.
79 Maglia, G., Javed, M. H., Allemann, R. K. (2003) Hydride transfer during catalysis by dihydrofolate reductase from *Thermotoga maritima*, *Biochem. J.* 374, 529–535.
80 Pu, J., Ma, S., Gao, J., Truhlar, D. G. (2005) Small temperature dependence of the kinetic isotope effect for the hydride transfer reaction catalyzed by Escherichia coli dihydrofolate reductase, *J. Phys. Chem. B.* 109, 8551–8556.
81 Scrutton, N. S., Basran, J., Sutcliffe, M. J. (1999) New insights into enzyme catalysis. Ground state tunnelling driven by protein dynamics, *Eur. J. Biochem.* 264, 666–671.
82 Scrutton, N. S. (1999) Enzymes in the quantum world, *Biochem. Soc. Trans.* 27, 767–779.
83 Sutcliffe, M. J., Scrutton, N. S. (2002) A new conceptual framework for enzyme catalysis. Hydrogen tunneling coupled to enzyme dynamics in flavoprotein and quinoprotein enzymes, *Eur. J. Biochem.* 269, 3096–3102 and many references cited therein.
84 Maglia, G., Allemann, R. K. (2003) Evidence for environmentally coupled hydrogen tunneling during dihydrofolate reductase catalysis, *J. Am. Chem. Soc.* 125, 13372–13373.
85 Schowen, R. L. (2002) Hydrogen tunneling. Goodbye to all that, *Eur. J. Biochem.* 269, 3095.
86 Borgis, D., Hynes, J. T. (1991) Molecular-Dynamics Simulation For a Model Nonadiabatic Proton Transfer Reaction in Solution, *J. Chem. Phys.* 94, 3619–3628.
87 Borgis, D., Hynes, J. T. (1993) Dynamical theory of proton tunneling transfer rates in solution – general formulation, *Chem. Phys.* 170, 315–346.
88 Kuznetsov, A. M., Ulstrup, J. (1999) Proton and hydrogen atom tunneling in hydrolytic and redox enzyme catalysis, *Can. J. Chem.* 77, 1085–1096.
89 Warshel, A. (1984) *Proc. Natl. Acad. Sci. USA* 81, 444–448.
90 Marcus, R. A., Sutin, N. (1985) Electron transfer in chemistry and biology, *Biochem. Biophys. Acta* 811, 265–322.
91 Hatcher, E., Soudackov, A. V., Hammes-Schiffer, S. (2004) Proton-coupled electron transfer in soybean lipoxygenase, *J. Am. Chem. Soc.* 126, 5763–5775.
92 Cui, Q., Karplus, M. (2001) Triosephosphate isomerase: A theoretical comparison of alternative

pathways, *J. Am. Chem. Soc.* 123, 2284–2290.

93 HWANG, J.-K., WARSHEL, A. (1996) How important are quantum mechanical nuclear motions in enzyme catalysis?, *J. Am. Chem. Soc.* 118, 11745–11751.

94 ALHAMBRA, C., SÁNCHEZ, M. L., CORCHADO, J. C., GAO, J., TRUHLAR, D. J. (2001) Quantum mechanical tunneling in methylamine dehydrogenase, *Chem. Phys. Lett.* 347, 512–518.

95 TRUHLAR, D. G., GAO, J., ALHAMBRA, C., GARCIA-VILCOA, M., CORCHADO, J. C., NCHEZ, M. L. S., VILLA, J. (2002) The incorporation of quantum effects in enzyme kinetics modeling, *Acc. Chem. Res.* 35, 341–349.

96 FAULDER, P. F., TRESADERN, G., CHOHAN, K. K., SCRUTTON, N. S., SUTCLIFFE, M. J., HILLIER, I. H., BURTON, N. A. (2001) QM/MM Studies show substantial tunneling for the hydrogen-transfer reaction in methylamine dehydrogenase, *J. Am. Chem. Soc.* 123, 8604–8605.

97 GARCIA-VILOCA, M., TRUHLAR, D. G., GAO, J. (2003) Reaction-path energetics and kinetics of the hydride transfer reaction catalyzed by dihydrofolate reductase, *Biochemistry* 42, 13558–13575.

98 NESHEIM, J. C., LIPSCOMB, J. D. (1996) Large kinetic isotope effects in methane oxidation catalyzed by methane monooxygenase: Evidence for C-H bond cleavage in a reaction cycle intermediate, *Biochemistry* 35, 10240–10247.

99 WHITTAKER, M. M., BALLOU, D. P., WHITTAKER, J. W. (1998) Kinetic isotope effects as probes of the mechanism of galactose oxidase, *Biochemistry* 37, 8426–8436.

100 BASRAN, J., SUTCLIFFE, M. J., SCRUTTON, N. S. (1999) Enzymatic H-transfer requires vibration-driven extreme tunneling, *Biochemistry* 38, 3218–3222.

101 BASRAN, J., PATEL, S., SUTCLIFFE, M. J., SCRUTTON, N. S. (2001) Importance of barrier shape in enzyme-catalyzed reactions – vibrationally assisted tunneling in tryptophan tryptophyl-quinone-dependent amine dehydrogenase, *J. Biol. Chem.* 276, 6234–6242.

102 BASRAN, J., SUTCLIFFE, M. J., SCRUTTON, N. S. (2001) Deuterium isotope effects during carbon–hydrogen cleavage by trimethylamine dehydrogenase, *J. Biol. Chem.* 276, 24581–24587.

103 HARRIS, R. J., MESKYS, R., SUTCLIFFE, M. J., SCRUTTON, N. S. (2000) Kinetic studies of the mechanism of carbon-hydrogen bond breakage by the heterotetrameric sarcosine oxidase of *Arthrobacter sp.* 1-IN, *Biochemistry* 39, 1189–1198.

104 CHOWDHURY, S., BANERJEE, R. (2000) Evidence for quantum mechanical tunneling in the coupled cobalt-carbon bond homolysis-substrate radical generation reaction catalyzed by methylmalonyl-CoA mutase, *J. Am. Chem. Soc.* 122, 5417–5418.

105 ABAD, J. L., CAMPS, F., FABRIAS, G. (2000) Is hydrogen tunneling involved in acylCoA desaturase reactios? The case of a D9 desaturase that transforms (*E*)-11-tetradecenoic acid into (*Z*,*E*)-9,11-tetradeienoic acid, *Angew. Chem. Int. Ed.* 122, 3279–3281.

106 GRANT, K. L., KLINMAN, J. P. (1989) Evidence that both protium and deuterium undergo significant tunneling in the reaction catalyzed by bovine serum amine oxidase, *Biochemistry* 28, 6597–6605.

107 JONSSON, T., EDMONDSON, D. E., KLINMAN, J. P. (1994) Hydrogen tunneling in the flavoenzyme monoamine oxidase B, *Biochemistry* 33, 14871–14878.

108 JONSSON, T., GLICKMAN, M. H., SUN, S. J., KLINMAN, J. P. (1996) Experimental evidence for extensive tunneling of hydrogen in the lipoxygenase reaction – implications for enzyme catalysis, *J. Am. Chem. Soc.* 118, 10319–10320.

109 KOHEN, A. (2006) Kinetic isotope effects as probes for hydrogen tunneling in enzyme catalysis, in *Isotope Effects in Chemistry and Biology*, KOHEN, A., LIMBACH, H. H. (Eds.), pp. 743–764, Taylor & Francis, CRC Press, New York.

13
Hydrogen Tunneling in Enzyme-catalyzed Hydrogen Transfer: Aspects from Flavoprotein Catalysed Reactions

Jaswir Basran, Parvinder Hothi, Laura Masgrau, Michael J. Sutcliffe, and Nigel S. Scrutton

13.1
Introduction

Enzymes are extremely efficient catalysts that can achieve rate enhancements of up to 10^{21} over the uncatalyzed reaction rate [1]. Our quest to understand the physical basis of this catalytic power – pivotal to our understanding of biological reactions and our exploitation of enzymes in chemical, biomedical and biotechnological processes – is challenging, and has involved sustained and intensive research efforts for over 100 years (for reviews see Refs. [2–6]). However, our understanding of how enzymes achieve phenomenal rate enhancements is far from complete. Recent years have witnessed new and important activity in this area, and these studies include roles for protein 'motion' [6–8], low barrier hydrogen bonds (for example see Refs. [9–11]), active site preorganization (for reviews see Refs. [4, 12]) and in particular the role of quantum mechanical tunneling in enzymic hydrogen transfer (for reviews see Refs. [13–16]). Understanding factors that drive this H-tunneling reaction is the key to understanding a large number of reactions in biology; C–H bond cleavage occurs in ∼50% of all biological reactions, and all of these are likely to involve tunneling to some degree.

Studies of H-transfer by quantum tunneling focused initially on deviations from values predicted by semiclassical models (in which zero point energies, but not tunneling, have been taken into account) – namely kinetic isotope effects (KIEs), Swain–Schaad relationships [17] $(\exp\left(\frac{\ln(k_H/k_T)}{\ln(k_D/k_T)}\right) > 3.26$, where k_H, k_D, and k_T are the rates of transfer for protium, deuterium and tritium, respectively) or Arrhenius prefactor ratios ($\gg 1$ for a reaction proceeding purely by tunneling, <1 for moderate tunneling). Early examples in which H-tunneling was inferred from measurements of KIEs include the quinoprotein bovine serum amine oxidase [18], the Zn^{2+}-dependent yeast alcohol dehydrogenase [19] and horse liver alcohol dehydrogenase [20], and the flavin-dependent monoamine oxidase [21]. These studies were shown to be consistent with the so-called Bell tunnel correction model of semiclassical transfer, which invokes tunneling (that is transfer occurs *through* the

Hydrogen-Transfer Reactions. Edited by J. T. Hynes, J. P. Klinman, H.-H. Limbach, and R. L. Schowen
Copyright © 2007 WILEY-VCH Verlag GmbH & Co. KGaA, Weinheim
ISBN: 978-3-527-30777-7

energy barrier separating reactant from product) just below the classical transition state [22]. This correction model accommodates small corrections to the rate of a reaction and predicts inflated KIEs and Arrhenius prefactor ratios less than unity (the so-called Kreevoy criteria for tunneling [23]) when KIEs are measured as a function of temperature. Such nonclassical behavior is expected for a light particle such as the H-nucleus (for transfer over short distances): the de Broglie wavelength is 0.63 Å for protium and 0.45 Å for deuterium (assuming an energy of 20 kJ mol^{-1}), and this positional uncertainty gives rise to a significant probability of H-transfer by tunneling. Recent studies from our own group [24–28] and that of Klinman [29, 30] have now indicated that the simple Bell-correction model cannot adequately account for observed KIEs in a number of enzyme systems. This has led to full tunneling models, akin to the established models for electron transfer, in which protein and/or substrate fluctuations are required to generate a configuration compatible with tunneling (see for example Refs. [14, 30, 31]). These full tunneling models are consistent with the strong temperature dependence of reaction rates, the variable temperature dependence of KIEs and the observed range of the Arrhenius prefactor ratio.

These new theoretical frameworks, which incorporate quantum mechanical tunneling coupled to protein motion, supported by experimental observations, have emerged in recent years [31–34]. The reaction itself (that is the breaking and making of bonds) can be modeled computationally based on the hybrid quantum mechanical/molecular mechanical (QM/MM) formulism, in which those atoms involved in the reaction are treated quantum mechanically and the rest of the system treated classically using molecular mechanics (for a review see [35]). Alternatively, the "quantum Kramers" method [15, 36, 37], which treats the whole enzymatic system using a simplified quantum mechanical formulism (note that current computers are not sufficiently powerful to treat the whole system with a full quantum mechanical formulism, hence the need for a simpler model), has been applied to small organic systems. The nature of the tunneling event itself is generally studied computationally using either a method based on variational transtition state theory with multidimensional tunneling corrections (VTST/MT) developed by Truhlar and coworkers (for a review see Refs. [38, 39]) or a method in which the electronic quantum effects are incorporated with an empirical valence bond potential and the hydrogen nucleus is represented as a multidimensional vibrational wavefunction developed by Hammes-Schiffer and coworkers (for a review see Ref. [39]). Given that protein 'motion' is thought to be an important factor in driving quantum tunneling in enzymic H-transfer reactions, methodology has also been developed for identifying computationally, residues important in creating reaction-promoting vibrations in enzymes [40, 41].

Our own studies, in which we have investigated H-tunneling in a number of quinoprotein and flavoprotein enzymes, have provided evidence consistent with H-transfer by quantum tunneling from the vibrational ground state of the reactive C–H bond of the substrate, and *either* H-tunneling in which the KIE is temperature independent – we interpret this to correspond to the absence of gated motion (that is no 'compression' of the transfer distance by substrate and/or protein fluctuations) *or* H-transfer in which the KIE is temperature dependent – we interpret this to correspond to the involvement of gated motion. Our work [26] has also highlighted the importance of energy barrier shape in determining the rates of H-transfer, and the concomitant values of KIEs, obtained in experimental studies. We have recently reviewed our work on quinoprotein enzymes elsewhere [42]. In this chapter we review our studies of H-tunneling in flavoprotein enzymes.

13.2
Stopped-flow Methods to Access the Half-reactions of Flavoenzymes

The flavoprotein enzymes are ideally suited to studies of H-transfer during substrate oxidation using stopped-flow methods. Analysis using the steady-state approach is often compromised by the inability to focus on a single chemical step, owing to the existence of multiple barriers for binding, product release and a number of chemical steps, each of which may contribute to the overall catalytic rate. Using the stopped-flow method, the chemical step can often be isolated and the true kinetics of C–H bond breakage determined without complications arising from other events in the catalytic sequence. With flavoprotein enzymes, the reactions catalyzed are conveniently divided into reductive and oxidative half-reactions. Enzyme reduction occurs by breakage of substrate or coenzyme C–H bonds. The kinetics of bond breakage are conveniently followed by absorbance spectrophotometry since the reaction is concomitant with reduction of the redox centre. Thus, the alternative redox states of the flavin center provide a readily available spectroscopic probe for following the kinetics of C–H bond breakage. The oxidative half-reaction usually involves long-range electron transfer to acceptor proteins (for example cytochromes, copper proteins or other flavoproteins). Again, the absorbance changes associated with oxidation of the flavin provide a readily available signal for monitoring H-transfer to the oxidizing substrate. The ability to interrogate each half-reaction by stopped-flow methods simplifies substantially the kinetic analysis and this makes these enzymes attractive targets in studies of H-transfer employing KIEs as probes of enzymic H-tunneling. Our flavoenzyme work has focused on trimethylamine dehydrogenase (TMADH), heterotetrameric sarcosine oxidase (TSOX), morphinone reductase (MR) and pentaerythritol tetranitrate (PETN) reductase. We have also determined high resolution crystallographic structures for MR [43], PETN reductase [44] and TMADH [45].

13.3
Interpreting Temperature Dependence of Isotope Effects in Terms of H-Tunneling

As mentioned in Section 13.1, the temperature dependent behavior of KIEs (that is temperature dependent versus temperature independent) is a key experimental result when considering the nature of the tunneling. Based on the phenomenological model provided by the Marcus-like framework for H-tunneling [14, 30, 31], the

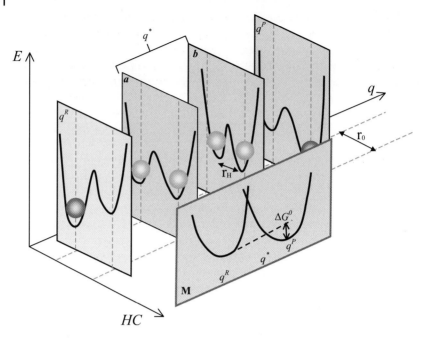

Figure 13.1. Representation of the model for the hydrogen transfer reaction used to interpret the experimental data (see text, and Refs. [6, 84, 85], for more details); some of the parameters in Eqs. (13.1)–(13.3) are shown. The three axes are: E, energy; q, environmental coordinate (from which the transferred hydrogen atom is excluded); HC, hydrogen coordinate. The four vertical panels show the potential energy curve as a function of the hydrogen coordinate for three values of the environmental coordinate: q^R is for the reactant, q^* is for the transition state and q^P is for the product. The gray spheres represent the groundstate vibrational wavefunction of the hydrogen nucleus. The panel labeled **M** shows a Marcus-like view of the free energy curves as functions of this environmental coordinate. The motions of the environment (related to the first exponential in Eq. (13.1)) modulate the symmetry of the double well, thus allowing the system to reach a configuration with (nearly) degenerate quantum states ($q = q^*$), from which the hydrogen is able to tunnel (F.C. Term in Eqs. (13.1) and (13.2)). The difference between panels a and b is a gating motion that reduces the distance between the two wells along the HC axis (r_H) away from its equilibrium value (r_0). This motion increases the probability of tunneling at the (nearly) degenerate configuration q^* (active dynamics term in Eqs. (13.1) and (13.2)).

nature of the temperature dependence of the KIE can be interpreted as follows: (i) H-tunneling in which the KIE is temperature dependent corresponds to the involvement of gated motion (that is motion along HC in Fig. 13.1), and (ii) H-tunneling in which the KIE is temperature independent corresponds to either the absence of, or at least no detectable contributions from, gated motion (that is no significant motion along HC in Fig. 13.1). In other words, two types of motion are important in enzymic tunneling: (i) those which facilitate attaining a nuclear configuration compatible with tunneling (that is a configuration with degenerate quantum states) – termed "passive dynamics", and (ii) those which enhance the probability of tunneling once (i) has been attained – termed "active dynamics". It is, however, not possible to completely decouple active and passive dynamics, since a given motion can contribute to both types of H-tunneling. However, such a role for gating in H-tunneling is not universally supported. For example: (i) the experimentally accessible temperature range is rather narrow, thus it is not always possible to show unambiguously that a given KIE is truly temperature independent; (ii) computational studies on a model system [46] have suggested that it is possible to have a temperature independent KIE in the presence of gating motion; and (iii) recent computational studies using the ensemble averaged variational transition state theory with multi-dimensional tunneling (EA-VTST/MT) framework [47] illustrate how, even with different tunneling behavior for H and D, nearly temperature independent KIEs can be observed. Thus, it is neither possible to map directly from the kinetic data to a detailed picture of the concomitant changes (motions) at the atomic level, nor, hence, to the nature of the (free) energy barrier separating reactants from products. Moreover, in cases where the KIE is temperature dependent (of which, from our own work, there are only two examples to date [26, 28]), a more traditional explanation is that the reaction takes place partly via the over-the-barrier route and partly by tunneling. In principle, this over/through the barrier explanation and/or the dominance of active gating is consistent with a temperature dependent KIE, and current methods for studying the KIE cannot unequivocally disentangle the contribution of each to the tunneling reaction.

The realization that tunneling might be driven by thermally induced vibrations [48] in the protein scaffold (that is a thermally fluctuating energy surface), as described by the theoretical model of Kuznestov and Ulstrup [31] (analogous to electron transfer theory [49]) and illustrated in Fig. 13.1, has been a major step forward in recent years. This model has been adopted by Knapp and Klinman [14, 30], and is of the form:

$$k_{\text{tunnel}} = (\text{const.}) \times \left[\exp\left\{\frac{-(\Delta G^\circ + \lambda)^2}{4\lambda RT}\right\}\right] \times (\text{F.C. Term}) \\ \times (\text{Active Dynamics Term}) \quad (13.1)$$

Here, k_{tunnel} is the tunneling rate constant; const. an isotope-independent term describing electronic coupling; the term in square parentheses is an environmental energy term relating the driving force of the reaction, ΔG°, and the reorganizational energy, λ (R is the gas constant and T the temperature in K); the F.C. Term is the Frank–Condon nuclear overlap along the hydrogen coordinate and arises from the overlap between the initial and the final states of the hydrogen's wavefunction. In the simplest limit, when only the lowest vibrational level is occupied for the nuclear wavefunction of the hydrogen, the F.C. Term is independent of temperature; otherwise, the F.C. Term will be temperature dependent. Temperature dependent 'gating' (or 'active') dynamics, which can be likened to a 'squeezing' of the potential energy barrier, can modulate the F.C. Term. From Eq. (13.1), the KIE can be expressed as:

$$\text{KIE} = \frac{\int \text{F.C.Term}_H \times \text{ActiveDynamicsTerm}_H}{\int \text{F.C.Term}_D \times \text{ActiveDynamicsTerm}_D} \qquad (13.2)$$

Thus, the KIE is temperature independent if the F.C. Term (which is distance independent) dominates, and temperature dependent if there is a significant contribution from a gating motion.

To better understand how Eq. (13.2) can be used to describe active and passive dynamics, the case study of soybean lipoxygenase-1 [14, 30] will be discussed. In this study, Knapp and Klinman [14, 30] adopted the fully nonadiabatic approach of Kuznestov and Ulstrup to analyse their experimental data on hydrogen atom (H•) transfer. It should be noted that H• transfer in soybean lipoxygenase-1 is non-adiabatic, whereas the quinoprotein systems discussed in this review, all of which involve proton transfer, are adiabatic in nature. In Knapp and Klinman's studies of H• transfer, the KIE (Eq. (13.2)) is written as [14]:

$$\text{KIE} = \frac{\int_{r_1}^{r_0} \exp(-m_H \omega_H r_H^2/2\hbar) \times \exp(-E_X/k_B T)\, dX}{\int_{r_1}^{r_0} \exp(-m_D \omega_D r_D^2/2\hbar) \times \exp(-E_X/k_B T)\, dX} \qquad (13.3)$$

Where k_B is Boltzmann's constant, r_0 is the equilibrium, and r_1 the final, separation of the reactant and product potential wells along the hydrogen coordinate, ω_H and ω_D the frequencies of the reacting bond, and m_H and m_D the masses of the transferred particle for protium and deuterium, respectively. The H/D transfer distance, r_H/r_D, is reduced by the distance the gating unit moves, r_X ($r_{H/D} = r_0 - r_X$) The energetic cost of gating (E_X) is given by [14]:

$$E_X = \frac{1}{2}\hbar\omega_X X^2 = \frac{1}{2}m_X \omega_X^2 r_X^2 \qquad (13.4)$$

Here the gating coordinate (X) is related to the gating oscillation (ω_X), the mass of the gating unit (m_X), and r_X as follows:

$$X = r_X \sqrt{m_X \omega_X / \hbar} \qquad (13.5)$$

This model predicts that if the gating term dominates (that is $\hbar\omega_X < k_B T$), the observed KIE can be temperature dependent, since this leads to different transfer distances for the heavy and light isotope [14]. In this regime the A_H/A_D value is predicted to be less than unity. Alternatively, if the Frank–Condon term dominates (that is $\hbar\omega_X > k_B T$) the KIE will be temperature independent. In this latter scenario, occupation of excited vibrational levels could result in some temperature dependence [30]. However, the Boltzmann distribution at 298 K suggests that tunneling should be predominantly from the vibrational ground state of the nuclear wavefunction of hydrogen. In the regime where $\hbar\omega_X \sim k_B T$ gating plays some role in modulating the tunneling probability, temperature dependent KIEs are observed and the A_H/A_D values decrease (compared with the regime where the Frank–Condon term dominates), and may approach unity [14].

Returning to the general case (Eq. (13.1)), studies of the temperature dependence of the KIE cannot disentangle unequivocally the contribution of each term to the tunneling reaction. Compounding the conceptual problem further, Warshel recently claimed [50, 51] that his work demonstrates that dynamical effects do not enhance enzyme catalysis over the equivalent reaction in solution, and he has suggested that the main contribution to catalysis comes from the fact that the barrier is lowered by electrostatic effects [52]. A key role for dynamics has, however, been advocated by others from theoretical studies and experimental observations, for example Bruno and Bialek [32], Klinman [53], Benkovic [7, 54–56], Hammes-Schiffer [41, 57–59] and Schwartz [15, 40, 48, 60–63]. The term "dynamics" is used by different authors to refer to very different events. For example, Schwartz uses "promoting vibrations" to refer to vibrations on the sub-picosecond timescale [61] that are coupled to the reaction coordinate and result in changes in the (quantum) free energy barrier. Hammes-Schiffer uses "promoting motions" to refer to motions averaged over Schwartz's faster "promoting vibrations". These "promoting motions" occur on the much longer timescale of the chemical reaction being catalysed; these thermally averaged motions affect the free energy. The term "dynamical motions" refers to those that influence barrier re-crossing [6]. A similar definition has been used by Warshel [50, 52]. Also, the more general context of atomic and molecular motion has been used by others (for example Karplus [38]). Thus, the jury is still very much out as to exactly what properties of the enzyme give rise to the temperature dependent behavior of the KIE.

The notion that enzymes have evolved to optimize H-tunneling by acquiring strategies during evolution to increase the probability of transfer remains controversial (see for example a recent News Feature in Nature [64]). In one study employing an adenosylcobalamin-dependent diol dehydratase model reaction it is argued by Finke and Doll [65, 66] that this B_{12}-dependent enzyme (which breaks a cobalt–carbon bond) exploits the same level of quantum mechanical tunneling that is available in the reaction occurring in the absence of enzyme (that is there is no 'compressive' motion that preferentially enhances H-tunneling in the enzyme over the reaction in solvent). Moreover, Siebrand and Smedarchina [67] have questioned the statistical significance of data reported over a relatively narrow temperature range for reactions of wild-type and mutant lipoxygenase – these data were originally presented as evidence for gated motion in this enzyme [30]. The same authors also argue, on theoretical grounds, that flexible proteins are "ill-equipped to cause strong local compression". One needs to be mindful of these issues, but our view is that it is not logical to generalize on the basis of a small number of studies and that a case-by-case analysis is appropriate.

13.4
H-Tunneling in Morphinone Reductase and Pentaerythritol Tetranitrate Reductase

We have used the formulism of Knapp and Klinman (see Section 13.3) to interpret the anomalous temperature dependences for H-transfer in FMN-containing morphinone reductase (MR) and the homologous pentaerythritol tetranitrate (PETN) reductase. In particular, we have studied the reactions of (i) PETN reductase with NADPH [28], (ii) MR with NADH [28], and (iii) MR in the oxidative half-reaction with 2-cyclohexenone [28], using stopped-flow and steady-state kinetic methods with protiated and deuterated nicotinamide coenzymes.

13.4.1
Reductive Half-reaction in MR and PETN Reductase

The temperature dependent behavior of the primary KIE for flavin reduction in MR and PETN reductase by nicotinamide coenzyme indicates that quantum mechanical tunneling plays a major role in hydride transfer. In PETN reductase, the KIE is essentially independent of temperature in the experimentally accessible range, and this contrasts with strongly temperature dependent reaction rates (Table 13.1). The data are consistent with a tunneling mechanism governed by passive dynamics and from the vibrational ground state of the reactive C–H/D bond. In MR, both the reaction rates with NADH and the KIE are dependent on temperature, and analysis using the Eyring equation (that is a plot of $\ln(k/T)$ versus $1/T$) sug-

Table 13.1. Tunneling regimes and associated parameters in various flavoprotein enzymes.

Enzyme	Substrate	A_H/A_D	ΔH^H (kJ mol^{-1})	ΔH^D (kJ mol^{-1})	KIE[a]	Passive[b]	Gated[2]	Reference
MR[c]	2-cyclohexenone	3.7	17.6 ± 0.9	17.1 ± 0.9	3.5 TI[a]	✔		28
MR[c]	NADH	0.126	35.3 ± 0.5	43.5 ± 0.8	3.9 TD[a]		✔	28
PETNR[c]	NADPH	4.1	36.4 ± 0.9	36.6 ± 0.9	4.1 TI[a]	✔		28
TMADH[c] (H172Q)	Trimethylamine	7.8	41.2 ± 2.6	41.7 ± 2.6	4.6 TI[a]	✔		25
TMADH[c] (Y169F)	Trimethylamine	2.5	42.1 ± 0.9	45.1 ± 1.6	8.7 TD[a]		✔	25
TSOX[c]	Sarcosine	5.8	39.4 ± 0.9	40.0 ± 1.2	7.3 TI[a]	✔		27

[a] TI = temperature independent KIE, TD = temperature dependent KIE; KIE values for enzyme–substrate combinations displaying a temperature dependent (TD) KIE (that is reactions involving gated motion [14, 30]) are given at 298 K. [b] The terms "passive" (that is the KIE is [almost] temperature independent) and "gated" (that is the KIE is temperature dependent) dynamics are taken from the work of Knapp and Klinman [14, 30]. See text for a discussion of the current limitations of this, and other, interpretations of factors affecting the temperature (in)dependence of KIEs.
[c] Enzyme abbreviations: MR, morphionone reductase; PETNR, pentaerythritol tetranitrate reductase; TMADH, trimethylamine dehydrogenase; TSOX, heterotetrameric sarcosine oxidase.

gests that hydride transfer has a major tunneling component, which unlike in PETN reductase, is gated by thermally induced vibrations in the protein (Table 13.1). We have suggested that PETN reductase is relatively more rigid than MR, consistent with gating being less dominant in PETN reductase, which in turn predicts that the KIE would be more temperature dependent in MR than in PETN reductase. Also, the active site of PETN reductase might be more optimally configured for hydride transfer than that of MR, thus requiring little (or no) vibrational assistance through gated motion. In other words, the active site of PETN reductase is ideally set up to transfer a hydride ion from NADPH to FMN, and nuclear reorganization associated with H-tunneling (that is passive dynamics) is the major dynamic component. We have compared the high resolution crystal structures of MR [43] and PETN reductase [44] in an attempt to provide insight into why gating is potentially more important in MR. Analysis of the structures of each enzyme suggest a key factor could be double stranded anti-parallel β-sheet D, against which the NAD(P)H coenzyme is thought to bind [43]. This region harbors arginine residues, important in the recognition of the 2'phosphate of NADPH (PETN reductase) and a glutamate residue required to form a H-bond with the 2'OH group of NADH (MR). The position of this sheet diverges at Leu-133 (PETN reductase)/Val-138 (MR) and converges again at Ile-141 (PETN reductase)/Gly-146 (MR). Also, there is an insertion of a glycine residue (Gly-133) in MR immediately before the start of β-sheet D. These differences are consistent with MR being more mobile at physiological temperatures in this region than PETN reductase, which in turn might assist in a 'squeezing' or 'compression' of the energy barrier in MR. This suggestion is consistent with the temperature factors for MR (all C_α temperature factors > 40; PDB [68] accession code 1GWJ) and PETN reductase (all C_α temperature factors < 20; PDB accession code 1GVQ) in this region. The next stage of our work to test this hypothesis is to obtain structural information for the coenzyme complexes at high resolution and to perform a more detailed theoretical analysis involving QM/MM, variational transition state theory with multidimensional tunneling and molecular dynamics studies.

13.4.2
Oxidative Half-reaction in MR

The oxidative half-reaction of MR with the substrate 2-cyclohexenone and NADH at saturating concentrations is fully rate-limiting in steady-state turnover; this has enabled us to investigate potential tunneling regimes in this part of the reaction cycle. Reduction of 2-cyclohexenone involves hydride transfer from $FMNH_2$ and protonation, and thus two H-transfer reactions are involved (Fig. 13.2). We have demonstrated that the KIE for hydride transfer from reduced flavin to the α/β unsaturated bond of 2-cyclohexenone is independent of temperature, contrasting with strongly temperature dependent reaction rates. A large solvent isotope effect (SIE) accompanies the oxidative half-reaction, which is also independent of temperature in the experimentally accessible range and double isotope effects indicate that hydride transfer from the flavin N5 atom to 2-cyclohexenone, and the protonation of

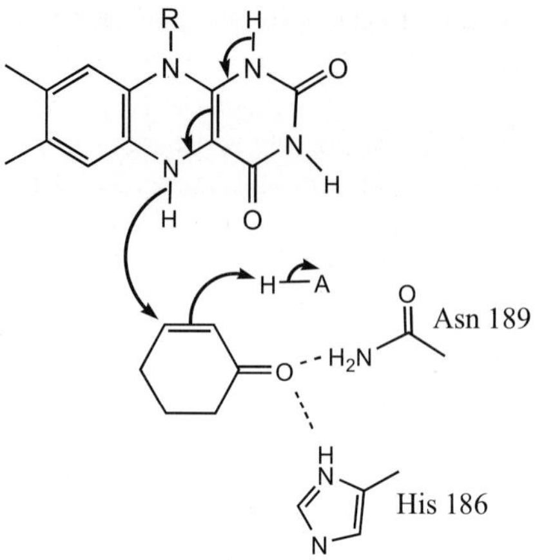

Figure 13.2. Proposed scheme for the oxidative half-reaction of morphinone reductase. The identity of the proton donor in the oxidative half-reaction is not known.

2-cyclohexenone, are concerted. Both the temperature independent KIE and SIE suggest that (i) gated motion is not required to compress the energy barrier, and (ii) this reaction proceeds by ground state quantum tunneling. Our work with MR is therefore the first to show that *both* passive *and* active dynamics are a feature of H-tunneling within the same native enzyme; in the reductive half-reaction we suggest barrier compression is required to facilitate hydride transfer from NADH to FMN, whereas in the oxidative half-reaction the active site is configured to catalyze hydride and proton transfer in a concerted fashion without vibrational assistance through gated motion.

13.5
H-Tunneling in Flavoprotein Amine Dehydrogenases: Heterotetrameric Sarcosine Oxidase and Engineering Gated Motion in Trimethylamine Dehydrogenase

The flavoprotein amine dehydrogenases have proven to be good model systems for studies of enzymic H-tunneling. Klinman and Edmondson provided early evidence for tunneling in mammalian monoamine oxidase, and the data were interpreted in terms of the Bell tunneling correction model [21]. More recent studies with heterotetrameric sarcosine oxidase (TSOX; [27]) and trimethylamine dehydrogenase (TMADH; [25]) indicate that the Bell tunneling correction model is inappropriate for these enzymes and that KIE data are consistent with the more recent full tunneling models.

Although H-transfer in flavoprotein amine dehydrogenases has been shown to occur by tunneling, the mechanisms of amine oxidation by flavoproteins remain controversial. Over the years mechanisms involving the following have been considered: (i) proton abstraction by an active site base to generate a carbanion species [69], (ii) an aminium radical cation species [70], (iii) H-atom abstraction by an active site radical species [71, 72] and (iv) nucleophilic attack by the substrate nitrogen on the flavin C4a atom, followed by proton abstraction by an active site base [73] or the flavin N5 atom [74] (analogous to a similar mechanism proposed for D-amino acid oxidase [75]).

13.5.1
Heterotetrameric Sarcosine Oxidase

TSOX is a diflavin enzyme containing FAD (the site of substrate oxidation) and 8α-(N^3-histidyl)-FMN (the site of oxygen reduction). Treatment of TSOX with sulfite provides the means for selective formation of a flavin-sulfite adduct with the covalent 8α-(N^3-histidyl)-FMN [27]. Formation of the sulfite-flavin adduct suppresses internal electron transfer between the noncovalent FAD and the covalent FMN and thus enables detailed characterization of the kinetics of FAD reduction by sarcosine using stopped-flow methods. The rate of FAD reduction was found to display a simple hyperbolic dependence on sarcosine concentration, and studies in the pH range 6.5 to 10 indicate there are no kinetically influential ionizations in the enzyme–substrate complex. A plot of the limiting rate of flavin reduction/the enzyme–substrate dissociation constant ($k_{\text{lim}}/K_{\text{d}}$) versus pH is bell-shaped and characterized by two macroscopic pK_a values of 7.4 ± 0.1 and 10.4 ± 0.2, indicating two kinetically influential ionizations in the free enzyme or free substrate which remain to be assigned. The KIE for breakage of the substrate C–H bond is 7.3, and the value is independent of temperature (and pH – see below) in the experimentally accessible range; in contrast, reaction rates are strongly dependent on temperature (Table 13.1). The lack of a temperature dependence on the kinetic isotope effect suggests gated motion is not dominant in this reaction.

13.5.2
Trimethylamine Dehydrogenase

13.5.2.1 Mechanism of Substrate Oxidation in Trimethylamine Dehydrogenase

Over the years there have been a number of mechanistic proposals for substrate oxidation by TMADH. An early proposal considered a carbanion mechanism in which an active site base deprotonates a substrate methyl group to form a substrate carbanion [69]; reduction of the flavin was then achieved by the formation of a carbanion–flavin N5 adduct, with subsequent formation of the product imine and dihydroflavin. A number of active site residues were identified as potential bases in such a reaction mechanism. Directed mutagenesis and stopped-flow kinetic studies, however, have been used to systematically eliminate the participation of these residues in a carbanion-type mechanism [76–79], thus indicating that a proton abstraction mechanism initiated by an active site residue does not occur in TMADH. Early proposals also invoked the trimethylammonium cation as the reactive species in the enzyme-substrate complex, owing to the high pK_a (9.81) of free trimethylamine. This was used to argue against mechanisms requiring trimethylamine base (or more explicitly a substrate nitrogen lone pair). However, more recent stopped-flow studies with trimethylamine and perdeuterated trimethylamine, which have taken advantage of force constant effects on the ionization of the trimethylammonium cation, have now established that trimethylamine base is in fact the reactive species in the enzyme–substrate complex [80]. Based on this finding, and the absence of residues in the active site that function as a base during amine oxidation, a mechanism involving addition of trimethylamine base at the C4a position of the flavin and abstraction of a substrate proton by the N5 atom of the flavin has been proposed [80]. This mechanism (Fig. 13.3(a)) is analogous to that proposed previously for monoamine oxidase A (MAO A) based on QSAR analysis with *para*-substituted benzylamines [74], and is consistent also with mechanisms arising from studies of model chemistry [73]. The proposed mechanism is likewise consistent with computational studies of TMADH that have indicated the C4a position is an electrophilic center [45], and with studies of inactivation of TMADH by phenylhydrazine where modification of the flavin occurs at the C4a position [81]. That said, other mechanistic possibilities exist, for example the equivalent of the aminyl radical cation mechanism proposed for MAO, but to date evidence for a protein-based radical to support such a mechanism in TMADH has not been obtained.

A key issue that arises from the demonstration that trimethylamine base is the reactive form of the substrate is the mechanism by which the pK_a for the ionisation of trimethylammonium cation is perturbed from 9.81 to ∼6.5 in the enzyme–substrate complex. The pH dependence of flavin reduction by trimethylamine and perdeuterated trimethylamine has been investigated in detail, and two kinetically influential ionizations have been identified [76]. The first ionization is perturbed by ∼0.5 pH units to higher pH when trimethylamine is replaced by perdeuterated trimethylamine, indicating that this kinetically influential ionization is attributed to deprotonation of substrate. The shorter C–D bond in the perdeuterated substrate results in a larger charge density, and is thus electron supplying relative to C–H. This has the effect of stabilizing the N–H bond, and thus elevating the pK_a for the ionization of substrate. The second ionization is attributed to residue His-172 in the active site of TMADH; the ionization is lost in the H172Q mutant enzyme [78], and is perturbed in a Y169F mutant enzyme (Y169 forms a hydrogen bond to His-172 in native enzyme) (Fig. 13.4). Studies of the pH dependence of the H172Q mutant TMADH indicated that the substrate pK_a in the enzyme-substrate complex is perturbed and elevated, indicating that His-172 is (partially) responsible for the lowering of the substrate pK_a (by ∼1.5 pH units) when bound to enzyme [78]. Residue Tyr-60 also plays a major role: this residue is one of three aromatic side chains (Tyr-60, Trp-264 and Trp-355) involved in binding substrate through amino-aromatic interactions [79, 82]. Replacement of Tyr-60 by phenylalanine elevates the substrate pK_a in the enzyme-substrate complex by about 1.3 pH units to a value of ∼8.8. In the double mutant (H172Q, Y60F) the substrate pK_a is raised even further to ∼9.3, which is close to that of free trimethylamine base [76]. Combined, the data indicate key roles for His-172 and Tyr-60 in

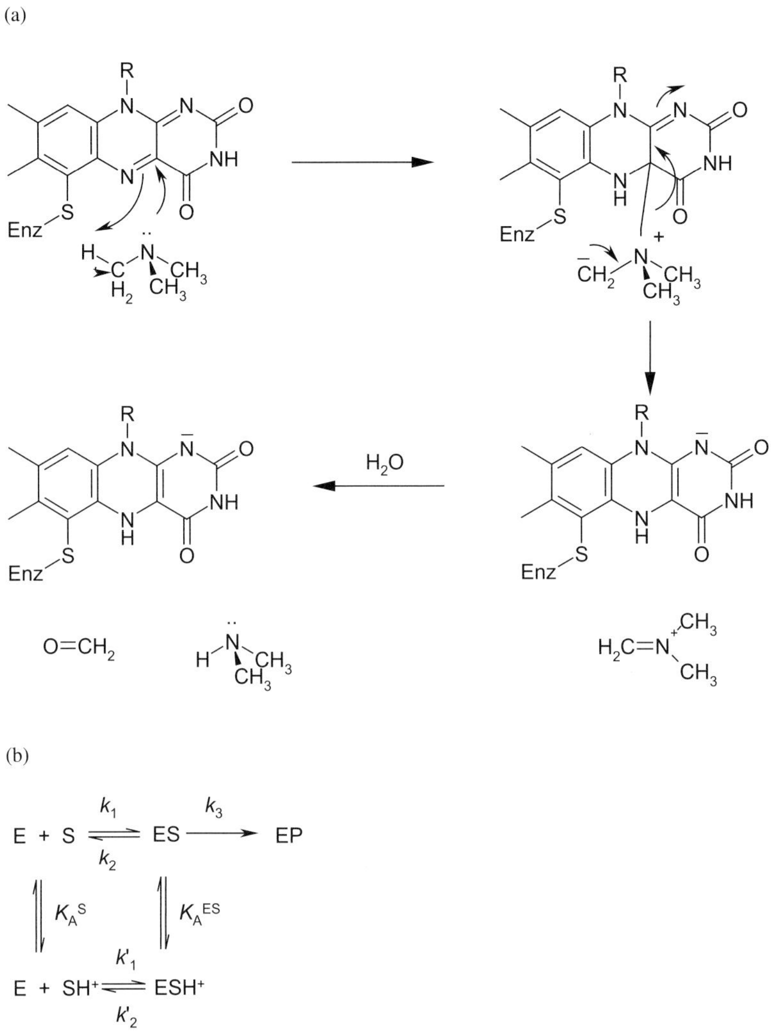

Figure 13.3. (a) A proposed mechanism for the oxidation of trimethylamine by TMADH [83]. (b) Kinetic scheme for the reaction of H172Q mutant TMADH with trimethylamine.

stabilizing the basic form of the substrate in the enzyme active site, thus facilitating catalysis at physiological pH values where the lone pair on the substrate nitrogen is required to initiate substrate oxidation.

13.5.2.2 H-Tunneling in Trimethylamine Dehydrogenase

We have investigated the effects of compromising mutations on tunneling in TMADH. Evidence from isotope studies (see below) supports the view that cataly-

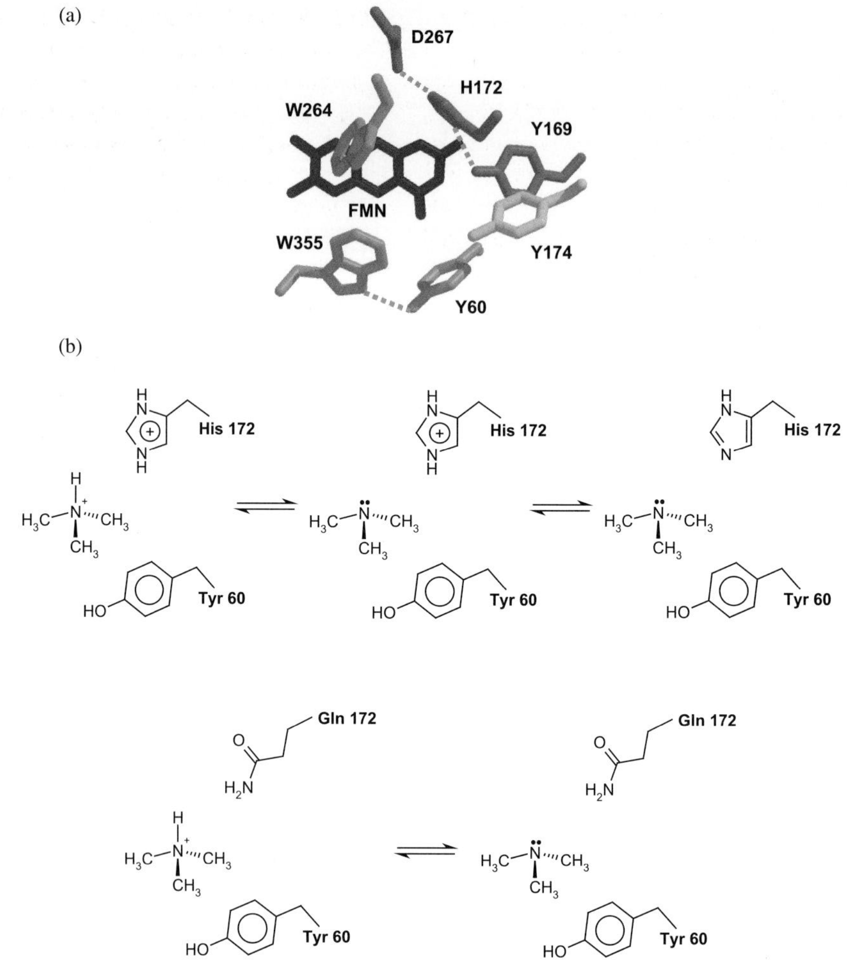

Figure 13.4. (a) Structure of the active site of TMADH showing ionizable residues and the aromatic 'bowl' (Tyr-60, Trp-264 and Trp-355) that interacts with the three methyl groups of the substrate through amino–aromatic interactions. Residues His-172 and Tyr-60 play key roles in stabilising the trimethylamine base in the enzyme–substrate complex. (b) Kinetically influential ionizations in the fast phase of the reductive half-reaction of TMADH. Upper sequence, ionizations in native TMADH; lower sequence, the single ionization in the H172Q mutant enzyme.

sis by TMADH proceeds from a Michaelis complex involving trimethylamine base and not, as thought previously, trimethylammonium cation (Fig. 13.3(a)). Stopped-flow studies and analysis of tunneling regimes in this enzyme are not straight-forward owing to the presence of four kinetically influential ionizations in the reduction of the 6-S-cysteinyl FMN of TMADH by substrate [two in the enzyme–substrate complex (pK_a 6.5 and 8.2), one attributable to free trimethylamine (pK_a 9.8) and one attributed to the free enzyme ($pK_a \sim 10$) which remains unassigned].

In native TMADH, reduction of the flavin by substrate (perdeuterated trimethylamine) is influenced by two ionizations in the Michaelis complex with pK_a values of 6.5 and 8.2 and maximal activity is realized in the alkaline region [83]. The latter ionization has been attributed to residue His-172 (through studies of the H172Q mutant enzyme; [78]) and, more recently, the former to the ionization of substrate itself [83]. Stopped-flow kinetic studies with trimethylamine as substrate have indicated that mutation of His-172 to Gln reduces the limiting rate constant for flavin reduction approximately 10-fold [78]. A kinetic isotope effect accompanies flavin reduction by H172Q TMADH, the magnitude of which varies significantly with solution pH. With trimethylamine, flavin reduction by H172Q TMADH is controlled by a single macroscopic ionization (pK_a 6.8 ± 0.1). This ionization is perturbed (pK_a 7.4 ± 0.1) in reactions with perdeuterated trimethylamine and is responsible for the apparent variation in the KIE with solution pH. The isotope dependence of this pK_a value is of interest. The evidence suggests that this pK_a represents the deprotonation of the substrate molecule itself [$(CH_3)_3NH^+ \rightarrow (CH_3)_3N$] on moving from low to high pH. It is anticipated that perdeuteration of the substrate will affect this ionization since: (i) the shorter C–D bond results in a larger charge density, and thus it is electron supplying (i.e. stabilizing the N–H bond) relative to C–H; (ii) the perdeuterated substrate has a greater reduced mass for the $(CD_3)_3N$–H stretching vibration, and therefore lies lower in the asymmetric potential energy well (although the impact of this would be very small). Thus, the $(CH_3)_3N$–H bond dissociates more readily than the $(CD_3)_3N$–H bond, accounting for the elevated macroscopic pK_a value seen with perdeuterated substrate in our kinetic studies. Figure 13.3(b) summarizes the prototropic control on flavin reduction in the enzyme–substrate complex. In Fig. 13.3(b) it is assumed that the rate of breakdown of the ES complex to EP is slow relative to the dissociation steps, so that the dissociation steps remain in thermodynamic equilibrium. Clearly, as a result of the elevated pK_a value seen with perdeuterated substrate there is a greater concentration of the ESH^+ (unreactive) complex (that is the lower branch of Fig. 13.3(b)). The effect of this partitioning between ES and ESH^+ forms of the enzyme–substrate complex is that the observed KIE is inflated over the intrinsic value that would be realized if the concentration of the ES species were equivalent (at a given pH value) for both perdeuterated and protiated substrate. Only at pH values of 9.5 and above (where the group identified in the plot of k_3 versus pH is fully ionized, and where the rate of flavin reduction is maximal), is the intrinsic isotope effect realized, owing to the enzyme being in the ES form for both protiated and perdeuterated substrate. In this regime, the KIE approaches a constant value of ∼4.5. In the enzyme–substrate complex, the pK_a for the ionisation of trimethylamine (6.8) is more acidic than that of free trimethylamine (9.8). Consequently, in the Michaelis complex, the ionisation of substrate is substantially perturbed leading to a stabilisation of trimethylamine base by ∼10 kJ mol^{-1}. We have shown by targeted mutagenesis and stopped-flow studies that this reduction of the pK_a is a consequence of electronic interaction with residues Tyr-60 and His-172 and these two residues are therefore key for optimising catalysis in the physiological pH range. Formation of a Michaelis complex with trimethylamine base is consistent with a mechanism of amine oxidation that we advanced in our previous computational and kinetic studies which involves nucleophilic attack by the substrate nitrogen atom on the electrophilic C4a atom of the flavin isoalloxazine ring (Fig. 13.3).

Substrate bond breakage by wild-type TMADH is too fast to be followed using the stopped-flow method in the regime where both His-172 and trimethylamine in the enzyme–substrate complex are deprotonated (that is ∼pH 10). Consequently, our tunneling studies have focused on the compromised mutant enzymes H172Q and Y169F (∼10-fold and 40-fold reduction in the limiting rate constant for flavin reduction compared with wild-type enzyme, respectively).

With H172Q TMADH flavin reduction is controlled by a single macroscopic ionization for substrate ionization (pK_a 6.8), which is slightly elevated compared with the value obtained for wild-type owing to the stabilizing electronic effects of the His-172 side-chain in the latter. With H172Q TMADH this ionization is perturbed (pK_a 7.4) in reactions with perdeuterated trimethylamine and is responsible for the apparent variation in the KIE with solution pH. At pH 9.5, where the substrate is fully ionized in the Michaelis complex, the KIE is independent of temperature in the range 277 to 297 K, whilst the reaction rates are still strongly dependent on temperature (Table 13.1); this is consistent with H-transfer by tunneling from the vibrational ground state of the reactive bond in a mechanism that is not dependent on gated motion (a possible alternative explanation is that there is still some gating, but this is not visible in the KIE temperature dependence as seen, for example, by Mincer and Schwartz [46]). With Y169F TMADH, the situation is different: the rate of flavin reduction is ∼4-fold more compromised than in H172Q TMADH and in the case of Y169F TMADH the KIE is dependent on temperature (Table 13.1). In this case, the temperature dependence of the KIE is consistent with the need for gated motion to facilitate the tunneling process.

13.6
Concluding Remarks

Quantum tunneling of hydrogen has emerged over recent years as a means by which enzymes catalyze reactions involving hydrogen transfer. The increasing body of experimental and computational evidence suggests that H-tunneling is likely to be adopted extensively by enzymes. The temperature dependent behavior of kinetic isotope effects has revealed that enzymes can catalyze reactions by "pure" quantum tunneling. KIEs give insight at the macroscopic level, but not at the atomic level. Computational studies are used to give insight into the atomic details of the mechanisms used by enzymes that invoke quantum tunneling. Protein dynamics (both 'active' and 'passive') drives these tunneling reactions, and a picture of how enzymes achieve this is beginning to emerge. However, many of the details at the atomic level remain to be discovered, and these will be probed in greater detail as more refined kinetic (for example at cryogenic temperatures) and computational techniques advance.

Acknowledgments

The authors are very grateful to Linus Johannissen, Kamaldeep Chohan, Richard Harris, Shila Patel, Adrian Mulholland and Kara Ranaghan for their valuable contributions to, and discussions about, the work presented. The BBSRC, EPSRC, University of Leicester and Wellcome Trust are thanked for providing financial support.

References

1. C. Lad, N. H. Williams, R. Wolfenden, *Proc. Natl. Acad. Sci. USA* **2003**, *100*, 5607–5610.
2. K. E. Neet, *J. Biol. Chem.* **1998**, *273*, 25527–25578.
3. W. R. Cannon, S. J. Benkovic, *J. Biol. Chem.* **1998**, *273*, 26257–26260.
4. A. Warshel, *J. Biol. Chem.* **1998**, *273*, 27035–27038.
5. W. W. Cleland, P. A. Frey, J. A. Gerlt, *J. Biol. Chem.* **1998**, *273*, 25529–25532.
6. S. J. Benkovic, S. Hammes-Schiffer, *Science* **2003**, *301*, 1196–1202.
7. C. E. Cameron, S. J. Benkovic, *Biochemistry* **1997**, *36*, 15792–15800.
8. P. T. Rajagopalan, S. Lutz, S. J. Benkovic, *Biochemistry* **2002**, *41*, 12618–21268.
9. P. A. Frey, S. A. Whitt, J. B. Tobin, *Science* **1994**, *264*, 1927–1930.
10. J. A. Gerlt, P. G. Gassman, *Biochemistry* **1993**, *32*, 11943–11952.
11. W. W. Cleland, M. M. Kreevoy, *Science* **1994**, *264*, 1887–1890.
12. W. R. Cannon, S. F. Singleton, S. J. Benkovic, *Nat. Struct. Biol.* **1996**, *3*, 821–833.
13. M. J. Sutcliffe, N. S. Scrutton, *Eur. J. Biochem.* **2002**, *269*, 3096–3102.
14. M. J. Knapp, J. P. Klinman, *Eur. J. Biochem.* **2002**, *269*, 3113–3121.
15. D. Antoniou, S. Caratzoulas, C. Kalyanaraman, J. S. Mincer, S. D. Schwartz, *Eur. J. Biochem.* **2002**, *269*, 3103–3112.
16. Z.-X. Liang, J. P. Klinman, *Curr. Opin. Struct. Biol.* **2004**, *14*, 648–655.
17. A. Kohen, J. H. Jensen, *J. Am. Chem. Soc.* **2002**, *124*, 3858–3864.
18. K. L. Grant, J. P. Klinman, *Biochemistry* **1989**, *28*, 6597–6605.
19. Y. Cha, C. J. Murray, J. P. Klinman, *Science* **1989**, *243*, 1325–1330.
20. B. J. Bahnson, D. H. Park, K. Kim, B. V. Plapp, J. P. Klinman, *Biochemistry* **1993**, *32*, 5503–5507.
21. T. Jonsson, D. E. Edmondson, J. P. Klinman, *Biochemistry* **1994**, *33*, 14871–14878.
22. R. P. Bell, *The Tunnel Effect in Chemistry*, Chapman and Hall, London, 1980, pp. 51–140.
23. Y. H. Kim, M. M. Kreevoy, *J. Am. Chem. Soc.* **1992**, *114*, 7116–7123.
24. J. Basran, M. J. Sutcliffe, N. S. Scrutton, *Biochemistry* **1999**, *38*, 3218–3222.
25. J. Basran, M. J. Sutcliffe, N. S. Scrutton, *J. Biol. Chem.* **2001**, *276*, 24581–2587.
26. J. Basran, S. Patel, M. J. Sutcliffe, N. S. Scrutton, *J. Biol. Chem.* **2001**, *276*, 6234–6242.
27. R. J. Harris, R. Meskys, M. J. Sutcliffe, N. S. Scrutton, *Biochemistry* **2000**, *39*, 1189–1198.
28. J. Basran, R. J. Harris, M. J. Sutcliffe, N. S. Scrutton, *J. Biol. Chem.* **2003**, *278*, 43973–43982.
29. A. Kohen, R. Cannio, S. Bartolucci, J. P. Klinman, *Nature* **1999**, *399*, 496–499.
30. M. J. Knapp, K. Rickert, J. P. Klinman, *J. Am. Chem. Soc.* **2002**, *124*, 3865–3874.
31. A. M. Kuznetsov, J. Ulstrup, *Can. J. Chem.* **1999**, *77*, 1085–1096.
32. W. J. Bruno, W. Bialek, *Biophys. J.* **1992**, *63*, 689–699.

33 D. Borgis, J. T. Hynes, *J. Phys. Chem.* **1996**, *100*, 1118–1128.
34 D. Antoniou, S. D. Schwartz, *Proc. Natl. Acad. Sci. U.S.A.* **1997**, *94*, 12360–12365.
35 J. Gao, M. Thompson, *Methods and Applications of Combined Quantum Mechanical and Molecular Mechanical Methods*, American Chemical Society, Washington DC, 1998.
36 D. Antoniou, S. D. Schwartz, *J. Chem. Phys.* **1999**, *110*, 7359–7364.
37 D. Antoniou, S. D. Schwartz, *J. Chem. Phys.* **1999**, *110*, 465–472.
38 M. Garcia-Viloca, J. Gao, M. Karplus, D. G. Truhlar, *Science* **2004**, *303*, 186–195.
39 S. Hammes-Schiffer, *Curr. Opin. Struct. Biol.* **2004**, *14*, 192–201.
40 J. S. Mincer, S. D. Schwartz, *J. Phys. Chem. B* **2003**, *107*, 366–371.
41 P. K. Agarwal, S. R. Billeter, P. T. Rajagopalan, S. J. Benkovic, S. Hammes-Schiffer, *Proc. Natl. Acad. Sci. USA* **2002**, *99*, 2794–2799.
42 L. Masgrau, J. Basran, P. Hothi, M. J. Sutcliffe, N. S. Scrutton, *Arch. Biochem. Biophys.* **2004**, *428*, 41–51.
43 T. Barna, H. L. Messiha, C. Petosa, N. C. Bruce, N. S. Scrutton, P. C. Moody, *J. Biol. Chem.* **2002**, *277*, 30976–30983.
44 T. M. Barna, H. Khan, N. C. Bruce, I. Barsukov, N. S. Scrutton, P. C. Moody, *J. Mol. Biol.* **2001**, *310*, 433–447.
45 P. Trickey, J. Basran, L.-Y. Lian, Z.-W. Chen, J. D. Barton, M. J. Sutcliffe, N. S. Scrutton, F. S. Mathews, *Biochemistry* **2000**, *39*, 7678–7688.
46 J. S. Mincer, S. D. Schwartz, *J. Chem. Phys.* **2004**, *120*, 7755–60.
47 J. Z. Pu, S. H. Ma, J. L. Gao, D. G. Truhlar, *J. Phys. Chem. B* **2005**, *109*, 8551–8556.
48 D. Antoniou, S. D. Schwartz, *J. Chem. Phys.* **1998**, *108*, 3620–3625.
49 R. A. Marcus, N. Sutin, *Biochim. Biophys. Acta* **1985**, *811*, 265–322.
50 A. Warshel, J. Villa-Freixa, *J. Phys. Chem. B* **2003**, *107*, 12370–12371.
51 M. H. Olsson, P. E. Siegbahn, A. Warshel, *J. Am. Chem. Soc.* **2004**, *126*, 2820–2828.
52 J. Villa, A. Warshel, *J. Phys. Chem. B* **2001**, *105*, 7887–7907.
53 Z. X. Liang, T. Lee, K. A. Resing, N. G. Ahn, J. P. Klinman, *Proc. Natl. Acad. Sci. USA* **2004**, *101*, 9556–9561.
54 C. J. Falzone, P. E. Wright, S. J. Benkovic, *Biochemistry* **1994**, *33*, 439–442.
55 D. M. Epstein, S. J. Benkovic, P. E. Wright, *Biochemistry* **1995**, *34*, 11037–11048.
56 M. J. Osborne, J. Schnell, S. J. Benkovic, H. J. Dyson, P. E. Wright, *Biochemistry* **2001**, *40*, 9846–9859.
57 S. R. Billeter, S. P. Webb, P. K. Agarwal, T. Iordanov, S. Hammes-Schiffer, *J. Am. Chem. Soc.* **2001**, *123*, 11262–11272.
58 P. K. Agarwal, S. R. Billeter, S. Hammes-Schiffer, *J. Phys. Chem. B* **2002**, *106*, 3283–3293.
59 S. Hammes-Schiffer, *Biochemistry* **2002**, *41*, 13335–13343.
60 S. Caratzoulas, S. D. Schwartz, *J. Chem. Phys.* **2001**, *114*, 2910–2918.
61 D. Antoniou, S. D. Schwartz, *J. Phys. Chem. B* **2001**, *105*, 5553–5558.
62 S. Caratzoulas, J. S. Mincer, S. D. Schwartz, *J. Am. Chem. Soc.* **2002**, *124*, 3270–3276.
63 S. D. Schwartz, *J. Phys. Chem. B* **2003**, *107*, 12372.
64 P. Ball, *Nature* **2004**, *431*, 396–397.
65 K. M. Doll, B. R. Bender, R. G. Finke, *J. Am. Chem. Soc.* **2003**, *125*, 10877–10884.
66 K. M. Doll, R. G. Finke, *Inorg. Chem.* **2003**, *42*, 4849–4856.
67 W. Siebrand, Z. Smedarchina, *J. Phys. Chem.* **2004**, *108*, 4185–4195.
68 H. M. Berman, J. Westbrook, Z. Feng, G. Gilliland, T. N. Bhat, H. Weissig, I. N. Shindyalov, P. E. Bourne, *Nucleic Acids Res.* **2000**, *28*, 235–242.
69 R. J. Rohlfs, R. Hille, *J. Biol. Chem.* **1994**, *269*, 30869–79.
70 R. B. Silverman, *Progr. Brain Res.* **1995**, *106*, 23–31.
71 D. E. Edmondson, *Xenobiotica* **1995**, *25*, 735–53.

72 S. Rigby, R. Hynson, R. Ramsay, A. W. Munro, N. S. Scrutton, *J. Biol. Chem.* **2005**, *280*, 4627–4631.

73 J. Kim, M. Bogdan, M. PS, *J. Am. Chem. Soc.* **1993**, *115*, 10591–10595.

74 J. R. Miller, D. E. Edmondson, *Biochemistry* **1999**, *38*, 13670–83.

75 R. Miura, C. Setoyama, Y. Nishina, K. Shiga, H. Mizutani, I. Miyahara, K. Hirotsu, *J. Biochem.* **1997**, *122*, 825–855.

76 J. Basran, M. J. Sutcliffe, N. S. Scrutton, *J. Biol. Chem.* **2001**, *276*, 42887–92.

77 J. Basran, M. J. Sutcliffe, R. Hille, N. S. Scrutton, *J. Biol. Chem.* **1999**, *274*, 13155–13161.

78 J. Basran, M. J. Sutcliffe, R. Hille, N. S. Scrutton, *Biochem. J.* **1999**, *341*, 307–314.

79 J. Basran, M. Mewies, F. S. Mathews, N. S. Scrutton, *Biochemistry* **1997**, *36*, 1989–98.

80 J. Basran, M. J. Sutcliffe, N. S. Scrutton, *J. Biol. Chem.* **2001**, *276*, 24581–24587.

81 J. Nagy, W. C. Kenney, T. P. Singer, *J. Biol. Chem.* **1979**, *254*, 2684–2688.

82 A. R. Raine, C. C. Yang, L. C. Packman, S. A. White, F. S. Mathews, N. S. Scrutton, *Protein Sci.* **1995**, *4*, 2625–2628.

83 J. Basran, M. J. Sutcliffe, N. S. Scrutton, *J. Biol. Chem.* **2001**, *276*, 42887–42892.

84 A. Kohen, J. P. Klinman, *Acc. Chem. Res.* **1998**, *31*, 397–404.

85 A. Kohen, J. P. Klinman, *Chem. Biol.* **1999**, *6*, R191–R198.

14
Hydrogen Exchange Measurements in Proteins

Thomas Lee, Carrie H. Croy,* Katheryn A. Resing,
and Natalie G. Ahn*

14.1
Introduction

Hydrogen exchange between protons in macromolecules and bulk solvent provides a powerful method to investigate global and local conformational changes, folding and stability, macromolecular interactions, and conformational mobility.

This chapter describes techniques for monitoring hydrogen exchange processes that occur between acidic protons bound to proteins and hydrogen isotopes within bulk solvent (e.g., D_2O). Current methodologies are most appropriate for measuring slower exchange rates (half life $\sim$ 1 s or longer), which favor backbone amide hydrogens and certain slowly changing side chain hydrogens. In unstructured peptides, differences in exchange rate at various amide residues are primarily determined by inductive effects from surrounding residues. In proteins, the rate of exchange is controlled by additional environmental factors affecting proton acidity, as well as accessibility of protons to solvent molecules which catalyze abstraction.

14.1.1
Hydrogen Exchange in Unstructured Peptides

Amide hydrogens within peptides which lack secondary structure are constitutively exposed to solvent, thus their exchange rates are mainly governed by the concentration of available catalyst and inductive effects of side chains. Hydrogen exchange in peptides can be described by a chemical exchange rate constant (k_{ch}) for each hydrogen, which depends on an "intrinsic" rate constant of exchange (k_{int}) multiplied by the concentration of base or acid catalysts, which in water are represented by OH^- or H_3O^+ ions ($k_{ch} = k_{int}$ [catalyst]) (Fig. 14.1(a)). The chemical exchange rate constant for freely exposed amide hydrogens in peptides is minimal near pH 2.5 ($k_{ch} = 10^{-1}$–10^{-2} min^{-1} at 5 °C). Rates increase by 10-fold with each pH unit above or below this minima, due, respectively, to base-catalyzed or acid-catalyzed

---

* These authors contributed equally to the preparation of this review.

Hydrogen-Transfer Reactions. Edited by J. T. Hynes, J. P. Klinman, H.-H. Limbach, and R. L. Schowen
Copyright © 2007 WILEY-VCH Verlag GmbH & Co. KGaA, Weinheim
ISBN: 978-3-527-30777-7

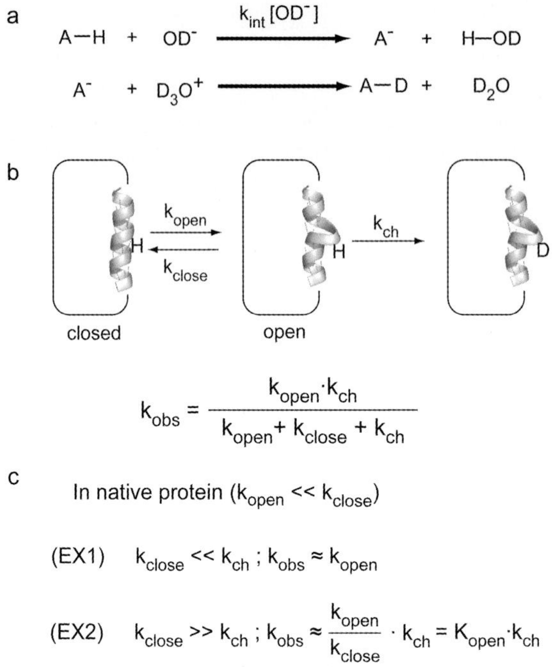

14 Hydrogen Exchange Measurements in Proteins

a

$$A-H + OD^- \xrightarrow{k_{int}[OD^-]} A^- + H-OD$$

$$A^- + D_3O^+ \longrightarrow A-D + D_2O$$

b

closed ⇌ open (with k_{open} forward, k_{close} reverse); open → D (with k_{ch})

$$k_{obs} = \frac{k_{open} \cdot k_{ch}}{k_{open} + k_{close} + k_{ch}}$$

c

In native protein ($k_{open} \ll k_{close}$)

(EX1) $k_{close} \ll k_{ch}$; $k_{obs} \approx k_{open}$

(EX2) $k_{close} \gg k_{ch}$; $k_{obs} \approx \dfrac{k_{open}}{k_{close}} \cdot k_{ch} = K_{open} \cdot k_{ch}$

Figure 14.1. Hydrogen exchange mechanisms: (a) Amide exchange at neutral pH involves base catalyzed proton abstraction and acid catalyzed transfer of deuterium from solvent. Measurable isotope effects on the amide hydrogen and a lack of a solvent isotope effect indicate that proton abstraction is rate limiting [1]. (b) Schematic diagram of the two-state model for hydrogen exchange in native proteins and kinetic prediction. The rate of observed hydrogen exchange (k_{obs}) depends on the fraction of conformers in the open state (e.g., breaking hydrogen bond) and the chemical exchange rate ($k_{ch} = k_{int}[\text{catalyst}]$), where catalyst is OH^- or buffer. (c) In native proteins, the rate of opening is assumed to be much slower than the rate of closing. The observed rates of hydrogen exchange lie on a continuum described by EX1 and EX2 conditions, as described in the text.

proton abstraction [1]. Effects of pH, temperature, and neighboring residues on amide hydrogen exchange rate have been carefully measured in model peptides by the Englander laboratory [1–3]. From these measurements, chemical exchange rate constants can be calculated for amide residues between any combination of adjacent amino acid residues.

In addition to catalyst concentration, the chemical exchange rates of amide hydrogens in peptides are influenced by surrounding amino acid residues. The intrinsic rate of exchange is dependent on local inductive effects which alter the pK_a of the exchangeable proton, as well as steric effects from side chain groups which alter solvent accessibility of the amide hydrogens. A recent study also found that the effects of neighboring side chain on hydrogen exchange can be explained by residual structures of unfolded peptides which affect peptide backbone solvation and protection of amide groups from solvent [4].

14.1.2
Hydrogen Exchange in Native Proteins

Exchange rates for amide hydrogens in native proteins can be much slower than the corresponding rates in peptides, typically differing by 10^2–10^9-fold. Protein folding, steric blocking, and hydrogen bonding interactions may reduce the local concentration of solvent catalyst by interfering with solvent accessibility, and may also alter chemical exchange rates by local perturbation of the pK_a. Thus, protein tertiary and electrostatic interactions, as well as dynamic motions that lead to protein fluctuations or local unfolding, may strongly influence the exchange rates of amide hydrogens in native proteins.

A two state kinetic mechanism, also referred to as the breathing or local unfolding model, has been a widely accepted model for hydrogen exchange processes [5, 6]. In this model, exchange reactions in native state proteins occur at or near protein–solvent interfaces at rates defined by the chemical exchange rate of unstructured peptides. However, reaction rates in the native protein are reduced by incomplete solvent accessibility, represented by the fraction of open or unfolded (open-state) versus closed or folded (closed-state) conformers (Fig. 14.1(b)). Thus, exchange rates can be influenced by transient protein fluctuations or flexibility, in addition to protein conformation. When the chemical exchange rate constant is much larger than the rate constant of structural closing (k_{close}), exchange occurs quickly after conversion of closed to open state ("EX1 regime", Fig. 14.1(c)). At this limit, the observed rate constant of exchange approaches the rate constants for structural opening ($k_{obs} \sim k_{open}$). When the chemical exchange rate is slower than the rate for structural closing, conversion between open and closed states reaches equilibrium before exchange occurs ("EX2 regime", Fig. 14.1(c)). At this limit, k_{obs} approximates the product of the chemical exchange rate constant and the equilibrium constant for structural opening ($k_{obs} \sim k_{ch} K_{eq} = k_{ch} k_{open}/k_{close}$).

Thermodynamic parameters for hydrogen exchange events can be estimated from exchange rates measured under EX2 conditions. The rate of exchange for a given amide hydrogen in a protein can be reported as a protection factor, P, which equals the ratio between the theoretical chemical exchange rate in an unstructured peptide and the observed exchange rate ($P = k_{ch}/k_{obs} \approx k_{close}/k_{open} = 1/K_{eq}$). The logarithm of the protection factor is proportional to the apparent free energy for exchange, $\Delta G_{HX} = -RT \ln(K_{eq}) \approx RT \ln(P)$, which reflects the change in state that a protein must achieve to enable structural opening and subsequent hydrogen exchange.

Rates measured under EX2 conditions are proportional to the chemical exchange rate, which is in turn proportional to the catalyst concentration. This is not true under EX1 conditions where the observed rate reports structural opening, independent of chemical exchange. Therefore, measuring the pH dependence of the observed rate of hydrogen exchange allows one to distinguish between EX1 (pH independent) versus EX2 (pH dependent) mechanisms. In studies of native state proteins, the EX2 mechanism usually predominates under neutral pH conditions, where k_{close} is typically greater than k_{ch}. Under alkaline pH conditions, k_{ch} may exceed k_{close}, thus k_{obs} approaches the EX1 regime. Measurements of k_{obs} at varying pH can be fit to a two-state kinetic model [$k_{obs} = (k_{open}k_{int}[OH^-])/(k_{close} + k_{int}[OH^-])$] from which individual rate constants for opening and closing can be reported [7]. The rate constant of opening can be directly and accurately estimated from these measurements. However, estimates of k_{close} depend on the value of k_{int} in the native protein, which may deviate from the measurements made in model peptides due to effects of tertiary environment. Therefore, measurements of k_{close} should be interpreted carefully.

In certain cases, biexponential rather than monoexponential time courses of hydrogen exchange from single backbone amide hydrogens may be observed, reflecting heterogeneous protein conformers that interconvert slowly. Such behavior can be observed in hydrogen exchange measurements of protein folding, under conditions where significant levels of open exchanging forms of a protein exist prior to initiation of the exchange reaction (e.g., $K_{eq} > 0.01$), and are not in equilibrium with closed or structured forms (i.e., the conversion of open to closed conformers is slower than the chemical exchange rate; $k_{close} < k_{ch}$). Rates of exchange from the pre-existing open forms are mixed with rates of exchange from closed forms, occurring with EX2 or EX1 kinetics [8, 9]. Examples include studies of protein folding intermediates of cytochrome c [10], where hydrogen exchange time courses were fit to a general biexponential pre-steady state rate equation, varying k_{ch} by changing pH or pulse strength, in order to estimate k_{open}, k_{close}, and ΔG_{HX} [9, 10]. Deviations in hydrogen exchange from monoexponential behavior have also been used as evidence for noninterconverting supramolecular forms of amyloid fibril molecules in solution [11].

14.1.3
Hydrogen Exchange and Protein Motions

In addition to conformation and folding, hydrogen exchange measurements reveal information about internal motions of the folded state. Conversions in structure between open and closed conformers are assumed to be completely reversible in native proteins [12–14]. Such motions can be represented by a continuum between different hydrogen exchange mechanisms: Under the EX1 regime, where chemical exchange occurs rapidly after conversion from the closed to open conformers, motions can be described by local unfolding, involving timescales from milliseconds to seconds in native proteins [15]. A typical signature of EX1 exchange events is the correlation between exchange rates from individual hydrogens within a localized region of a protein. Rate limitation by k_{open} leads to correlated exchange rates because the rate of reclosing is slow enough to allow exchange to occur from multiple hydrogens [16, 17].

Recent examples illustrate the determination of structural opening rates in a native protein from hydrogen exchange rates measured in the EX1 regime [15, 18]. Hydrogen exchange in the turkey ovalbumin third domain protein was monitored by NMR at varying pH. Observed rates were fit versus pH to a two-state kinetic model $[k_{obs} = k_{open}k_{int}[OH^-]/(k_{close} + k_{int}[OH^-])]$ from which estimates of $k_{open} = 0.3$–2.8 s^{-1} and $k_{close} = 0.8 \times 10^3$–$1.5 \times 10^4$ s^{-1} were obtained for individual, slow exchanging amide hydrogens, and $k_{open} = 40$–200 s^{-1} and $k_{close} = 4$–400 s^{-1} were obtained for fast exchanging amides. A caveat with this method is that the EX1 exchange reactions were carried out at or above pH 10, conditions where many proteins are destabilized or unfolded. Therefore, protein stability and activity at high pH need to be examined carefully and confirmed when applying this method. While the approach is not suitable for all proteins, it provides a simple and valuable strategy to derive rate constants for conversion between open and closed protein conformers.

Under EX2 conditions, where the rate of structural closing occurs faster than the rate of chemical exchange, individual hydrogens exchange in an uncorrelated manner, even when exposed to solvent in a single concerted motion. In this case, k_{obs} reports the equilibrium between open versus closed conformers, from which the free energy of interconversion can be estimated. Motions that give rise to hydrogen exchange can be described by local fluctuations which are assumed to occur on millisecond to microsecond timescales [19]. Local fluctuations allowing hydrogen exchange events are assumed to involve breaking of hydrogen bonds, allowing solvent penetration into proteins and direct contact with solvent catalysts. In studies of hydrogen exchange by NMR, such fluctuations appear to occur over limited regions in proteins, typically less than 10 amino acids in length [15, 20–22].

In summary, the ability of hydrogen exchange measurements to report local protein changes at short timescales provides a powerful tool to measure solution behavior of proteins relevant to stability and conformational mobility.

14.2
Methods and Instrumentation

14.2.1
Hydrogen Exchange Measured by Nuclear Magnetic Resonance (NMR) Spectroscopy

NMR relaxation studies of nuclei in ^{15}N- and ^{13}C-labeled proteins, as well as heteronuclear Overhauser effects (NOE), have provided a wealth of insight into protein motions that occur in solution on picosecond–nanosecond and microsecond–millisecond timescales (reviewed in Refs. [23–25]). Hydrogen exchange NMR experiments provide complementary strategies to probe slower dynamic processes. Hydrogen exchange measurements are most often carried out by solvent exchange, reporting information about protons with slow or intermediate exchange rate constants (<0.1 s^{-1}). An alternative hydrogen exchange NMR strategy uses water magnetization transfer experiments, enabling rapidly exchanging protons, with exchange rate constants ∼0.1–100 s^{-1}, to be observed. We briefly outline each experiment.

For measurement of slower rates, solvent exchange is initiated either by dissolving lyophilized proteins into D$_2$O, or by solvent exchange into D$_2$O using rapid gel filtration. Proton signals are then monitored by collecting spectra at varying times after solvent exchange [26, 27]. A common collection method for ^{15}N-labeled protein would be a heteronuclear single quantum coherence (HSQC) experiment. The extent of exchange versus time can be measured by the reduction in the areas of the proton resonance peaks, normalizing intensities against the average intensity of multiple long-lived cross-peaks. The normalized data are then fit by nonlinear least squares to an exponential decay equation.

Hydrogen exchange rates can be measured at specific residues when resonances for backbone amide protons are assigned. This is typically done by double labeling proteins with ^{13}C and ^{15}N amino acids and carrying out standard three-dimensional experiments, which include HNCO and HN(CA)CO for backbone carbonyl assignments [28–31]; HNCA and HN(CO)CA for Cα assignments [31, 32]; HNCACB and CBCA(CO)NH for Cβ assignments [33, 34], and all the above for amide resonance assignments. The dead time of the experiment is greater than 5 min, thus protons with slow exchange rate constants (<0.001 s^{-1}) are selectively monitored. These are assumed to be hydrogen bonded within regions of secondary structure, and typically represent more than half of the total amide hydrogens. Protons with faster rates of exchange, such as those found on surface exposed backbone atoms, cannot be monitored by methods involving change in solvent.

Exchange can also be performed by measuring protons transferred from H$_2$O into deuterium-exchanged proteins, a strategy often used in protein folding experiments. For example, an early NMR protocol for monitoring slowly exchanging protons during folding, developed simultaneously by the Englander and Baldwin laboratories [17, 35], involved complete deuteration of unfolded proteins at low pH, initiation of folding followed by pulse labeling with H$_2$O at increased pH, and quenching exchange by decreasing pH. Transient states that showed partial exchange with hydrogen provided unequivocal support for the existence of intermediate states during refolding.

Magnetization transfer techniques provide the means to measure rates at rapidly exchanging protons [36–38]. No change of solvent is involved; instead, water is excited with a selective 180° pulse which saturates the hydrogen spin states in solvent molecules, and exchange is detected by the appearance of proton resonances as magnetization from the solvent transfers to sites on the protein. Early magnetization transfer studies were often ambiguous due to signal contributions from other transfer mechanisms, such as NOEs arising from nonselective magnetization of Cα hydrogens whose chemical shifts are coincident with water, or exchange-relayed NOEs arising from nearby protons that undergo rapid exchange. Multiple pulse methods were therefore developed to suppress these distracting contributions. The Phase-Modulated CLEAN chemical EXchange (CLEANEX-PM) sequence [39] works by dephasing those protein NMR resonances that would give rise to distracting NOEs, while maintaining coherence of solvent protons. Amide proton peaks appear as exchange occurs during the mixing time, and increases in peak height versus time are used to determine k_{obs}. Hwang and colleagues were able to measure accurate chemical exchange rates on submillisecond timescales using CLEANEX-PM and 2D-Fast HSQC (FHSQC) detection of ^{15}N-labeled staphylococcal nuclease. Tugarinov and Kay later demonstrated that CLEANEX-FHSQC experiments can be modified using a 3D-TROSY-HNCO type sequence in order to probe exchange processes in larger proteins [40, 41]. Magnetization transfer experiments on ^{13}C–^{15}N labeled malate synthase G were performed to reveal millisecond timescale motions occurring within the protein upon binding glyoxylate or pyruvate.

NMR allows hydrogen exchange rates to be measured at specific amide hydrogens, enabling protection factors to be determined and related directly to the structural environment. However, determination of exchange rates with high resolution requires assignments of spectral peaks, which can be time consuming and require high concentrations of soluble protein. Furthermore, coverage of amide hydrogens is hampered by exchange rates outside the measurable range.

14.2.2
Hydrogen Exchange Measured by Mass Spectrometry

Mass spectrometry represents a complementary approach to NMR, providing faster analysis with high sequence coverage at lower protein amounts, although usually with reduced sequence resolution. MS enables detection at subpicomole sensitivity for hydrogen exchange measurements, which for many proteins is close to physiological concentration.

Mass spectrometry detects hydrogen exchange in proteins by increased polypeptide mass, corresponding to 1 Da for each exchange event of a proton for solvent deuteron [42–44]. Initial studies measured exchange into intact proteins, for example by monitoring intact protein mass of ubiquitin to measure deuterium incorporation under native versus denatured conditions [45]. Later, strategies were developed to improve resolution by quenching exchange at varying times followed by proteolysis, and analysis of peptide weighted average mass [46, 47]. The sensitivity of mass spectrometers allows exchange reactions to be performed at submicromolar protein concentrations.

Mass spectrometry is optimal for monitoring exchange into amide hydrogens. The exchange reaction is typically initiated by dilution of protein sample in water by ~10–20-fold into D$_2$O. Reaction times are between seconds and hours, and reaction temperatures can easily be varied. Thus, rate constants for fast, intermediate and slow amides can be distinguished over 3 orders of magnitude (~0.00005–0.05 s^{-1}). A larger range of rate constants can be captured by incorporating quench-flow methods to monitor exchange times down to 5 ms [48]. Exchange is quenched by rapidly lowering pH (pD$_{read}$ ~ 2.4) and temperature (0 °C) in order to reduce back-exchange, and the protein is diluted into water and rapidly (1–5 min) digested by proteolysis. Pepsin is used most often for proteolysis, due to its acidic pH optimum and stability against autolysis, although use of other proteases has also been reported [49, 50]. Proteolysis provides medium resolution that varies according to the lengths of peptide digestion products. Because pepsin cleavage specificities are not predictable, the cleavage sites must be inferred from peptide sequencing. This is performed most often by MS/MS, although accurate mass tag analysis, post-source decay, and C-terminal carboxypeptidase Y digestion have also been used. Often, coverage greater than 90% of amide hydrogens can be achieved.

Weighted average masses of peptides are corrected for in-exchange during quenching and proteolysis ("artifactual in-exchange") and back-exchange, from which the number of deuterons incorporated versus time is determined and fit to exponential models [51]. Because multiple exchange reactions occur on each peptide, the number of exponential terms theoretically equals the number of amides. However, in practice, rates of individual amides often cannot be distinguished and are instead grouped into classes with fast (>0.05 s^{-1}), intermediate (0.001–0.05 s^{-1}), and slow (<0.001 s^{-1}) averaged rate constants, the latter includes those that are nonexchanging over the observation period. Examples of mass spectra along with a multi-exponential fitted curve are shown in Fig. 14.2 [52].

A feature unique to mass spectrometry datasets is that populations of molecules that exchange with greatly different rates but interchange slowly can be distinguished by the appearance of bimodal mass/charge distributions [53]. Thus, denatured populations within protein samples can be quantified by the appearance of peptides that show isotopic distributions at lower average mass, reflecting EX2 exchange by native forms, and higher average mass, reflecting rapid exchange by unfolded forms.

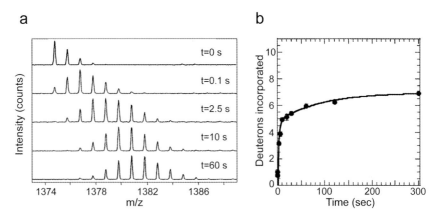

Figure 14.2. Quantifying deuterium incorporation into a protein detected by HX-MS. (a) MALDI-TOF mass spectra showing the incorporation of deuterium into a peptide from IκBα, EVIRQVKGDLAF (MH^{+1} = 1374.77), after indicated times of hydrogen/deuterium exchange. Apparent deuteration at each time point is measured by calculating the weighted average mass of the peptide and subtracting the weighted average mass at $t = 0$. Average weighted mass is calculated as the $(\sum_i (m/z)_i I_i)/\sum_i I_i$, where $(m/z)_i$ is the mass/charge for each isotopic peak and I_i is the intensity of that isotopic peak. The number of deuterons incorporated into peptides at each time point are corrected for back-exchange and artifactual in-exchange. (b) Time course of deuteration, fit to a sum of three exponentials reporting fast (30 s^{-1}), intermediate (0.2 s^{-1}), and slow (0.01 s^{-1}) rate constants. Adapted with permission from Croy et al. [52]. Reprinted with permission from Croy C.H., Bergqvist S., Huxford T., Ghosh G., Komives E.A., *Protein Sci.*, **2004**, *13*, 1767–1777. Copyright 2004 Cold Spring Harbor Laboratory Press.

Mass spectral analysis is usually carried out by electrospray ionization (ESI), which generates multiply charged forms of peptides, or matrix assisted laser desorption/ionization (MALDI), which generates singly charged forms. These ionization methods require different sample preparation protocols. ESI enables peptide separation by reversed-phase HPLC in-line with the mass detector. Peptides are desalted on the column, leading to loss of deuteration at side-chains which rapidly back-exchange to water. As a result, amide hydrogens are observed almost exclusively. By embedding the column and injector in an ice/water slurry (0 °C), and carrying out proteolysis in the sample loop or on a protease column, back-exchange of amide hydrogens to water during the liquid chromatographic (LC) separation can be minimized to ~20% or less [54].

Matrix-assisted laser desorption ionization (MALDI) analysis involves mixing peptide analytes with matrix solution, typically α-cyano-4-hydroxycinnamic acid, and drying the mixture on a MALDI target prior to laser desorption and time of flight (TOF) detection. A 10–20 fold dilution into water is performed at the time of quenching in order to reduce deuteration of side-chains and preferentially monitor amide hydrogens [55, 56]. Back-exchange of amide hydrogens can be maintained between 35 and 50% by preparing the matrix solution in a mixture of acetonitrile, ethanol, and trifluoroacetic acid at pH 2.2–2.5, pre-chilling MALDI targets, and rapidly drying samples under vacuum [55].

Advantages of MALDI over ESI are that the spectra are simpler because singly charged ions are primarily observed, and HPLC is not required, therefore protocols and analyses are easier to perform. Advantages of ESI are that back-exchange is lower, ionization suppression is lower compared to MALDI, and LC separations reduce spectral overlap between peaks, increasing protein sequence coverage and allowing exchange reactions to be performed for longer times.

ESI and MALDI interfaces can be coupled to various types of mass detectors including quadrupole, ion-trap, time-of-flight (qTOF), and Fourier transform ion cyclotron resonance (FT-ICR) mass analyzers. Mass spectral resolution, defined as observed mass divided by the full-width at half maximum peak height (FWHM) varies between instruments (quadrupole ~800, ion-trap ~3000, TOF ~10 000, FT-ICR ~70 000). ESI-quadrupole-TOF and MALDI-TOF instruments both provide baseline resolution of isotopic peaks and high data quality. The higher resolution of FT-ICR enables faster LC runs, thus reducing back-exchange, and enables analysis of large proteins or multiprotein samples with more complex peptide composition [57]. Recent protocols have automated hydrogen exchange measurements by LC-ESI-MS using a quadrupole ion trap instrument, or by MALDI-TOF using an autosampler for target plate spotting, allowing experiments to be performed with higher throughput [58, 59].

14.2.3
Hydrogen Exchange Measured by Fourier-transform Infrared (FT-IR) Spectroscopy

Infrared detection represents one of the first strategies used to monitor protein hydrogen exchange [60]. Most IR studies measure the integrated intensity of the amide II band (~1550 cm^{-1}), assigned to a mode that couples N–H bond bending with C–N stretching. As amide protons are exchanged for deuterons, the wavenumber maximum decreases to ~1450 cm^{-1} (amide II'), which mainly represents C–N stretching vibrations that are no longer coupled with N–D bending (reviewed in Refs. [61, 62]). The number of amide hydrogens that exchange over a given period, at slow or intermediate exchange rates, averaged over the entire protein, may be determined by fitting the normalized amide II intensity [$(A_{II,t} - A_{II,\infty})/A_{II,H2O}$] against time using a multi-exponential decay equation.

FT-IR measurements yield information about global hydrogen exchange into polypeptides, and can be used to reveal changes in tertiary structure or domain interactions in large protein complexes. A recent example measured overall hydrogen exchange changes which occur upon photoactivation of rhodopsin by monitoring the isotopic shift in the amide II band [63]. The authors used time resolved HX-FT-IR to monitor conformational changes which occur within the 7-helix core of rhodopsin upon photoactivation, an event that allows binding of the ligand transducin. In order to monitor the solvent-protected core, the authors first allowed the accessible amides in the protein to exchange in D$_2$O for 48 h prior to photoactivation, after which 67% of the amide II intensity remained. Upon photoactivation, they observed a new population of exchangeable protons which consisted of fast (time constant ~30 min) and slow (~11 h) components. Although they were not able to identify the regions in the protein that accounted for the increased exchange or measure whether these regions correlated to the appearance and decay of known rhodopsin intermediates, they were able to conclude that photoactivation caused buried portions of the core to become more accessible. A caveat of any method that monitors global hydrogen exchange is that exchange rates cannot be treated as occurring uniformly along the polypeptide backbone, thus information about exchange behavior in localized regions may be lost. This is illustrated by a study examining hydrogen exchange in different forms of alcohol dehydrogenase and comparing FT-IR versus mass spectrometry [64]. The "corresponding state" theory postulates that homologous thermophilic and psychrophilic proteins will show similar flexibility when measured at their physiological temperatures [65]. Global hydrogen exchange measured by FT-IR spectroscopy appeared to refute this prediction, showing overall lower exchange in psADH than htADH. However, analysis of individual peptides by HX-MS showed enhanced exchange in psADH compared to htADH, localized to regions involved in substrate and cofactor binding. Thus, localized flexibility within the active site is more likely relevant to catalysis than global motions.

The amide I band region (1620–1680 cm^{-1}) can be used to monitor different classes of secondary structure, where absorbance is dominated by C=O vibrations. Second derivative peaks in this region are characteristic of secondary structure environments (β-sheet ~ 1638/1628 cm^{-1}; α-helix ~ 1655 cm^{-1}; β-turn ~ 1686/1679 cm^{-1}; 3_{10} helix ~ 1660 cm^{-1}), and correlations between changes in these peaks with time of exposure in D$_2$O can be used to assign exchange rates to various secondary structures, thus improving resolution [66–68]. For example, a recent study of brain-derived neurotrophic factor (BDNF) binding to its receptor (trkB) used

HX-FT-IR to monitor the global solvent accessibility of receptor in its unbound versus bound state, and then interpreted the observed hydrogen exchange differences by examining changes in secondary structure and thermal stability from the amide I′ bands [66]. BDNF was metabolically labeled with ^{13}C- and ^{15}N-amino acids, resulting in isotopic shifts within the amide I′ region. This enabled observation of both BDNF and trkB in the complex without spectral overlap, a method referred to as isotope-edited FT-IR. Amide I′ band wavenumbers demonstrated that both proteins contained ∼50% β-sheet structure, and that a small percentage of loop structure in ligand-free trkB was converted into β-turn structure upon ligand binding. The overall stability of trkB decreased upon complex formation, indicated by a new amide I′ band appearing at 1613 cm^{-1}, a wavenumber associated with thermally denatured proteins. Correspondingly, HX into trkB increased significantly upon ligand binding, as observed by a 75% increase in amide II′ band intensity, indicating that a large number of amides showed increased solvent accessibility in the complexed form. Raman spectroscopy showed significant conformational changes at six disulfide bonds upon ligand–receptor complex formation, consistent with loosening of tertiary structure, increased HX, and increased sensitivity to thermal denaturation.

Attenuated total reflection Fourier transform infrared spectroscopy (ATR-FT-IR) involves depositing a thin film of protein on an internal reflection element, initiating deuteration by flushing the sample with nitrogen gas saturated with D_2O, and monitoring loss of signal from the amide II band. Advantages over normal transmission FT-IR are that the experiment requires lower amounts of analyte (1–100 µg), and can be performed within seconds after initiation, enabling amides exchanging with intermediate to fast rates to be observed. The solid support also enables sample orientation. By combining ATR with linear dichroism, secondary structures in membrane transporter proteins and orientation of helices relative to the membrane plane could be determined, allowing hydrogen exchange measurements to be selectively recorded from integral membrane-bound helices [69, 70].

14.3
Applications of Hydrogen Exchange to Study Protein Conformations and Dynamics

14.3.1
Protein Folding

Hydrogen exchange in proteins reports secondary and tertiary structures as well as loss of structure due to unfolding. From such measurements, protein folding intermediates and their kinetic and thermodynamic properties can be monitored. Most applications to examine protein folding use kinetic pulse labeling or equilibrium hydrogen exchange labeling strategies.

In pulse labeling [71], an unfolded and fully deuterated protein, prepared in D_2O at high denaturant concentration, is diluted into H_2O to initiate refolding. The diluting solution is maintained at moderately low pH and temperature (e.g., pH 6,

10 °C) in order to slow chemical exchange rates at solvent accessible amides (half-life ~ s). At various times during refolding, the pH is elevated (pH 8.5–10, half-life ~ ms) to initiate pulsed back-exchange for short, fixed times (e.g. 50 ms) and then exchange is quenched by rapidly lowering the pH to 2.5. Backbone amide hydrogens in unstructured regions undergo back-exchange during the short pulse, whereas amide hydrogens in structured regions do not, providing a snapshot of folding in local regions at varying times.

A recent study by Krishna et al. [10] provides an excellent example using hydrogen exchange by pulse labeling to investigate protein folding intermediates and their kinetics. Folding reactions with denatured cytochrome c were carried out with a fixed 100 ms prelabeling period at pH 6, followed by 50 ms pulses at varying pHs ranging from 7.5 to 10. NMR measurements of fractional back-exchange versus pH were fit to pre-steady state solutions, from which individual rate constants for opening and closing could be estimated (Section 14.1.2, Fig. 14.1(b)). The results revealed protection of N-terminal and C-terminal helices from back-exchange, indicating their secondary structure formation within 100 ms. In contrast, other regions of the protein showed partial or no protection. The free energies of unfolding (ΔG_{HX}) estimated from protection factors measured in the N and C-terminal regions were greater than those estimated for model helices in solution, suggesting that additional tertiary interactions formed within these helices which were more similar to the native state. Rates of structural closing indicated that folding in these regions occurred on a millisecond timescale, much slower than the nanosecond timescale observed for stable helices in solution. Importantly, these regions showed the highest free energy of unfolding in cytochrome c, and were proposed to represent the last step in a sequential unfolding pathway ([72]; see below), presumably due to steric hindrance and interactions between side chains. Rate constants of structural opening (k_{open}) in these regions were similar, suggesting that the two helices unfold in a concerted manner, and that unfolding begins at the ends of each helix. Overall, the results provided evidence for protein folding intermediates containing secondary and tertiary structures and resembling a partially constructed native state.

Hydrogen exchange by equilibrium labeling monitors proteins at varying equilibria between folded and unfolded states, typically under conditions of added denaturants or lowered pH. Using this approach, protein folding intermediates of cytochrome c were monitored by measuring exchange rates and calculating ΔG_{HX} versus increasing concentrations of guanidinium chloride (GdmCl) [72, 73]. A linear dependence of unfolding free energy on denaturant concentration [$\Delta G_U(\text{GdmCl}) \approx \Delta G_{HX}(\text{GdmCl})$] was assumed ($\Delta G_{HX}(\text{GdmCl}) = \Delta G_{HX}(0) - m[\text{GdmCl}]$), where the slope ($m$) increases with the degree of denaturant-sensitive surface exposed during unfolding. At low [GdmCl], hydrogen exchange is dominated by local fluctuations and ΔG_{HX} is independent of denaturant ($m \approx 0$). As GdmCl increases, global unfolding is enhanced and the slope increases in magnitude until linearity is observed between ΔG_{HX} and [GdmCl]. Extrapolation to [GdmCl] = 0 provided estimates of $\Delta G_{HX}(0)$.

It was observed in cytochrome c that groups of amino acids (usually within the

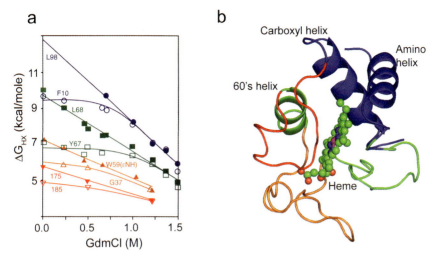

Figure 14.3. HX reveals cooperative unfolding in cytochrome c. (a) A plot of ΔG_{HX} versus guanidium chloride (GdmCl). Residues in cytochrome c with similar slopes at high GdmCl indicate similar free energies of global partial unfolding. Four of the five identified groups of unfolding units, termed "foldons" (red, orange, green, blue) are illustrated in this figure. The data suggest that residues in each foldon participate in a cooperative unfolding unit and that cytochrome c unfolds in a sequential manner from the least stable foldon (red) to the most stable foldon (blue). (b) Foldons of cytochrome c, color coded against the X-ray structure in order of increasing stability (red, orange, green, blue). Adapted with permission from Bai et al. [72]. Reprinted with permission from Bai Y., Sosnick T.R., Mayne L., Englander S.W., *Science*, **1995**, *269*, 192–197. Copyright 1995 AAAS.

same secondary structure) were observed to have similar slope and intercepts, suggesting their participation as a cooperative unit ("foldon") with the same $\Delta G_{HX}(0)$. The results suggested that cytochrome c has five distinct foldons (named nested yellow, red, yellow, green, blue), which were proposed to unfold in a stepwise and sequential manner from lowest to highest $\Delta G_{HX}(0)$ (N < R < Y < G < B) (Fig. 14.3). The sequential unfolding model was further examined by introducing a point mutation (E62G) into the Y foldon, which disrupted salt bridge formation and destabilized the protein by 0.8 kcal mol^{-1} [73]. The mutation decreased $\Delta G_{HX}(0)$ by ∼0.8 kcal mol^{-1} in Y, R, and B foldons, but had little or no effect on N or R foldons. Changes in $\Delta G_{HX}(0)$ would be expected only for the Y foldon, if unfolding of each unit occurred independently. Therefore the results confirmed that the five units in cytochrome c unfold via a sequential, stepwise pathway.

Hydrogen exchange by equilibrium labeling has also been used to study protein molten globules, a term that describes compact, partially folded species which lack complete tertiary interactions of the native state. In general, molten globules are viewed as an interconverting ensemble of structures, rather than a single conformational species. Creative means have been employed to observe such folding intermediates and to glean information about their behavior. Various proteins are known to exist in discrete molten globule states under mildly denaturing or acidic pH conditions [74]. Their isolation and characterization by NMR structural analysis is complicated by the propensity of these partially folded states to aggregate, leading to extreme line broadening and poor chemical shift dispersion. Use of hydrogen exchange can circumvent these problems, by allowing exchange to occur with the molten globule state, and then refolding the protein to the native structure. Well-resolved NMR spectra are obtained which report regions of solvent accessibility in the intermediate state.

In this way, hydrogen exchange from α-lactalbumin was initiated in D_2O under pH conditions that favored the molten globule [74–76]. After varying times, the reaction was quenched and the protein was freeze-dried until NMR acquisition. Redissolving the protein in pH 5.5 buffer containing Ca^{2+} promoted the native structure of α-lactalbumin, from which 2D-HSQC data were collected. Amides within A-, B-, and C-helices were protected from exchange and identified by retention of ^1H signal, evidence that these secondary structures were maintained in both the molten globule and the native state of the protein. Furthermore, the relative stability of these regions could be evaluated by titrating denaturant into the sample and measuring protection factors, demonstrating reduced stability of the molten globule state ($P = 100$) compared to the native state ($P > 10^7$) [76]. Thus, by characterizing "molten globule" states, insights into kinetically transient intermediates in folding pathways were obtained.

A practical application of folding analysis by hydrogen exchange is to screen proteins for crystallization optimization. A bottleneck in structural genomics is producing suitably diffracting protein crystals. Unstructured regions in proteins with higher flexibility cause structural heterogeneity in proteins, preventing crystallization. The ability of hydrogen exchange measurements to successfully reveal regions that interfere with crystallization was tested, using ESI-MS [77, 78]. Sites of proteolytic cleavage and regions of rapid deuteration (10 s) were mapped in 24 *Thermotoga maritima* proteins with known crystallization behavior. Regions of high deuteration were found to correlate with regions disordered in X-ray structures. Truncations were then made to remove rapidly exchanging regions in two proteins. Crystallization trials showed that one protein crystallized more readily in its truncated form (76 crystals out of 1920 attempts) than its corresponding full-length form (2 crystals out of 2400 attempts) and diffracted well (maximum resolution = 1.9 Å). Thus, hydrogen exchange measurements may be useful for distinguishing regions of flexibility from stable cores, facilitating protein analysis.

14.3.2
Protein–Protein, Protein–DNA Interactions

Comparing the extent of hydrogen exchange between bound versus unbound proteins provides a facile method for probing protein–protein interactions, macromolecular complexes, and protein–ligand interactions. Decreased hydrogen exchange (increased protection factors) around a binding interface can be interpreted by steric exclusion of bulk solvent from the binding interface, or alternatively, stabilization and reduced flexibility of the binding interface upon ligand binding.

Both NMR and MS techniques have been used to analyze protein–protein interactions by hydrogen exchange, which reports interfaces between proteins as well as conformational changes induced upon protein complex formation. Early hydrogen exchange labeling studies were carried out by 2D-NMR to determine sites for antibody binding to horse cytochrome c [79]. Later, the same approach was used to map regions of association in a complex formed between yeast ferricytochrome c and cytochrome c peroxidase [80]. Although the X-ray structure of this complex had previously revealed an interface for protein–protein interactions, significant conformational differences between free versus complexed proteins were not observed. In contrast, the HX-NMR study revealed long distance conformational changes. In addition to expected chemical shifts of residues near the interface and around the exposed heme cofactor, hydrogen exchange protection was observed within the C-terminus of ferricytochrome c, a region located far from the protein–protein interface.

A practical concern in mapping protein–protein interactions by HX-MS is the need for relatively high binding affinities. In order to identify protected interfaces and/or conformational changes upon binding, bound forms should ideally represent ~75% of total protein. Given typical protein concentrations used in HX-MS experiments (1–5 μM), a 1:1 molar ratio of proteins is sufficient to achieve this level of binding when K_d is less than ~0.1 μM. However, lower affinities require higher molar ratios of one of the two proteins in order to maintain sufficient complex formation, which often leads to saturating signals and/or obscures peptides from the less abundant protein. In such cases, use of synthetic peptides in place of proteins, which are known to mimic protein interactions, allows examination of low affinity binding interactions.

A recent example used HX-MS to locate binding sites on mitogen activated protein kinases (MAPKs) for substrate docking motifs, conserved regions in substrates and other proteins which enhance affinity and specificity of binding interactions [81]. Hydrogen exchange protection of MAP kinases p38 and ERK2 by peptide ligand revealed the location of homologous binding sites for a basic residue docking motif ($K_d \sim 20$ μM), confirming previous conclusions made from X-ray cocrystal structures of p38. In addition, a surface hydrophobic pocket in ERK2, protected by a separate hydrophobic docking motif ($K_d \sim 10$ μM), was found to be located in a region distal to the site of the basic residue docking motif (Fig. 14.4). Disruption of protein binding by site-directed mutagenesis confirmed the importance of the surface hydrophobic pocket as the relevant binding site for the hydrophobic docking motif. Together with the kinase catalytic cleft, these sites for docking motif binding revealed a new view of how tripartite interactions with enzyme confer substrate recognition.

Two HX-MS studies investigated sites of interaction between regulatory (R) and catalytic (C) subunits of cAMP dependent protein kinase (PKA) where the affinity was in the nanomolar range, enough to maintain a stable complex between intact proteins [82, 83]. Kinase activity in PKA is inhibited by R and C subunit binding, and de-repressed when cAMP binds to R, which leads to subunit dissociation and release of the active C subunit. Hydrogen exchange by MALDI-MS was used to

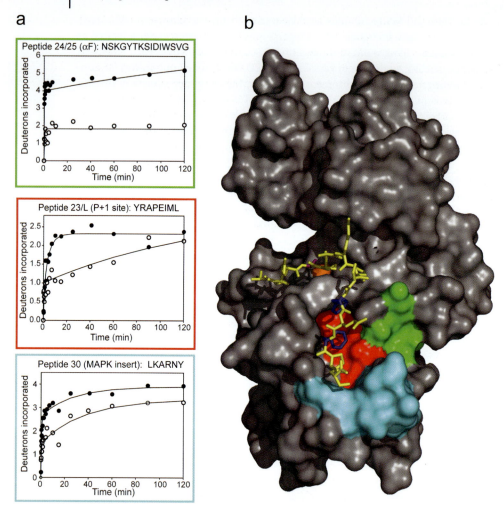

Figure 14.4. HX reveals the binding sites for substrate docking motifs on ERK2. (a) Three regions in ERK2 showed significant decreases in hydrogen exchange upon binding a docking motif peptide (open circles) compared to the unbound state of ERK2 (closed circles). (b) The regions represented in panel (a) are mapped onto the surface of ERK2 (red, green, and cyan). A model of substrate interactions with ERK2 (yellow) showing a sequence (PRSPAKLSFQFPS) from Elk1. Ser3 represents the site of phosphorylation, which is facilitated by a conserved Phe-X-Phe docking motif at res. 9–11. The model constrained the Ser hydroxyl (purple) to lie within hydrogen bonding distance of the catalytic Asp nucleophile (orange). The Phe-X-Phe motif (dark blue) was allowed to interact with a hydrophobic pocket that was revealed by HX-MS and confirmed by site-directed mutagenesis. Adapted with permission from Lee et al. [81]. Reprinted from Lee T., Hoofnagle A.N., Kabuyama Y., Stroud J., Min X., Goldsmith E.J., Chen L., Resing K.A., Ahn N.G., Mol. Cell, **2004**, 14, 43–45. Copyright 2004 Elsevier.

compare free R subunit, R–C holoenzyme complex in the absence of cAMP, and R + C in the presence of cAMP. Binding of C resulted in protection of two alpha helices within a subdomain of the R subunit. This was located within a region adjacent to, but not overlapping with, residues previously shown by mutagenesis to be important for binding, suggesting that the mutations reduce affinity by perturbing direct binding interactions. Unexpectedly, C subunit binding also led to increased protection at a distal site in R that contained the cAMP binding pocket. Conversely, cAMP binding increased solvent accessibility within the helical subdomain in R comprising the R–C interface. Based on this evidence, the authors proposed that, in addition to sites of direct protein interaction, hydrogen exchange protection revealed transmission of conformational information between binding sites that were relevant to ligand allostery.

Further resolution of the PKA binding interaction sites based on hydrogen exchange data were explored using a computational docking program. The docking program, DOT [84], was used to compute 117 billion potential docking structures based on X-ray coordinates of R and C subunits. The top 100 000 lowest energy structures were then filtered using hydrogen exchange information. For example, when amide exchange data showed two protected sites within a given region in R, docked structures would only be selected when two Cα atoms from this region were within 10 Å of the R–C subunit interface. After filtering, 15 structures remained, each showing a similar R–C interface, but differing in the rotational orientation between subunits. Further clustering of these structures yielded a subpopulation used to model the holoenzyme complex, which predicted R–C contacts consistent with previous mutational studies. Thus, data obtained from HX-MS are able to refine protein docking models generated computationally.

Hydrogen exchange has also been used to probe specific interactions between proteins and DNA. Kalodimos et al. [85] characterized the mechanism of interactions between Lac headpiece (HP) protein and DNA operator sequences by HX-NMR. Protection factor estimates in LacHP indicated greater structural stability in the DNA bound state ($P \sim 10^7$) compared to the unbound state ($P \sim 10^2$). Exchange rates were then measured under EX1 conditions by varying pH, in order to estimate k_{open} and k_{close} in LacHP, which were assumed to reflect respective rate constants of dissociation and association with DNA. The measurements revealed three distinct ranges for k_{open} within the hinge helix, containing the DNA binding site (~ 0.20 h^{-1}), within Asn50 and the C-terminus of helix III (~ 0.1 h^{-1}), and within the rest of LacHP (~ 0.02 h^{-1}). These were interpreted to reflect sequential steps in the pathway of protein–DNA dissociation. Residues in each region showed similar values of k_{open}, suggesting that each group of residues dissociates from DNA as a cooperative unit. Importantly, independent measurements of the macroscopic dissociation rate constant (~ 0.05 h^{-1}, [86]) corresponded to the slowest and rate limiting value of k_{open}. Parallel measurements of k_{close} revealed that protein–DNA association is initiated by interactions between the recognition helix (helix II) in LacHP and the major groove of DNA, occurring with a rate constant ~ 40 s^{-1}, which is consistent with the macroscopic rate constant for association. Overall, the data provided a model for LacHP–DNA association and dissociation, in which the rate limiting step for binding involves proper orientation between DNA and the LacHP helix-turn-helix, followed by association of the hinge helix with the minor groove of DNA. Greater flexibility of LacHP in the unbound state may enable initial DNA sliding, which is then followed by specific recognition of operator sequences in the major groove, formation of the hinge helix, and recognition of the minor groove. In contrast, LacHP–DNA dissociation is initiated by hinge helix unfolding which disrupts stable interactions in the major groove. This study illustrates the ability of HX measurements to elucidate details of protein–DNA interactions which cannot easily be monitored by conventional biochemical methods.

14.3.3
Macromolecular Complexes

An important advantage of HX-MS is its ability to study large proteins and macromolecular complexes. The size limitation is mainly governed by how well peptides in proteolytic digests can be uniquely identified and resolved from each other. Studies on macromolecular structures have been carried out using ESI-MS coupled to reverse phase-HPLC [87–89] as well as MALDI-MS without LC separation [90]. Examples described in this section illustrate analyses of heterotetramers, conformational changes in oligomeric capsid assemblies and formation of heterogeneous polymeric fibrils.

A recent study compared nucleosome subcomplexes containing histone H3 or its centromeric variant, CENP-A, in an effort to explore differential packing in nucleosomes targeted to noncentromeric versus centromeric DNA [87]. Heterotetramers formed between CENP-A and histone H4 showed markedly reduced hydrogen exchange within two regions of H4 compared to the heterotetramers formed between histones H3 and H4, suggesting selective protection and subunit compaction by CENP-A association. Bimodal exchange behavior in 13 overlapping peptides was observed in H3-H4, revealing multiple discrete structural states. In agreement, analytical ultracentrifugation showed a smaller Stokes radius in CENP-A-H4 than H3-H4, implying greater compactness in the centromeric complex. Chimeras that substituted regions in H3 located at the H4 interface with homologous regions in CENP-A were then made and found to confer greater subunit compaction as well as centrometric targeting. The results support a model in which CENP-A directs centromeric targeting in part by conferring structural rigidity to nucleosome complexes.

Viral capsids consist of large oligomeric assemblies, usually composed of single subunit types. High resolution models of viral capsids are often constructed from X-ray structures of individual subunits configured against capsids imaged by cryo-electron microscopy. However, these provide little information about intersubunit interactions and conformational changes that occur as the capsids change their form. Hydrogen exchange by ESI-MS of the ~3.6 MDa brome mosaic virus

(BMV) particle was performed to examine the behavior of the BMV capsid protein [89]. Reversible swelling of viral particles occurs when the pH is increased from 5 to 7, which has been linked to capsid stability and disassembly. After correcting for pH effects on chemical exchange, increased deuterium incorporation was observed at high pH, consistent with overall loosening of the protein assembly upon expansion. In contrast, solvent protection at the C terminus of the capsid protein was unaffected by pH, indicating that strong intersubunit interactions were still maintained. Hydrogen exchange by MALDI-MS was used to study the expansion of bacteriophage P22 capsids which occurs upon DNA packaging and viral maturation [90]. P22 maturation was initiated by raising temperature, identifying regions in the capsid protein that varied in flexibility between procapsid and mature capsid assemblies. In addition, a pulse exchange method was used to carry out in-exchange during the expansion phase followed by back-exchange from the expanded form. Regions of the capsid protein that were solvent exposed only during the expansion period were identified by retention of deuterium, and interpreted as domains that were transiently exposed, then buried or stabilized in the mature form.

Structures of fibril forming proteins have proven challenging to study at high-resolution, due to the size and heterogeneity of the protein aggregates. However, HX-MS provides a method well suited to distinguish different aggregate forms. Studies of amyloid aggregation have explored protection patterns in amyloid Aβ peptide (1-40), a proteolytic product of amyloid precursor protein, in unstructured monomeric, protofibril, and mature fibril forms [91–93]. Amyloid Aβ protofibrils are oligomeric forms that are relevant to assemblies in Alzheimer's disease, but are transiently observed and challenging to characterize. Hydrogen exchange monitored for 2 days showed strong solvent protection in protofibrils, covering 40% of backbone amides, in contrast to unstructured monomers which exchanged rapidly throughout the polypeptide. Strong protection of mature amyloid fibrils was also observed, covering 60% of backbone amides, indicating a significant core structure in protofibrils that approximates mature fibrils.

Another study examined conformational changes in prion proteins during transition from soluble to aggregated states [93]. The soluble Het-S protein contains a well-ordered globular domain (residues 1–230) and an unstructured C-terminus (residues 240–289). Hydrogen exchange by MALDI-MS revealed that the C-terminal region is solvent accessible in the monomer but undergoes substantial solvent protection upon amyloid formation, suggesting that this region represents the site involved in aggregation.

14.3.4
Protein–Ligand Interactions

Hydrogen exchange has been used to map protein binding sites for ligands as small as metal ions. An early HX-MS study examined effects of Ca^{2+} binding on recoverin, a sensor that enhances the lifetime of photoexcited rhodopsin by inhibiting rhodopsin kinase at high intracellular Ca^{2+} concentrations. A "calcium-myristoyl switch" model proposed that Ca^{2+} binding promotes membrane association of recoverin by exposing an N-terminal myristoyl group. Neubert et al. [94] monitored hydrogen exchange on recoverin in myristoylated versus unmyristoylated states and in the presence and absence of Ca^{2+} using ESI mass spectrometry. The results showed protection of a hydrophobic cleft in myristoylated recoverin that was abolished upon addition of Ca^{2+}, supporting a model in which the myristoyl group interacts with the hydrophobic core in the absence of metal ion, but extends into the solvent after Ca^{2+} binding. Interestingly, protection from hydrogen exchange was observed in only two of four helix-loop-helix "EF-hand" peptides, suggesting that only two EF-hands coordinate Ca^{2+}. These conclusions were subsequently confirmed by NMR solution structures of myristoylated recoverin in the presence and absence of Ca^{2+} [95, 96].

Hydrogen exchange also proves a powerful tool to identify sites on proteins targeted by pharmaceutical compounds. Eg5, a member of the kinesin superfamily, is essential for microtubule spindle formation and centrosome separation during mitosis, and a viable candidate for antimitotic drug development. Using hydrogen exchange mass spectrometry, Brier et al. [97] examined binding of S-trityl-L-cysteine, a potent inhibitor of this enzyme ($IC_{50} \sim 1.0$ μM). Global deuteration of intact protein for 120 min showed reduced exchange at ~ 30 backbone amide hydrogens in inhibitor-bound versus free Eg5. Hydrogen exchange followed by proteolysis with pepsin revealed these amides to be located within the loop L5-α2 helix and β5-α3 regions, a domain containing the microtubule motor. Substitution of 21 residues in the motor domain of human Eg5 with equivalent residues from *Neurospora Crassa* Eg5 eliminated the effect of S-trityl-L-cysteine on ATPase activity, confirming the motor domain as the relevant binding site for the inhibitor. This study demonstrates that HX-MS can be used to rapidly probe interactions between small molecules and receptors prior to the high resolution structural determination.

Development of high throughput methods for measuring protein–ligand binding is important in fields such as pharmaceutical therapeutics. "Stability of Unpurified Proteins from Rates of hydrogen EXchange" (SUPREX), is a MALDI-MS hydrogen exchange approach designed to rapidly quantify protein folding energy [59]. In this method, ligand interactions are assumed to increase protein folding energy, thus, ligand binding should increase the amount of denaturant needed to unfold the protein, which is indicated when the isotopic envelope shifts to high mass [98]. Hydrogen exchange rates in the presence versus absence of ligand are measured in the EX2 regime at increasing concentrations of denaturant, from which apparent dissociation constants and free energies for ligand binding can be calculated. For high-throughput applications, hydrogen exchange can be monitored at a fixed denaturant concentration (single point SUPREX). Powell and Fitzgerald [99] successfully identified high affinity ligands for a model protein (S protein), by screening combinatorial libraries containing peptides with affinities varying between $K_d \sim 34$ nM to 1000 μM. When fully automated, single point SUPREX was proposed to enable screening of 100 000 compounds per day.

14.3.5
Allostery

Allosteric interactions between ligand binding sites play important roles in protein regulation. These involve changes in conformation or dynamics that may occur over long distances, which can be difficult to infer from static X-ray structures. Classically, allostery has been explained by transitions between discrete conformational states, which modulate binding affinity through changes in enthalpy [100, 101]. An alternative mechanism, proposed by Cooper and Dryden [102–104], suggested instead that "dynamic allostery" may involve modulation of affinity via changes in the distribution of fluctuations around the mean structure of a protein. Thus, allosteric regulation of binding may be attributed to entropic effects, e.g., due to altered side chain mobilities not observable by X-ray crystallography. The ability of hydrogen exchange measurements to report internal motions or conformational mobility within localized regions has led to increased use of this approach for documenting long distance communication in proteins. Due to protein size limitations with NMR, mass spectrometry is generally preferred for hydrogen exchange analyses of multidomain proteins, where allosteric interactions more often occur.

Hemoglobin (Hb) represents a classic model for protein allostery, and structural and thermodynamic changes responsive to ligand (oxygen) binding were successfully examined by hydrogen exchange in tritiated water, using scintillation counting to monitor exchange [105–107]. Each of the 4 subunits in Hb ($\alpha 2\beta 2$) contains a heme group, and sequential binding of oxygen to each subunit enhances the binding affinity to unoccupied sites. The basis of allostery has been modeled from X-ray structures of deoxy (T-state) versus oxy (R-state) forms of Hb, in which movement of Fe(II) into the plane of the heme upon oxygen binding is propagated to the $\alpha\beta$ dimer interface and quaternary structure, affecting the binding affinity at each of the other subunits. Studies by the Englander laboratory [105–110] examined global hydrogen exchange into Hb with tritiated water, revealing changes in rate constants for amide exchange between different states of Hb. Later, higher resolution was obtained by digesting the protein with pepsin and separating the resulting peptides by HPLC [111]. In these studies, five amide hydrogens at the N-terminus of Hbα showed 9-fold higher exchange rates in the R- versus T-state, while four amide hydrogens at the C-terminus of Hbβ showed 190-fold higher exchange rates in the R- versus T-state. Allostery is indicated by the reduction in free energy upon oxygen binding (T to R) of -1.2 kcal mol^{-1} monomer^{-1} for Hbα and -2.85 kcal mol^{-1} monomer^{-1} for Hbβ. The free energy of allosteric destabilization upon oxygen binding (-8.1 kcal mol^{-1} at 0 °C) is consistent with the value measured independently in binding studies (-8.2 kcal mol^{-1} at 5 °C). Hydrogen exchange also revealed that ligand binding to Hbα subunits altered exchange within unbound Hbβ subunits, and vice versa, evidence for cross-subunit communications within the T-state Hb. Importantly, cross-subunit effects were not observed with the HbM (Milwaukee) mutant (Hbβ–Val67Glu) in which Fe(II) is 0.2 Å away from the heme plane, suggesting that cross-subunit communications depend on the movement of Fe(II) with respect to the heme, rather than oxygen binding [110].

Analysis of thrombin illustrates the ability of hydrogen exchange mass spectrometry to reveal long distance interactions between an enzyme active site and distal regions in the protein. Thrombin regulates procoagulant pathways via proteolytic cleavage of fibrinogen, into fibrin, which then self-assembles to form insoluble fibrin clots. Binding of thrombomodulin (TM) alters the substrate specificity of thrombin to favor protein C, which initiates anticoagulant pathways. Biochemical studies indicated that TM binds to anion binding exosite 1 (ABE1), a site in thrombin distant from the active site. It was proposed that TM binding alters substrate specificity through allosteric regulation of the active site conformation, however, no conformational changes were apparent in X-ray cocrystal structures of TM-thrombin [112]. However, hydrogen exchange experiments showed that TM binding reduced exchange within a surface loop adjacent to the active site as well as in the ABE1 [84]. Furthermore, binding of a competitive peptide inhibitor reduced hydrogen exchange at ABE1 as well as the active site [113]. The results provide evidence for reciprocal interactions between the TM binding site (ABE1) and the catalytic site via mechanisms that are more likely to involve changes in conformational mobility than structural perturbations.

Hydrogen exchange measurements thus appear to be quite sensitive to perturbations in exchange within regions distal from ligand binding sites, and similar effects have been observed in many proteins. However, so far, the evidence for causality between binding and changes in hydrogen exchange are largely correlative, and the physical basis of effects reported by hydrogen exchange measurements, as well as their importance towards allosteric function, remains to be determined. Further studies are needed to explore this intriguing hypothesis by characterizing mutations in proteins that both eliminate allostery and disrupt effects on hydrogen exchange at long distances. These studies also illustrate an important caveat when examining sites for ligand binding interactions, that reduced hydrogen exchange does not always reflect steric effects from direct ligand binding interactions, and therefore require confirmation by independent methods.

14.3.6
Protein Dynamics

An exciting application of hydrogen exchange is to explore the relevance of global and local protein flexibility and dynamics in enzyme catalysis. It is often postulated that protein motions underlie conformational and entropic contributions to catalysis; however, motions in enzymes have been difficult to assay. Although NMR measurements of chemical relaxation and order parameters report local dynamics of proteins on timescales ranging from picoseconds to nanoseconds (see Section 14.2.1), the relevance of extremely fast motions in proteins to binding or catalysis is not clear (with the possible exception of electron or hydrogen tunneling). Hydrogen exchange can be used to probe local dynamics of proteins at timescales ranging from microseconds to seconds (Section 14.1.3), which overlaps timescales relevant to enzyme catalysis.

An early study that correlated enzyme rate with hydrogen exchange behavior compared active versus inactive forms of the signaling enzyme, MAP kinase kinase 1 (MKK1) [114]. Two sets of activating mutations in MKK1 were examined, one which replaced regulatory phosphorylation sites with acidic residues, and another which deleted 8 residues from the N-terminus. Either mutation elevated the specific activity of MKK1 by 40–80-fold, and combining these mutations synergistically elevated activity by up to 600-fold. Localized changes in hydrogen exchange were examined in the single and combinatorial mutants, and compared to inactive WT enzyme. Noteworthy was a significant increase in exchange occurring within the N-terminal ATP-binding domain. This was attributed to enhanced flexibility upon activation and, due to overlapping proteolytic products, the sites of increased hydrogen exchange were able to be pinpointed to single amides in this domain. Such behavior was more consistent with fluctuations or breathing motions of the enzyme, which transiently enhance solvent accessibility in these regions, because local unfolding would be expected to alter hydrogen exchange at adjacent amides. Importantly, both sets of mutations individually increased the hydrogen exchange in the N-terminal domain and, in each case, the magnitude of the increase was intermediate to that observed in the combinatorial mutant. Thus, the increased hydrogen exchange in this region correlated qualitatively with increased specific activity. The authors hypothesized that enhanced flexibility within the N-terminal domain is important for MKK1 activation, perhaps to promote ATP binding, phosphoryl transfer, or ADP release.

Local dynamic changes upon mutation that correlate with changes in enzyme activity were also suggested in glutathione transferase, GSTM1, which catalyzes the addition of glutathione to 1-chloro-2,4-dinitrobenzene (CDNB) with product release as the rate-limiting step [115]. Elevated activity of mutant GSTM1-Tyr115Phe was attributed to increased product release rates, whereas values of k_{chem} for glutathione transfer and K_d for CDNB binding were similar to those of WT enzyme. The high resolution X-ray structure of GSTM1-Tyr115Phe was identical to WT enzyme, except for loss of a hydrogen bond between Tyr115 and a neighboring backbone amide. However, comparison of GSTM1-WT versus GSTM1-Tyr115Phe by hydrogen exchange showed increased exchange rates, not only in the region containing the Tyr115Phe mutation, but also in an adjacent region containing the channel for product release. The results suggested that increased protein motions near the channel may explain how the mutation enhances the rate of product release.

Other studies have used hydrogen exchange to reveal protein conformational and dynamic changes following covalent modifications of proteins. Hoofnagle et al. [116] monitored changes in conformational mobility in ERK2 following kinase phosphorylation, which increases the enzyme specific activity >1000-fold. Altered hydrogen exchange rates were observed upon phosphorylation in regions located >10 Å from the site of covalent modification. X-ray structures of unphosphorylated versus phosphorylated ERK2 revealed no conformational differences that could explain the altered hydrogen exchange rates, leading these investigators to hypothesize that the effects on hydrogen exchange reflect changes in conformational mobility or flexibility upon phosphorylation. Interestingly, regions of ERK2 that showed increased exchange upon kinase activation were those that would be expected to undergo motions during catalysis. Follow-up studies comparing active and inactive ERK2 using site-directed spin labeling (SDSL) and electron paramagnetic resonance spectroscopy (EPR) supported the conclusion that increased hydrogen exchange rates in these regions reflect increased flexibility in a manner correlated with enhanced turnover [117].

Perhaps the strongest evidence substantiating the connection between hydrogen exchange measurements and catalytic function has been made from correlations between exchange rates and hydrogen tunneling in related enzymes with differing temperature optima. Through a mechanism termed "environmentally coupled hydrogen tunneling", heavy atom motions in the active site control the probability of hydrogen transfer in the thermophilic alcohol dehydrogenase (htADH) from *B. stearothermophilus* [118]. Previous studies of temperature-dependent kinetic isotope effects showed discontinuities in Arrhenius plots which were attributed to a change in the properties of tunneling at elevated temperatures (Fig. 14.5(a)) [119]. Such results could be explained by invoking environmental coupling, in which hydrogen transfer is enhanced at high temperatures by increased sampling of a conformational space that is inaccessible at low temperature, thus reducing the reaction barrier width and enhancing coupling between reactant and product states.

One test of the model was to demonstrate a transition in conformational mobility accompanying increased tunneling at high temperatures, which would reflect coupling to enzyme dynamics. HX-MS was carried out on htADH at 10–65 °C, and averaged hydrogen exchange rate constants versus $1/T$ were evaluated for each peptide [120]. Overall, averaged hydrogen exchange rate constants increased with temperature, and most varied linearly with $1/T$, as expected by Arrhenius behavior. But, unexpectedly, peptides that formed the cofactor and substrate binding sites underwent discrete transitions that increased temperature dependences above 30 °C (Fig. 14.5(b) and (c)). Thus, regions in the active site undergo a transition in motion or conformation, which coincides with the change in the activation energy of k_{cat} for hydride transfer and tunneling, indicating a direct correlation between hydride transfer and localized protein mobility in htADH. These results substantiate the ability of HX-MS to report changes in enzyme mobility, revealing locations in proteins that are relevant to catalytic function.

Studies of enzyme orthologs with different temperature optima provide strong evidence for the importance of dynamics on enzyme catalysis, and their examination by HX has been insightful [64]. As mentioned above (Section 14.2.3), studies comparing HX behavior of thermophilic versus psychrophilic alcohol dehydrogenase by FT-IR versus mass spectrometry provided evidence that localized motions within discrete regions of the molecule are more likely relevant to catalysis than global motions. Similar conclusions were reached by hydrogen exchange NMR studies of thermophilic versus mesophilic rubredoxin. Structural comparisons suggest that the enhanced thermal stability of proteins from thermophilic organisms

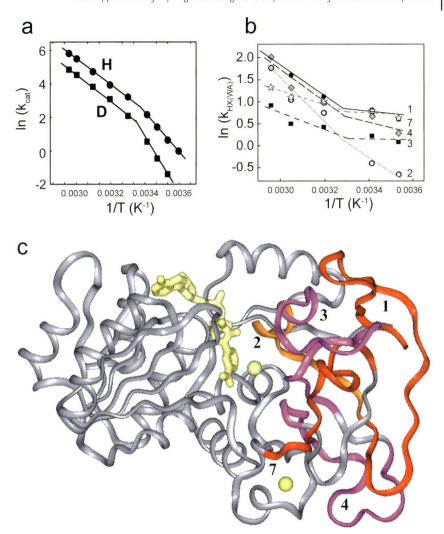

Figure 14.5. HX reveals a temperature-dependent transition in mobility. (a) Arrhenius plot for the oxidation of protonated (circles) or deuterated (squares) benzyl alcohol by htADH. The discontinuity at 30 °C indicates a transition in activation energy for the reaction. (b) Weighted averaged HX rate constant ($k_{HX(WA)}$) for peptides from htADH plotted versus $1/T$ shows discontinuities at 30 °C in five peptides. The weighted averaged kHX is defined as $(Ak_1 + Bk_2 + Ck_3)/NH$ where NH is the total number of amide hydrogens in the peptide, and A, B, and C are the number of amide hydrogens exchanging with rate constants k_1, k_2, and k_3, respectively. (c) The X-ray structure of htADH monomer where the five peptides from panel (b) (coded by number) are mapped onto the structure. All five peptides are located within the substrate binding domain, suggesting that localized changes in protein mobility proximal to the active site are linked to changes in the catalytic rate. Adapted with permission from Liang et al. [120]. Reprinted from Liang Z.X., Lee T., Resing K.A., Ahn N.G., Klinman J.P., *Proc. Natl. Acad. Sci. U.S.A.*, **2004**, *101*, 9556–9561. Copyright 2004, National Academy of Sciences, U.S.A.

may be caused by increased accumulation of salt bridges and other intramolecular interactions [121, 122]. To address whether global conformational rigidity may also account for increased thermostability, hydrogen exchange rates at fast exchanging amide protons ($k_{ex} > 0.2$ s^{-1}) in thermophilic versus mesophilic rubredoxin were measured at varying pH and temperature by CLEANEX-PM [123]. At room temperature, mesophilic rubredoxin underwent significantly faster hydrogen exchange in regions surrounding the catalytic metal binding site and a three-stranded β-sheet compared to its thermophilic ortholog, suggesting higher flexibility at equivalent temperatures. In contrast, reduced hydrogen exchange occurred within a multiple turn domain in mesophilic compared to thermophilic rubredoxin. This was explained in part by ion pair interactions found only in the thermophilic enzyme; however, the exchange rates in this region showed no temperature dependence in thermophilic rubredoxin, unlike the mesophilic ortholog. These results suggest that thermal stability in the thermophilic compared to the mesophilic enzyme are due to localized rather than global conformational rigidity of the protein.

14.4
Future Developments

Further developments promise to broaden the technical capabilities and scope of problems that can be addressed by hydrogen exchange measurements. New TROSY techniques that yield high resolution spectra for large proteins, new methods for measuring residual dipolar couplings, and cryoprobe technologies are extending the range of protein sizes and concentrations that can be observed by NMR, from which improved protocols for HX-NMR are expected to follow.

Major advances in hydrogen exchange measurements by mass spectrometry are needed to increase spatial resolution, by localizing deuteration events at specific amides and allowing measurement of individual protection factors. Increased resolution can be obtained to some degree by comparing the exchange rates between peptides with overlapping sequences. These are often generated by incomplete proteolysis by pepsin, and may also be enhanced using multiple proteases to produce variable patterns of cleavage [49, 114]. However, exchange rates can rarely be measured at more than a few amides.

Several attempts have been made to achieve higher resolution by MS/MS, examining fragment ions for incremental mass shifts that report deuteration at specific residues, and requiring that the sum of deuterated sites equals the total deuterium incorporation into the peptide (reviewed in Ref. [124]). However, so far, the results have been mixed. A significant problem with MS/MS is that the high energy required to induce fragmentation often leads to nonspecific exchange along the peptide, a phenomenon termed scrambling. Some studies have shown significant scrambling behavior, particularly using model peptides deuterated at specific residues. However, other studies reported deuteration of b-ion fragments at individual amides without scrambling [125, 126], which in each case correlated well with corresponding hydrogen exchange rates determined by NMR. Residue-specific deuteration has also been reported, upon inducing collision activated dissociation of an intact protein in the nozzle-skimmer region of the ESI source [127].

Most likely, scrambling occurs in a nonrandom fashion, determined by mechanisms for gas phase peptide bond cleavage. The most common cleavage mechanism involves formation of an oxazolone b ion and y ion, via attack of the backbone carbonyl oxygen at the $n-1$ position at a protonated peptide bond [128–132]. Very rapid cleavages probably occur at sites protonated during the ionization processes, while slower cleavages probably involve interchange of protons by vibrational motion (mobile proton model). Thus, it is likely that scrambling is problematic at some peptide bonds, but not others. Recent computational algorithms based on classical kinetics and the mobile proton hypothesis have successfully simulated peptide MS/MS fragmentation generated in a 3D ion trap mass spectrometer [133]. Predictive models for fragmentation may be useful for improving the design of fragmentation experiments, in order to define conditions that minimize intramolecular proton interchange.

Further improvement of hydrogen exchange methods will involve methods for improving throughput as well as better computational tools to facilitate data reduction. Such goals for hydrogen exchange experiments can be useful for screening compound libraries for protein binding, providing complementary information to high throughput structure determination by NMR and X-ray crystallography. In contrast to high resolution analyses, which show limited compatibility with many proteins and are often carried out under restrictive conditions, HX-MS shows applicability towards a wide range of proteins, accommodating greater diversity in protein sizes, solubilities, concentrations, and solution conditions. Furthermore, HX-MS is readily adaptable to high-throughput protocols. A prototype for automated acquisition and analysis in HX-MS studies was reported by Hamuro et al., who described a fully-automated sample handling system that carries out 100 labeling and quenching reactions, an automated LC system that performs in-line proteolysis and RP separation prior to MS detection, and a software analysis package which can rapidly analyze time courses of exchange [58]. Further development of these methods as well as approaches such as SUPREX (Section 14.3.1) will broaden applications of HX-MS for gaining insight into protein structure and dynamics.

References

1 G. P. CONNELLY, Y. BAI, M. F. JENG, S. W. ENGLANDER, Proteins, 1993, 17, 87–92.
2 Y. BAI, J. S. MILNE, L. MAYNE, S. W. ENGLANDER, Proteins, 1993, 17, 75–86.
3 R. S. MOLDAY, S. W. ENGLANDER, R. G. KALLEN, Biochemistry, 1972, 11, 150–158.
4 F. AVBELJ, R. L. BALDWIN, Proc. Natl. Acad. Sci. U.S.A., 2004, 101, 10967–10972.
5 A. HVIDT, C. R. Trav. Lab. Carlsberg, 1964, 34, 299–317.
6 S. W. ENGLANDER, N. R. KALLENBACH, Quart. Rev. Biophys., 1983, 16, 521–655.
7 T. SIVARAMAN, A. D. ROBERTSON, Methods Mol. Biol., 2001, 168, 193–214.

8 H. Qian, S. I. Chan, *J. Mol. Biol.*, **1999**, *286*, 607–616.
9 M. M. Krishna, L. Hoang, Y. Lin, S. W. Englander, *Methods*, **2004**, *34*, 51–64.
10 M. M. Krishna, Y. Lin, L. Mayne, S. W. Englander, *J. Mol. Biol.*, **2003**, *334*, 501–513.
11 K. Yamaguchi, H. Katou, M. Hoshino, K. Hasegawa, H. Naiki, Y. Goto, *J. Mol. Biol.*, **2004**, *338*, 559–571.
12 Y. Bai, J. J. Englander, L. Mayne, J. S. Milne, S. W. Englander, *Methods Enzymol.*, **1995**, *259*, 344–356.
13 J. Clarke, L. S. Itzhaki, *Curr. Opin. Struct. Biol.*, **1998**, *8*, 112–118.
14 S. W. Englander, N. W. Downer, H. Teitelbaum, *Annu. Rev. Biochem.*, **1972**, *41*, 903–924.
15 C. B. Arrington, A. D. Robertson, *J. Mol. Biol.*, **2000**, *296*, 1307–1317.
16 A. Miranker, C. V. Robinson, S. E. Radford, R. T. Aplin, C. M. Dobson, *Science*, **1993**, *262*, 896–900.
17 H. Roder, G. A. Elove, S. W. Englander, *Nature*, **1988**, *335*, 700–704.
18 C. B. Arrington, A. D. Robertson, *Biochemistry*, **1997**, *36*, 8686–8691.
19 G. Hernandez, F. E. Jenney, Jr., M. W. Adams, D. M. LeMaster, *Proc. Natl. Acad. Sci. U.S.A.*, **2000**, *97*, 3166–3170.
20 K. Kuwajima, R. L. Baldwin, *J. Mol. Biol.*, **1983**, *169*, 299–323.
21 G. Wagner, C. I. Stassinopoulou, K. Wuthrich, *Eur. J. Biochem.*, **1984**, *145*, 431–436.
22 H. Maity, W. K. Lim, J. N. Rumbley, S. W. Englander, *Protein Sci.*, **2003**, *12*, 153–160.
23 L. E. Kay, *Nat. Struct. Biol.*, **1998**, *5*, 513–517.
24 R. Ishima, D. A. Torchia, *Nat. Struct. Biol.*, **2000**, *7*, 740–743.
25 J. Cavanagh, W. J. Fairbrother, A. G. Palmer III, N. J. Skelton, *Protein NMR Spectroscopy: Principles and Practice*, Academic Press, San Diego, 1996.
26 B. K. John, D. Plant, R. E. Hurd, *J. Magn. Reson., Ser. A*, **1992**, *101*, 113–117.
27 L. E. Kay, P. Keifer, T. Saarinen, *J. Am. Chem. Soc.*, **1992**, *114*, 10663–10665.
28 S. Grzesiek, A. Bax, *J. Magn. Reson.*, **1992**, *96*, 432–440.
29 H. Matsuo, E. Kupce, H. J. Li, G. Wagner, *J. Magn. Reson., Ser. B*, **1996**, *111*, 194–198.
30 R. T. Clubb, V. Thanabal, G. Wagner, *J. Magn. Reson.*, **1992**, *97*, 213–217.
31 L. E. Kay, M. Ikura, R. Tschudin, A. Bax, *J. Magn. Reson.*, **1990**, *89*, 496–514.
32 A. Bax, M. Ikura, *J. Biomol. NMR*, **1991**, *1*, 99–104.
33 M. Wittekind, L. Müller, *J. Magn. Reson., Ser. B*, **1993**, *101*, 201–205.
34 S. Grzesiek, A. Bax, *J. Am. Chem. Soc.*, **1992**, *114*, 6291–6293.
35 J. B. Udgaonkar, R. L. Baldwin, *Nature*, **1988**, *335*, 694–699.
36 G. Gemmecker, W. Jahnke, H. Kessler, *J. Am. Chem. Soc.*, **1993**, *115*, 11620–11621.
37 S. Mori, M. Johnson, J. M. Berg, P. C. M. van Zijl, *J. Am. Chem. Soc.*, **1994**, *116*, 11982–11984.
38 S. Grzesiek, A. Bax, *J. Biomol. NMR*, **1993**, *3*, 627–638.
39 T. L. Hwang, P. C. van Zijl, S. Mori, *J. Biomol. NMR*, **1998**, *11*, 221–226.
40 V. Tugarinov, L. E. Kay, *J. Am. Chem. Soc.*, **2003**, *125*, 13868–13878.
41 V. Tugarinov, L. E. Kay, *J. Mol. Biol.*, **2003**, *327*, 1121–1133.
42 A. N. Hoofnagle, K. A. Resing, N. G. Ahn, *Annu. Rev. Biophys. Biomol. Struct.*, **2003**, *32*, 1–25.
43 L. S. Busenlehner, R. N. Armstrong, *Arch. Biochem. Biophys.*, **2005**, *433*, 34–46.
44 S. J. Eyles, I. A. Kaltashov, *Methods*, **2004**, *34*, 88–99.
45 V. Katta, B. T. Chait, *Rapid Commun. Mass Spectrom.*, **1991**, *5*, 214–217.
46 Z. Zhang, D. L. Smith, *Protein Sci.*, **1993**, *2*, 522–531.
47 R. S. Johnson, K. A. Walsh, *Protein Sci.*, **1994**, *3*, 2411–2418.
48 H. Yang, D. L. Smith, *Biochemistry*, **1997**, *36*, 14992–14999.
49 L. Cravello, D. Lascoux, E. Forest,

Rapid Commun. Mass Spectrom., **2003**, *17*, 2387–2393.

50 J. J. Englander, C. Del Mar, W. Li, S. W. Englander, J. S. Kim, D. D. Stranz, Y. Hamuro, V. L. Woods, Jr., *Proc. Natl. Acad. Sci. U.S.A.*, **2003**, *100*, 7057–7062.

51 A. N. Hoofnagle, K. A. Resing, N. G. Ahn, *Methods Mol. Biol.*, **2004**, *250*, 283–298.

52 C. H. Croy, S. Bergqvist, T. Huxford, G. Ghosh, E. A. Komives, *Protein Sci.*, **2004**, *13*, 1767–1777.

53 D. M. Ferraro, N. D. Lazo, A. D. Robertson, *Biochemistry*, **2004**, *43*, 587–594.

54 K. A. Resing, A. N. Hoofnagle, N. G. Ahn, *J. Am. Soc. Mass Spectrom.*, **1999**, *10*, 685–702.

55 J. G. Mandell, A. M. Falick, E. A. Komives, *Anal. Chem.*, **1998**, *70*, 3987–3995.

56 J. G. Mandell, A. M. Falick, E. A. Komives, *Proc. Natl. Acad. Sci. U.S.A.*, **1998**, *95*, 14705–14710.

57 J. Lanman, P. E. Prevelige, Jr., *Curr. Opin. Struct. Biol.*, **2004**, *14*, 181–188.

58 Y. Hamuro, S. J. Coales, M. R. Southern, J. F. Nemeth-Cawley, D. D. Stranz, P. R. Griffin, *J. Biomol. Technol.*, **2003**, *14*, 171–182.

59 S. Ghaemmaghami, M. C. Fitzgerald, T. G. Oas, *Proc. Natl. Acad. Sci. U.S.A.*, **2000**, *97*, 8296–8301.

60 H. Lenormant, E. R. Blout, *Nature*, **1953**, *172*, 770–771.

61 A. Barth, C. Zscherp, *Quart. Rev. Biophys.*, **2002**, *35*, 369–430.

62 R. B. Gregory, A. Rosenberg, *Methods Enzymol.*, **1986**, *131*, 448–508.

63 P. Rath, W. J. DeGrip, K. J. Rothschild, *Biophys. J.*, **1998**, *74*, 192–198.

64 Z. X. Liang, I. Tsigos, V. Bouriotis, J. P. Klinman, *J. Am. Chem. Soc.*, **2004**, *126*, 9500–9501.

65 P. Zavodszky, J. Kardos, Svingor, G. A. Petsko, *Proc. Natl. Acad. Sci. U.S.A.*, **1998**, *95*, 7406–7411.

66 T. Li, J. Talvenheimo, L. Zeni, R. Rosenfeld, G. Stearns, T. Arakawa, *Biopolymers*, **2002**, *67*, 10–19.

67 H. H. de Jongh, E. Goormaghtigh, J. M. Ruysschaert, *Biochemistry*, **1997**, *36*, 13593–13602.

68 H. H. de Jongh, E. Goormaghtigh, J. M. Ruysschaert, *Biochemistry*, **1997**, *36*, 13603–13610.

69 V. Grimard, C. Vigano, A. Margolles, R. Wattiez, H. W. van Veen, W. N. Konings, J. M. Ruysschaer, E. Goormaghtigh, *Biochemistry*, **2001**, *40*, 11876–11886.

70 C. Vigano, M. Smeyers, V. Raussens, F. Scheirlinckx, J. M. Ruysschaert, E. Goormaghtigh, *Biopolymers*, **2004**, *74*, 19–26.

71 P. S. Kim, R. L. Baldwin, *Biochemistry*, **1980**, *19*, 6124–6129.

72 Y. Bai, T. R. Sosnick, L. Mayne, S. W. Englander, *Science*, **1995**, *269*, 192–197.

73 H. Maity, M. Maity, S. W. Englander, *J. Mol. Biol.*, **2004**, *343*, 223–233.

74 C. Redfield, *Methods Mol. Biol.*, **2004**, *278*, 233–254.

75 J. Baum, C. M. Dobson, P. A. Evans, C. Hanley, *Biochemistry*, **1989**, *28*, 7–13.

76 V. Forge, R. T. Wijesinha, J. Balbach, K. Brew, C. V. Robinson, C. Redfield, C. M. Dobson, *J. Mol. Biol.*, **1999**, *288*, 673–688.

77 D. Pantazatos, J. S. Kim, H. E. Klock, R. C. Stevens, I. A. Wilson, S. A. Lesley, V. L. Woods, Jr., *Proc. Natl. Acad. Sci. U.S.A.*, **2004**, *101*, 751–756.

78 G. Spraggon, D. Pantazatos, H. E. Klock, I. A. Wilson, V. L. Woods, Jr., S. A. Lesley, *Protein Sci.*, **2004**, *13*, 3187–3199.

79 Y. Paterson, S. W. Englander, H. Roder, *Science*, **1990**, *249*, 755–759.

80 Q. Yi, J. E. Erman, J. D. Satterlee, *Biochemistry*, **1994**, *33*, 12032–12041.

81 T. Lee, A. N. Hoofnagle, Y. Kabuyama, J. Stroud, X. Min, E. J. Goldsmith, L. Chen, K. A. Resing, N. G. Ahn, *Mol. Cell*, **2004**, *14*, 43–55.

82 G. S. Anand, C. A. Hughes, J. M. Jones, S. S. Taylor, E. A. Komives, *J. Mol. Biol.*, **2002**, *323*, 377–386.

83 G. S. Anand, D. Law, J. G. Mandell, A. N. Snead, I. Tsigelny, S. S.

Taylor, L. F. Ten Eyck, E. A. Komives, *Proc. Natl. Acad. Sci. U.S.A.*, **2003**, *100*, 13264–13269.

84 J. G. Mandell, A. Baerga-Ortiz, S. Akashi, K. Takio, E. A. Komives, *J. Mol. Biol.*, **2001**, *306*, 575–589.

85 C. G. Kalodimos, R. Boelens, R. Kaptein, *Nat. Struct. Biol.*, **2002**, *9*, 193–197.

86 C. G. Kalodimos, G. E. Folkers, R. Boelens, R. Kaptein, *Proc. Natl. Acad. Sci. U.S.A.*, **2001**, *98*, 6039–6044.

87 B. E. Black, D. R. Foltz, S. Chakravarthy, K. Luger, V. L. Woods, Jr., D. W. Cleveland, *Nature*, **2004**, *430*, 578–582.

88 J. Lanman, T. T. Lam, S. Barnes, M. Sakalian, M. R. Emmett, A. G. Marshall, P. E. Prevelige, Jr., *J. Mol. Biol.*, **2003**, *325*, 759–772.

89 L. Wang, L. C. Lane, D. L. Smith, *Protein Sci.*, **2001**, *10*, 1234–1243.

90 R. Tuma, H. Tsuruta, J. M. Benevides, P. E. Prevelige, Jr., G. J. Thomas, Jr., *Biochemistry*, **2001**, *40*, 665–674.

91 I. Kheterpal, H. A. Lashuel, D. M. Hartley, T. Walz, P. T. Lansbury, Jr., R. Wetzel, *Biochemistry*, **2003**, *42*, 14092–14098.

92 I. Kheterpal, S. Zhou, K. D. Cook, R. Wetzel, *Proc. Natl. Acad. Sci. U.S.A.*, **2000**, *97*, 13597–13601.

93 A. Nazabal, S. Dos Reis, M. Bonneu, S. J. Saupe, J. M. Schmitter, *Biochemistry*, **2003**, *42*, 8852–8861.

94 T. A. Neubert, K. A. Walsh, J. B. Hurley, R. S. Johnson, *Protein Sci.*, **1997**, *6*, 843–850.

95 J. B. Ames, R. Ishima, T. Tanaka, J. I. Gordon, L. Stryer, M. Ikura, *Nature*, **1997**, *389*, 198–202.

96 T. Tanaka, J. B. Ames, M. Kainosho, L. Stryer, M. Ikura, *J. Biomol. NMR*, **1998**, *11*, 135–152.

97 S. Brier, D. Lemaire, S. Debonis, E. Forest, F. Kozielski, *Biochemistry*, **2004**, *43*, 13072–13082.

98 K. D. Powell, M. C. Fitzgerald, *Biochemistry*, **2003**, *42*, 4962–4970.

99 K. D. Powell, M. C. Fitzgerald, *J. Comb. Chem.*, **2004**, *6*, 262–269.

100 E. J. Goldsmith, *Faseb J*, **1996**, *10*, 702–708.

101 A. D. Robertson, *Trends Biochem. Sci.*, **2002**, *27*, 521–526.

102 A. Cooper, D. T. Dryden, *Eur. Biophys. J.*, **1984**, *11*, 103–109.

103 A. Cooper, *Curr. Opin. Chem. Biol.*, **1999**, *3*, 557–563.

104 A. J. Wand, *Nat. Struct. Biol.*, **2001**, *8*, 926–931.

105 J. J. Englander, J. R. Rogero, S. W. Englander, *J. Mol. Biol.*, **1983**, *169*, 325–344.

106 J. Ray, S. W. Englander, *Biochemistry*, **1986**, *25*, 3000–3007.

107 G. Louie, J. J. Englander, S. W. Englander, *J. Mol. Biol.*, **1988**, *201*, 765–772.

108 S. W. Englander, J. J. Englander, R. E. McKinnie, G. K. Ackers, G. J. Turner, J. A. Westrick, S. J. Gill, *Science*, **1992**, *256*, 1684–1687.

109 J. J. Englander, G. Louie, R. E. McKinnie, S. W. Englander, *J. Mol. Biol.*, **1998**, *284*, 1695–1706.

110 J. J. Englander, J. N. Rumbley, S. W. Englander, *J. Mol. Biol.*, **1998**, *284*, 1707–1716.

111 J. R. Rogero, J. J. Englander, S. W. Englander, *Methods Enzymol.*, **1986**, *131*, 508–517.

112 P. Fuentes-Prior, Y. Iwanaga, R. Huber, R. Pagila, G. Rumennik, M. Seto, J. Morser, D. R. Light, W. Bode, *Nature*, **2000**, *404*, 518–525.

113 C. H. Croy, J. R. Koeppe, S. Bergqvist, E. A. Komives, *Biochemistry*, **2004**, *43*, 5246–5255.

114 K. A. Resing, N. G. Ahn, *Biochemistry*, **1998**, *37*, 463–475.

115 S. G. Codreanu, J. E. Ladner, G. Xiao, N. V. Stourman, D. L. Hachey, G. L. Gilliland, R. N. Armstrong, *Biochemistry*, **2002**, *41*, 15161–15172.

116 A. N. Hoofnagle, K. A. Resing, E. J. Goldsmith, N. G. Ahn, *Proc. Natl. Acad. Sci. U.S.A.*, **2001**, *98*, 956–961.

117 A. N. Hoofnagle, J. W. Stoner, T. Lee, S. S. Eaton, N. G. Ahn, *Biophys. J.*, **2004**, *86*, 395–403.

118 M. J. Knapp, J. P. Klinman, *Eur. J. Biochem.*, **2002**, *269*, 3113–3121.

119 A. Kohen, R. Cannio, S. Bartolucci, J. P. Klinman, *Nature*, **1999**, *399*, 496–499.

120 Z. X. Liang, T. Lee, K. A. Resing,

N. G. Ahn, J. P. Klinman, *Proc. Natl. Acad. Sci. U.S.A.*, **2004**, *101*, 9556–9561.

121 K. S. Yip, T. J. Stillman, K. L. Britton, P. J. Artymiuk, P. J. Baker, S. E. Sedelnikova, P. C. Engel, A. Pasquo, R. Chiaraluce, V. Consalvi, *Structure*, **1995**, *3*, 1147–1158.

122 N. Frankenberg, C. Welker, R. Jaenicke, *FEBS Lett.*, **1999**, *454*, 299–302.

123 G. Hernandez, D. M. LeMaster, *Biochemistry*, **2001**, *40*, 14384–14391.

124 J. K. Hoerner, H. Xiao, A. Dobo, I. A. Kaltashov, *J. Am. Chem. Soc.*, **2004**, *126*, 7709–7717.

125 M. Y. Kim, C. S. Maier, D. J. Reed, M. L. Deinzer, *J. Am. Chem. Soc.*, **2001**, *123*, 9860–9866.

126 Y. Deng, P. Pan, D. L. Smith, *J. Am. Chem. Soc.*, **1999**, *121*, 1966–1967.

127 S. J. Eyles, P. S. Speir, G. H. Kruppa, L. M. Gierasch, I. A. Kaltashov, *J. Am. Chem. Soc.*, **2000**, *122*, 495–500.

128 V. H. Wysocki, G. Tsaprailis, L. L. Smith, L. A. Breci, *J. Mass Spectrom.*, **2000**, *35*, 1399–1406.

129 L. A. Breci, D. L. Tabb, J. R. Yates, 3rd, V. H. Wysocki, *Anal. Chem.*, **2003**, *75*, 1963–1971.

130 G. Tsaprailis, H. Nair, W. Zhong, K. Kuppannan, J. H. Futrell, V. H. Wysocki, *Anal. Chem.*, **2004**, *76*, 2083–2094.

131 C. Gu, G. Tsaprailis, L. Breci, V. H. Wysocki, *Anal. Chem.*, **2000**, *72*, 5804–5813.

132 E. A. Kapp, F. Schutz, G. E. Reid, J. S. Eddes, R. L. Moritz, R. A. O'Hair, T. P. Speed, R. J. Simpson, *Anal. Chem.*, **2003**, *75*, 6251–6264.

133 Z. Zhang, *Anal. Chem.*, **2004**, *76*, 6374–6383.

15
Spectroscopic Probes of Hydride Transfer Activation by Enzymes

Robert Callender and Hua Deng

15.1
Introduction

Dehydrogenases, reductases and a number of other enzymes, such as UDP-glucose epimerase, utilize NAD or NADP as an enzymatic cofactor and catalyze the oxidation/reduction of various substrates, facilitating the usually reversible stereospecific hydride transfer from the C4 position of the 1,4 dihydronicotinamide ring of NAD(P)H to substrate. The reaction catalyzed by lactate dehydrogenase and a schematic drawing of the putative hydride transfer reaction that takes place are shown in Fig. 15.1.

The transition state paradigm states that enzymes achieve their high catalytic power by stabilizing the transition state and/or destabilizing (or activating) the ground state to reduce the reaction barrier. The forces that bring about transition state stabilization or ground state destabilization can be deduced typically from an examination of the distortions that enzymes bring about on bound substrates, or substrate mimics, or from the structures of bound transition state mimics. For example, stabilization of the transition state almost always brings the ground state of bound substrates structurally closer to the transition state compared to when unbound, even if the distortion may be small. Activation of the ground state typically shows up in larger distortions.

Our purpose here is to review spectroscopic approaches, optical and vibrational, applied to the determination of enzyme structure and dynamics. We focus on hydride transfer reactions in protein catalysis. Vibrational spectroscopy is especially useful in the study of the molecular mechanism of enzymes because it is structurally specific and is of high resolution; bond distortions as small as 0.01–0.001 Å can be discerned by vibrational spectroscopy. It is at this level of atomic resolution that enzyme induced bond distortions usually manifest themselves. In addition, both enthalpic and entropic factors can be characterized by vibrational spectroscopy, sometimes in quantitative terms. Although most of the chapter is concerned with the structures of static protein–ligand complexes, the dynamics of how these complexes are formed and depleted has recently become a viable topic for scientific

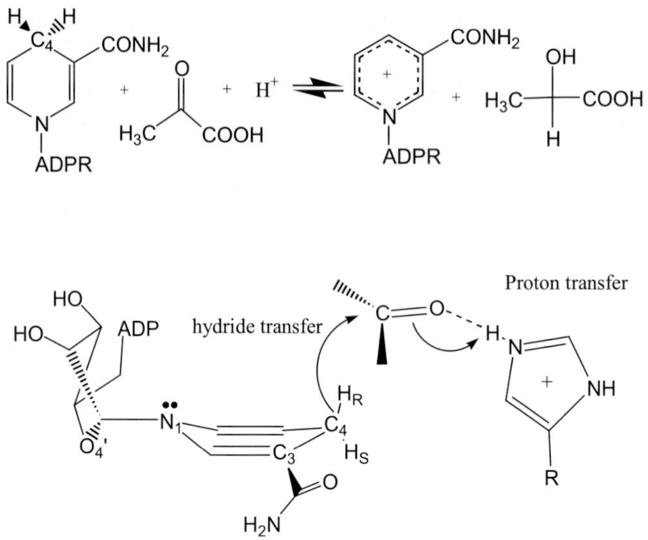

Figure 15.1. Reaction catalyzed by LDH and active site contacts in LDH. It shows an A-side hydride transfer reaction from NADH to a substrate C=O which is accompanied by a proton transfer, either sequentially or simultaneously, from His195. The 1,4 dihydronicotinamide is in an "anti" conformation (relative to ribose) and its amide arm at C3 is in transoid conformation, and pro-R hydrogen at the C4 is in a pseudoaxial position.

research. Aspects of the dynamical nature of enzymes revealed by optical/vibrational spectroscopies are also then discussed here.

The reason that vibrational spectroscopy yields useful information on a quantitative basis is that the observed vibrational frequencies arise from interatomic force constants; these are a measure of the distribution of the electrons within the molecule [1–3]. This electronic distribution is disturbed, in some cases, by protein–ligand interactions, in direct proportion to the degree of interaction. For example, the strengths of hydrogen bonds in model systems can be determined directly from linear free energy correlations between frequency measurements and the ΔH of formation of the H-bonds. Besides the determination of electron density in a chemical bond, vibrational spectroscopy can also be used to relate molecular structure/conformation to the size of interactions between molecular groups, especially when complemented by vibrational analysis based on quantum chemical calculations. Thus, this spectroscopic method is well suited for the determination of the substrate activation using enzyme–ligand complexes that simulate the ground state or transition state of the enzyme catalyzed reaction. The vibrational approach to the study of enzyme–ligand interactions has been reviewed previously [4–7], and the reader may find these reviews quite useful and interesting.

The chapter is organized as a series of summaries of actual uses of optical/vibrational spectroscopies based on selected experiments of specific enzymes. We relate the results to the molecular mechanism. We feel this is the most productive way of providing an overview of spectroscopic probes of our enzyme class. Much of our past work on the NAD(P) linked enzymes, for example, concerns lactate dehydrogenase and dihydrofolate reductase, and we use these systems and others as selected specific examples in reviewing how various spectroscopic methods yield useful information on enzymatic catalysis within this enzyme group.

15.2
Substrate Activation for Hydride Transfer

For concomitant proton transfer to a substrate C=O oxygen or C=N nitrogen (see Fig. 15.1), the enzyme can activate the substrate for catalysis by increasing the partial positive charge on the carbon of the to-be-reduced substrate C=O or C=N bond. This can be achieved by forming a strong hydrogen bond, or other form of electrostatic interaction, to the oxygen of the C=O (C=N) bond, thereby polarizing it and stabilizing the polar nature of the transition state, or by protonation of the C=N bond. Both forms of activation have been found.

15.2.1
Substrate C=O Bond Activation

C=O stretch mode is quite strong in Raman or IR spectra and in some cases quantitative correlation between hydrogen bonding energy and C=O stretch frequency can be determined [8, 9]. Thus, it is ideally suited for studies of the substrate C=O bond activation in the enzyme complexes. Two dehydrogenases, LDH and LADH, have been studied in detail by vibrational spectroscopy.

15.2.1.1 Hydrogen Bond Formation with the C=O Bond of Pyruvate in LDH
Lactate dehydrogenase (LDH) accelerates the oxidation of lactate by NAD^+ to pyruvate and NADH by about 10^{14}-fold relative to a corresponding model reaction [10]. This reversible reaction involves the direct transfer of a hydride ion, H^- to the pro-R, *re-face* of the nicotinamide moiety of NAD^+ from the C2 carbon of L-lactate forming NADH and pyruvate with a high degree of stereochemical fidelity [11]. The nature of the LDH/NADH•pyruvate complex was originally studied by examining an adduct complex, E/NAD-pyr, that is formed by the addition of the C3 carbon of pyruvate enol to the C4 position of the nicotinamide ring of NAD^+ in the presence of LDH [12, 13]. It is believed that the interactions between the enzyme and pyruvate in the E/NADH•pyr central complex are largely maintained in the adduct complex, in spite of the additional covalent linkage present in the adduct complex.

The frequency of the C=O stretch of pyruvate in solution lies at 1710 cm^{-1}. The bond order, hence stretch frequency, of this polar bond is modulated by electrostatic interactions. The stretch frequency is therefore a monitor of these interactions, which are believed to be quite important in stabilizing the transition state of LDH. It is obviously therefore of interest to determine the stretch frequency for

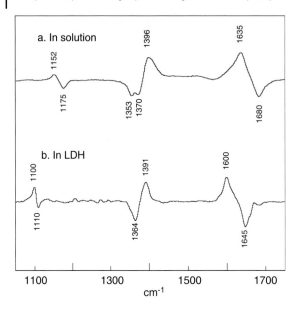

Figure 15.2. Isotope edited difference Raman spectra between $^{13}C={}^{18}O$ pyruvate and $^{13}C=O$ pyruvate in solution and in LDH/NAD-pyr adduct.

pyruvate bound to the enzyme. Although this is just one C=O moiety in the enzyme–substrate complex among hundreds of others, it is possible to determine the C=O stretch of the bound pyruvate unambiguously using isotope editing approaches. In this approach, ternary E/NAD-pyr complexes are formed using pyruvate with and without ^{13}C (and/or ^{18}O) label on the C=O bond under identical conditions and their spectra are taken separately and then subtracted. The difference spectrum only shows the vibrational modes that are affected by the isotope labeling as positive and negative peak pairs, while all other protein and ligand peaks not affected by the isotope label are cancelled out. Isotope edited difference Raman studies showed that the frequency of the carbonyl stretch of pyruvate shifts downward by 35 cm^{-1} relative to its solution value upon forming a complex with pig heart LDH (Fig. 15.2). This downward shift in frequency is associated with a strong bond polarization where a significant single bond, $^+$C–O$^-$, resonance form is mixed into the mostly double bond of the carbonyl upon binding. The net interaction energy between the protein and the carbonyl moiety that produces this bond polarization is about −14 to −17 kcal mol^{-1} [14]. Further studies with a number of mutant LDHs revealed that there is a good correlation between the hydride transfer rate and the pyruvate C=O stretch frequency shifts in the E/NAD-pyr complex [15]; Fig. 15.3. The reasonable agreement observed in Fig. 15.3 can be rationalized by supposing that the enthalpy of the transition state is stabilized to an even greater extent than the ground state, so that the net reaction barrier is lowered by the electrostatic interaction. This is true because the carbonyl moiety is substantially more

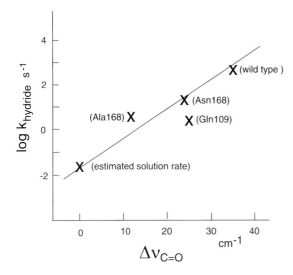

Figure 15.3. The log of the rate of hydride transfer step (in s^{-1}) versus the observed change in C=O stretch frequency of pyruvate upon binding to LDH/NADH.

polarized in the transition state, with more negative character on the oxygen. Also, it seems likely that the proton located on the imidazole in the ground state is now closer to the oxygen in the transition state. For example, *ab initio* calculations on hydride transfer reaction in a model system place the proton closer to the oxygen than to the imidazole ring [16]. This analysis is in accord with a previous study that showed that the equilibrium towards enol-pyruvate from pyruvate is increased by $10^{5.5}$-fold at the active site in LDH compared to solution by the stabilization of C–O$^-$ character into the pyruvate's C=O. All these results prompted the suggestion that some $10^{5.5}$ fold of LDH's 10^9-fold rate enhancement arises from the C=O $\cdots$ His-195$^+$ interaction [10].

15.2.1.2 Hydrogen Bond Formation with the C=O Bond of Substrate in LADH

Horse liver alcohol dehydrogenase (LADH) catalyzes the reactions of aldehydes and their corresponding alcohols with the coenzymes NADH and NAD$^+$. Activation of substrate complexes via polarization of substrate C=O bond has been observed in LADH by vibrational spectroscopy. Two enzyme complexes have been studied by difference Raman measurements, the E/NADH•DABA complex [17, 18] and the E/NADH•CXF complex [19]. DABA is a poor substrate while CXF is a substrate analog. X-ray crystallography has shown that the polarization of the substrate C=O bond is mainly due to a coordination to the active site Zn^{++} ion [20, 21]. For example, polarization of the C=O bond of DABA in the LADH complex was found to be substantial, half way between a single and double bond as compared to DABA in solution [18].

Vibrational studies revealed substantial features of how CXF binds to LADH/

NADH [19]. *Ab initio* normal mode analyses of the Raman data in solution and in the enzyme complex were carried out and empirical correlations were established that yielded bond stretch versus interaction enthalpy relationships. The total binding Gibbs free energy between CXF and LADH/NADH complex is about 7 kcal mol^{-1}. Since this is made up of both enthalpic and entropic components, and the entropic $T\Delta S$ term of the Gibbs free energy is almost certainly unfavorable, the binding enthalpy is greater than 7 kcal mol^{-1}. It was found that 4 kcal mol^{-1} of the enthalpic binding energy is from the amide moiety. The binding of the carbonyl group of CXF to the catalytic zinc and the hydrogen bond between the oxygen and the hydroxy group of Ser-48 would account for about 5.5 kcal mol^{-1} to the enthalpic term. The binding of the amide N–H moiety is destabilizing, although it would be even more destabilizing if not for the cation–π interaction at the binding site [19].

15.2.2
Substrate C=N Bond Activation

The intensity of the C=N stretch mode is also quite strong in Raman or IR spectra. Activation of the C=N bond for hydride transfer can be realized by formation of hydrogen bonding to the C=N nitrogen or by protonation of the C=N nitrogen. These two cases can be distinguished conclusively by vibrational spectroscopy, as shown below.

15.2.2.1 N5 Protonation of 7,8-Dihydrofolate in DHFR

Dihydrofolate reductase catalyzes the reduction of 7,8-dihydrofolate (H$_2$folate) to 5,6,7,8-tetrahydrofolate (H$_4$folate) by facilitating the addition of a proton to N5 of H$_2$folate and the transfer of a hydride ion from the pro-R side of NADPH C4 to C6 (see Fig. 15.4). Despite extensive kinetic, site directed mutagenesis, X-ray crystallographic, and theoretical molecular modeling studies that have been performed on this enzyme, the reaction mechanism of DHFR is still under debate. In fact, the electronic nature of the ground state within the active site in the productive DHFR•NADPH•H$_2$folate complex is unclear, and this is key to an understanding of the reaction mechanism of DHFR. For example, X-ray crystallographic studies have revealed that the only ionizable group near the pteridine ring is a carboxylic acid, equivalent to Asp27 in *E. coli* DHFR and there is no protein residue in the immediate vicinity of H$_2$folate N5 that can act as a general acid/base or form a hydrogen bond to N5 [22–24]. Thus, activation of the H$_2$folate substrate, if it exists, is unlikely to be realized by a strong hydrogen bond to C6=N5. It is possible that there are one or more binding site water molecules interacting with the pterin ring and that the proton that ends up on N5 arrives from the binding site carboxyl group via these water molecules. In this regard, the rate of hydride transfer from NADPH to H$_2$folate has a pK_a of around 6.5 compared with the pK_a of 2.6 for N5 of DHF in solution [25, 26].

The possibility of activating the H$_2$folate substrate in DHFR/NADPH•H$_2$folate complex by the protonation of N5 for the subsequent hydride transfer was pro-

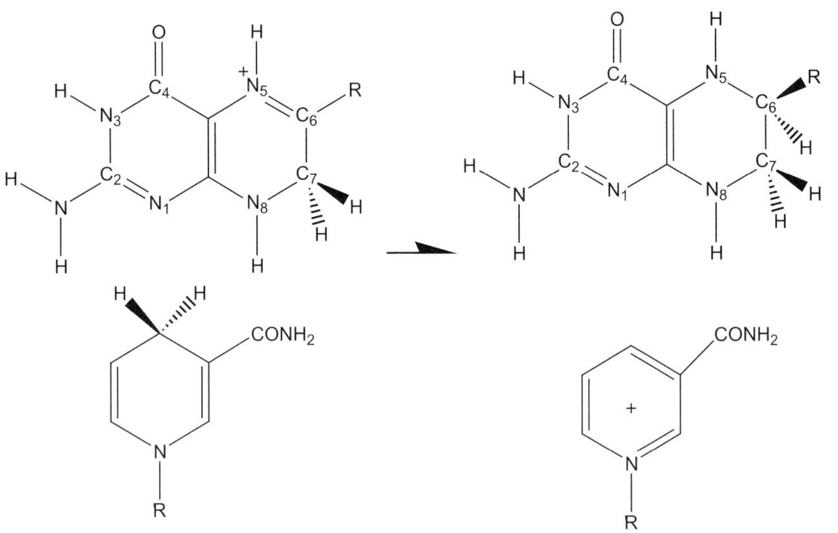

Figure 15.4. Reaction catalyzed by DHFR.

posed [23] and later supported by the Raman difference spectroscopic studies of the DHFR/NADP$^+$•DHF complex, which is a mimic of the DHFR/NADPH•H$_2$folate Michaelis complex [27]. In the Raman spectrum of DHFR/NADP$^+$•H$_2$folate complex, two N5=C6 stretch 'marker' bands indicating unprotonated (1650 cm^{-1}) or protonated (1675 cm^{-1}) N5 were identified. The assignments were based on isotope labeling, comparisons to the spectra of solution models, and the positioning of the bands. Based on these assignments, N5 in the binary DHFR/H$_2$folate complex was found to be unprotonated at near neutral pH values. In the ternary DHFR/NADP$^+$•H$_2$folate complex, however, another band at 1675 cm^{-1} was observed (Fig. 15.5). A titration study, using the 1650 and protonated 1675 cm^{-1} marker bands as indicators for unprotonated and protonated species respectively, showed that the pK_a of N5 is raised from 2.6 in solution [25] to 6.5 in this complex [27] (Fig. 15.6). Further studies based on *ab initio* vibrational analysis of the Raman data confirmed the original assignments and further indicated that the immediate environment of N5 in the DHFR•NADP$^+$•H$_2$folate complex is quite hydrophobic [28]. The activation of the H$_2$folate due to N5 protonation can be rationalized by the reduced electron clouds near the C6=N5 carbon [28, 29].

In summary, vibrational spectroscopic studies have shown that in the enzyme catalyzed hydride transfer reaction, the substrate C=O or C=N bond in the Michaelis complex may first be activated in two different ways for the subsequent hydride transfer. For a substrate that contains a C=O bond to be reduced by the enzyme, strong electron withdrawing interaction due to hydrogen bonding (in LDH) or electrostatic interaction (in LADH) can polarize this bond to reduce the electron cloud near the carbonyl carbon to facilitate the hydride transfer. This is consistent with theoretical studies that one of the driving forces for the hydride transfer is the large

15 Spectroscopic Probes of Hydride Transfer Activation by Enzymes

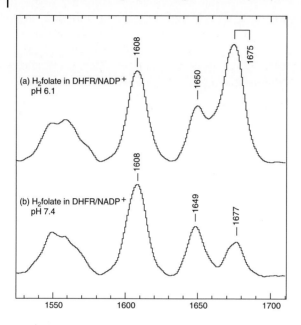

Figure 15.5. Difference Raman spectra between DHFR/NADP$^+$•H$_2$folate and DHFR/NADP$^+$ at pH 6.1 and pH 7.4.

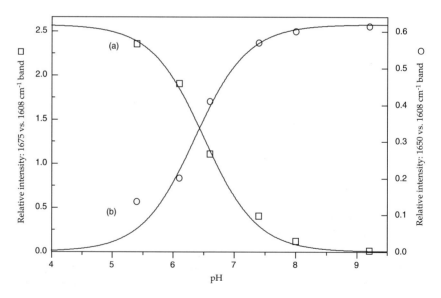

Figure 15.6. pH dependence of the intensities of the C=NH band (1675 cm^{-1}) and the C=N band (1650 cm^{-1}) of H$_2$folate in the DHFR/NADP$^+$•H$_2$folate complex. The pK_a = 6.5.

partial positive charge on the carbonyl carbon caused by the strong hydrogen bonding on the C=O bond [30]. For a substrate that contains a C=N bond, protonation of the nitrogen can be used by the enzyme to reduce the electron cloud near the Schiff base carbon to facilitate the hydride transfer. In this case, hydrogen bonding to N–H would be electron donating and thus not required for the activation of the substrate.

15.3
NAD(P) Cofactor Activation for Hydride Transfer by Enzymes

Previous theoretical and enzyme kinetic studies have identified a number of factors that could be important in activating the NAD(P) cofactor. Several of these factors can be studied by spectroscopy to determine if and how the cofactor NAD(P) is activated within the Michaelis complex to facilitate the hydride transfer.

15.3.1
Ring Puckering of Reduced Nicotinamide and Hydride Transfer

Theoretical calculations suggested that the features of the transition state structure of the hydride transfer reaction include a ring puckering of the dihydronicotinamide ring of NADH, which renders the transferred hydrogen at the psuedoaxial position [16, 30, 31]. The formation of a boat conformation may contribute to a reduction of the reaction barrier by as much as 4–6 kcal mol^{-1} with C4 and N1 out of the ring plane by 10–15 degrees, as predicted by semiempirical AM1 studies of LDH [30]. Thus, if such a form of activation is significant, a pro-R hydrogen will be observed at the pseudoaxial position in A-side enzymes (in which the pro-R hydrogen is transferred) and pro-S hydrogen at the pseudoaxial position in B-side enzymes (in which the pro-S hydrogen is transferred).

The frequencies of the carbon–hydrogen stretching mode of the pro-R and pro-S C4-H bonds of NADH are a sensitive indicator of the nicotinamide planarity; they have been determined in solution and when bound to pig heart lactate dehydrogenase (LDH) by isotope edited Raman spectroscopy [32]; Fig. 15.7. This was achieved by specifically deuterating the C4 pro-R or pro-S hydrogens of NADH and determining the frequencies of the resulting C4–D stretches by Raman difference spectroscopy. The frequencies of the two C4–D stretching modes for the two bonds are essentially the same for the unliganded coenzyme in solution. On the other hand, the position of the pro-S [4-^{2}H]NADH stretch shifts upwards by about 23–30 cm^{-1} in its binary complex with lactate dehydrogenase relative to that observed in solution, while that for the bound pro-R [4-^{2}H]NADH is relatively unchanged. Semiempirical quantum mechanical calculations (MINDO/3, MNDO and AM1) suggested that the orientation of the amide arm, puckering of the reduced nicotinamide ring, and external charge or dipole are among the factors that can affect the C4–H stretch frequency. Within the range of the study, the positions of the C4–D stretches may be understood as the result of two conformational

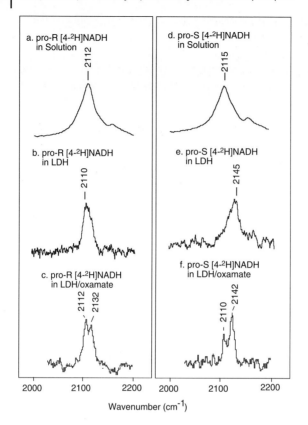

Figure 15.7. Raman difference spectra of (a) 100 mM pro-R [4-^{2}H]NADH at 4 °C, $\Gamma_{1/2} = 45$ cm^{-1}; (b) pro-R [4-^{2}H]NADH in LDH (LDH•NADH = 1.5/5 mM) at 4 °C, $\Gamma_{1/2} = 16$ cm^{-1}; (c) pro-R [4-^{2}H]NADH complexed with LDH•oxamate (LDH/NADH•oxamate = 1.5/5/5 mM) at 4 °C, $\Gamma_{1/2} = 6$ cm^{-1} for either the 2112 or 2132 cm^{-1} band; (d) 100 mM pro-S [4-^{2}H]NADH at 4 °C, $\Gamma_{1/2} = 45$ cm^{-1}; (e) pro-S [4-^{2}H]NADH in LDH (LDH•NADH = 1.5/5 mM) at 4 °C, $\Gamma_{1/2} = 22$ cm^{-1}; (f) pro-S [4-^{2}H]NADH complexed with LDH•oxamate (LDH/NADH•oxamate = 1.5/5/5 mM) at 4 °C $\Gamma_{1/2} = 6$ cm^{-1} for either the 2110 or 2124 cm^{-1} band. All spectra were obtained by subtracting the corresponding NADH spectrum taken under the same conditions.

changes of the nicotinamide ring that occur when NADH forms a binary complex with LDH: the rotation of the amide group from a solution *cisoid* to *transoid, in situ*, (which results in a blue shift of the average of the two C–D stretch frequencies) and the adoption of a 'half-boat' of the dihydronicotinamide ring of NADH when bound to the enzyme from an essentially planar solution structure (this results in a higher C–D stretch frequency of the pseudoequitorial C–D bond). The estimated angle of the C4 ring carbon with respect to the other carbon atoms is around 10–15 degrees, with the pro-R hydrogen at a pseudo-axial position and the pro-S hydrogen at a pseudo-equatorial position. Since LDH is a so-called A-side enzyme (transfers the pro-R hydrogen), the ground state structural distortions imposed on the cofactor appear to populate preferentially the correct ring geometry for the hydride transfer reaction.

Similar difference Raman studies have been carried out on two other enzymes in their binary and ternary (Michaelis mimics) complexes with pro-R or pro-S specifically labeled NAD(P)H: the A-side specific dihydrofolate reductase (DHFR) and the B-side specific glycerol-3-phosphate dehydrogenase (G3PDH) [33]. DHFR shows quite similar behavior to that of LDH. In the ternary complex data involving B-side enzyme G3PDH, only a single band is observed and, hence, only a single conformer. The nearly equal frequencies for the C4 deuteron of NADH bound in the G3PDH ternary complex suggest that the ring is essentially planar. For all three ternary complexes, the bandwidths of the C4–D stretch modes have narrowed by a factor of two from those of the binary complexes, indicating the flexibility of the nicotinamide is further limited upon binding of the inhibitor. This is discussed further in Section 15.3.3.

15.3.2
Effects of the Carboxylamide Orientation on the Hydride Transfer

Theoretical studies suggest that a transoid amide significantly reduces the activation energy for hydride transfer compared to a cisoid amide, even though the cisoid conformation is more stable in solution [31]. In addition, there is a clear preference for the amide carbonyl oxygen to be located on the same face as the transferring hydride when the amide is transoid. This orientation preference has been attributed to electrostatic attraction between the amide oxygen and the partially positively charged hydride acceptor that is absent with a cisoid amide [31]. Thus, hydride transfer activation can be achieved by orientating the amide arm to the transoid conformation.

Several vibrational frequencies of the reduced nicotinamide are sensitive to the orientation of the amide. Besides the C4–D stretch modes, discussed in Section 15.3.1, the coupled ring C=C stretch and the C=O stretch modes of the reduced nicotinamide ring are also sensitive to the amide orientation. Vibrational analysis based on *ab initio* calculations suggests that the stretch motions of the two C=C bonds in the reduced nicotinamide ring and the C=O stretch of the amide are highly coupled to form three vibrational modes [32]. The vibrational mode formed by the anti-phase combination of the two C=C stretches is the highest frequency mode and is not sensitive to the orientation of the amide. The vibrational mode with the lowest frequency of the three modes is very sensitive to the amide orientation, shifts up by nearly 40 cm^{-1} when the amide is changed from cisoid to transoid. Since this mode intensity is quite strong in the Raman spectrum of NADH, it can be conveniently used for the determination of the amide orientation. According to such criteria, the amide of NADH in LDH and cytoplasmic malate dehydrogenase binary complexes is transoid while in the mitochondrial malate dehydrogenase it is cisoid [34, 35].

Apparently, the orientation of the amide can be controlled by active site contacts in the enzymes. Difference Raman studies of an NAD analog APAD, in which the amide group of NAD is replaced by an acetyl group, showed that the hydrogen bonding on the acetyl C=O bond in the LDH complex is 4–5 kcal mol^{-1} stronger than in aqueous solution [36].

15.3.3
Spectroscopic Signatures of "Entropic Activation" of Hydride Transfer

Entopic effects may play a role in enzymatic catalysis by lowering the $T\Delta S$ component of ΔG between the ground state and the transition state along the reaction coordinate for the enzyme catalyzed reaction, relative to the reaction pathway in solution (cf. Jencks, [37]). This may come about either by lowering the number of available states in the ground state (ground state entropic loss) or by increasing the number of available states in the transition state (transition state entropic gain). This concept has recently been developed quantitatively using numerical simulations [30, 38]. These computational studies emphasize bringing the substrates from the ground state to a state closely resembling the structure of the transition state that is called the "near attack conformation" (NAC's). Several states may be accessible from the ground state enzyme–substrate complex, with some states approaching the NAC structure and, hence, being poised for bond breaking/making to occur while others are not catalytically productive. The enzymatic rate enhancement is then proportional to the fraction of time that the complex spends in the NAC conformation.

In the Raman difference spectroscopic studies of the C4 deuterated NADH bound to LDH, it was observed that the width of the C4–D stretches of [4,4-D2]NADH decreases by a factor of ~2.5 upon the formation of a ternary complex with LDH and oxamate [36] (Fig. 15.7). Such a change in the Raman spectrum of bound NADH from the binary to the ternary complex suggests that the protein conformational change that accompanies loop closure and the formation of the ternary Michaelis complex involves a 'stiffening' of the active site in addition to the changes in electrostatic interactions reviewed in the discussion above. The most likely explanation for the unusual width of the C4–D stretch bands in water and in the binary complex is that the reduced dihydronicotinamide ring adopts various boat conformations (as discussed in Section 15.3.1) since quantum mechanical calculations suggest there is little energy difference between various boat conformations, including the planar conformation [16, 32, 39]. Since the frequency of the C–D stretch is sensitive to this angle [32], a heterogeneous mixture of various boat forms results in the observed broad bands. Band narrowing arises from the selection of a particular ring conformation forced by the formation of the E/NADH●oxamate Michaelis mimic complex.

The data thus suggest that the protein conformational change driven by substrate binding decreases the number of nicotinamide ring conformations that are energetically available to the coenzyme in the ternary E/NADH●substrate complex compared to the E/NADH binary complex and presumably aligns the C4–H (pro-R) bond along the direction of the reaction coordinate in the transition state. This suggests that entropic effects are involved in substrate binding and probably influence catalytic activity. Indeed, Burgner and Ray [10] estimated that the immobilization of reactants at the active site of LDH contributes at least a thousand-fold (ca 4.2 kcal mol^{-1}) to the lowering of the transition state barrier for hydride transfer to and from NAD to substrate on the basis of a series of reactions catalyzed by LDH.

The approximate change in entropy between the binary LDH/NADH and the ternary LDH/NADH•oxamate complexes, as monitored by the changes observed on the C4–H bond, is estimated by assuming that the number of available states, Ω, is proportional to the heterogeneous bandwidth of the observed Raman band. For example, taking the bandwidth of the C4–D stretch in the ternary complex as arising from essentially homogeneous line broadening processes and using this to fit the profile of this normal mode in the binary data, we find that three bands of essentially equal strength are needed. Thus, there are three states available to the ring in the binary complex as opposed to one in the ternary complex. Since $S = R \ln \Omega$, $T\Delta S = RT \ln 3$ or 0.7 kcal mol^{-1}. Of course, this just involves the entropy change associated with the available states of the coenzyme that affect the C–D stretch at the C4 position, and the total entropic change is likely to be larger [36].

15.3.4
Activation of C–H bonds in NAD(P)$^+$ or NAD(P)H

It has been argued that the activation of the oxidized nicotinamide in the ground state complex should force the C4 of NAD(P)$^+$ from sp2- to sp3-like. In this case, the C4–H bond order, and thus the bond stretch frequency, should be reduced [11, 31]. A relatively large equilibrium deuterium and tritium isotope effect ($\sim$1.1) on binding of C-4H labeled NAD$^+$ to LDH was detected. If the entire isotope effect is assumed to be due to the frequency change of the C4–H stretch upon NAD$^+$ binding to LDH, the predicted downward shift of the C4–H stretch frequency is $\sim$100 cm^{-1} for NAD$^+$ bound to LDH compared to free in solution [11]. By the same argument, activation of the NAD(P)H C4–H bond that is not involved in the hydride transfer should result in a bond order increase, thus a blue shift of the stretch frequency, by the distortion of the electronic nature of C4 from sp3 to sp2. It is less clear how the C4–H stretch frequency should change for the C4–H bond involved in the hydride transfer since this would depend on the nature of the hydride transfer process: two-electron transfer followed by proton transfer, or vice versa, or all at the same time.

In any case, the stretch frequency of the C4–H bond is a direct monitor of its bond order and therefore presents an opportunity to test the concepts above. On the NADH side, large shifts in the frequency of the C4–D stretch are *not* observed when NAD(P)H binds to the enzymes studied thus far (the A-side specific lactate dehydrogenase (LDH) and dihydrofolate reductase (DHFR) and the B-side specific glycerol-3-phosphate dehydrogenase (G3PDH)), as would be predicted [33]. The observed frequency shifts are generally not larger than the heterogeneously broadened band of the cofactor's solution spectrum. In addition, the C4–H frequencies in these three enzymes also do *not* indicate a direct relationship between frequency and activity. This suggests that the rate enhancements accomplished by NAD(P)H dependent enzymes do not reside in the enzymes ability to activate C4–H in the ground state by changing the electronic nature of the C4–H bond. With regards to the NAD(P)$^+$ side of the reaction, the C4–D stretch mode is essentially unaffected when NAD(P)$^+$ binds to either LDH or DHFR [33]. Neither the position nor the bandwidth is changed by binding upon formation of enzyme NAD(P)$^+$. Small shifts in frequency are observed for NAD(P)$^+$ in the ternary complexes made with LDH and DHFR compared to the respective binary complexes, but it is hard to discern any pattern since the shift is positive in one protein (DHFR) and negative in the other (LDH). ^{13}C NMR studies of the ^{13}C4 labeled NAD$^+$ bound to UDP-galactose 4-epimerase showed that the ^{13}C4 signal was shifted downfield by 3.4 ppm, consistent with electron withdrawing from C4 in the enzyme [40]. Studies on a series of NAD$^+$ analog *N*-alkylnicotinamides, whose rates of reduction by cyanoborohydride in aqueous solution differ by three orders of magnitude, did show a clear correlation between the C4 chemical shift reduction rate (3.4 ppm downfield chemical shift correlated to 3200-fold rate increase). However, the Raman studies found only small changes in the C4–D stretch frequencies of the same series of *N*-alkylnicotinamides (less than 1 cm^{-1}, our unpublished observations). Thus, it seems clear that C4 can be activated without affecting the electronic nature of the C4–H bond [33].

15.4
Dynamics of Protein Catalysis and Hydride Transfer Activation

While the static structures of a substantial number of Michaelis complexes are known for many enzymes, little is known of the nature or degree regarding protein conformational flexibility that is required to form the Michaelis complex and to accommodate the changing chemical nature of the bound ligand as the system evolves along the reaction coordinate. Just how a ligand binds to a protein, the specific pathway(s), the time ordering of events, and the atoms and groups of atoms involved in the binding process, is largely uncharacterized. These dynamical processes cover a wide range of time scales, from picoseconds, involving small scale displacement of atoms or molecular groups, to nanoseconds/microseconds involving activated motions of atomic groups (such as motions of loops either on the surface of the protein or buried), to millisecond times or longer for activated motions of domains. In general, time scales shorter than milliseconds have been difficult to access experimentally, and this has been a major reason for the lack of characterization of enzyme associated dynamics.

Over the past years, a number of approaches to the study of atomic motion within proteins have been developed and are undergoing continued development. These include NMR relaxation spectroscopies, line shape analysis of spectral bands, and others. Our purpose here is to discuss the uses that investigators have developed for the optical/vibrational spectroscopic study of protein dynamics. We further narrow this perspective to examine briefly experimental studies of the 'pump–probe' type that have been carried out on the NAD(P) linked enzymes. In this approach, the chemistry of the system is initiated by some perturbant, and the evolving structural changes are followed by a suitable probe. This is the typical arrangement for conventional stopped-flow experiments, the problem being that the normal resolution of this approach is no better than around one millisecond. This limitation is being remedied to an extent by the development of fast mixers that can achieve a resolution of 10–50 μs (see e.g. Ref. [41]). For the NAD(P) enzymes, the rather limited number of studies that access time scales faster than one millisecond have centered on the use of T-jump relaxation spectroscopy. This approach was developed in the 1960s [42] and has recently found new and fruitful uses. Here the temperature of a chemical system (any chemical system that is in equilibrium) is quickly raised by 10–30 °C, typically now by irradiating, for example, a protein solution by a pulse of near-IR laser light tuned to weak water absorption bands. As the chemical system evolves to a new equilibrium point defined by the new temperature, structural changes are monitored by spectroscopic probes. The approach is capable of achieving a resolution of about 10 ps [43]. Optical/vibrational spectroscopies can follow events and structural changes on such fast time scales since their characteristic time scales are sub-picosecond, even femtosecond. The methodology is therefore suitable to study the dynamics of enzymatic catalysis over multiple time scales from picoseconds to minutes [44].

15.4.1
The Approach to the Michaelis Complex: the Binding of Ligands

The binding of substrate in LDH is ordered and follows the formation of LDH/NADH binary complex. The substrate binding pocket is somewhat deep into the protein [45]. It supplies the catalytically crucial His195 which acts as a general acid/base in the catalyzed reaction and also polarizes pyruvate's C=O bond and stabilizes the transition state, which contains a highly polarized –C–O⁻-like bond; the preformed pocket also 'solvates' the substrate's carboxyl group by supplying Arg171 [15]. Once the substrate reaches a position close enough to the enzyme's active site, a number of events take place. A surface loop, or flap, of the polypeptide chain, residues 98–110 (often referred to as the 'mobile loop'), closes over the active site entrance, the key residue Arg109 located on the loop is brought deep into the active site in close contact with the C=O bond of bound substrate, water leaves the binding pocket, the enzyme tightens around the bound substrate and NADH bringing them, as well as key protein residues, close together in a proper geometry for the catalyzed reaction to occur, much of which is outlined above. The rate limiting step in the kinetics of hydride transfer catalyzed by LDH is loop closure, which occurs on the 1–10 ms time scale [46]. Spectroscopic probes of the kinetics of structural changes in the NAD(P) linked enzymes on the atomic scale include the absorption and emission of NAD(P)H, which occurs in the near-UV and visible, respectively, and the absorptions of the various substrate specific vibrational bands. The sensitivities to structure of various IR bands have been discussed above. In addition, it can be possible to introduce strategically placed chromophores (typically Trp residues) at strategic places within a specific protein of interest; the folding kinetics of LDH [47] was monitored in just this way, as were the structural changes involved with loop closure [48].

The dynamics of the binding of NADH to LDH has been studied by laser induced T-jump spectroscopy and NADH emission [49]; the binding process was found to occur via a multiple-step process over multiple time scales from nanoseconds to milliseconds. The kinetics of substrate binding has been investigated in several studies [46, 50, 51] (and S. McClendon, N. Zhadin, and R. Callender, unpublished). Most of these studies have employed the substrate (pyruvate) mimic, oxamate, which is an unreactive inhibitor of LDH. For example, a laser induced T-jump relaxation study of oxamate binding to LDH/NADH monitoring the changes in emission of bound NADH is shown in Fig. 15.8 [52]. Several steps occurring at 1.5, 53, and 119 µs (the fast ca. 15 ns and slow ca. 5 ms transients in Fig. 15.8 are instrument response). A kinetic scheme of the binding process is built up by performing a series of studies varying the amounts of free [LDH/NADH] and [oxamate] (cf., Ref. [53], which we have recently performed (S. McClendon, N. Zhadin, and R. Callender, unpublished). NADH emission is a very direct probe of substrate (or inhibitor mimic) binding since the emission yield of NADH bound to LDH decreases by a factor of more than ten when the ternary complex is formed.

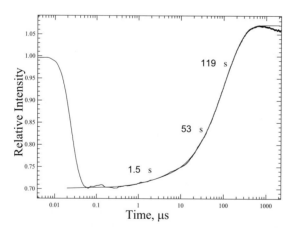

Figure 15.8. The kinetic response of NADH emission at 450 nm stimulated by energy transfer from excited Trp residues as a result of a T-jump from 20 to 41 °C for the reaction of LDH/NADH with oxamate. The kinetics were fitted to a three exponential function with time constants of 1.5, 53, and 119 µs, which is overlaid on the data. The relative initial concentration of LDH, NADH, and oxamate were 100, 200, and 500 µM, respectively. All samples were in 100 mM sodium phosphate, pH = 7.2.

15.4 Dynamics of Protein Catalysis and Hydride Transfer Activation

Holbrook and coworkers [51] first studied the binding kinetics of oxamate with LDH/NADH using a LDH with Tyr-237 nitrated. Putting their results together (cf., Ref. [46] with our current results yields the following preliminary kinetic scheme for the binding of substrate to LDH to form the Michaelis complex (at 20 °C):

$$\text{LDH/NADH} + \text{oxamate} \underset{2000\ \text{s}^{-1}}{\overset{37\ \mu\text{M}^{-1}\text{s}^{-1}}{\rightleftharpoons}} \text{LDH}^1/\text{NADH}\bullet\text{oxamate}$$

$$\underset{4000\ \text{s}^{-1}}{\overset{4000\ \text{s}^{-1}}{\rightleftharpoons}} \text{LDH}^{2(\text{loopOpen})}/\text{NADH}\bullet\text{oxamate}$$

$$\underset{180\ \text{s}^{-1}}{\overset{700\ \text{s}^{-1}}{\rightleftharpoons}} \text{LDH}^{3(\text{loopClosed})}/\text{NADH}\bullet\text{oxamate}$$

In these studies, the temperature dependence of the rates can be tracked so that the thermodynamics of the various steps are determined. One of the most interesting findings is that the on rate to form the initial encounter complex (termed LDH^1/NADH•oxamate in the kinetic scheme above) actually *slows down* with increasing temperature. This implies a temperature dependent activation enthalpy which is explained by the 'melting' or exposure of hydrophobic residues to water in the transition state of the encounter complex formation (cf., Ref. [54]). In order to understand numerically the size of the change in the rate of formation of the encounter complex, about 30–40 residues would have to become unfolded in the transition state. This suggests that the structure of the LDH/NADH binary complex is quite labile comformationally, perhaps designed this way so as to facilitate proper binding of ligands.

On the other hand, the relative motions of key residues at the substrate binding site of LDH moving against the substrate in the LDH/NAD-pyruvate adduct complex, a complex that closely resembles the productive Michaelis complex of this enzyme, except that the substrate is effectively trapped at the active site by the noncovalent contacts between protein and the NADH-like moiety of NAD-pryuvate (see above), were probed on the 10 ns to 10 ms time scale using laser induced temperature jump relaxation spectroscopy while employing isotope edited IR absorption spectroscopy as a structural probe (Fig. 15.9, [52]). The frequencies of NAD-pryuvate adduct's C=O stretch and –COO$^-$ antisymmetric stretch will shift substantially should any relative motion of the polar moieties at the active site (His-195, Asp-168, Arg-109, and Arg-171) occur. Apart from the 'melting' of a few residues on the protein's surface, although the measurements were made with a high degree of accuracy, no kinetics were observed on any time scale in experiments on the bound NAD-pyruvate adduct, even for final temperatures close to the unfolding transition of the protein. This is contrary to simple physical considerations and models. These results were interpreted to mean that, once a productive protein–substrate complex is formed, the binding pocket becomes very rigid, with very little, if any, motion apart from the mobile loop, and that loop opening involves concomitant movement of the substrate out of the binding pocket.

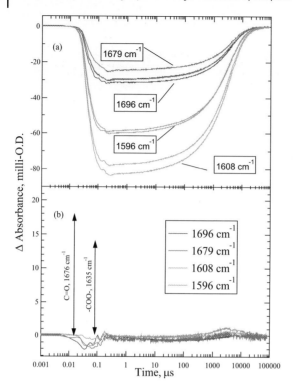

Figure 15.9. (a) Kinetic IR response of LDH/NAD-pyr in 5 mM deuterated sodium phosphate buffer pH* = 7.0 for a T-jump from 6 to 19 °C at 1608, 1596, 1679 and 1696 cm^{-1} for LDH/NAD-pyr (sample side, dashed line) and LDH/NAD-[^{13}C1,^{13}C2]pyr (reference side, solid lines). The path length is 50 μm. The total protein optical absorbances at 1652 cm^{-1} (maximum of the amide-peak of LDH) is 561 mOD, sample side, and 782 mOD, reference. (b) The difference transients formed by taking the difference between sample and reference signals, where the two responses have been scaled to take into account the difference in concentration. The vertical bars represent the peak absorption of the C=O and −COO$^-$ stretch bands at 1676 and 1635 cm^{-1}, respectively, of the LDH/NAD-pyr adduct as deduced from static isotope edited IR spectroscopy scaled for path length and concentration.

15.4.2
Dynamics of Enzymic Bound Substrate–Product Interconversion

In general, the conversion of enzyme–substrate to enzyme–product occurs on the 1 ms time scale over a very wide range of enzymes [38]. Hence, if we are to understand how chemistry is catalyzed by enzymes, dynamical process on shorter time scales require investigation. For the NAD(P) systems, a very preliminary study was performed on LDH using laser induced T-jump relaxation spectroscopy [44].

In this study, at high initial concentrations of lactate and LDH/NAD$^+$, an equilibrium was established minimally consisting of:

15.4 Dynamics of Protein Catalysis and Hydride Transfer Activation

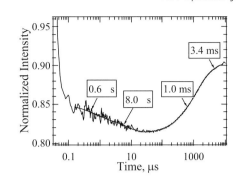

Figure 15.10. Time resolved fluorescence at 450 nm (NADH emission) of a reactive LDH/substrates mixture in response to a T-jump of 10 to 23 °C. The $\lambda_{ex} = 290$ nm. The sample initially contained 100 μN LDH, 100 μM NAD^+, and 10 mM lactate and was allowed to reach equilibrium before the T-jump experiments.

$$LDH/NADH + pyruvate \Leftrightarrow LDH/NADH\bullet pyruvate$$
$$\Leftrightarrow LDH/NAD^+\bullet lactate \text{ (plus free lactate)}$$

The LDH/NADH•pyruvate ternary complex concentration is quite low, and it was found that the concentration of LDH/NADH + pyruvate equals approximately that of the LDH/NAD$^+$•lactate. A temperature increase tips the equilibrium from right to left. Figure 15.10 shows the time-resolved fluorescence emission of NADH at 450 nm in response to a T-jump from 10 to 23 °C. There are two instrument response times: one near 30 ns, which is the pulse width of the laser irradiation heating the sample, and the second is diffusion of heat out from the laser interaction volume that occurs around 15 ms (the latter response is not shown). Fitting the data (solid line) with a function of multiexponentials yielded four rates, as indicated on Fig. 15.10, in addition to these instrument response functions.

The data of Fig. 15.10 are remarkable in showing that there exists a number of fast processes previously unresolved in studies of enzymic dynamics. *It is clear that the minimal number of steps shown above is not sufficient to explain the kinetic data.* In general, the minimal number of species in a kinetic model of the data is equal to one more than the number of observed relaxation rates. Hence, the data can be fit by four processes involving five species. We believe that none of the transients correspond to the reaction LDH/NADH → LDH + NADH because we have measured this reaction [49], and none of the events of the binding reaction correspond to any of signals shown in Fig. 15.10. The dissociation of pyruvate from LDH/NADH would be observed on the millisecond time scale. Taking a binding constant of 0.27 mM and the bimolecular rate constant of 8.33×10^5 M^{-1} s^{-1} yields a $k_{obs} = k_{binding}([LDH\bullet NDH] + [pyruvate]) + k_{release}$ of 3 ms. Tentatively, we assign the 3.4 ms transient to this event. Such an assignment is reinforced by the increasing signal associated with the 3.4 ms transient since there is a large emission increase as pyruvate comes off from LDH/NADH.

Acknowledgments

Supported by research grants GM068036, EB001958.

Abbreviations

LDH, lactate dehydrogenase; DHFR, dihydrofolate reductase; G3PDH, glycerol-3-phosphate dehydrogenase; LADH, liver alcohol dehydrogenase; H_2folate, 7,8-dihydrofolate; H_4folate, 5,6,7,8-tetrahydrofolate; H_2biopterin, 7,8-dihydrobiopterin; NADH, reduced β-nicotinamide adenine dinucleotide; NAD^+, oxidized β-nicotinamide adenine dinucleotide; NADPH, reduced β-nicotinamide adenine dinucleotide phosphate; $NADP^+$, oxidized β-nicotinamide adenine dinucleotide phosphate; DABA, p-(dimethylamino)benzaldehyde; CXF, N-cyclohexylformamide; NAD-pyr, adducts formed in the presence of LDH from pyruvate and NAD^+; T-jump, temperature-jump.

References

1 BELLAMY, L. J. (1975) *Advances in Infrared Group Frequencies*, Vol. 1, Chapman and Hall, London.
2 BELLAMY, L. J. (1980) *The Infrared Spectra of Complex Molecules: Advances in Infrared Group Frequencies*, Vol. 2, 2nd edn., Chapman and Hall, London.
3 COLTHUP, N. B., DALY, L. H., WIBERLEY, S. E. (1990) *Introduction to Infrared and Raman Spectroscopy*, 3rd edn., Academic Press, San Diego.
4 CAREY, P. R. (1999) Raman spectroscopy, the sleeping giant in structural biology, awakes, *J. Biol. Chem.* 274, 26625–26628.
5 DENG, H., CALLENDER, R. (1999) Raman spectroscopic studies of the structures, energetics, and bond distortions of substrates bound to enzymes, *Methods Enzymol.* 308, 176–201.
6 DENG, H., CALLENDER, R. (2001) in *Infrared and Raman Spectroscopy of Biological Materials*, GREMLICH, H.-U., YAN, B. (Eds.), pp. 477–513, Marcel Dekker, New York.
7 CALLENDER, R., DENG, H. (1994) Nonresonance Raman difference spectroscopy: a general probe of protein structure, ligand binding, enzymatic catalysis, and the structures of other biomacromolecules, *Annu. Rev. Biophys. Biomol. Struct.* 23, 215–245.
8 WHITE, A. J., WHARTON, C. W. (1990) Hydrogen-Bonding in Enzymatic Catalysis, *Biochem. J.* 270, 627–637.
9 TONGE, P. J., CAREY, P. R. (1992) Forces, bond lengths, and reactivity: fundamental insight into the mechanism of enzyme catalysis, *Biochemistry* 31, 9122–9125.
10 BURGNER II, J. W., RAY, W. J. (1984) On the Origin of Lactate Dehydrogenase Induced Rate Effect, *Biochemistry* 23, 3636–3648.
11 ANDERSON, V. E., LaREAU, R. D. (1988) Hydride Transfer Catalyzed by Lactate Dehydrogenase Displays Absolute Stereospecificity at the C4 of the Nicotinamide Ring, *J. Am. Chem. Soc.* 110, 3695–3697.
12 BURGNER II, J. W., RAY, W. J. (1978) Mechanistic Study of the Addution of Pyruvate to NAD Catalyzed by Lactate Dehydrogenase, *Biochemistry* 17, 1654–1661.
13 BURGNER II, J. W., RAY, W. J. (1984) The Lactate Dehydrogenase Catalyzed Pyruvate Adduct Reaction:

Simultaneous General Acid-Base Catalysis Involving an Enzyme and an External Catalysis, *Biochemistry 23*, 3626–3635.

14 DENG, H., ZHENG, J., BURGNER, J., CALLENDER, R. (1989) Molecular properties of pyruvate bound to lactate dehydrogenase: a Raman spectroscopic study, *Proc. Natl. Acad. Sci. U.S.A. 86*, 4484–4488.

15 DENG, H., ZHENG, J., CLARKE, A., HOLBROOK, J. J., CALLENDER, R., BURGNER, J. W., 2nd. (1994) Source of catalysis in the lactate dehydrogenase system. Ground-state interactions in the enzyme-substrate complex, *Biochemistry 33*, 2297–2305.

16 WU, Y., HOUK, K. N. (1991) Theoretical Evaluation of Conformational Preferences of NAD+ and NADH: An Approach to Understanding the Stereospecificity of NAD+/NADH-Dependent Dehydrogenases, *J. Am. Chem. Soc. 113*, 2353–2358.

17 JAGODZINSKI, P. W., PETICOLAS, W. L. (1981) Resonance Enhanced Raman Identification of the Zinc-Oxygen Bond in a Horse Liver Alcohol Dehydrogenase-Nicotinamide Adenine Dinucleotide-Aldehyde Transient Chemical Intermediate, *J. Am. Chem. Soc. 103*, 234–236.

18 CALLENDER, R., CHEN, D., LUGTENBURG, J., MARTIN, C., REE, K. W., SLOAN, D., VANDERSTEEN, R., YUE, K. T. (1988) The Molecular Properties of p-Dimethylamino Benzaldehyde Bound to Liver Alchohol Dehydrogenase: a Raman Spectroscopic Study, *Biochemistry 27*, 3672–3681.

19 DENG, H., SCHINDLER, J. F., BERST, K. B., PLAPP, B. V., CALLENDER, R. (1998) A Raman spectroscopic characterization of bonding in the complex of horse liver alcohol dehydrogenase with NADH and N-cyclohexylformamide, *Biochemistry 37*, 14267–14278.

20 CEDERGREN-ZEPPEZAUER, E., SAMAMA, J. P., EKLUND, H. (1982) Crystal structure determinations of coenzyme analogue and substrate complexes of liver alcohol dehydrogenase: binding of 1,4,5,6-tetrahydronicotinamide adenine dinucleotide and trans-4-(N,N-dimethylamino)cinnamaldehyde to the enzyme, *Biochemistry 21*, 4895–4908.

21 RAMASWAMY, S., SCHOLZE, M., PLAPP, B. V. (1997) Binding of formamides to liver alcohol dehydrogenase, *Biochemistry 36*, 3522–3527.

22 MCTIGUE, M. A., DAVIES, J. F., II, KAUFMAN, B. T., KRAUT, J. (1992) Crystal Structures of Chicken Liver Dihydrofolate Reductase Complexed with NADP+ and Biopterin, *Biochemistry 31*, 7264–7273.

23 BYSTROFF, C., OATLEY, S. J., KRAUT, J. (1990) Crystal Structure of *Escherichia coli* Dihydrofolate Reductase: The Binding of NADP+ Holoenzyme and the Folate•NADP+ Ternary Complex. Substrate Binding and a Model for the Transition State, *Biochemistry 29*, 3263–3277.

24 KLON, A. E., HEROUX, A., ROSS, L. J., PATHAK, V., JOHNSON, C. A., PIPER, J. R., BORHANI, D. W. (2002) Atomic structures of human dihydrofolate reductase complexed with NADPH and two lipophilic antifolates at 1.09 a and 1.05 a resolution, *J. Mol. Biol. 320*, 677–693.

25 MAHARAJ, G., SELINSKY, B. S., APPLEMAN, J. R., PERLMAN, M., LONDON, R. E., BLAKLEY, R. L. (1990) Dissociation Constants of Dihydrofolic Acid and Dihydrobiopterin and Implications for Mechanistic Models for Dihydrofolate Reductase, *Biochemistry 29*, 4554–4560.

26 FIERKE, C. A., JOHNSON, K. A., BENKOVIC, S. J. (1987) Construction and Evaluation of the Kinetic Scheme Associated with Dihydrofolate Reductase from *Escherichia coli*, *Biochemistry 26*, 4085–4092.

27 CHEN, Y.-Q., KRAUT, J., BLAKLEY, R. L., CALLENDER, R. (1994) Determination by Raman Spectroscopy of the pKa of N5 of Dihydrofolate Bound to Dihydrofolate Reductase: Mechanistic Implications, *Biochemistry 33*, 7021–7026.

28 DENG, H., CALLENDER, R. (1998) The Structure of Dihydrofolate when

Bound to Dihydrofolate Reductase, *J. Am. Chem. Soc.* 120, 7730–7737.

29 GREADY, J. E. (1985) Theoretical Studies on the Activation of the Pterin Cofactor in the Catalytic Mechanism of Dihydrofolate Reductase, *Biochemistry* 24, 4761–4766.

30 ALMARSSON, O., BRUICE, T. C. (1993) Evaluation of the factors influencing reactivity and stereospecificity in NAD(P)H dependent dehydrogenase enzymes, *J. Am. Chem. Soc.* 115, 2125–2138.

31 WU, Y.-D., LAI, D. K. W., HOUK, K. N. (1995) Transition Structures of Hydride Transfer Reactions of Protonated Pyridinium Ion with 1,4-Dihydropyridine and Protonated Nicotinamide with 1,4-Dihydronicotinamide, *J. Am. Chem. Soc.* 117, 4100–4108.

32 DENG, H., ZHENG, J., SLOAN, D., BURGNER, J., CALLENDER, R. (1992) A vibrational analysis of the catalytically important C4–H bonds of NADH bound to lactate or malate dehydrogenase: ground-state effects, *Biochemistry* 31, 5085–5092.

33 CHEN, Y.-Q., VAN BEEK, J., DENG, H., BURGNER, J., CALLENDER, R. (2002) Vibrational Structure of NAD(P) Cofactors Bound to Three NAD(P) Dependent Enzymes: an Investigation of Ground State Activation, *J. Phys. Chem. B.* 106, 10733–10740.

34 DENG, H., ZHENG, J., SLOAN, D., BURGNER, J., CALLENDER, R. (1989) Classical Raman spectroscopic studies of NADH and NAD+ bound to lactate dehydrogenase by difference techniques, *Biochemistry* 28, 1525–1533.

35 DENG, H., BURGNER, J., CALLENDER, R. (1991) Raman spectroscopic studies of NAD coenzymes bound to malate dehydrogenases by difference techniques, *Biochemistry* 30, 8804–8811.

36 DENG, H., BURGNER, J., CALLENDER, R. (1992) Raman Spectroscopic Studies of the Effects of Substrate Binding on Coenzymes Bound to Lactate Dehydrogenase, *J. Am. Chem. Soc.* 114, 7997–8003.

37 JENCKS, W. P. (1980) in *Molecular Biology, Biochemistry, and Biophysics*, CHAPEVILLE, F., HAENNI, A.-L. (Eds.), pp. 3–25, Springer Verlag, New York.

38 BRUICE, T. C., BENKOVIC, S. J. (2000) Chemical basis for enzyme catalysis, *Biochemistry* 39, 6267–6274.

39 CUMMINS, P., GREADY, J. (1989) Mechanistic Aspects of Biological Redox Reactions Involving NADH 1: ab initio Quantum Chemical Structure of the 1-Methyl-nicotinamide and 1-Methyl-Dihydronicotinamide Coenzyme Analogues, *J. Mol. Struct. (Theochem)* 183, 161–174.

40 BURKE, J. R., FREY, P. A. (1993) The Importance of Binding Energy in Catalysis of Hydride Transfer by UDP-Galactose 4-Epimerase: A 13C and 15N NMR and Kinetic Study, *Biochemistry* 32, 13220–13230.

41 JAMIN, M., YEH, S. R., ROUSSEAU, D. L., BALDWIN, R. L. (1999) Submillisecond unfolding kinetics of apomyoglobin and its pH 4 intermediate, *J. Mol. Biol.* 292, 731–740.

42 EIGEN, M., DE MAEYER, L. D. (1963) in *Techniques of Organic Chemistry*, FRIESS, S. L., LEWIS, E. S., WEISSBERGER, A. (Eds.), pp. 895–1054, Interscience, New York.

43 DYER, R. B., GAI, F., WOODRUFF, W., GILMANSHIN, R., CALLENDER, R. H. (1998) Infrared Studies of Fast Events in Protein Folding, *Acc. Chem. Res.*, 31, 709–716.

44 CALLENDER, R., DYER, R. B. (2002) Probing protein dynamics using temperature jump relaxation spectroscopy, *Curr. Opin. Struct. Biol.* 12, 628–633.

45 HOLBROOK, J. J., LILJAS, A., STEINDEL, S. J., ROSSMANN, M. G. (1975) in *The Enzymes*, BOYER, P. D. (Ed.), pp. 191–293, Academic Press, New York.

46 DUNN, C. R., WILKS, H. M., HALSALL, D. J., ATKINSON, T., CLARKE, A. R., MUIRHEAD, H., HOLBROOK, J. J. (1991) Design and Synthesis of New Enzymes based upon the Lactate dehydrogenase Framework, *Philos. Trans. R. Soc. (London) Ser.B* 332, 177–185.

47 ATKINSON, T., BARSTOW, D., CHIA, W., CLARKE, A., HART, K., WALDMAN, A., WIGLEY, D., WILKS, H., HOLBROOK,

J. J. (1987) Mapping Motion in Large Proteins by Single Tryptophan Probes Inserterd by Site-Directed Mutagenesis: Lactate Dehydrogenase, *Biochem. Soc. Trans. 15*, 991–993.

48 WALDMAN, A. D. B., W., H. K., CLARKE, A. R., WIGLEY, D. B., BARSTOW, D. A., ATKINSON, T., CHIA, W. N., HOLBROOK, J. J. (1988) The Use of a Genetically Engineered Tryptophan to Identify the Movement of a Domain of B. Stearothermophilus Lactate Dehydrogenase with the Process which Limits the Steady-State Turnover of the Enzyme, *Biochem. Biophys. Res. Commun. 150*, 752–759.

49 DENG, H., ZHADIN, N., CALLENDER, R. (2001) Dynamics of protein ligand binding on multiple time scales: NADH binding to lactate dehydrogenase, *Biochemistry 40*, 3767–3773.

50 PARKER, D. M., JECKEL, D., HOLBROOK, J. J. (1982) Slow structural changes shown by the 3-nitrotyrosine-237 residue in pig heart [Tyr(3NO2)237] lactate dehydrogenase, *Biochem. J. 201*, 465–471.

51 CLARKE, A. R., WALDMAN, A. D. B., HART, K. W., HOLBROOK, J. J. (1985) The Rates of Defined Changes in Protein Structure During the Catalytic Cycle of Lactate Dehydrogenase, *Biochim. Biophys. Acta S29*, 397–407.

52 GULOTTA, M., DENG, H., DYER, R. B., CALLENDER, R. H. (2002) Toward an understanding of the role of dynamics on enzymatic catalysis in lactate dehydrogenase, *Biochemistry 41*, 3353–3363.

53 CANTOR, C. R., SCHIMMEL, P. R. (1980) *Biophysical Chemistry*, Vol. 2, W. H. Freeman and Company, San Francisco.

54 FERSHT, A. (1999) *Structure and Mechanism in Protein Science: a Guide to Enzyme Catalysis and Protein Folding*, W.H. Freeman, New York.

Part IV
Hydrogen Transfer in the Action of Specific Enzyme Systems

Although the previous section of this volume refers to numerous enzyme reactions, the focus has been on the general properties of hydrogen tunneling and enzyme dynamics. In the next four chapters, specific enzyme systems are the focus. In the chapter by Tittmann and co-workers, an emphasis is on the C-H activation in thiamin diphosphate, i.e., loss of the C2-H proton to yield the reactive zwitterionic intermediate. The high pK_a of the C2-H in solution implicates specific interactions within the enzyme that increase the rate of proton loss. Through the use of NMR to follow the H/D exchange kinetics, the authors show how functional groups within the cofactor, as well as specific active site protein side chains lead to large increases in the rate of deuterium exchange. One notable effect is that of an allosteric effector in yeast pyruvate dehydrogenase that increases the rate of H/D exchange by 3 orders of magnitude. The contribution by Benkovic and Hammes-Schiffer describes the *E. coli* dihydrofolate reductase (DHFR), which has been one of the major "players" in discerning the role of protein dynamics, and more recently, tunneling in hydride transfer reactions. Both NMR and X-ray crystallographic studies have implicated multiple conformations for DHFR, in particular a closed vs. occluded form, binding substrates and products, respectively. The key question has been how motions within specific regions of the protein correlate with these conformations and facilitate the hydride transfer from NADPH to dihydrofolate. The use of site-specific mutagenesis, coupled to detailed kinetic studies and computational analyses, implicates a network of residues whose correlated motions are coupled to the efficiency of H-transfer. These authors point out that the type of motions that they are measuring are not "dynamically coupled", but rather are equilibrium, thermally averaged conformational changes that change the active site structure in such a way as to favor the hydride transfer from NADPH to dihydrofolate. A focus on hydrogen-atom transfers occurs in the chapter by Banerjee and co-workers on the H-transfer reactions catalyzed by B-12 enzymes. One very interesting aspect of these B-12 enzymes is the coupling of cleavage of the cobalt-carbon bond of the B-12 cofactor to the hydrogen abstraction from substrate, i.e., significant isotope effects are observed on the formation of the cleaved Co(II) form of the cofactor. Stopped flow studies have allowed the measurement of very large, temperature-dependent kinetic isotope effects. QM/MM modeling of these data support a very substantial contribution of tunneling to the reaction coordinate.

In concluding, the authors contrast their data with recent model studies, which indicate isotope effects of a similar size and temperature dependence to the enzyme. Whereas the bulk of the currently available experimental studies increasingly implicate specific roles for a protein in facilitating tunneling, this comparison raises the question of the exact role of methylmalonyl-CoA mutase in catalyzing its tunneling reaction. The contribution by Stein is a departure from the above chapters, with its focus on the important class of proton transfers to and from heteroatoms (illustrated for enzymes catalyzing an addition of water to their substrates). The determination of the size of solvent isotope effects is emphasized as a diagnostic tool to interrogate whether protein-aided acid or base catalysis is occurring; as discussed, by extension of such studies to mixtures of H_2O and D_2O the number of transferred protons can be inferred.

16
Hydrogen Transfer in the Action of Thiamin Diphosphate Enzymes

Gerhard Hübner, Ralph Golbik, and Kai Tittmann

16.1
Introduction

The coenzyme thiamin diphosphate (ThDP, I in Scheme 16.1), the biologically active form of vitamin B_1, is used by different enzymes that perform a wide range of catalytic functions, such as the oxidative and nonoxidative decarboxylation of α-ketoacids, the formation of acetohydroxyacids and ketol transfer between sugars.

In these reactions, the C2-atom of ThDP must be deprotonated to allow this atom to attack the carbonyl carbon of the different substrates. In all ThDP-dependent enzymes this nucleophilic attack of the deprotonated C2-atom of the coenzyme on the substrates results in the formation of a covalent adduct at the C2-atom of the thiazolium ring of the cofactor (IIa and IIb in Scheme 16.1). This reaction requires protonation of the carbonyl oxygen of the substrate and sterical orientation of the substituents. In the next step during catalysis either CO_2, as in the case of decarboxylating enzymes, or an aldo sugar, as in the case of transketolase, is eliminated, accompanied by the formation of an α-carbanion/enamine intermediate (IIIa and IIIb in Scheme 16.1). Dependent on the enzyme this intermediate reacts either by elimination of an aldehyde, such as in pyruvate decarboxylase, or with a second substrate, such as in transketolase and acetohydroxyacid synthase. In these reaction steps proton transfer reactions are involved. Furthermore, the α-carbanion/enamine intermediate (IIIa in Scheme 16.1) can be oxidized in enzymes containing a second cofactor, such as in the α-ketoacid dehydrogenases and pyruvate oxidases. In principal, this oxidation reaction corresponds to a hydride transfer reaction.

In the next section, the mechanism of the C2-H deprotonation of ThDP in enzymes is considered, followed by a discussion of the proton transfer reactions during catalysis. Finally, the oxidation mechanism of the α-carbanion/enamine intermediate in pyruvate oxidase is discussed.

Hydrogen-Transfer Reactions. Edited by J. T. Hynes, J. P. Klinman, H.-H. Limbach, and R. L. Schowen
Copyright © 2007 WILEY-VCH Verlag GmbH & Co. KGaA, Weinheim
ISBN: 978-3-527-30777-7

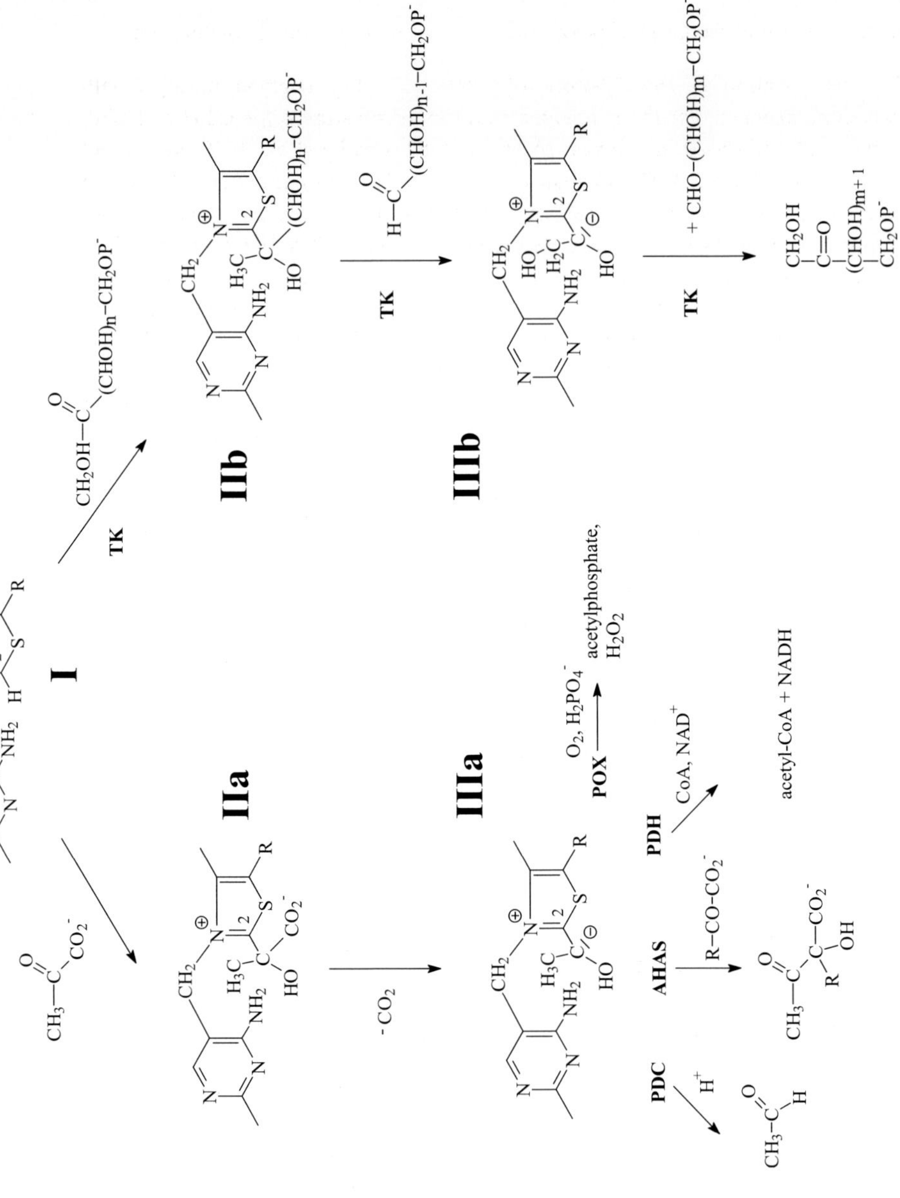
I
IIa
IIb
IIIa
IIIb
TK
- CO2
POX
O2, H2PO4-
acetylphosphate, H2O2
PDH
CoA, NAD+
acetyl-CoA + NADH
AHAS
R–CO–CO2-
PDC
H+

Scheme 16.1

16.2
The Mechanism of the C2-H Deprotonation of Thiamin Diphosphate in Enzymes

The deprotonation of the C2-atom of ThDP is a key reaction in all ThDP-dependent enzymes. For the reaction with different substrates, the C2-H of ThDP, having a pK_a of 17–20 in solution (Breslow, 1962; Crosby et al., 1970; Kemp and O'Brien, 1970; Kluger, 1987; Washabaugh and Jencks, 1988), must be activated by the enzyme environment. ^{13}C NMR investigations on pyruvate decarboxylase containing ^{13}C-C2-labelled ThDP exclude the existence of a C2-carbanion of ThDP in detectable amounts under physiological conditions (Kern et al., 1997). Therefore, in the enzyme-catalysed reaction, the addition of the carbonyl group of any substrate to the C2-atom of ThDP requires essentially a fast dissociation of the C2-proton. In order to determine the rate of this deprotonation, the proton/deuterium exchange kinetics (H/D exchange) of the proton bound to the C2-atom of ThDP in the enzymes pyruvate decarboxylase, transketolase, pyruvate oxidase, and in the pyruvate dehydrogenase multienzyme complex have been examined by ^{1}H NMR.

The exchange reactions were initiated by dilution of a sample solution containing the enzyme (active site concentration 0.1–0.5 mM) with D$_2$O at a mixing ratio of 1+1 in a chemical quenched-flow device. The exchange reactions were stopped by addition of DCl and trichloroacetic acid. In addition, this procedure causes a rapid and complete denaturation and precipitation of the protein and a release of the cofactor. After separation of the denatured protein by centrifugation, the ^{1}H NMR spectra of the supernatant containing the ThDP can be recorded (Kern et al., 1997). These spectra show the C2-H signal of ThDP at 9.55 ppm. For quantification, this signal can be compared with the C6'-H signal at 7.85 ppm as a nonexchanging standard (Fig. 16.1). Under the experimental conditions used, the

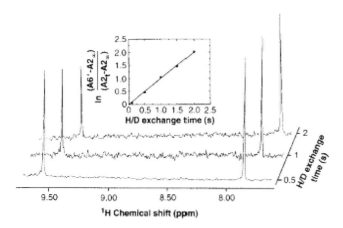

Figure 16.1. Kinetics of H/D exchange of the C2-H in pyruvate decarboxylase from *Saccharomyces cerevisiae*. The ^{1}H NMR spectra are expansions showing the ThDP signals C2-H at 9.55 ppm and C6'-H at 7.85 ppm as a nonexchanging standard for quantification.

H/D exchange follows a pseudo-first order reaction. It must be noticed that the H/D exchange rate constant at the C2-atom of enzyme-bound ThDP represents the lower limit of the C2-H deprotonation, because this observed value reflects not only the deprotonation, but also the exchange rate constant of the base responsible for the deprotonation with solvent protons. The observed rate constant of the H/D-exchange k_{obs} is composed of the rate constant of deuteration k_D of the carbanion intermediate, and that of its reprotonation k_H according to the equation $k_{obs} = \frac{1}{2} \cdot (k_H + k_D)$ (Tittmann, 2000). The fractionation factor ϕ of this reaction is 1.03 (Tittmann, 2000).

16.2.1
Deprotonation Rate of the C2-H of Thiamin Diphosphate in Pyruvate Decarboxylase

Pyruvate decarboxylase catalyses the nonoxidative decarboxylation of pyruvate yielding CO_2 and acetaldehyde. The H/D exchange rate constant of the C2-H of ThDP bound to the homotetrameric and allosterically regulated pyruvate decarboxylase from *Saccharomyces cerevisiae* is accelerated by three orders of magnitude compared with that of free ThDP (Table 16.1). However, this rate constant is still one order of magnitude too small to allow the enzyme catalysis to proceed at the observed catalytic constant of 10 s^{-1} at 4 °C for each active site. Since the value of the catalytic constant represents the rate constant of the allosteric enzyme in the activated state, the H/D exchange of the C2-H of ThDP was investigated in the presence of the allosteric activator pyruvamide (Hübner et al., 1978). In this case, the deprotonation rate constant is at least three orders of magnitude higher than that of the nonactivated enzyme (Table 16.1). The C2-H dissociation of the co-

Table 16.1. Cofactor activation in pyruvate decarboxylase

Sample	Rate constant (s^{-1})
Free ThDP	$9.5 \pm 0.4 \times 10^{-4}$
Free 4′-deamino-ThDP	$1.2 \pm 0.1 \times 10^{-3}$
Yeast pyruvate decarboxylase (wild type)	$9.7 \pm 0.9 \times 10^{-1}$
Yeast pyruvate decarboxylase (wild type), pyruvamide activated	≥ 600
Yeast pyruvate decarboxylase E51Q variant	$7.6 \pm 0.6 \times 10^{-2}$
Yeast pyruvate decarboxylase (wild type), reconstituted with 4′-deamino-ThDP	$3.4 \pm 0.1 \times 10^{-5}$
Pyruvate decarboxylase (wild type) from *Zymomonas mobilis*	110 ± 20
Pyruvate decarboxylase (E473D) from *Zymomonas mobilis*	104 ± 20
Pyruvate decarboxylase (D27E) from *Zymomonas mobilis*	117 ± 13
Pyruvate decarboxylase (H113K) from *Zymomonas mobilis*	96 ± 24

enzyme ThDP is not rate-limiting in the activated pyruvate decarboxylase from *Saccharomyces cerevisiae*, whereas it is indeed rate-limiting in the nonactivated enzyme. This indicates that the allosteric activation in the yeast enzyme is accomplished by an increase in the C2-H dissociation rate of the enzyme-bound ThDP. This model was substantiated by measuring the H/D exchange of C2-H of ThDP in pyruvate decarboxylase from *Zymomonas mobilis*, which shows no allosteric activation by the substrate (Bringer-Meyer et al., 1986). According to this model, the deprotonation rate constant of the cofactor in pyruvate decarboxylase from *Zymomonas mobilis* exceeds its catalytic constant of 17 s^{-1} at 4 °C for each active site (Table 16.1) and is not altered by pyruvamide.

The crystal structure of pyruvate decarboxylase from *Saccharomyces cerevisiae* (Arjunan et al., 1996; Dyda et al., 1993) shows that the side chain of a glutamate is at a short distance from the N1'-atom of the pyrimidine ring of ThDP, indicating the formation of a hydrogen bond. On the other hand, studies involving ThDP analogs bound in various ThDP-dependent enzymes point to an essential requirement of the N1'-atom and the 4'-amino group for the catalytic activity (Golbik et al., 1991; Schellenberger, 1990; Schellenberger et al., 1997). In order to demonstrate the cofactor activation by the protein environment Glu51 in pyruvate decarboxylase from *Saccharomyces cerevisiae* was mutated to glutamine and, additionally, the 4'-amino group of the coenzyme was eliminated. The E51Q mutant enzyme binds ThDP as strongly as the wild type enzyme, shown by characteristic changes in the near-UV circular dichroism spectrum (Killenberg-Jabs et al., 1997). However, the residual catalytic activity of the variant was only 0.04% of that of the wild type enzyme. The slow dissociation rate constant of the C2-H of ThDP in the *Saccharomyces cerevisiae* pyruvate decarboxylase E51Q variant (Table 16.1) suggests that this glutamate is essentially involved in the proton abstraction of the enzyme-bound ThDP. Furthermore, pyruvate decarboxylase from *Saccharomyces cerevisiae* was reconstituted with 4'-deamino-ThDP to unravel the function of the 4'-amino group of the coenzyme. This modification of the cofactor results in an inactive enzyme after reconstitution and in a markedly decreased H/D exchange rate constant of C2-H of the analog compared with the enzyme containing the entire coenzyme (Table 16.1). These findings point to an essential function of the 4'-amino group in the deprotonation step.

In order to investigate the functional contributions of side chains of the active site to the H/D exchange of ThDP (beside the substitution of the conserved glutamate interacting with the N1'-atom of ThDP) all putative residues were mutated in this location in pyruvate decarboxylase from *Zymomonas mobilis*. As shown in Table 16.1, these mutations have no influence on the C2-H deprotonation rate constant of the enzyme-bound cofactor.

In order to further characterize the key role of the 4'-amino group of ThDP for cofactor activation, the influence of the chemical environment at the active site of pyruvate decarboxylase from *Zymomonas mobilis* on the electronic properties of the 4'-amino group was studied by two-dimensional proton-nitrogen correlated NMR spectroscopy (Tittmann et al., 2005a). Chemical shift analysis and its pH dependence indicate that the acceleration of C2 deprotonation by 5 orders of magnitude is not mainly of thermodynamic nature caused by a significant increase in basicity of the 4′-amino group, but rather of kinetic nature caused by an optimal spatial orientation of the activated amino group towards the C2 hydrogen enforced by the adopted *V* conformation of the cofactor in the active site.

16.2.2
Deprotonation Rate of the C2-H of Thiamin Diphosphate in Transketolase from *Saccharomyces cerevisiae*

In order to investigate whether the mechanism of ThDP activation is a common phenomenon in other ThDP-dependent enzymes, the H/D exchange of the C2-H of ThDP was measured in transketolase from *Saccharomyces cerevisiae* catalyzing the ketol transfer between aldo- and ketosugars. As observed for the above mentioned pyruvate decarboxylase, the C2-H deprotonation of ThDP is not rate-limiting in the wild type enzyme (Table 16.2). The crystal structure of transketolase from *Saccharomyces cerevisiae* (Lindqvist et al., 1992; Nikkola et al., 1994) shows the importance of the N1′-atom of the pyrimidine ring of the cofactor, which is located at a hydrogen bond distance from the side chain of Glu418. A mutation of this glutamate to alanine results in an enzyme with only 0.1% of the activity measured for the wild type enzyme and a slow H/D exchange rate constant. However, the mutation of the closest base to the C2-atom of ThDP in this transketolase, realized in the variant H481A, does not change the rate constant of C2-H deprotonation (Table 16.2). Therefore, a mechanism assuming His481 as the base for C2-proton abstraction could be essentially ruled out.

In order to unravel the function of both the 4′-amino group and the N1′-atom of the coenzyme in transketolase from *Saccharomyces cerevisiae*, the apoenzyme was reconstituted with either the 4′-deamino-ThDP, or the N1′ → C-substituted ThDP analog (N1′-deaza-ThDP, N3′-pyridyl-ThDP). Both modifications of the cofactor result in inactive enzymes and a markedly decreased H/D exchange rate constant of C2-H compared with the enzyme containing the natural coenzyme (Table 16.2). Structural changes of the respective complexes of the wild type enzyme with the analogs were not detectable by X-ray crystallography (König et al., 1994). These results establish the essential function of both the 4′-amino group, and the N1′-atom in the deprotonation reaction of the coenzyme in this enzyme as well.

Table 16.2. Cofactor activation in transketolase

Sample	Rate constant (s^{-1})
Free ThDP	$3.0 \pm 0.1 \times 10^{-3}$
Free N1′ → C substituted ThDP	$1.6 \pm 0.1 \times 10^{-4}$
Free 4′-deamino-ThDP	$3.2 \pm 0.1 \times 10^{-3}$
Transketolase (wild type)	61 ± 2
Transketolase E418A variant	$3.7 \pm 0.1 \times 10^{-1}$
Transketolase H481A variant	61 ± 2
Transketolase (wild type) reconstituted with 4′-deamino-ThDP	$9.5 \pm 0.1 \times 10^{-5}$
Transketolase (wild type) reconstituted with N1′ → C substituted ThDP	$1.6 \pm 0.2 \times 10^{-4}$

The same coenzyme binding pattern and no structural changes in the protein component were detectable for the mutant enzymes of transketolase from *Saccharomyces cerevisiae* and their complexes with coenzyme analogs studied by X-ray crystallography (König et al., 1994; Wikner et al., 1994). Summarizing, it can be ruled out that the differences in the H/D exchange rate constants of transketolase from *Saccharomyces cerevisiae* are a result of a different solvent accessibility of a base involved in the proton abstraction mechanism of ThDP.

16.2.3
Deprotonation Rate of the C2-H of Thiamin Diphosphate in the Pyruvate Dehydrogenase Multienzyme Complex from *Escherichia coli*

In the pyruvate dehydrogenase complex, catalyzing the oxidative decarboxylation of pyruvate with NAD^+ and coenzyme A as cosubstrates and yielding acetyl-coenzyme A, the ThDP-containing E1 component catalyzes the rate-limiting step of the overall reaction (Akiyama and Hammes, 1980; Bates et al., 1977) and, for this reason, represents an ideal target for regulation. Due to the reversible binding of ThDP to the complex the H/D exchange experiments were carried out in the presence of equimolar amounts of ThDP. Considering the very slow H/D exchange rate constant of the small amount of free ThDP in the reaction mixture ($k_{obs} = 3 \times 10^{-3}$ s^{-1} at pH 7.0 and 4 °C), a value of 16 s^{-1} at pH 7.0 and 4 °C was calculated for the enzyme-bound ThDP in bacterial E1, which exceeds the catalytic constant of 2 s^{-1} measured under the same experimental conditions.

On the basis of the crystal structure of a *Bacillus stearothermophilus* pyruvate dehydrogenase subcomplex formed between the heterotetrameric E1 and the peripheral subunit binding domain of E2 with an evident structural dissymmetry of the two active sites, a direct active center communication via an acidic proton tunnel has been proposed (Frank et al., 2004). According to this, one active site is in a closed state with an activated cofactor even before a substrate molecule is engaged, whereas the activation of the second active site is coupled to decarboxylation in the first site. Our own kinetic NMR studies on human PDH E1 (unpublished) support the model suggested, but similar studies on related thiamin enzymes, such as pyruvate decarboxylase, transketolase or pyruvate oxidase reveal that half-of-the-sites reactivity is a unique feature of ketoacid dehydrogenases. In line with this, X-ray crystallography studies on intermediates in transketolase catalysis indicated an active site occupancy close to unity in both active sites (Fiedler *et al.*, 2002 and G. Schneider, personal communication).

16.2.4
Deprotonation Rate of the C2-H of Thiamin Diphosphate in the Phosphate-dependent Pyruvate Oxidase from *Lactobacillus plantarum*

The pyruvate oxidase from *Lactobacillus plantarum* catalyzes the oxidative decarboxylation of pyruvate and the formation of acetylphosphate, CO_2 and H_2O_2 in the presence of oxygen and phosphate (Götz and Sedewitz, 1990; Sedewitz et al., 1984a; Sedewitz et al., 1984b). Each subunit of the homotetrameric enzyme binds

Table 16.3. Cofactor activation in pyruvate oxidase

Sample	Rate constant (s^{-1})
Free ThDP in 50 mM phosphate buffer	$9.5 \pm 0.4 \times 10^{-4}$
Holo pyruvate oxidase in 50 mM phosphate buffer	314 ± 12
Holo pyruvate oxidase without phosphate	20 ± 0.8
Holo pyruvate oxidase reconstituted with 5-carba-5-deaza-FAD in 50 mM phosphate buffer	8 ± 0.3
Apo-ThDP-Mg^{2+} pyruvate oxidase in 50 mM phosphate buffer	$1.0 \pm 0.05 \times 10^{-2}$
Pyruvate oxidase E59A variant in 50 mM phosphate buffer	0.49 ± 0.12

one ThDP and one FAD in the presence of Mn^{2+} or Mg^{2+}. The presence of these metal ions provided both FAD, and ThDP alone can form binary complexes with the apoenzyme (Risse et al., 1992). The binary complexes, however, are enzymatically inactive in the native overall oxidation reaction. The deprotonation of the C2-H of ThDP is also an important catalytic step in the pyruvate oxidase reaction. As shown in Table 16.3, the H/D exchange rate constant of the C2-H of ThDP in the apoenzyme–ThDP binary complex is very slow and would not allow catalysis at the rate constant observed. FAD binding to this binary complex accelerates the H/D exchange rate constant by four orders of magnitude compared to that of free ThDP (Table 16.3) and exceeds the enzyme's catalytic constant of 2 s^{-1} at 4 °C. This fast H/D exchange in the native holoenzyme does not appear to be mediated by a direct interaction of the FAD with the C2-H of the enzyme-bound ThDP, but rather by interactions with functional groups of the protein that are operative only in the holoenzyme. This interpretation is consistent with the observation that the rate constant of the H/D exchange of ThDP is only marginally reduced compared to that of the native holoenzyme in the ternary complex with 5-carba-5-deaza-FAD (Table 16.3). In the crystal structure, the distance of the closest FAD atom to C2-H of ThDP is 11 Å (Muller and Schulz, 1993; Muller et al., 1994). Based on the structural homology of the ThDP binding site to other ThDP-dependent enzymes, it may be assumed that Glu59 is the residue mediating this activation in pyruvate oxidase by interacting with the N1'-atom of ThDP. This principle would be in analogy to pyruvate decarboxylase and transketolase (see sections above) displaying the same type of interaction. The slow H/D exchange rate constant in the E59A variant of pyruvate oxidase confirms this presumed function of Glu59 (Table 16.3).

Interestingly, the presence of the second substrate phosphate increases the rate constant of the H/D exchange of ThDP in the holoenzyme of pyruvate oxidase from *Lactobacillus plantarum* 16-fold compared to that measured in a phosphate-free buffer (Table 16.3). At present, this behaviour of the enzyme observed in the presence of phosphate cannot be interpreted in molecular detail.

16.2.5
Suggested Mechanism of the C2-H Deprotonation of Thiamin Diphosphate in Enzymes

The data measured for the deprotonation of the C2-H of ThDP in different enzymes show that its N1'-atom and 4'-amino group are essential for coenzyme activation. In addition, a fast deprotonation of C2-H requires an interaction of a conserved acidic group with the N1'-atom of the pyrimidine moiety of ThDP. As described in Section 16.2.1 a fast deprotonation of enzyme-bound ThDP is not mainly a result of an increased basicity of the 4'-amino group shown by proton-nitrogen correlated NMR spectroscopy, but is rather caused by an optimal orientation of the amino group towards the C2 hydrogen. This kinetic control of the C2 deprotonation is in agreement with the previously performed experiments with ^{13}C2-labeled ThDP showing that the enzyme-bound cofactor does not generate a significant population of the C2 carbanion intermediate (Kern et al., 1997). A significant tautomerization of the aminopyrimidine part to the 1'–4' imino tautomer, as detected for covalent intermediates in pyruvate decarboxylase from *Saccharomyces cerevisiae* (Jordan et al., 2002; Nemeria et al., 2004) and the E1 component of the pyruvate dehydrogenase complex from *Escherichia coli* (Jordan et al., 2003), is not a prerequisite for a fast deprotonation (Scheme 16.2).

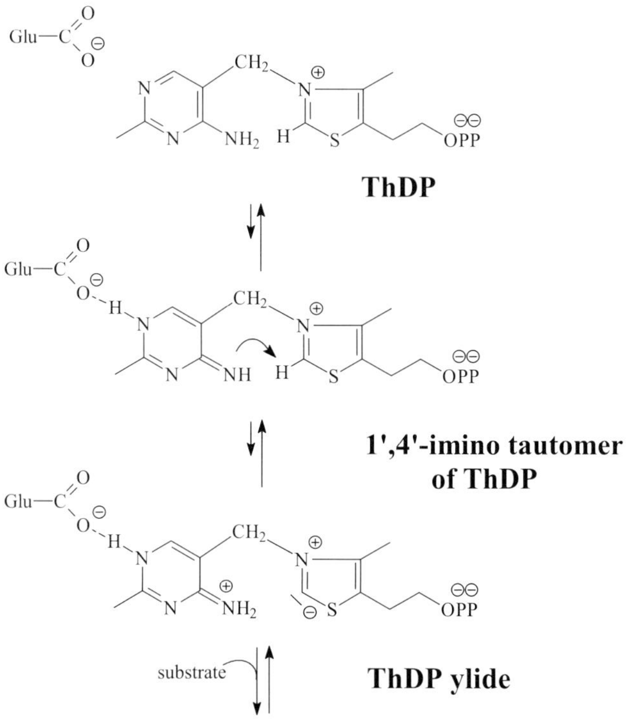

Scheme 16.2

16.3
Proton Transfer Reactions during Enzymic Thiamin Diphosphate Catalysis

Whereas the mechanism of the C2-H deprotonation of ThDP has been shown to be identical in all ThDP-dependent enzymes investigated, the following steps in catalysis of the different enzymes require different protonation and deprotonation reactions of the intermediates formed along the process. In order to identify side chains involved in proton transfer steps, the distribution of reaction intermediates during catalysis of any wild type enzyme can be compared with that of active site mutant enzymes. Rate constants for single steps in catalysis can be calculated from

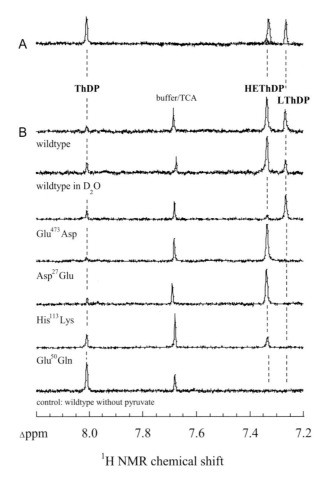

Figure 16.2. (A) C6′-H ^{1}H NMR fingerprint region of ThDP (I in Scheme 16.1), 2-lactyl-ThDP (II in Scheme 16.1) and 2-(α-hydroxyethyl)-ThDP (protonated III in Scheme 16.1). (B) Intermediate distribution of the covalent intermediates formed during the nonoxidative decarboxylation of pyruvate by pyruvate decarboxylase from *Zymomonas mobilis* and its active site variants.

16.3 Proton Transfer Reactions during Enzymic Thiamin Diphosphate Catalysis

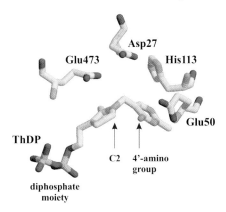

Figure 16.3. Crystal structure of active site residues of pyruvate decarboxylase from *Zymomonas mobilis* with the enzyme-bound cofactor ThDP in its typical V-conformation.

the distribution of reaction intermediates and compared for the wild type enzyme as well as for variants carrying mutations in the active site. This leads to a functional assignment of the corresponding side chains (Tittmann et al., 2003). Fortunately, in ThDP-dependent enzymes the intermediates (ThDP-C2 adducts) can be discriminated using 1H NMR (Tittmann et al., 2003). The chemical shifts of the C6′-H singlets of the aminopyrimidine moiety can be used as a fingerprint region for the discrimination of all C2-derived covalent ThDP adducts (Fig. 16.2A). They can be separated from the respective enzymes working at steady state conditions by acid quench treatment (Tittmann et al., 2003).

Pyruvate decarboxylase from *Zymomonas mobilis* was investigated to determine the function of interacting groups in the interconversion of the reaction intermediates as a first example of this method (Tittmann et al., 2003). This enzyme, showing four amino acid side chains located at a short distance from the cofactor ThDP (Fig. 16.3), catalyzes the nonoxidative decarboxylation of pyruvate yielding acetaldehyde and carbon dioxide. The minimal catalytic scheme (Scheme 16.3) comprises the reversible, noncovalent binding of the substrate to the Michaelis complex, carbon–carbon bond formation between the C2-atom of ThDP and the carbonyl carbon of pyruvate to yield enzyme-bound 2-lactyl-ThDP (LThDP), the subsequent decarboxylation to the α-carbanion/enamine of 2-(α-hydroxyethyl)-ThDP (HEThDP), and finally, the liberation of acetaldehyde. Although a covalent binding of pyruvate to ThDP can be assumed to be reversible, the rate constant of this step calculated from the intermediate distribution at steady state reflects mainly the forward reac-

$$\text{E-ThDP} + \text{CH}_3\text{COCOO}^- \xrightleftharpoons{K_1} \text{E-ThDP}*\text{CH}_3\text{COCOO}^- \xrightarrow{k_2} \text{E-LThDP}^- \xrightarrow{k_3} \text{E-HEThDP}^- + \text{H}^+ \xrightarrow{k_4} \text{E-ThDP} + \text{CH}_3\text{CHO}$$
$$- \text{CO}_2$$

Scheme 16.3

Table 16.4. Microscopic rate constants of catalysis in pyruvate decarboxylase wildtype and variants

	k_{cat} (s^{-1})	Rate constant (s^{-1})		
		C–C bonding	CO_2 release	Acetaldehyde release
ZmPDC wt	150 ± 5	2650 ± 210	397 ± 20	265 ± 13
ZmPDC wt (D_2O)	100 ± 4	685 ± 70	530 ± 45	150 ± 14
ZmPDC Glu^{473}Asp	0.10 ± 0.004	0.60 ± 0.08	0.13 ± 0.01	1.2 ± 0.2
ZmPDC Asp^{27}Glu	0.05 ± 0.002	>5	>5	0.051 ± 0.002
ZmPDC His^{113}Lys	0.24 ± 0.03	>25	>25	0.25 ± 0.03
ZmPDC Glu^{50}Gln	0.04 ± 0.003	0.07 ± 0.01	>7	0.09 ± 0.01
ScPDC wt	45 ± 2	294 ± 20	105 ± 6	105 ± 6

tion of this microscopic step because of the large forward commitment factor of LThDP decarboxylation (Sun et al., 1995).

The analysis of the intermediate distribution of pyruvate decarboxylase and its active site variants at steady state (Fig. 16.2B, Table 16.4) revealed two independent proton relay systems working in this enzyme. Furthermore, the kinetic solvent isotope effect measured for the wild type enzyme indicates the binding of pyruvate to the C2-atom of ThDP and the release of the reaction product acetaldehyde (both processes require proton transfer steps) to be the mainly affected steps in catalysis. The proton relay, involving the amino group of the cofactor and the interaction of the Glu50 with the N1'-atom of ThDP, influences not only the C2-H deprotonation of the coenzyme, but also the carbon–carbon bond formation between the C2-carbanion of ThDP and the carbonyl-carbon of the substrate pyruvate and the release of the reaction product acetaldehyde (Table 16.4). According to the primary isotope effect measured, it can be assumed that this proton relay protonates the carbonyl-oxygen of the substrate as a prerequisite for the addition of its carbonyl-carbon to the C2-carbanion of the coenzyme (step 2 in Scheme 16.4). This proton relay and a second proton relay consisting of the side chains of His113 and Asp27 are involved in product release (Table 16.4). This reaction step requires a protonation of the α-carbanion/enamine formed after decarboxylation of LThDP and a deprotonation of the α-OH group of HEThDP (step 4 in Scheme 16.4). The His113/Asp27 dyad seems to be responsible for the protonation of the α-carbanion of HEThDP, since a mutation of Asp27 to alanine leads to a variant of this pyruvate decarboxylase catalyzing the formation of acetolactate as a result of the perturbation of the protonation reaction and the accumulation of the α-carbanion/enamine of HEThDP. A second pyruvate has the chance to attack the accumulated α-carbanion of HEThDP in this protein. This reaction is usually catalyzed by acetohydroxyacid synthase (AHAS), a ThDP-dependent enzyme in which the His/Asp dyad is missing.

Interestingly, if the pyruvate decarboxylase reaction is carried out in a mixture of H_2O/D_2O, a deuterium discrimination at the C1-atom of acetaldehyde will be ob-

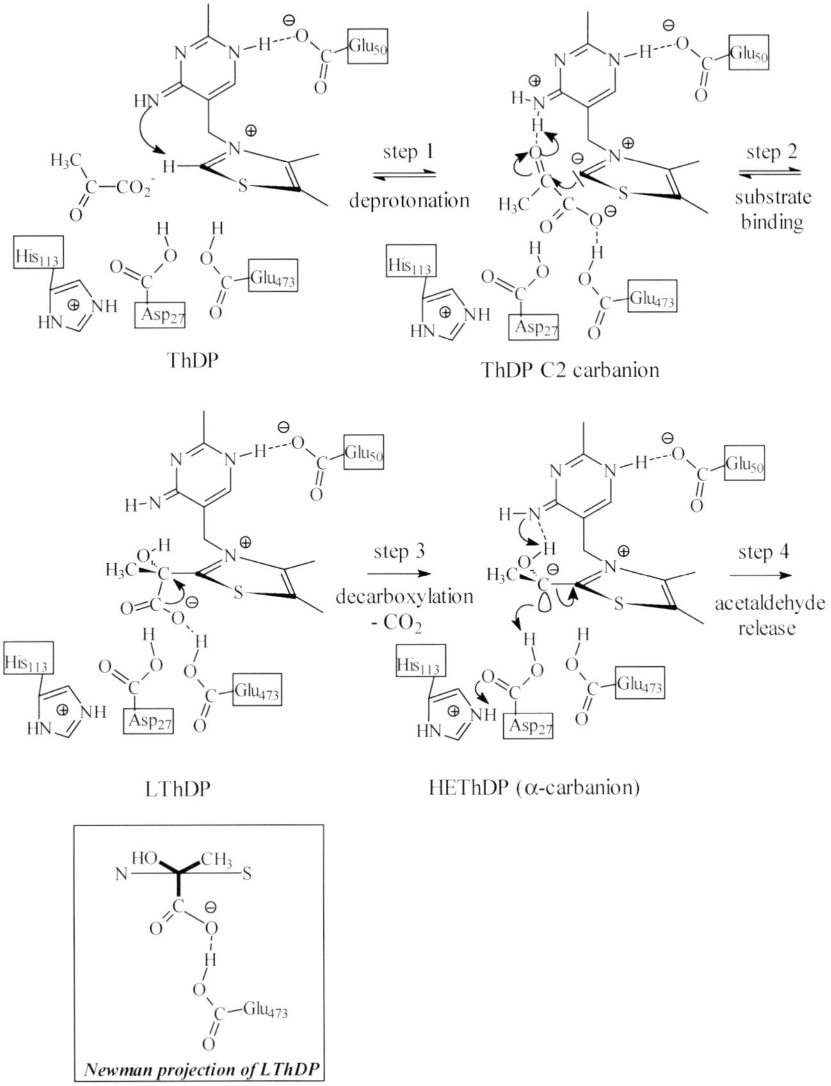

16.3 Proton Transfer Reactions during Enzymic Thiamin Diphosphate Catalysis

Scheme 16.4

served (Ermer et al., 1992). This discrimination of deuterium, indicating a specific protonation of the α-carbanion/enamine intermediate by an active site residue, can be attributed in its magnitude either to a protonation by a functional group with low fractionation factor, or to a kinetic isotope effect of the protonation involving an asymmetric transition state.

A mutation of Glu50, perturbing the proton relay Glu50 – N1′-atom – 4′-amino group, drastically decreases the rate constant of acetaldehyde release (Table 16.4). Therefore, the deprotonation of the α-OH group of HEThDP is very likely catalyzed by the 4′-imino group of its 1′,4′-imino tautomeric form and would indicate this group of the cofactor as the preferred acid/base catalyst for this reaction.

Similar results were obtained in our studies on the thiamin-dependent indolepyruvate decarboxylase, where protonation and deprotonation reactions are also catalyzed by a Glu-cofactor proton shuttle and a His-Asp-Glu relay (Schütz et al., 2005).

16.4
Hydride Transfer in Thiamin Diphosphate-dependent Enzymes

Hydride transfer reactions can be expected in those ThDP-dependent enzymes catalyzing the oxidative decarboxylation of α-ketoacids, such as α-ketoacid dehydrogenase complexes and pyruvate oxidases. The electron/proton transfer in α-ketoacid dehydrogenase complexes cannot be ascribed unambiguously to the reaction of the α-carbanion/enamine of HEThDP (formed by the E1 component of the complex) with the lipoate of the E2 component leading to enzyme-bound 2-acetyl-ThDP. An alternative mechanism leading to 2-acetyl-lipoate is imaginable, which involves an initial C–S bond formation and S–S bond fission together with proton reorganisation. On the other hand, in pyruvate oxidases the target for the oxidation reaction by enzyme-bound FAD can be attributed unambiguously to the α-carbanion/enamine of HEThDP. In this section, the mechanism of the electron transfer in pyruvate oxidase from *Lactobacillus plantarum* is discussed. This enzyme is a homotetrameric protein containing one ThDP and one FAD per monomer and catalyzes the conversion of pyruvate to acetylphosphate, CO_2 and H_2O_2 in the presence of phosphate and oxygen.

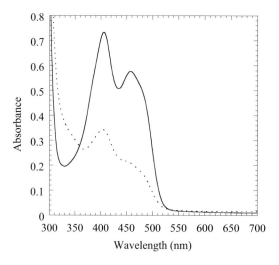

Figure 16.4. Absorption spectra of oxidized (solid line) and reduced (dotted line) pyruvate oxidase from *Lactobacillus plantarum*.

For investigating pyruvate oxidase the spectroscopic properties of the isoalloxazine system of FAD are an excellent probe for monitoring steps relevant to enzyme catalysis directly. In pyruvate oxidase from *Lactobacillus plantarum* the lowest π–π^* transition can be used to distinguish between oxidized and reduced FAD (Fig. 16.4). In this enzyme the reductive half-reaction, including the catalytic steps of substrate binding, decarboxylation and electron transfer from the α-carbanion of HEThDP to FAD (Scheme 16.5), can be monitored after complete removal of oxygen in the mixture before the reaction is started with pyruvate. A back electron transfer from FADH$_2$ to 2-acetyl-ThDP is negligible, because in the absence of oxygen pyruvate completely reduces enzyme-bound FAD to FADH$_2$. The progress curves of these single-turnover experiments measured at different concentrations of pyruvate (Fig. 16.5) reveal rate constants for the catalytic steps of the reductive half reaction of $k_{on} = 6.5 \times 10^4$ M^{-1} s^{-1} and $k_{off} = 20$ s^{-1} for the reversible binding of pyruvate, $k_{dec} = 112$ s^{-1} for the decarboxylation, and $k_{red} = 422$ s^{-1} for the reduction of FAD, respectively (Tittmann et al., 2000). In addition, time-resolved absorption spectra of enzyme-bound FAD indicate that blue or red flavosemiqui-

$$E\begin{matrix}FAD\\ThDP\end{matrix} + CH_3COCOO^- \underset{k_{off}}{\overset{k_{on}}{\rightleftharpoons}} E\begin{matrix}FAD\\lactyl\text{-}ThDP\end{matrix} \xrightarrow{k_{dec}} E\begin{matrix}FAD\\HETHDP\end{matrix} \xrightarrow{k_{red}} E\begin{matrix}FADH_2\\acetyl\text{-}ThDP\end{matrix}$$

Scheme 16.5

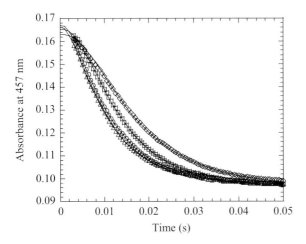

Figure 16.5. Reduction of oxidized pyruvate oxidase by different concentrations of the substrate pyruvate under anaerobic conditions in 0.2 M potassium phosphate buffer, pH 6.0. Pyruvate concentration: 2.5 mM (open diamond), 5 mM (open square), 20 mM (open circle), and 50 mM (open triangle).

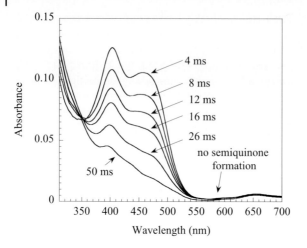

Figure 16.6. Time-resolved absorption spectra of enzyme-bound FAD in pyruvate oxidase from *Lactobacillus plantarum* during the first turnover reduction by pyruvate under anaerobic conditions.

none species are not kinetically stabilized at pre-steady state (Fig. 16.6) and at steady state (data not shown) of the catalytic reaction. The absorbance appears to be composed of oxidized and fully reduced flavin species only.

Kinetic solvent isotope effect experiments were performed to address the question, whether the reduction in pyruvate oxidase proceeds in a two-step single electron transfer or possibly via a hydride transfer after protonation of the α-carbanion/enamine of HEThDP. Single-wavelength stopped-flow experiments were carried out with pyruvate and phosphate at saturating concentrations and air-saturated buffer using different mole fractions of deuterium oxide at pL 6.0. No kinetic solvent isotope effect for the reduction of FAD was found (Fig. 16.7). However, the catalytic constant representing the rate-limiting steps of catalysis shows a kinetic solvent isotope effect of 1.8 and a linear dependence of the proton inventory curve. In summary, the reduction of enzyme-bound FAD by the HEThDP intermediate proceeds via a two-step single electron transfer. Recent results obtained from ^{1}H NMR measurements on pyruvate oxidase suggest the decarboxylation of the substrate to be the rate-limiting step of catalysis (Tittmann et al., 2005b).

A mechanism that involves protonation of the α-carbanion/enamine of HEThDP and subsequent hydride transfer, which has been proposed for several flavine-dependent enzymes (Pollegioni et al., 1997), is unlikely since no kinetic solvent isotope effect is evident for this catalytic step (Fig. 16.7). In accordance, after replacement of FAD by 5-carba-5-deaza-FAD, a FAD analog not catalyzing a transfer of single electrons but functioning as hydride acceptor, no reduction is observed by the HEThDP intermediate in pyruvate oxidase from *Lactobacillus plantarum* (Tittmann et al., 1998).

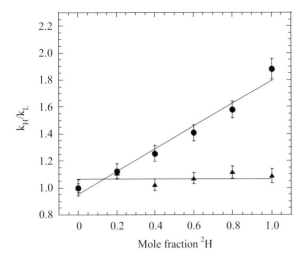

Figure 16.7. Proton inventory for both the reduction of pyruvate oxidase by the substrate pyruvate (filled triangle) and for the catalytic constant (open square).

A direct α-carbanion mechanism involving the formation of a covalent adduct is very unlikely due to the spatial orientation of the two cofactors and the long distance between ThDP and FAD. The distance between the C2-atom of ThDP and the C4α-atom of FAD in pyruvate oxidase from Lactobacillus plantarum is about 11 Å as derived from the X-ray crystallography structure (Muller et al., 1993; Muller and Schulz, 1993). The results are in good agreement with the findings that Cα-deprotonated HEThDP reduces a flavin analog in a two-step electron-transfer reaction in solution. Moreover, it has been concluded from the electron-transfer properties of free HEThDP that it may act as a two-electron donor with potentials of $E^0_{ox(1)} = -0.97$ V and $E^0_{ox(2)} = -0.56$ V (Nakanishi et al., 1997a; Nakanishi et al., 1997b).

Despite these clear indications for a radical two-step transfer mechanism, populated flavin radicals could neither be detected during the first turnover electron transfer (Fig. 16.6), nor at steady state or during anaerobic reduction (Fig. 16.4). It is unclear how the electrons are transferred between HEThDP and FAD. Based on the X-ray crystallography structure, it has been supposed that a stepwise electron transfer may be related to the benzene rings of Phe479 of the one and/or Phe121 of the other subunit (Muller and Schulz, 1993; Muller et al., 1994). On the other hand, the electrons may simply tunnel from HEThDP to FAD as proposed by Dutton and coworkers (Page et al., 1999). Since the redox potentials of the enzyme-bound HEThDP in pyruvate oxidase from Lactobacillus plantarum is still unknown, it is difficult to calculate the theoretical rate constant of the electron transfer from HEThDP to FAD according to Dutton (Page et al., 1999). However, it is very unlikely that the rate of reduction reflects the intrinsic rate of electron transfer. The absence of a kinetic solvent isotope effect argues against a proton transfer being the rate-limiting step. Possibly, rather, a reduction-associated distortion of FAD impeded by the protein environment is rate-limiting for the reduction process.

References

AKIYAMA, S. K., HAMMES, G. G. (1980), Elementary steps in the reaction mechanism of the pyruvate dehydrogenase multienzyme complex from *Escherichia coli*: kinetics of acetylation and deacetylation, *Biochemistry* **19**, 4208–4213.

ARJUNAN, P., UMLAND, T., DYDA, F., SWAMINATHAN, S., FUREY, W., SAX, M., FARRENKOPF, B., GAO, Y., ZHANG, D., JORDAN, F. (1996), Crystal structure of the thiamin diphosphate-dependent enzyme pyruvate decarboxylase from the yeast *Saccharomyces cerevisiae* at 2.3 Å resolution, *J. Mol. Biol.* **256**, 590–600.

BATES, D. L., DANSON, M. J., HALE, G., HOOPER, E. A., PERHAM, R. N. (1977), Self-assembly and catalytic activity of the pyruvate dehydrogenase multienzyme complex of *Escherichia coli, Nature* **268**, 313–316.

BRESLOW, R. (1962), Part II. Cocarboxylase: Mode of action, reaction mechanism, and role of the active acetate, *Ann. N. Y. Acad. Sci.* **98**, 445–452.

BRINGER-MEYER, S., SCHIMZ, K. L., SAHM, H. (1986), Pyruvate decarboxylase from *Zymomonas mobilis*. Isolation and partial characterization, *Arch. Microbiol.* **146**, 105–110.

CROSBY, J., STONE, R., LIENARD, G. E. (1970), Mechanism of thiamine-catalyzed reactions. Decarboxylation of 2-(1-carboxy-1-hydroxyethyl)-3,4-dimethylthiazolium-chloride, *J. Am. Chem. Soc.* **92**, 2891–2900.

DYDA, F., FUREY, W., SWAMINATHAN, S., SAX, M., FARRENKOPF, B., JORDAN, F. (1993), Catalytic centers in the thiamin diphosphate dependent enzyme pyruvate decarboxylase at 2.4-Å resolution, *Biochemistry* **32**, 6165–6170.

ERMER, J., SCHELLENBERGER, A., HÜBNER, G. (1992), Kinetic mechanism of pyruvate decarboxylase. Evidence for a specific protonation of the enzymic intermediate, *FEBS Lett.* **299**, 163–165.

FIEDLER, E., THORELL, S., SANDALOVA, T., GOLBIK, R., KÖNIG, S., SCHNEIDER, G. (2002), Snapshot of a key intermediate in enzymatic thiamin catalysis: crystal structure of the α-carbanion of (α,β-dihydroxyethyl)-thiamin diphosphate in the active site of transketolase from *Saccharomyces cerevisiae, Proc. Nat. Acad. Sci. USA* **99**, 591–595.

FRANK, R. A., TITMAN, C. M., PRATAP, J. V., LUISI, B. F., PERHAM, R. N. (2004), A molecular switch and proton wire synchronize the active sites in thiamine enzymes, *Science* **306**, 818–820.

GOLBIK, R., NEEF, H., HÜBNER, G., KÖNIG, S., SELIGER, B., MESHALKINA, L. E., KOCHETOV, G. A., SCHELLENBERGER, A. (1991), Function of the aminopyrimidine part in thiamine pyrophosphate enzymes, *Bioorg. Chem.* **19**, 10–17.

GÖTZ, F., SEDEWITZ, B. (1990), Physiological role of pyruvate oxidase in the aerobic metabolism of *Lactobacillus plantarum*, in *Biochemistry and Physiology of Thiamin Diphosphate Enzymes* (BISSWANGER, H., ULLRICH, J., eds.), pp. 286–293. VCH, Weinheim.

HÜBNER, G., WEIDHASE, R., SCHELLENBERGER, A. (1978), The mechanism of substrate activation of pyruvate decarboxylase: A first approach, *Eur. J. Biochem.* **92**, 175–181.

JORDAN, F., NEMERIA, N. S., ZHANG, S., YAN, Y., ARJUNAN, P., FUREY, W. (2003), Dual catalytic apparatus of the thiamin diphosphate coenzyme: acid-base via the $1',4'$-iminopyrimidine tautomer along with its electrophilic role, *J. Am. Chem. Soc.* **125**, 12732–12738.

JORDAN, F., ZHANG, Z., SERGIENKO, E. (2002), Spectroscopic evidence for participation of the $1',4'$-imino tautomer of thiamin diphosphate in catalysis by yeast pyruvate decarboxylase, *Bioorg. Chem.* **30**, 188–198.

KEMP, D. S., O'BRIEN, J. T. (1970), Base catalysis of thiazolium salt hydrogen exchange and its implications for enzymatic thiamine cofactor catalysis, *J. Am. Chem. Soc.* **92**, 2554–2555.

KERN, D., KERN, G., NEEF, H., TITTMANN, K., KILLENBERG-JABS, M., WIKNER, C., SCHNEIDER, G. HÜBNER, G. (1997), How thiamine diphosphate is activated in enzymes, *Science* **275**, 67–70.

KILLENBERG-JABS, M., KÖNIG, S., EBERHARDT, I., HOHMANN, S., HÜBNER, G. (1997), Role of Glu51 for cofactor binding and catalytic activity in pyruvate decarboxylase from yeast studied by site-directed mutagenesis, *Biochemistry* **36**, 1900–1905.

KLUGER, R. (1987). Thiamin diphosphate: A mechanistic update on enzymic and nonenzymic catalysis of decarboxylation. *Chem. Rev.* **87**, 863–876.

KÖNIG, S., SCHELLENBERGER, A., NEEF, H., SCHNEIDER, G. (1994), Specificity of coenzyme binding in thiamin diphosphate-dependent enzymes. Crystal structures of yeast transketolase in complex with analogs of thiamin diphosphate, *J. Biol. Chem.* **269**, 10879–10882.

LINDQVIST, Y., SCHNEIDER, G., ERMLER, U., SUNDSTRÖM, M. (1992), Three-dimensional structure of transketolase, a thiamine diphosphate dependent enzyme, at 2.5 Å resolution, *EMBO J.* **11**, 2373–2379.

MULLER, Y. A., LINDQVIST, Y., FUREY, W., SCHULZ, G. E., JORDAN, F., SCHNEIDER, G. (1993), A thiamin diphosphate binding fold revealed by comparison of the crystal structures of transketolase, pyruvate oxidase and pyruvate decarboxylase, *Structure* **1**, 95–103.

MULLER, Y. A., SCHULZ, G. E. (1993), Structure of the thiamine- and flavin-dependent enzyme pyruvate oxidase, *Science* **259**, 965–967.

MULLER, Y. A., SCHUMACHER, G., RUDOLPH, R., SCHULZ, G. E. (1994), The refined structures of a stabilized mutant and of wild-type pyruvate oxidase from *Lactobacillus plantarum*, *J. Mol. Biol.* **237**, 315–335.

NAKANISHI, I., ITOH, S., SUENOBU, T., FUKUZUMI, S. (1997a), Electron transfer properties of active aldehydes derived from thiamin coenzyme analogues, *Chem. Commun.* **19**, 1927–1928.

NAKANISHI, I., ITOH, S., SUENOBU, T., INOUE, H., FUKUZUMI, S. (1997b), Redox behavior of active aldehydes derived from thiamin coenzyme analogs, *Chem. Lett.* 707–708.

NEMERIA, N. S., BAYKAL, A., EBENEZER, J., ZHANG, S., YAN, Y., FUREY, W., JORDAN, F. (2004). Tehtrahedral intermediates in thiamin diphosphate-dependent decarboxylations exist as a 1′,4′-imino tautomeric form of the coenzyme, unlike the Michaelis complex or the free coenzyme, *Biochemistry* **43**, 6565–6575.

NIKKOLA, M., LINDQVIST, Y., SCHNEIDER, G. (1994), Refined structure of transketolase from *Saccharomyces cerevisiae* at 2.0 Å resolution, *J. Mol. Biol.* **238**, 387–404.

PAGE, C. C., MOSER, C. C., CHEN, X., DUTTON, P. L. (1999), Natural engineering principles of electron tunnelling in biological oxidation – reduction, *Nature* **402**, 47–52.

POLLEGIONI, L., BLODIG, W., GHISLA, S. (1997), On the mechanism of D-amino acid oxidase. Structure/linear free energy correlations and deuterium kinetic isotope effects using substituted phenylglycines, *J. Biol. Chem.* **272**, 4924–4934.

RISSE, B., STEMPFER, G., RUDOLPH, R., MÖLLERING, H., JAENICKE, R. (1992), Stability and reconstitution of pyruvate oxidase from *Lactobacillus plantarum*: Dissection of the stabilizing effects of coenzyme binding and subunit interaction, *Protein Sci.* **1**, 1699–1709.

SCHELLENBERGER, A. (1990), Die Funktion der 4′-Aminopyrimidin-Komponente im Katalysemechanismus von Thiaminpyrophosphatenzymen aus heutiger Sicht, *Chem. Ber.* **123**, 1489–1494.

SCHELLENBERGER, A., HÜBNER, G., NEEF, H. (1997), Cofactor designing in functional analysis of thiamin diphosphate enzymes, *Methods Enzymol.* **279**, 131–146.

SCHÜTZ, A., GOLBIK, R., KÖNIG, S., HÜBNER, G., TITTMANN, K. (2005), Intermediate and transition states in thiamin diphosphate-dependent decarboxylases. A kinetic and NMR study on wild-type indolepyruvate decarboxylase and variants using indolepyruvate, benzoylformate, and pyruvate as substrates, *Biochemistry* **44**, 6164–6179.

SEDEWITZ, B., SCHLEIFER, K. H., GÖTZ, F. (1984a), Physiological role of pyruvate oxidase in the aerobic metabolism of *Lactobacillus plantarum*, *J. Bacteriol.* **160**, 462–465.

SEDEWITZ, B., SCHLEIFER, K. H., GÖTZ, F. (1984b), Purification and biochemical characterization of pyruvate oxidase from *Lactobacillus plantarum*, *J. Bacteriol.* **160**, 273–278.

SUN, S. X., DUGGLEBY, R. G., SCHOWEN, R. L. (1995), Linkage of catalysis and regulation in enzyme action – carbon isotope effects, solvent isotope effects, and proton inventories for the unregulated pyruvate decarboxylase of *Zymomonas mobilis*, *J. Am. Chem. Soc.* **117**, 7317–7322.

TITTMANN, K. (2000), Untersuchungen zu Katalysemechanismen von Flavin- und Thiamindiphosphat-abhängigen Enzymen. Aktivierung von Thiamindiphosphat in Enzymen. Katalysemechanismus der Pyruvatoxidase *aus Lactobacillus plantarum*, PhD Thesis, Martin-Luther-University Halle-Wittenberg.

TITTMANN, K., GOLBIK, R., GHISLA, S., HÜBNER, G. (2000), Mechanism of elementary catalytic steps of pyruvate oxidase from *Lactobacillus plantarum*, *Biochemistry* **39**, 10747–10754.

TITTMANN, K., GOLBIK, R., UHLEMANN, K., KHAILOVA, L., SCHNEIDER, G., PATEL, M., JORDAN, F., CHIPMAN, D. M., DUGGLEBY, R. G., HÜBNER, G. (2003), NMR analysis of covalent intermediates in thiamin diphosphate enzymes, *Biochemistry* **42**, 7885–7891.

TITTMANN, K., NEEF, H., GOLBIK, R., HÜBNER, G., KERN, D. (2005a), Kinetic control of thiamin diphosphate activation in enzymes studied by proton-nitrogen correlated NMR spectroscopy, *Biochemistry* **44**, 8697–8700.

TITTMANN, K., PROSKE, D., SPINKA, M., GHISLA, S., RUDOLPH, R., HÜBNER, G., KERN, G. (1998), Activation of thiamin diphosphate and FAD in the phosphate-dependent pyruvate oxidase from *Lactobacillus plantarum*, *J. Biol. Chem.* **273**, 12929–12934.

TITTMANN, K., WILLE, G., GOLBIK, R., WEIDNER, A., GHISLA, S., HÜBNER, G. (2005b), Radical phosphate transfer mechanism for the thiamin diphosphate- and FAD-dependent pyruvate oxidase from *Lactobacillus plantarum*. Kinetic coupling of intercofactor electron transfer with phosphate transfer to acetyl-thiamin diphosphate via transient FAD semiquinone/hydroxyethyl-ThDP radical pair, *Biochemistry* **44**, 13291–13303.

WASHABAUGH, M. W., JENCKS, W. P. (1988), Thiazolium C(2)-proton exchange: structure-reactivity correlations and the pK_a of thiamin C(2)-H revisited, *Biochemistry* **27**, 5044–5053.

WIKNER, C., MESHALKINA, L., NILSSON, U., NIKKOLA, M., LINDQVIST, Y., SUNDSTRÖM, M., SCHNEIDER, G. (1994), Analysis of an invariant cofactor-protein interaction in thiamin diphosphate-dependent enzymes by site-directed mutagenesis. Glutamic acid 418 in transketolase is essential for catalysis, *J. Biol. Chem.* **269**, 32144–32150.

17
Dihydrofolate Reductase: Hydrogen Tunneling and Protein Motion

Stephen J. Benkovic and Sharon Hammes-Schiffer

Dihydrofolate reductase (DHFR, EC 1.5.1.3) is an essential enzyme required for normal folate metabolism in prokaryotes and eukaryotes. Its role is to maintain necessary levels of tetrahydrofolate to support the biosynthesis of purines, pyrimidines and amino acids. Many compounds of pharmacological value, notably methotrexate and trimethoprim, work by inhibition of DHFR. Their clinical importance justified the study of DHFR in the rapidly evolving field of enzymology. Today, there is a vast amount of published literature (ca. 1000 original research articles) on the broad subject of dihydrofolate reductase contributed by scientists from diverse disciplines. We have selected kinetic, structural, and computational studies that have advanced our understanding of the DHFR catalytic mechanism with special emphasis on the role of the enzyme–substrate complexes and protein motion in the catalytic efficiency achieved by this enzyme.

17.1
Reaction Chemistry and Catalysis

DHFR catalyzes the reduction of 7,8-dihydrofolate (H_2F) to 5,6,7,8-tetrahydrofolate (H_4F) using nicotinamide adenine dinucleotide phosphate (NADPH) as a cofactor (Fig. 17.1). Specifically, the pro-R hydride of NADPH is transferred stereospecifically to the C6 of the pterin nucleus with concurrent protonation at the N5 position [1]. Structural studies of DHFR bound with substrates or substrate analogs have revealed the location and orientation of H_2F, NADPH and the mechanistically important side chains [2]. Proper alignment of H_2F and NADPH is crucial in enhancing the rate of the chemical step (hydride transfer). *Ab initio*, mixed quantum mechanical/molecular mechanical (QM/MM), and molecular dynamics computational studies have modeled the hydride transfer process and have deduced optimal geometries for the reaction [3–6]. The optimal C–C distance between the C4 of NADPH and C6 of H_2F was calculated to be ~2.7 Å [5, 6], which is significantly smaller than the initial distance of 3.34 Å inferred from X-ray crystallography [2].

One proposed chemical mechanism involves a keto–enol tautomerization (Fig.

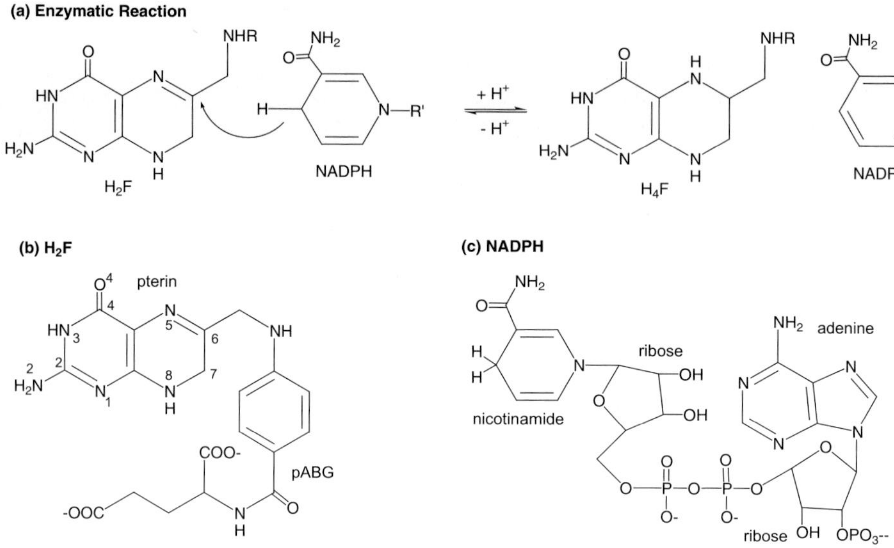

(a) Enzymatic Reaction

(b) H$_2$F

(c) NADPH

Figure 17.1. (a) The reaction catalyzed by DHFR. R and R' denote functional groups of H$_2$F and NADPH, respectively. (b) Structure of dihydrofolate (H$_2$F). pABG denotes the *para*-amino benzoyl group of H$_2$F. (c) Structure of cofactor NADPH. (Reproduced from Ref. [38].)

17.2) driven by the low dielectric environment of the DHFR active site [7–9]. In this pathway, H$_2$F binds initially as the 4-oxo tautomer and expels most of the water molecules from the active site. The resulting decrease in dielectric constant raises the pK_a of Asp-27 and serves as the driving force for the tautomerization. The enol tautomer (4-hydroxy) results in a complex poised for chemical reaction. Hydride transfer then proceeds along with concerted proton transfers involving O4, N5 and N3 positions of the pterin and Asp-27 with the product H$_4$F formed as the 4-oxo tautomer. The hydride transfer rate (k_{hyd}) displays strong pH dependence with an experimentally determined maximal value of 950 s^{-1} and a pK_a value of 6.5 [10]. However, there is some dispute as to whether the measured pK_a reflects that of Asp-27, that of the N5 of H$_2$F, or both [7, 9].

These mechanistic issues have been studied via several different theoretical approaches. Cummins et al. assessed the energetically most likely substrate and enzyme protonation sites and pathways by performing QM/MM calculations [11]. In addition to explaining control of the likely protonation site by a structurally conserved water molecule that hydrogen bonds to both the carboxyl of Asp-27 and the O4 of the pterin, their results support a mechanism in which Asp-27 is protonated first, followed by direct protonation of the keto form of the pterin at the N5 position for H$_2$F reduction. Analysis of the hydrogen-bonding distances between water molecules and the N5 position in classical molecular dynamics simulations has been used to postulate a mechanism in which the hydride transfer occurs before

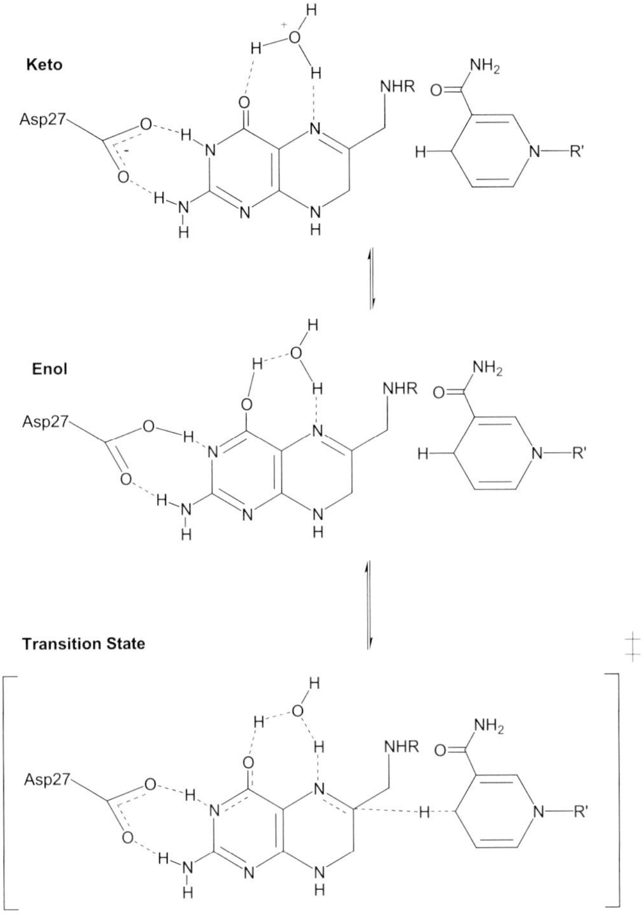

Figure 17.2. The keto/enol tautomerization that plays a role in one of the proposed chemical mechanisms for hydride transfer.

the proton transfer [12, 13]. Recent calculations [14] have provided more direct evidence that the reaction proceeds through an initial proton transfer followed by a hydride transfer. These calculations illustrated that the free energy barrier for hydride transfer is more than 30 kcal mol^{-1} greater for nonprotonated H_2F than for protonated H_2F. Most molecular dynamics simulations of the hydride transfer reaction have assumed that the protonation of H_2F occurs prior to hydride transfer.

17.1.1
Hydrogen Tunneling

Theoretical simulations indicate that hydrogen tunneling plays an important role in the hydride transfer reaction catalyzed by DHFR. A hybrid quantum/classical

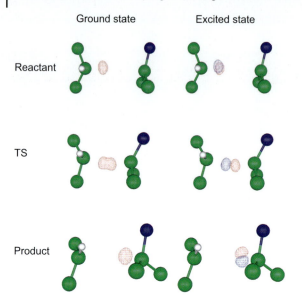

Figure 17.3. Three-dimensional vibrational wavefunctions representing the transferring hydride for reactant, transition state, and product configurations obtained from hybrid quantum/classical molecular dynamics simulations of the hydride transfer reaction catalyzed by DHFR. On the donor side, the donor carbon atom and its first neighbors are shown, and on the acceptor side the acceptor carbon atom and its first neighbors are shown. (Reproduced from Ref. [15].)

molecular dynamics method that includes electronic and nuclear quantum effects, as well as the motion of the entire solvated enzyme, has been applied to this hydride transfer reaction [6, 15, 16]. In this approach, the transferring hydrogen nucleus is represented by a three-dimensional vibrational wavefunction (Fig. 17.3), and the free energy profile for the hydride transfer reaction is generated as a function of a collective reaction coordinate. Nuclear quantum effects such as zero point energy and hydrogen tunneling were found to lower the free energy barrier by 2–3 kcal mol^{-1} [15]. The calculated primary deuterium kinetic isotope effect was consistent with the experimental value of 3, and hydrogen tunneling in the direction along the donor–acceptor axis was found to be significant [15]. The subsequent application of an alternative semiclassical tunneling method led to similar conclusions [14].

Mixed labeling experiments with specifically isotopically substituted 4R- and 4S-NADPH cofactors established the primary and secondary kinetic isotope effects and their temperature dependence for the hydride transfer reaction. Indeed, resulting data could be rationalized only by a reaction model featuring an extensive tunneling contribution that is environmentally coupled. The difference in the observed and calculated intrinsic kinetic isotope effects requires a commitment factor arising from dissecting the pre-steady state hydride step into kinetic steps, one the actual hydride transfer step itself and the other a motion of the protein and/or nicotinamide associated with the hydride transfer step [17].

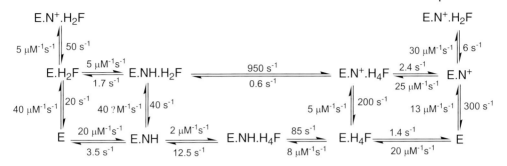

Figure 17.4. The pH-independent kinetic scheme for DHFR catalysis at 25 °C. This scheme pertains to the reaction at pH less than 7. E is DHFR; NH is NADPH; N^+ is $NADP^+$; H_2F is dihydrofolate; and H_4F is tetrahydrofolate. (Reproduced from Ref. [38].)

17.1.2
Kinetic Analysis

DHFR catalysis follows the kinetic scheme described in Fig. 17.4. Pre-steady state kinetic experiments augmented by equilibrium binding measurements were used to elucidate this scheme, which correctly predicts the full-time course kinetics as a function of substrate concentrations and pH [10]. The preferred pathway that produces a maximal rate of H_4F turnover involves DHFR cycling between five kinetically observable species. Initially, rapid hydride transfer within the Michaelis complex (E•NH•H_2F) produces the product complex (E•N^+•H_4F) from which $NADP^+$ (N^+) dissociates to produce the E•H_4F binary complex. A NADPH (NH) molecule then associates to form a reduced ternary complex (E•NH•H_4F) from which H_4F dissociates, returning the E•NH complex poised for another round of turnover. The steady state rate is controlled by the rate of H_4F release from the reduced ternary complex (E•NH•H_4F). The presence of both reduced ligands in this complex causes negative cooperativity and an elevation in the off rates of both ligands. This complete kinetic analysis of DHFR catalysis served as the basis for comparative studies with numerous site directed mutants. Similar kinetic schemes were developed for these mutant DHFRs, thereby allowing an in-depth and unambiguous analysis of the function of these mutated regions [10, 18–26] when combined with structural considerations furnished by X-ray and NMR methods.

17.2
Structural Features of DHFR

From the standpoint of protein structure, DHFR possesses prominent structural features that guide the substrate and cofactor through the preferred catalytic pathway (Fig. 17.5). X-ray crystallographic studies have shown that the *E. coli* DHFR is

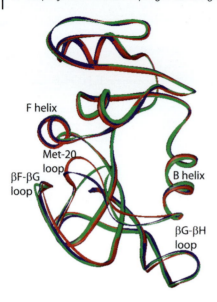

Figure 17.5. Structural features of E.coli DHFR. The figure is an overlay of three crystallographically observed Met-20 loop conformations. They are open (blue), closed (red) and occluded (green). (Reproduced from Ref. [38].)

a monomeric enzyme and has several secondary structural elements [2, 8, 27]. An eight stranded β-sheet (βA-βH) and four α helices are interspersed with loop regions that connect these structural elements. The protein structure can be divided into two subdomains, namely the adenosine binding subdomain and the loop subdomain [18]. The loop subdomain contains three flexible loops: the Met-20 loop, the βF-βG loop, and the βG-βH loop. Space between the two subdomains forms the active site cleft where NADPH and H_2F bind at a 45° angle to each other. The subdomains rotate open and closed during passage through the preferred catalytic pathway.

17.2.1
The Active Site of DHFR

Key hydrophobic contacts exist between the *para*-amino benzoyl glutamate (pABG) group of H_2F and active site residues Leu-28, Phe-31, Ile-50 and Leu-54 that could be enhanced during the hydride transfer step [2, 21, 26]. After reduction to H_4F, the resulting ring pucker at the pterin C6 position causes the disruption of these van der Waals contacts. This provides a basis for discrimination between H_2F and H_4F. The side chains of Leu-28 and Leu-54 are separated by 8 Å on opposite sides of the active site but interact through the bound H_2F. This coupling was revealed by double mutational analysis of the two residues, as shown in Fig. 17.6. Kinetic constants for key steps were measured for both the single mutants (L28Y and

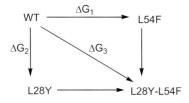

$$\Delta G_x = -RT\ln(k_{wt}/k_{mut})$$

$$\Delta G_I = \Delta G_3 - (\Delta G_1 + \Delta G_2)$$

Figure 17.6. Representative double mutational cycle involving mutations L28Y and L54F. Mutational effects are calculated as free energy changes (ΔG_x) relative to wild type. Nonaddi- tivity (ΔG_I) in the double mutational cycle is then calculated for any kinetic or thermodynamic parameter. (Reproduced from Ref. [38].)

L54F) and the double mutant (L28Y-L54F) (Table 17.1) and converted to changes in free energy (ΔG_x) [21]. If the residues act independently, the free energies for a given step in the two single mutants would sum to that for the double mutant with $\Delta G_I = 0$. Nonadditive free energies (nonzero ΔG_I) indicative of side chain

Table 17.1. Mutational effects on ligand binding (K_D) and hydride transfer rate (k_{hyd}). All data were obtained at pH = 7.5.

Mutation	K_D H$_2$F (µM)	K_D NADPH (µM)	k_{hyd} (s^{-1})	References
Wild type	0.21	0.33	220	10
H$_2$F contacts				
L28Y	0.11	0.15	109	21
L28F	0.15	0.40	4000	39
L54F	0.10	0.17	20	21
L28Y-L54F	11.0	0.60	77	21
L28F-L54F	0.20	3.5	126	26
Met-20 loop				
DL1	2.0	5.3	1.7	22
βF-βG loop				
G121V	0.36	14.2	1.4	20
G121S	NR	NR	40.1	20
D122N	0.38	0.92	9.4	23
D122S	0.37	1.1	5.9	23
D122A	0.39	1.3	4.0	23
βG-βH loop				
S148D	0.18	0.15	319	25
S148A	1.06	0.05	157	25
S148K	0.72	0.16	162	25

coupling were, however, observed for H_2F and NADPH binding and the hydride transfer rates. Additive effects were displayed for the rates of reverse hydride transfer and product dissociation, indicating that for these steps the side chains act independently. The conclusion is that the substrate complexes (E•NH and E•NH•H_2F) are in conformations in which the side chains of Leu-28 and Leu-54 can couple, whereas the product complexes (E•N^+•H_4F and E•NH•H_4F) are in conformations in which these side chains cannot couple. Of particular note is the fact that the forward and reverse hydride transfer steps exhibit different ΔG_I values, thereby excluding interaction by these residues in their common transition state. Collectively the data suggest that DHFR adopts different conformations involving Leu-28 and Leu-54 during the catalytic cycle.

Later efforts examined the additivity of mutational effects caused by the L54F mutation in combination with the L28F mutation [26]. The nonadditive effects observed with the related double mutational cycle involving the L28Y mutation were essentially absent, showing that a single functional group change (–OH in L28Y to –H in L28F) can control the additivity of mutational effects. In general, the effect of the second mutation within the active site, although deleterious, did not alter the thermodynamic and kinetic parameters to the extent anticipated if they were additive. Similar trends were observed for double mutational cycles involving paired mutations at active site residues Phe-31 and Ile-50 in conjunction with paired Leu-28 and Leu-54 mutations. This resiliency to mutation may be thought of as a built in "active site redundancy" that conserves function and may have evolutionary significance.

17.2.2
Role of Interloop Interactions in DHFR Catalysis

Sawaya and Kraut compiled structures representative of all five complexes and the TS in the kinetically preferred pathway by cocrystallizing the enzyme with substrate and cofactor analogs and subsequently tracing variations in loop and subdomain positions [2]. DHFR, as mentioned earlier, has three flexible loops that cover 45% of the sequence [2, 28]. Several structural studies have shown that the Met-20 loop adopts either a closed or occluded conformation (Fig. 17.5) depending on the ligands bound. In the closed conformation, the Met-20 loop residues interact with NADPH through hydrogen bonding to the carboxamide group and through van der Waals contacts with the ribose unit. This closed conformation is observed in surrogate substrate complexes (E•NH and E•NH•H_2F) and is stabilized by interactions with βF-βG loop residues. Specifically, the amide backbone of both Gly-15 and Glu-17 in the Met-20 loop forms hydrogen bonds with Asp-122 in the βF-βG loop. Mutagenesis of Asp-122 to Asn, Ser and Ala (in order of decreasing ability to hydrogen bond) showed two important trends [23]. First, a significant correlation was observed between decreased NADPH binding and lowered hydride transfer rates (Table 17.1). This finding suggests that the interactions of Asp-122 participate in the collective reaction coordinate leading to the TS. Secondly, the mutations in

Asp-122 altered the preferred catalytic pathway, indicating that loop interactions actively control ligand affinity and turnover.

The presence of H_2F or an analog has implications for the conformational state of the Met-20 loop. In the absence of a folate analog, the occupancy of the nicotinamide binding pocket by this moiety of the cofactor in the E•NH complex is $\sim 75\%$. When H_2F binds, the occupancy shifts to 100% and the nicotinamide ring is positioned favorably for hydride transfer. After the chemical reaction, the nicotinamide pocket occupancy falls to zero in the E•$NADP^+$•H_4F complex [2, 27]. Concomitant disruption of hydrogen bonds between the Met-20 and βF-βG loops and formation of new ones between the Met-20 and βG-βH loops result in the Met-20 loop adopting an occluded conformation in the product complexes. The occluded conformation excludes the nicotinamide ring from its binding pocket.

The closed/occluded Met 20 loop conformations were also assigned to various DHFR ligand complexes in solution through observation of chemical shift changes associated with alanine ^1H–^{15}N resonances [29]. The findings are in close agreement with those from crystallography, namely, the key substrate complexes E•NH and E•NH•H_2F are closed, and the key product complexes E•H_4F, E•N^+•H_4F and E•NH•H_4F are occluded. The HSQC spectrum of the E•N^+ binary complex revealed the presence of both closed and occluded forms, probably associated with fluctuation of the ribose-nicotinamide moiety into and out of the binding cleft. The chemical shift of the Ala 6 resonance shows that the pterin ring occupies the active site in all complexes (both closed and occluded conformations) and irrespective of whether the ring is planar or strongly puckered. On the other hand, if either the pterin or nicotinamide rings are non-planar, e.g. NADPH or H_4F, the conformation of the Met 20 loop is occluded with the nicotinamide ring excluded from its binding site. Consequently, during turnover, the enzyme must cycle between the closed and the occluded states at two steps in the reaction cycle: following hydride transfer it changes from closed to occluded, and after product release it changes from occluded to closed.

Site directed mutagenesis of βG-βH loop residues (residues 146-148) in which the ability to hydrogen bond with the Met-20 loop was systematically varied included single mutant enzymes S148D, S148A, S148K and the deletion mutant enzyme $\Delta(146\text{-}148)$. (Note that the amide backbone of Asn-23 in the Met-20 loop hydrogen bonds to Ser-148 in the βG-βH loop.) Kinetic studies indicated that the Met-20 and βG-βH loop interaction modulated the ligand off rates and degree of cycling through the preferred kinetic pathway. Collectively, these kinetic and structural studies implicated a series of conformational states characterized by extensive loop movements that guide the reaction cycle.

17.3
Enzyme Motion in DHFR Catalysis

Structures derived from X-ray crystallography provide static images of different protein conformations and lack dynamical information. Moreover, deviations due to crystal packing forces can cause a different picture to be presented than the true situation in solution. To address these issues, NMR studies were initiated to examine the solution structure of DHFR in the presence and absence of ligands [28, 30]. The dynamics of the Met-20 loop were investigated by analyzing the 2D NOESY spectra of the apo-enzyme. The time dependence of Trp-22 cross peaks was used as an indicator of Met-20 loop exchange between the closed and occluded states [28]. The frequency of oscillation of the Met-20 loop was found to be 35 s^{-1}, which is similar to the off rate of H$_4$F under steady state conditions, suggesting that loop movement may be a limiting step in substrate turnover. Replacement of the central portion of the Met-20 loop (residues Met-16 to Ala-19) with a single glycine (mutant DL1 in table 1) altered the loop dynamics and decreased substrate binding 10-fold and hydride transfer 400-fold [22].

Epstein et al. studied the backbone and tryptophan side-chain dynamics of the ^{15}N-labeled DHFR•folate complex [30]. Measurements of the ^{15}N spin–lattice (T_1) and spin–spin (T_2) relaxation times and ^{1}H–^{15}N heteronuclear Overhauser effects (NOEs) were made for the protonated backbone nitrogen atoms. Dynamic information for each residue was then calculated and expressed in terms of a generalized order parameter (S^2), an effective correlation time for internal motions (τ_e), and a ^{15}N exchange broadening contribution (R_{ex}). The values of S^2 lie between 0 and 1, where a lower S^2 value implies increased disorder. The average S^2 values of secondary structure elements are similar, but much lower S^2 values and larger τ_e values are observed in the loop regions, indicating increased motions on the ns–ps time scale. Four notable regions of enhanced flexibility are the Met-20 loop (residues 16-22), the adenosine binding loop (residues 67-69), the hinge regions (residues 38 and 88), and the βF-βG loop (residues 119-123) [30]. The DHFR•folate complex studied represents the product binary complex (DHFR•H$_4$F) in which the Met-20 loop occludes the nicotinamide binding site.

Subsequent work included DHFR•dhNADPH•folate and DHFR•NADP$^+$•folate species that represent the product ternary complexes (DHFR•NADP$^+$•H$_4$F and DHFR•NADPH•H$_4$F) and the Michaelis complex (DHFR•NADPH•H$_2$F), respectively [31]. Folate and dhNADPH (5,6-dihydro NADPH) serve as nonreactive substrate and cofactor analogs, respectively. Conformations of the Met-20 loop and occupancy of the nicotinamide binding pocket in each complex were identified using a diagnostic set of marker resonances. The ns–ps timescale backbone dynamics of the regions in the DHFR•folate and DHFR•dhNADPH•folate complexes (Met-20 loop in the occluded state) are largely attenuated in the DHFR•NADP$^+$•folate complex (Met-20 loop in the closed state), as indicated by increased S^2 and decreased τ_e values. The change in dynamics observed between the closed and occluded complexes can be rationalized from structural considerations [2]. In the occluded conformation, residues 14-16 of the Met-20 loop occupy the nicotinamide binding pocket while residues 16-22 and 119-123 (βF-βG loop) are solvent exposed, resulting in high flexibility. In the closed state, however, tight packing involving the Met-20 loop, the βF-βG loop, and the nicotinamide group results in reduced flexibility. Despite the suggestive nature of the data, the available NMR measurements cannot alone conclusively link the backbone motions to catalysis.

The relationship between dynamics and catalysis was probed by site directed mutagenesis of Met-20 and βF-βG loop residues guided in part by the NMR findings [20]. The mutant enzymes were characterized by building complete kinetic schemes as in Fig. 17.4. Mutagenesis of Gly-121 in the βF-βG loop weakened NADPH binding and slowed hydride transfer (Table 17.1). In particular, the G121V mutation caused hydride transfer to be reduced 200-fold and introduced a kinetically significant conformational change step in the ternary Michaelis complex preceding the hydride transfer step [20]. NMR studies of this mutant have shown that the closed Met 20 loop is destabilized so that key substrate complexes are in the occluded conformation. Structurally, this corresponds to insertion of the nicotinamide ring into its binding pocket. The rate of insertion has been reduced from a rate of >2000 s^{-1} in the wild type enzyme to 2 s^{-1} in the mutant. However, despite insertion into the binding pocket, the hydride transfer rate is still reduced [20]. Two interpretations are plausible: the population of productive ternary complexes at the active site is also reduced by the mutation and/or an important conformational motion along the reaction coordinate has been affected by the mutation.

In order to determine if the dynamic motions observed by NMR methods are correlated with one another and catalysis, classical molecular dynamics (MD) simulations were performed on various DHFR complexes. Radkiewicz et al. generated 10 ns MD simulations on the three ternary complexes in the preferred pathway [32]. The starting points for the simulations were crystal structures that represented the DHFR•NADPH•H_2F, DHFR•$NADP^+$•H_4F and DHFR•NADPH•H_4F complexes. During the simulations the flexible loop region underwent conformational changes while the protein retained its overall secondary and tertiary structure. Furthermore, these changes were dependent on the nature of the ligand bound. Residue–residue based maps of correlated motions representing fluctuations about an average structure were generated for all three complexes analyzed. These fluctuations are local to the average structure and typically occur on the femtosecond to picosecond timescale. In these maps, regions of correlated and anticorrelated motions were defined in which the two residues moved in concert, either in the same or in opposite directions, respectively. Strong correlated and anticorrelated motions involving spatially distinct regions in the protein structure were observed for the Michaelis complex (DHFR•NADPH•H_2F) only. These correlated motions appeared in many of the same regions of the protein observed in the dynamic NMR measurements. The absence of these correlated motions in the product complexes implied that they may be tied to catalysis, although it is not known which correlations are essential. Their possible relevance to catalysis is underscored by the fact that mutations made in regions of correlated motion have a deleterious effect on DHFR activity. Classical MD simulations on the G121V mutant, which has a 200-fold lower hydride transfer rate, revealed an absence of similar patterns of correlated motion [33]. The collective analysis of kinetic data for site directed mutants, data from NMR spectroscopy for regions of dynamic motions, MD simulations for correlated motions, and genomic content for sequence conservation across 36 species has led to the identification of a conserved set of residues that are catalytically relevant (Fig. 17.7) [6].

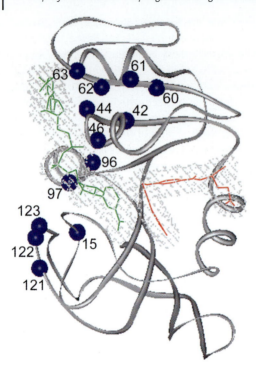

Figure 17.7. Chain of residues important for hydride transfer in E.coli DHFR. Substrate and cofactor structures are in red and green respectively with dot surfaces. Blue balls with adjacent numbering denote residue location. (Reproduced from Ref. [38].)

In addition, hybrid quantum/classical molecular dynamics simulations of hydride transfer catalyzed by DHFR provided evidence of a network of coupled motions extending throughout the protein and ligands [6, 15]. These coupled motions represent equilibrium thermally averaged conformational changes along the collective reaction coordinate, leading to configurations conducive to the reaction. These motions are averaged over the fast vibrational modes and reflect conformational changes that may occur on the much slower millisecond timescale of the overall hydride transfer reaction. The equilibrium molecular motions in this network are not dynamically coupled to the chemical reaction, but rather give rise to conformations in which the hydride transfer reaction is facilitated because of short transfer distances, suitable orientation of substrate and cofactor, and a favorable electrostatic environment for charge transfer. A portion of this network of coupled motions is illustrated in Fig. 17.8. Applications of two other computational approaches to hydride transfer catalyzed by DHFR have shown geometrical changes to be in agreement with this network of coupled motions [14, 34]. Moreover, subsequent

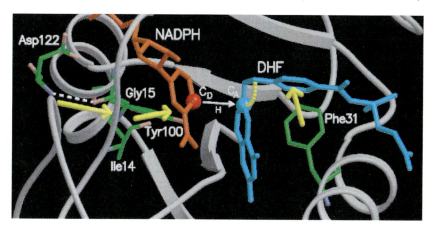

Figure 17.8. Schematic diagram of a portion of the network of coupled motions in DHFR. The yellow arrows and arc indicate the coupled motions. This picture does not represent a complete or unique network but rather illustrates the general concept of reorganization of the enzymatic environment to provide configurations conducive to the hydride transfer reaction. (Reproduced from Ref. [6].)

to the identification of Ile-14 as a key player in this network of coupled motions, NMR experiments confirmed the catalytic importance of this residue [35]. Specifically, the observed line broadening for Ile-14 in the measurement of methyl deuterium relaxation rates suggests motions on a microsecond or millisecond timescale.

Applications of the hybrid quantum/classical molecular dynamics method to mutant DHFR enzymes have provided further insights into this network of coupled motions. Hybrid simulations of the G121V mutant lead to a rate reduction that is consistent with the experimental rate measurements and suggest that the mutation may modify the network of coupled motions through structural perturbations, thereby increasing the free energy barrier and decreasing the reaction rate [36]. Recently, a more comprehensive analysis of coupled motions correlated to hydride transfer was applied to the triple mutant M42F-G121S-S148A (Fig. 17.9) and the associated single and double mutants [37]. This analysis indicates that each enzyme system samples a unique distribution of coupled motions correlated to hydride transfer. These coupled motions provide an explanation for the experimentally measured nonadditivity effects in the hydride transfer rates for these mutants. Moreover, this analysis illustrates that site-specific mutations distal to the active site can introduce subtle perturbations that impact the catalytic rate by altering the conformational sampling of the entire enzyme. Since distal regions of the enzyme are coupled to each other through long-range electrostatics and extended hydrogen bonding networks [16], the introduction of a site-specific mutation alters the thermal motions of the entire enzyme. Altering the thermal motions of the enzyme affects the probability of sampling conformations conducive to the catalyzed chemical reaction, thereby impacting the free energy barrier and the rate.

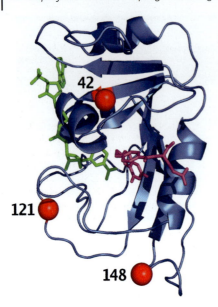

Figure 17.9. Three-dimensional structure of wild-type DHFR, with the NADPH cofactor shown in green and the DHF substrate shown in magenta. The residues involved in the triple mutant M42F-G121S-S148A are labeled with red spheres.

17.4
Conclusions

The participation of distant residues in catalysis is not unique to DHFR. Distant residues have been shown to determine the energetics of active site residues in serine proteases. Switching the substrate specificity of trypsin to chymotrypsin required a large set of multiple mutations at unexpected positions. In addition to catalysis, other physiologically important phenomena where protein motion may be involved are the allosteric behavior of enzymes and protein–protein interactions. In both cases an event such as the binding of a ligand can be signaled to regions far away from the binding site using protein motion as the medium of communication. Evidence for the biological significance of protein motion may be manifested in conservation of residues involved in key conformational changes. The combination of genomic analyses, kinetic studies, and theoretical calculations is continuing to provide insight into the role of protein motion in enzyme catalysis.

References

1 CHARLTON, P. A., YOUNG, D. W., BIRDSALL, B., FEENEY, J., ROBERTS, G. C. K. (1979) Stereochemistry of reduction of folic acid using dihydrofolate reductase, *J. Chem. Soc., Chem. Commun.* 922–924.

2 SAWAYA, M. R., KRAUT, J. (1997) Loop and subdomain movements in the

mechanism of *Escherichia coli* dihydrofolate reductase: crystallographic evidence, *Biochemistry* 36, 586–603.

3 WU, Y. D., HOUK, K. N. (1987) Theorectical transmission structures for hydride transfer to methyleneiminium ion from methylamine and dihydropyridine. On the nonlinearity of hydride transfers, *J. Am. Chem. Soc.* 109, 2226–2227.

4 WU, Y. D., LAI, D. K. W., HOUK, K. N. (1995) Transition structures of hydride transfer reactions of protonated pyridinium ion with 1,4-dihydropyridine and protonated nicotinamide with 1,4-dihydronicotinamide, *J. Am. Chem Soc.* 117, 4100–4108.

5 CASTILLO, R., ANDRES, J., MOLINER, V. (1999) Catalytic mechanism of dihydrofolate reductase enzyme: A combined quantum-mechanical/molecular mechanical characterization of transition state structure for the hydride transfer step, *J. Am. Chem Soc.* 121, 12140–12147.

6 AGARWAL, P. K., BILLETER, S. R., RAJAGOPALAN, P. T. R., BENKOVIC, S. J., HAMMES-SCHIFFER, S. (2002) Network of coupled promoting motions in enzyme catalysis, *Proc. Natl. Acad. Sci. USA* 99, 2794–2799.

7 CHEN, Y. Q., KRAUT, J., BLAKLEY, R. L., CALLENDER, R. (1994) Determination by Raman Spectroscopy of the pKa of N5 of dihydrofolate reductase: mechanistic implications, *Biochemistry* 33, 7021–7026.

8 LEE, H., REYES, V. M., KRAUT, J. (1996) Crystal structures of *Escherichia coli* dihydrofolate reductase complexed with 5-formyltetrahydrofolate (folinic acid), in two space groups: evidence for enolization of pteridine O4, *Biochemistry* 35, 7012–7020.

9 CANNON, W. R., GARRISON, B. J., BENKOVIC, S. J. (1997) Electrostatic characterization of enzyme complexes: evaluation of the mechanism of catalysis of dihydrofolate reductase, *J. Am. Chem. Soc.* 119, 2386–2395.

10 FIERKE, C. A., JOHNSON, K. A., BENKOVIC, S. J. (1987) Construction and evaluation of the kinetic scheme associated with dihydrofolate reductase from *Escherichia coli*, *Biochemistry* 26, 4085–4092.

11 CUMMINS, P. L., GREADY, J. E. (2001) Energetically most likely substrate and active-site protonation sites and pathways in the catalytic mechanism of dihydrofolate reductase, *J. Am. Chem. Soc.* 123, 3418–3428.

12 SHRIMPTON, P. J., ALLEMANN, R. K. (2002) Role of water in the catalytic cycle of *E.coli* dihydrofolate reductase, *Protein Sci.* 11, 1442–1451.

13 SHRIMPTON, P. J., MULLANEY, A., ALLEMANN, R. K. (2003) Functional role for Tyr31 in the catalytic cycle of chicken dihydrofolate reductase, *Proteins: Struct. Funct. Genet.* 51, 216–223.

14 GARCIA-VILOCA, M., TRUHLAR, D. G., GAO, J. (2003) Reaction-path energetics and kinetics of the hydride transfer reaction catalyzed by dihydrofolate reductase, *Biochemistry* 42, 13558–13575.

15 AGARWAL, P. K., BILLETER, S. R., HAMMES-SCHIFFER, S. (2002) Nuclear Quantum Effects and Enzyme Dynamics in Dihydrofolate Reductase Catalysis, *J. Phys. Chem. B.* 106, 3283–3293.

16 WONG, K., WATNEY, J. B., HAMMES-SCHIFFER, S. (2004) Analysis of electrostatics and correlated motions for hydride transfer in dihydrofolate reductase, *J. Phys. Chem. B.* 108, 12231–12241.

17 SIKORSKI, R. S., WANG, L., MARKHAM, K. A., RAJAGOPALAN, P. T. R., BENKOVIC, S. J., KOHEN, A. (2004) Tunneling and coupled motion in the *Escherichia coli* dihydrofolate reductase catalysis, *J. Am. Chem Soc.* 126, 4778–4779.

18 MILLER, G. P., BENKOVIC, S. J. (1998) Stretching exercises – flexibility in dihydrofolate reductase catalysis, *Chem. Biol.* 5, R105–R113.

19 BENKOVIC, S. J., FIERKE, C. A., NAYLOR, A. M. (1988) Insights into enzyme function from studies on mutants of dihydrofolate reductase, *Science* 239, 1105–1110.

20 CAMERON, C. E., BENKOVIC, S. J. (1997) Evidence for a functional role of the dynamics of glycine-121 of *Escherichia coli* dihydrofolate reductase

obtained from kinetic analysis of a site-directed mutant, *Biochemistry* 36, 15792–15800.
21 HUANG, Z., WAGNER, C. R., BENKOVIC, S. J. (1994) Nonadditivity of mutational effects at the folate binding site of *E.coli* dihydrofolate reductase, *Biochemistry* 33, 11576–11585.
22 LI, L., FALZONE, C. J., WRIGHT, P. E., BENKOVIC, S. J. (1992) Functional role of a mobile loop of *E.coli* dihydrofolate reductase in transition-state stabilization, *Biochemistry* 31, 7826–7833.
23 MILLER, G. P., BENKOVIC, S. J. (1998) Strength of an interloop hydrogen bond determines the kinetic pathway in catalysis by *Escherichia coli* dihydrofolate reductase, *Biochemistry* 37, 6336–6342.
24 MILLER, G. P., BENKOVIC, S. J. (1998) Deletion of a highly motional residue affects formation of the Michaelis complex for *Escherichia coli* dihydrofolate reductase, *Biochemistry* 37, 6327–6335.
25 MILLER, G. P., WAHNON, D. C., BENKOVIC, S. J. (2001) Interloop contacts modulate ligand cycling during catalysis by *Escherichia coli* dihydrofolate reductase, *Biochemistry* 40, 867–875.
26 WAGNER, C. R., HUANG, Z., SINGLETON, S. F., BENKOVIC, S. J. (1995) Molecular basis for nonadditive mutational effects in *Escherichia coli* dihydrofolate reductase, *Biochemistry* 34, 15671–15680.
27 BYSTROFF, C., OATLEY, S. J., KRAUT, J. (1990) Crystal structures of *Escherichia coli* dihydrofolate reductase the NADP+ holoenzyme and the folate NADP+ ternary complex. Substrate binding and a model for the transition state, *Biochemistry* 29, 3263–3277.
28 FALZONE, C. J., WRIGHT, P. E., BENKOVIC, S. J. (1994) Dynamics of a flexible loop in dihydrofolate reductase from *Escherichia coli* and its implication for catalysis, *Biochemistry* 33, 439–442.
29 VENKITAKRISHNAN, R. P., ZABOROWSKI, E., MCELHENY, D., BENKOVIC, S. J., DYSON, H. J., WRIGHT, P. E. (2004). Conformational Changes in the Active Site Loops of Dihydrofolate Reductase during the Catalytic Cycle, *Biochemistry* 43, 16046–16055.
30 EPSTEIN, D. M., BENKOVIC, S. J., WRIGHT, P. E. (1995) Dynamics of the dihydrofolate reductase – folate complex: Catalytic sites and regions known to undergo conformational change exhibit diverse dynamical features, *Biochemistry* 34, 11037–11048.
31 WRIGHT, P. E. (2002) personal communication.
32 RADKIEWICZ, J. L., BROOKS, C. L. I. (2000) Protein dynamics in enzymatic catalysis: Exploration of dihydrofolate reductase, *J. Am. Chem. Soc.* 122, 225–231.
33 ROD, T. H., RADKIEWICZ, J. L., BROOKS, C. L. I. (2003) Correlated motion and the effect of distal mutations in dihydrofolate reductase, *Proc. Natl. Acad. Sci. USA* 100, 6980–6985.
34 THORPE, I. E., BROOKS, C. L. I. (2003) Barriers to hydride transfer in wild type and mutant dihydrofolate reductase fom *E.coli*, *J. Phys. Chem. B.* 107, 14042–14051.
35 SCHNELL, J. R., DYSON, H. J., WRIGHT, P. E. (2004) Effect of cofactor binding and loop conformation on side chain methyl dynamics in dihydrofolate reductase, *Biochemistry* 43, 374–383.
36 WATNEY, J. B., AGARWAL, P. K., HAMMES-SCHIFFER, S. (2003) Effect of mutation on enzyme motion in dihydrofolate reductase, *J. Am. Chem Soc.* 125, 3745–3750.
37 WONG, K., SELZER, T., BENKOVIC, S. J. (2004) Impact of distal mutations on the network of coupled motions correlated to hydride transfer in dihydrofolate reductase, *Proc. Natl. Acad. Sci. USA*, 102, 6807–6812.
38 RAJAGOPALAN, P. T. R., BENKOVIC, S. J. (2002) Preorganization and protein dynamics in enzyme catalysis, *The Chemical Record* 2, 24–36.
39 WAGNER, C. R., THILLET, J., BENKOVIC, S. J. (1992) Complementary Pertubation of the Kinetic Mechanism and Catalytic Effectiveness of Dihydrofolate Reductase by Side-Chain Interchange, *Biochemistry* 31, 7834–7840.

18
Proton Transfer During Catalysis by Hydrolases

Ross L. Stein

18.1
Introduction

Hydrolases are a large family of enzymes whose members catalyze the hydrolytic cleavage of a variety of chemical bonds, including esters, epoxides, amides, acetals, acid anhydrides, and halides. The generalized reaction that is catalyzed by these enzymes is shown in Scheme 18.1 and illustrates the two chemical imperatives of all hydrolytic enzymes:

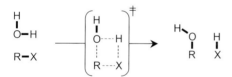

Scheme 18.1. Generalized reaction catalyzed by hydrolases. In this scheme, R–X is the completely generalized substrate for hydrolases, where R contains the electropositive carbon that undergoes nucleophilic attack and X is the electronegative leaving group.

- Activation of nucleophilic attack by proton abstraction.
- Stabilization of intermediate formation and/or leaving group departure by proton donation.

In this chapter we will examine the catalytic strategies that hydrolases employ to meet these chemical demands.

18.1.1
Classification of Hydrolases

Table 18.1 summarizes the scheme that the Enzyme Commission has adopted to classify hydrolases. Each major division or class of hydrolase designates the chem-

Hydrogen-Transfer Reactions. Edited by J. T. Hynes, J. P. Klinman, H.-H. Limbach, and R. L. Schowen
Copyright © 2007 WILEY-VCH Verlag GmbH & Co. KGaA, Weinheim
ISBN: 978-3-527-30777-7

Table 18.1. Summary of hydrolases and their mechanisms.

EC Number	Example	Mechanism
EC 3.1.0.0 Ester bonds		
3.1.1.7	acetylcholinesterase	acyl-Ser intermediate
3.1.3.48	protein-tyrosine-phosphatase	phospho-Cys intermediate
3.1.27.5	pancreatic ribonuclease	non-covalent intermediate
EC 3.2.0.0 Acetyl bonds of glycosides		
3.2.1.17	β-galactosidase	galactosyl-Glu intermediate
3.2.1.139	α-glucuronidase	direct, Glu- and Asp-assisted
EC 3.4.0.0 Amide bonds of proteins		
3.4.11.1	leucyl aminopeptidase	direct, diM^{++}-promoted water attack
3.4.17.1	carboxypeptidase A	direct, Zn^{++}-promoted water attack
3.4.21.1	chymotrypsin	acyl-Ser intermediate
3.4.22.2	papain	acyl-Cys intermediate
3.4.23.1	pepsin	direct, Asp-promoted water attack
3.4.24.27	thermolysin	direct, Zn^{++}-promoted water attack
EC 3.5.0.0 Carbon nitrogen bonds (other than peptide bonds)		
3.5.2.6	β-lactamase	acyl-Ser intermediate
3.5.3.3	creatinase	direct, His-promoted water attack
3.5.4.4	adenosine deaminase	direct, Zn^{++}-promoted water attack
EC 3.6.0.0 Acid anhydrides		
3.6.1.1	inorganic pyrophosphatase	direct, Mg^{++}-promoted water attack
3.6.1.7	acylphosphatase	direct, Arg-promoted water attack
EC 3.8.0.0 Halide bonds		
3.8.1.5	haloalkane dehalogenase	alkyl-Asp intermediate

ical bond that the enzymes of that particular class hydrolyze (e.g., EC 3.4.0.0, amide bonds of proteins). Subclasses that are shown in this table represent individual enzymes (e.g., EC 3.4.22.2, papain). Each of the enzymes in Table 18.1 represents a specific mechanistic type and is accompanied by a brief statement of the defining mechanistic feature for this enzyme. Note that neither this table nor this chapter is intended to be comprehensive; there are a number of classes of hydrolase enzyme that are not listed in the table and will not be considered in this chapter. The enzymes listed in Table 18.1 will be used as illustrative examples throughout this chapter.

18.1.2
Mechanistic Strategies in Hydrolase Chemistry

Two broad strategies will concern us here. First, we will need to understand the two types of general chemical mechanism that hydrolases use and the heavy atom rearrangements that each entails. The two chemical mechanisms lead to differences in enzyme kinetic mechanism. Operationally, the practicing enzymologist is often able to use observation of a particular kinetic mechanism to draw conclusions about chemical mechanism. The second broad strategic concern is how these enzymes have used proton bridging as a means to stabilize the transition states of these chemical mechanisms. Proton bridging, and the proton transfer which often accompanies it, is the principle topic of this chapter.

18.1.2.1 Heavy Atom Rearrangement and Kinetic Mechanism

Scheme 18.2 illustrates the two classes of reaction that hydrolases catalyze: single displacement reactions, in which the enzyme catalyzes the direct attack of water on a substrate, and double displacement reactions, in which the enzyme catalyzes the initial attack of an enzyme-bound nucleophile with expulsion of the leaving group

I. Single Displacement

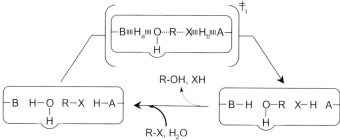

II. Double Displacement

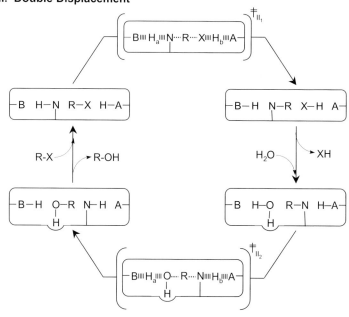

Scheme 18.2. Mechanistic strategies employed by hydrolases.

to form an intermediate which undergoes subsequent enzyme-catalyzed attack of water. In all but a very few cases, the enzyme-bound nucleophile is a side-chain heteroatom of an amino acid of the enzyme's primary sequence. One exception that I will discuss in this chapter is ribonuclease which proceeds through the intermediacy of a cyclic phosphate ester.

Both classes of reaction in Scheme 18.2 are shown as catalytic cycles to emphasize the often overlooked fact that hydrolase turnover depends on the exchange of products for substrate and water, and proton shuffling to 'prime' the system for another round of catalysis. We will see that this step can be kinetically significant, leading to rate-limiting product release or protein- conformational isomerization.

The double-displacement mechanism with its covalent intermediate can often be distinguished from single displacement reactions by kinetic experiments in which an appropriate nucleophilic species is added to reaction solutions of enzyme and substrate and allowed to compete with water for the covalent intermediate. The dependence of initial velocity on the concentrations of substrate and added nucleophile will produce a distinctive kinetic pattern (i.e., 'ping-pong' kinetics). Observation of such a pattern constitutes strong evidence suggesting that the hydrolase under study effects catalysis through a double-displacement mechanism.

18.1.2.2 Proton Bridging and the Stabilization of Chemical Transition States

In the reactions of Scheme 18.2, the principle chemical events are nucleophilic attack and leaving group expulsion. These processes invariably proceed through transition states that can be stabilized by the formation of appropriate proton bridges. For single displacement reactions, the single catalytic transition state is depicted in Scheme 18.2 as having two critical proton bridges. One of these proton bridges, mediated by H_a, activates the attacking water molecule, and the other proton bridge, mediated by H_b, stabilizes the departing leaving group. Without the stabilization that these proton bridges afford, a transition state of much higher energy would form. Proton bridges are formed in exactly analogous manner in transition states II_1 and II_2 of the double displacement reaction of Scheme 18.2. In addition to the generic bridges mediated by protons H_a and H_b, enzyme-specific proton bridging occurs and will be described in this chapter as the occasion arises.

Often, bridging protons of transition states will become fully transferred protons. In Scheme 18.2, protons H_a and H_b, which first participate in bridges, are ultimately transferred. This is not always the case for proton bridges; some that are formed do not result in transfer to a heteroatom different from the one on which it started. Again, these important exceptions are enzyme-specific and will be described in due course.

18.1.3
Focus and Organization of Chapter

My goal in this chapter is to describe the proton transfer reactions that occur during the course of hydrolase-catalyzed reactions and to provide some insight into how they contribute to the catalytic efficiency of these enzymes. Rather than present a case study for each of the hydrolases of Table 18.1, I have chosen instead to organize the chapter around proton transfer reactions themselves, presenting proton abstraction and proton donation, in turn. When we consider proton abstraction, we will first examine how it activates active-site nucleophiles in the first step of double displacement reactions, and then how it is utilized by hydrolases to activate the water molecule for nucleophilic attack in both the second step of double displacement reactions and single displacement reactions. When we turn to proton donation, we first consider how this reaction stabilizes the formation of the covalent intermediate of double displacement reactions and then how proton donation helps to promote leaving group departure in both single displacement and double displacement reactions.

Throughout this chapter, I illustrate key points using the specific enzyme examples from Table 18.1 together with data for enzymes related to those of the table. Since the stated goal of this chapter is to shed light on how proton transfer contributes to the catalytic efficiency of hydrolases, I will necessarily be concerned with the formation of proton bridges in catalytic transition states. Since, arguably, the only experimental technique that directly probes these bridges is the kinetic solvent deuterium isotope effect, much of the data cited here will be these effects, along with results from proton inventory studies. Solvent isotope effects are determined by measuring reaction rates in H_2O and D_2O and are expressed as ratios of the rate constants, typically k_H/k_D, while proton inventories are determined by measuring reaction rates in mixtures of isotopic waters and are expressed graphically as the dependence of rate on mole fraction of solvent deuterium. The numerical magnitude of the solvent isotope effect and the shape of the proton inventory are diagnostic of reaction mechanism and, upon detailed analysis, can often lead to novel mechanistic insights.

So-called "low-barrier hydrogen bonds" [1] may also play a role in catalysis effected by hydrolases, but their catalytic role is less secure [2–5] relative to the role of transition state proton bridges. It is indeed ironic that while the structural origins of catalytically-questionable LBHBs can be determined with certainty, the structural origins of catalytically-critical TSPBs are elusive. The implications of the latter point will be the topic of the final section of this chapter where we examine several cases in which the results of solvent isotope effect studies that may have been interpreted in the context of the hydrolase chemistry, in fact reflect rate-limiting product release or conformational isomerization of the enzyme.

18.2
Proton Abstraction – Activation of Water or Amino Acid Nucleophiles

18.2.1
Activation of Nucleophile – First Step of Double Displacement Mechanisms

The enzymes that we consider in this section all react through double-displacement mechanisms that require nucleophile activation by proton abstraction during the mechanism's first step. In all cases considered here, the nucleophile is an amino acid residue and proton transfer is through a general-base mechanism to another active site residue.

Acyl-transfer reactions comprise the largest single group of enzymic process that we will be considering in this chapter, and among these enzymes the most thoroughly studied are the serine proteases. *Chymotrypsin* and other serine proteases are acylated by a process that begins with attack of the active site serine hydroxyl on the carbonyl carbon of the substrate's scissile bond. This step is subject to general-base catalysis by the active site histidine residue (i.e., transfer of H_a in transition state II_1 of Scheme II), which itself might be assisted by proton-bridging to the active-site aspartate residue. In chymotrypsin, the residues Ser^{95}-His^{57}-Asp^{102} comprise what has been called the 'charge-relay system' or the 'catalytic triad'. Attack of serine leads to formation of a tetrahedral addition adduct, which decomposes with release of leaving group to form the acyl-enzyme. Thus, in contrast to the general picture of Scheme 18.2, the first step of the double-displacement mechanism for these enzymes is not concerted, proceeding through the single transition state II_1, but rather goes through a tetrahedral intermediate.

An informative study which probed acylation was conducted with the serine protease human leukocyte elastase, where a solvent isotope effect on k_c/K_m of 2.0 was observed for hydrolysis of Z-Val p-nitrophenyl ester [6]. Unlike the chymotrypsin-catalyzed hydrolysis of similar substrates (e.g., Ac-Tyr p-nitrophenyl ester) where acylation chemistry is rapid and k_c/K_m is rate-limited by the binding of substrate, producing solvent isotope effects of unity, the hydrolysis of Z-Val p-nitrophenyl ester by elastase is rate-limited by chemistry. More specifically, since acylation is rate-limited by attack of the active site serine to form the tetrahedral intermediate rather than departure of the p-nitrophenylate leaving group from the intermediate, the observed isotope effect reflects protolytic activation of the serine nucleophile for attack. Additional insight into this reaction step was gained from the proton inventory for this reaction which was observed to be linear, suggesting the existence of a single protonic bridge in the transition state. One-proton catalysis provides evidence against operation of the catalytic triad, which, if fully functional, would be expected to produce a 'bowl-shaped' proton inventory indicative of two proton catalysis [7–11]. However, as discussed below, Z-Val p-nitrophenyl ester is a minimal substrate for leukocyte elastase and does not fulfill the enzyme's structural requirements of a specific substrate. For such substrates, linear proton inventories are the rule [6, 8, 11–14].

Acetylcholinesterase is mechanistically related to serine proteases and involves acylation of Ser^{200} and contains the catalytic triad Ser^{200}-His^{440}-Glu^{327}. Precise data for acetylation by the natural substrate acetylcholine is difficult to obtain due to lack of a convenient assay method, so the mechanistically equivalent surrogate acetylthiocholine has frequently been used to probe mechanistic aspects of this enzyme [15]. To explore proton transfer reactions that accompany acetylation by this substrate, solvent isotope effect measurements and proton inventories on k_c/K_m have been conducted [16, 17]. The isotope effects are near unity and the proton inventories bowed-upwards, suggesting that the transition state for k_c/K_m is a virtual transition [18], partially rate-limited by the chemical steps of acylation and some physical process. Since acylation is likely rate-limited by attack of the active site hydroxyl rather than the leaving group departure from the tetrahedral intermediate, the isotope effect reflects, at least in part, general-base catalysis of nucleophilic attack by His^{440}. To probe active site chemistry more directly, the active site mutant in which Glu^{202} of the active site was replaced by Ala was studied. This mutant is accompanied by a 2–3 order of magnitude decrease in reactivity towards acetylthiocholine [16]; for this mutant, k_c/K_m reflects acylation only. The solvent isotope now becomes large and produces a linear proton inventory, suggesting one-proton, general-base catalysis by His^{440} during acetylation. This conclusion is consistent with solvent isotope effect studies of acetylation by unnatural anilide and phenyl ester substrates [19–24]. These results all suggest that if Glu^{327} participates in acetylation, it does not engage in formation of an isotope effect-generating proton bridge with the His^{440}; that is, it would stabilize the transition state for this reaction by a mechanism other than that of a general catalyst, perhaps electrostatically.

Acyl-transfer of β-lactam and acyclic substrates to active site Ser^{64} of the β-lactamase of *Enterobacter cloacae* P99 occurs with a unique mechanism for nucleophilic activation in which the phenolate of Tyr^{150}, stabilized in free enzyme by interaction with both the hydroxyl of Ser^{64} and by ionic interaction with protonated Ly^{367}, acts as general-base [25]. The unusual hydrogen bonding system which exists in free enzyme is thought to manifest as a ground state fractionation factor that is less than unity and account for the inverse solvent isotope effects around 0.7 that have been observed on k_c/K_m [25–27].

Acyl-transfer to the sulfhydryl protease *papain* occurs on the thiolate form of the active site cysteine Cys^{25}, which is thought to be stabilized by ionic interaction with the imidazolium form of an essential active site histidine His^{159} [28–30]. So, unlike serine hydrolases, papain and other cysteine acyl-transfer enzymes do not activate their nucleophiles by general-base catalysis but rather through a mechanism involving a pre-equilibrium process in which the sulfhydryl proton of Cys^{25} of the uncharged pair SH/Im is transferred to the imidazole of His^{159} to produce the ion pair S$^-$/ImH$^+$. Such a mechanism manifests itself in solvent isotope effects on k_c/K_m whose magnitudes are largely dependent both on the equilibrium constant for tautomerization of SH/Im and S$^-$/ImH$^+$, and the proton fractionation factor for sulfhydryl SH [29, 31]. In cases where SH/Im predominates, or is at least not in very low concentration relative to S$^-$/ImH$^+$, inverse isotope effects ranging from 0.5–0.9 are predicted. However, small normal solvent isotope effects can also be observed if leaving group expulsion from the tetrahedral intermediate is rate-limiting and the leaving group requires general-acid catalysis for expulsion. In these cases, the observed isotope effect will have contributions from both the fractionation factor of the ground state sulfhydryl and the fractionation factor of the catalytic proton bridge in the transition state.

Protein-tyrosine phosphatase has an active site Cys^{403} with an unusually low pK_a of 4.7 [32] that is thought to result from stabilization that the thiolate anion receives from an extensive network of H-bonding interactions, including one from the closest proton donor, Thr^{410} [33]. Nucleophilic attack on substrate phosphate esters is thought to be by the thiolate of Cys[403] and thus to occur without the need of general-base catalysis. This mechanism is supported by the observation of solvent isotope effects of unity on k_c/K_m [34].

18.2.2
Activation of Active-site Water

The oxygen atom of water is not especially reactive relative to other heteratoms that enzymes might employ as effective nucleophiles, such as the ε-amine of Lys, the unprotonated heterocyclic nitrogen of His, or the sulfhydryl of Cys. So, to promote attack on an electropositve carbon atom, hydrolases have had to devise strategies for activating water for nucleophilic attack. Two strategies are employed by hydrolases: proton abstraction from water by a basic amino acid residue acting as a general-base catalyst or coordination of the oxygen of water to an active-site metal to promote deprotonation and formation of metal-bound hydroxide.

18.2.2.1 Double-displacement Mechanisms – Second Step

Other than ribonuclease, the enzymes we will be considering here all proceed by mechanisms in which the substrate has undergone covalent interaction with an active site nucleophile to form an intermediate species. In this section, we will examine the role of proton transfer in the enzyme-promoted hydrolysis of this intermediate to generate the second product of the reaction and liberate free enzyme. The mechanisms used by these enzymes all involve activation of the attacking water molecule by an active site residue participating as a general-base.

Our discussion again begins with *chymotrypsin* and other serine proteases. Perhaps the earliest paper documenting the investigation of the proton transfer involved in the attack of water on an acyl-chymotrypsin was that of Myron Bender in 1962 where he reported a solvent isotope effect of 2.8 on k_c for the hydrolysis of Ac-Trp methyl ester [35]. Since k_c for this substrate and other esters is rate-limited by deacylation rather than acylation and, further, since deacylation is itself rate-limited by attack of water on the acyl-enzyme to form a tetrahedral intermediate rather than expulsion of carboxylate product from the tetrahedral intermediate, this value does indeed probe the protonic bridge that is established in the catalytic transition state between the attacking water and His[57]. Since 1962, results similar to these have been observed for a great many serine proteases and their ester substrates [11, 14, 36–41]. In addition, it has also been repeatedly observed that proton inventories of k_c for reaction of serine proteases with specific peptide substrates are "bowl-shaped" and can be fit by a simple model involving two proton catalysis. This has been interpreted to suggest that in the transition states for these reactions the catalytic triad is fully engaged with proton transfer occurring from the hydroxyl of Ser[95] to the imidazole of His[57] and a catalytic proton bridge being established between His[57] and Asp[102].

It is unclear, however, if the operation of the catalytic triad imparts any catalytic advantage. This was investigated in a systematic study of human leukocyte elastase [14, 42]. For *p*-nitroanilides of substrates R-Val-, R-Pro-Val-, R-Ala-Pro-Val-, and R-

Ala-Ala-Pro-Val- (R = methoxysuccinyl-), the first-order rate constant for acylation from within the Michaelis-complex increases as 0.06, 0.42, 43, and 43 s^{-1}, respectively, indicating that substrates of increasing length are better able to fulfill the enzyme's specificity requirements. When proton inventories were determined for hydrolyses of the acyl-enzymes derived from these substrates, it was found that while R-Val-HLE and R-Pro-Val-HLE hydrolyze by a mechanism of one-proton catalysis, presumably involving general-base catalysis by the active site His; R-Ala-Pro-Val-HLE and R-Ala-Ala-Pro-Val-HLE hydrolyze by a mechanism of two proton catalysis, presumably involving full engagement of the catalytic triad [14]. Curiously, deacylation rate-constants are essentially identical for the four substrates and equal 12 ± 1 s^{-1}. It still "remains to be established whether coupling of the charge relay system gives rise to any substantial acceleration or if, perhaps, it is a mere structural and dynamic coincidence that it is coupled with physiological substrates but not with small substrates" [43].

During hydrolytic deacetylation of *acetylcholinesterase*, His^{440} acts as a general-base catalyst to abstract a proton from a water molecule as it attacks the acetyl-Ser^{200} intermediate. This reaction generates a solvent isotope effect of around 2 and a linear proton inventory [17, 22, 44] suggesting, as we saw above for acylation of this enzyme, that if Glu^{327} participates in deacetylation, it does not engage in formation of an isotope effect-generating proton bridge with the His^{440}. The observation of single-proton catalysis during deacylation of the enzyme's natural substrate should be contrasted with observation of two-proton catalysis that has uniformly been observed when serine proteases act on their specific substrates. The significance of this observation for catalysis by acetylcholinesterase is currently unclear.

Like serine hydrolases, the thiol protease *papain* contains the catalytic triad Cys^{25}-His^{159}-Asp^{158} and thus offers the possibility of two-proton catalysis. To investigate the possible operation of this structural feature in catalysis, solvent isotope effects and proton inventories were determined for hydrolysis of the acyl-enzymes that form during reaction of papain with the methyl esters of N-benzoyl-Gly and N-(β-phenylpropionyl)Gly, and the p-nitrophenyl esters of Z-Gly, Z-Lys, and N-methoxycarbonyl-Phe-Gly [45, 46]. Solvent isotope effects ranged from 2.0 to 2.6 and were all accompanied by linear proton inventories suggesting processes of one proton catalysis in which His^{159} acts as a general-base to abstract the proton from the attacking water molecule and a catalytic proton bridge between His^{159} and Asp^{158} does not form. Thus, it appears that, at least for hydrolysis of these substrates, papain's catalytic triad is not fully operative. It could be that two-proton catalysis would be observed if longer peptide substrates were used that fulfilled the enzyme's specificity requirements.

Hydrolysis of the phospho-Cys^{403} intermediate that forms during reactions of *protein-tyrosine phosphatase* is accompanied by a solvent isotope effect of 1.5 [47, 48] and a linear proton inventory [47] suggesting a process of one-proton catalysis in which Asp^{356} acts as a general-base to abstract the proton from the attacking water molecule [49].

The second step of *ribonuclease* catalysis involves the hydrolysis of cyclic nucleoside 2′,3′-phosphate esters to 3′-phosphates. The nucleophilic water that attacks the phosphorus atom of the cyclic ester is activated by proton abstraction by His[119] acting as general-base catalyst [50]. This mechanism is supported by observation of a large solvent isotope effect of 3.1 on k_c/K_m for the ribonuclease-catalyzed hydrolysis of cytidine cyclic 2′,3′-monophosphate [51, 52]. Interestingly, this system generates a bowed-downward proton inventory that can be fit with a model involving two-proton catalysis [52]. The second proton 'in-flight' in the rate-limiting transition state for which these data provide evidence is likely the proton that is thought to be donated by the imidazolium of His[12] to the 2′-oxygen [50]. Observation of an identical proton inventory during hydrolysis of a substituted phenyl cyclic 3′,4′-monophosphate by the ribonuclease model β-cyclodextrin 6,6′-bis(imidazole) supports this mechanistic proposal [53].

β-galactosidase proceeds through the intermediacy of a galactosyl-enzyme whose hydrolytic decomposition rate-limits k_c for substrates with sufficiently reactive aglycone leaving groups where formation of this intermediate is rapid [54–57]. For this process, it is thought that Glu[461] acts as a general-base catalyst to abstract a proton from the attacking water molecule [55, 56]. In such a situation, one would anticipate observation of a large solvent isotope effect. Interestingly though, a solvent isotope effect on k_c of only 1.1 was measured for the β-galactosidase-catalyzed hydrolysis of 3,4-dinitrophenyl β-D-galactopyranoside [54]. It is unclear why such a small isotope effect is observed for this reaction.

Haloalkane dehalogenase catalyzes hydrolysis of carbon–halide bonds of alkyl halides by a double-displacement mechanism in which the carboxylate anion of Asp[124] (*Xanthobacter autotrophicus* GJ10) attacks the carbon of the carbon–halide bond and displaces the halide with formation of an alkyl ester intermediate [58–60]. Interestingly, decomposition of this intermediate is rate-limited by one of three steps, depending on the bacterial species from which the enzyme is isolated [60]: hydrolysis (*Sphingomonas paucimobilis*), halide release (*Xanthobacter autotrophicus*), or alcohol release (*Rhodococcus rhodochrous*). To probe the possibility of catalytically-important proton transfer during hydrolysis of the alkyl ester intermediate, pre-steady-state kinetic experiments were preformed for turnover of dibromomethane by the enzyme from *Xanthobacter autotrophicus* and reveal an isotope effect of 2.5 on the rate-constant and amplitude of a pre-state burst [59]. This was interpreted to reflect general-base catalysis of attack of water by His[289].

18.2.2.2 Single Displacement Mechanisms

The enzymes we will be considering in this section all catalyze their hydrolytic reactions through mechanisms that lack the intermediacy of any species other than simple, noncovalent Michaelis complexes. While transient intermediates may form, such as oxocarbenium species during catalysis by inverting glycosidases or tetrahedral intermediates during catalysis by aspartyl- and metallo-proteinases, they are of insufficient stability to accumulate in the steady state and thus have structures that are similar to the transition states that precede and follow them on the reaction pathway. For these reactions, the single displacement mechanism of Scheme 18.2, with its single transition state, is not an inaccurate depiction.

In this section, we will be concerned with understanding how proton H_a of this mechanism is abstracted from the attacking water molecule. Two mechanisms of activation will be seen to be used by these enzymes: activation by an amino acid general-base and activation by virtue of complexation of the catalytic water to a metal atom.

Activation by Amino Acid General-base α-*Glucuronidases* and other inverting glycosidases catalyze their hydrolytic reactions through an S_N1 mechanism with a transition state having substantial oxocarbenium ion character, as evidenced by large α-secondary deuterium kinetic isotope effects of the order of 1.1–1.2 [55, 56, 61]. In these reactions, the attacking water molecule is usually activated by an active site carboxylate moiety. α-Glucuronidases from two bacterial species have recently had their X-ray structures determined [62, 63], revealing key aspartyl residues that play the role of general-base catalyst in the displacement of aglycone by water. However, in some cases another residue is used. For example, a His has been implicated in the active site of an inverting exo-β(1 → 3)-glucanase [64]. Reactions of this enzyme proceed with large solvent isotope effects on k_c of the order of 2.5 indicating the establishment of a proton bridge in the catalytic transition state and perhaps suggesting a role of general-base catalyst for the His residue.

Pepsin catalyzes the hydrolysis of amide bonds of proteins using a unique mechanism in which water attack on the amide is activated by concerted action of two active site aspartyl carboxyl groups. Small normal solvent isotope effects are observed but cannot be unambiguously interpreted regarding their origins and how they relate to proton abstraction from water and/or proton donation to amine leaving group [65].

Creatinase catalyzes the hydrolysis of creatine to N-methyl-glycine and urea by a mechanism in which water is activated for attack on substrate guanidinium carbon by the imidazole of His[232] acting as general base [66].

Acylphosphatase is one of the smallest enzymes (98 residues; MW 11 kDa) and hydrolyzes acylphosphates, such as acetylphosphate, succinylphosphate, and β-asparatylphosphate [67]. While there is scant kinetic and no isotope effect data, structural and mutagenesis studies suggest a mechanism in which the amide functionality of Asn[41] aligns a water molecule for attack on the phosphorus atom of the substrate with substrate-assisted general-base catalysis of this attack by an anionic oxygen of the phosphate group. A pentacovalent phosphorous may form and carboxylate leaves unassisted.

Activation of Metal-bound Water In this mechanism, a metal ion, usually a zinc, complexes a water molecule and lowers its pK_a from 14 to values perhaps as low as 7 or 8, thus readily producing a nucleophilic metal-bound hydroxide anion [68, 69]. Proton transfer thus occurs in a pre-equilibrium step and does not participate, in the form of general acid–base catalysis, in the rate-limiting step.

This mechanism may best be exemplified by metal-dependent proteases and peptidases such as *thermolysin*, *leucine aminopeptidase*, and *carboxypeptidase A*. Of these, the most systematic study of proton transfer has been conducted with thermolysin and the mechanistically related metalloendoproteinase stromelysin [68–73]. For both of these enzymes, solvent isotope effects on k_c/K_m of about 0.75 have been observed [69–71] and support a mechanism in which the zinc-bound water molecule of free enzyme undergoes reaction with substrate to become part of a tetrahedral addition adduct with a fractionation factor of one. The inverse isotope effect thus originates from the fractionation factors of about 0.85 for each of the two protons of the zinc-bound water molecule.

Leucine aminopeptidase is interesting in that its active site contains two zinc atoms which together bind and activate the water molecule [74]. Despite this enzyme containing a dinuclear metal center at its active site, its mechanism, and specifically its mode of proton transfers reactions, appear to follow the general theme established by thermolysin and carboxypeptidase

Adenosine deaminase and other members of the family of nucleoside and nucleotide deaminases utilize zinc-bound water as the catalytic nucleophile to displace ammonia from the 6-position of purines or the 4-position of pyrimidines and in all cases display inverse solvent deuterium isotope effects ranging from 0.3 to 0.8 on k_c/K_m [75–80]. These effects are reminiscent of those observed for metalloproteases and have their origins, like those of the proteases, in fractionation factors for the protons of the bound water that are less than one.

18.3
Proton Donation – Stabilization of Intermediates or Leaving Groups

18.3.1
Proton Donation to Stabilize Formation of Intermediates

All of the enzymes to be considered here proceed through double-displacement mechanisms in which stabilization of the intermediate occurs with donation of a proton by an amino acid of the primary sequence of the enzyme.

In *chymotrypsin* and other serine proteases, the oxyanion hole exists as a means of stabilizing intermediates. The oxyanion hole is a structural unit at the active sites of these enzymes that is thought to participate in stabilizing acyl-transfer transition states via hydrogen bonds from two protein backbone amide groups to the oxyanion that exists during the formation and collapse of tetrahedral intermediates. The evidence for the existence of the oxyanion hole and hypotheses regarding its role in catalysis come chiefly from X-ray crystallographic studies of these enzymes and their stable complexes with inhibitors [81].

Like serine proteases, *acetylcholinesterase*, appears to possess an oxyanion hole, comprised of the peptide NH groups of Gly[118], Gly[119], and Ala[201] that stabilizes oxyanionic transition states and tetrahedral intermediates by hydrogen bonding [82]. Linear proton inventories on k_c suggest that these stabilizing interactions do not generate proton bridges that give rise to fractionation factors less than unity; if this were the case, the multiproton origin of the solvent isotope effect would produce bowed-downward proton inventories [17].

Papain also contains an oxyanion hole comprising hydrogen-bond donation from the amide side chain of Gln^{19} and the backbone NH of Cys^{25} [83]. Evidence for the existence of this structural features comes largely from X-ray crystallographic studies.

The phospho-Cys^{403} intermediate that forms during the course of reactions catalyzed by *protein-tyrosine phosphatase* is stabilized by hydrogen-bond donation from backbone residues of the so-called P-loop comprising Cys^{403}, Arg^{409}, and Thr^{410} as well as the guanidinium moiety of Arg^{409} [49]. These interactions also serve to stabilize the Michaelis complex.

During turnover of creatine by *creatinase*, the transition state for attack by water is stabilized by hydrogen bonding by carboxylate residues of Glu262 and Glu358 to the departing guanidinium group [66].

18.3.2
Proton Donation to Facilitate Leaving Group Departure

18.3.2.1 Double-displacement Mechanisms

For many of the reactions that have been considered here, leaving group departure during formation of the stable intermediate of double-displacement reactions requires protonation of the scissile heteroatom. This proton is thought to come either from solvent, if the active site is sufficiently accessible to solvent, or from a protonated active site residue serving as general-acid.

Chymotrypsin and other serine proteases undergo acylation by amide and anilide substrates with at least partial rate-limiting expulsion of the amine or aniline leaving group. ^{15}N-isotope effects of 1.010 ± 0.001 and 1.014 ± 0.001 on k_c/K_m for reaction of chymotrypsin with N-acetyl-Trp amide [84] and N-succinyl-Phe p-nitroanilide [85] are smaller than the theoretical limit of 1.044 and suggest that the observed transition state for these reactions may be a virtual transition state [18] reflecting both formation and decomposition of the tetrahedral addition adduct that exists as an intermediate between the Michaelis complex and acyl-enzyme. Thus, in studies which seek to examine proton bridging to the departing leaving group, care must be taken in the interpretation of the data, since it is likely compromised by partial rate-limiting water attack and the proton bridging that accompanies it. For example, the solvent isotope effect on the first-order rate constant for acylation of chymotrypsin by N-succinyl-Ala-Ala-Pro-Phe p-nitroanilide from the stage of the Michaelis complex is 3.1 [86]. From the previous analysis, it is likely that this value is a weighted average of the isotope effect on proton abstraction from the attacking water molecule and the isotope effect on proton donation to the departing leaving group.

For *acetylcholinesterase*, it is thought that the protonated imidazole of His^{440} acts as a general-acid to assist departure of aniline during hydrolysis of anilide substrates. Based on analysis of the ^{15}N isotope effect in light and heavy water, it was estimated that departure of o-nitroaniline during hydrolysis of o-nitroacetanilide is accompanied by a modest solvent isotope near 1.5, supporting the general-acid role of ImH^+-His^{440} [19].

Protein-tyrosine phosphatase catalyzes the hydrolysis of aryl phosphates with rate-limiting dephosphorylation for k_c. However, for alkyl phosphates, k_c is rate-limited by phosphorylation [34] and allows proton transfer in this process to be explored. Specifically for our present discussion, we would like to learn about proton transfer as involved in leaving group departure. For the hydrolysis of 4-phenylbutyl phosphate, a solvent isotope effect of 2.4 is observed along with a linear proton inventory. Since nucleophilic attack by the thiolate of Cys^{403} cannot generate this isotope effect (see above), it seems likely that the effect originates in the general-catalysis of leaving group departure from the tetrahedral intermediate. Mutagenesis and structural studies suggest that the general-acid is Asp^{356} [87, 88].

β-galactosidase from *E. coli* is thought to use Glu^{461} as a general-acid to donate a proton to the departing aglycone leaving group during transfer of β-D-galactopyranosyl to Glu^{537} to form the intermediate galactosyl-enzyme species [55, 56]. Some of the most compelling evidence for this comes from recent mutagenesis experiments in which the E461G enzyme was shown to have pH-rate dependences and solvent isotope effects predicted for this mutant [57]. Specifically for the latter property, the solvent isotope effect on k_c for hydrolysis of 4-nitrophenyl β-D-galactopyranoside of 1.7 for wild-type enzyme [54] was reduced to a value near unity with the E461G mutant [57]. Since k_c is thought to be rate-limited by galactosyl-enzyme formation, and it is known that nucleophilic attack on the anomeric carbon by Glu^{537} is not subject to general-catalysis [55, 56], the observed solvent isotope effect for the native enzyme must be due the establishment of a proton bridge between the protonated carboxyl moiety of Glu^{461} and the oxygen of the leaving group. In the E461G mutant, catalysis of this type is unavailable and no significant isotope is observed.

18.3.2.2 Single-displacement Mechanisms

The *α-glucuronidases* from the two bacterial species mentioned above have been suggested to have aspartyl residues as part of hydrogen-bonding networks that participate in proton donation to the departing aglycone moiety [62, 63].

For the mechanistically related enzymes *thermolysin* and *stromelysin*, a strong case has been made supporting the hypothesis that the amine of leaving groups during turnover of peptide substrates is fully protonated in the transition state and thus requires no assistance from a general acid [71].

Departure of urea leaving group during turnover of creatine by *creatinase* is facilitated by proton donation by the imidazolium group of His232, the same residue that extracted the proton from water to promote nucleophilic attack [66].

18.4
Proton Transfer in Physical Steps of Hydrolase-catalyzed Reactions

18.4.1
Product Release

For *thermolysin* and *stromelysin*, release of the carboxylate product is not a simple dissociation process but rather involves a ligand exchange reaction on zinc; that is, that step involves the attack of water on zinc to displace the carboxylate product.

Rate-limiting protolytic catalysis of this process is suggested by the observation of a solvent isotope effect of 1.6 on k_c for peptide hydrolysis by stromelysin [71].

Since release of carboxylate product is the reverse of the association of carboxylate-based inhibitors with metalloproteinases, one might anticipate that much could be learned about the former by study of the latter. For reaction of such inhibitors [73], as well as phosphorus-containing inhibitors [72], inhibitor displaces zinc-bound water to form the enzyme–inhibitor complex and generate isotope effects on the association rate constant of 1.6 to 1.9 [69, 72, 73]. These large isotope effects arise from proton catalytic bridges that form in the transition state for association of enzyme and inhibitor. Insufficient data exist to say whether this mechanism can be generalized to *leucine aminopeptidase* and *carboxypeptidase A*. In fact for both of these enzymes, solvent isotope effects ranging from 1.3 to 2.6 on k_c have been observed and interpreted to reflect general-acid catalysis of the amine leaving group departure by an active site carboxylate of a Glu residue [89, 90].

A solvent isotope effect of 3 was observed on k_c for *inorganic pyrophosphatase* and was originally interpreted to reflect the chemistry of pyrophosphate hydrolysis [91]. However, this interpretation required revision in the light of microscopic rate constants that were calculated from the results of studies that measured P_i-dependent formation of enzyme-bound pyrophosphate and measured rates of H_2O/P_i oxygen exchange [92]. Based on these studies it now appears that the large solvent isotope effect is for P_i dissociation and likely reflects changes in the energetics of hydrogen bonds between P_i and enzyme.

18.4.2
Protein Conformational Changes

Haloalkane dehalogenase from *Xanthobacter autotrophicus* GJ10 exhibits a solvent isotope effect of 3 and linear proton inventories for k_c for hydrolysis of dibromethane and 1,2-dichloroethane [58]. Stopped-flow fluorescence experiments under single-turnover conditions suggest that halide release limits k_c. This isotope effect was interpreted to reflect a conformational change of the protein that allows departure of halide.

Data is accumulating for *pepsin* and HIV protease to suggest that they may catalyze their reactions through a mechanism in which release of products leaves the enzyme in a conformation which must undergo isomerization before another round of catalysis can commence [65]. More work is needed here to eliminate other mechanistic possibilities.

References

1 CLELAND, W. W. (2000) *Arch. Biochem. Biophys.*, 382, 1–5.
2 WARSHEL, A., PAPAZYAN, A. (1996) *Proc. Natl. Acad. Sci. USA*, 93, 13665–13670.
3 ASH, E. L., SUDMEIER, J. L., DE FABO, E. C., BACHOVCHIN, W. W. (1997) *Science* 278, 1128–1132.
4 STRATTON, J. R., PELTON, J. G., KIRSCH, J. F. (2001) *Biochemistry*, 40, 10411–10416.

5 Poi, M. J., Tomaszewski, J. W., Yuan, C., Dunlap, C. A., Andersen, N. H., Gelb, M. H., Tsai, M.-D. (2003) *J. Mol. Biol.*, 329, 997–1009.

6 Stein, R. L. (1985) *J. Am. Chem. Soc.* 107, 5767–5775.

7 Elrod, J. P., Gandour, R. D., Hogg, J. L., Kise, M., Maggiora, G. M., Schowen, R. L., Venkatasubban, K. S. (1976) *Faraday Symp. Chem. Soc.*, 10, 145–153.

8 Venkatasubban, K. S., Schowen, R. L. (1984) *Crit. Rev. Biochem.*, 17, 1–44.

9 Mordy, C. W., Schowen, R. L. (1987) *Adv. Biosci.*, 65, 273–280.

10 Demuth, H. U., Heins, J., Fujihara, H., Mordy, C. W., Fischer, G., Barth, A., Schowen, R. L. (1987) *Stud. Org. Chem.*, 31, 439–446.

11 Schowen, R. L. (1988) *Mol. Struct. Energ.*, 9, 119–168.

12 Stein, R. L. (1983) *J. Am. Chem. Soc.* 105, 5111–5116.

13 Stein, R. L., Elrod, J. P., Schowen, R. L. (1983) *J. Am. Chem. Soc.* 105, 2446–2452.

14 Stein, R. L., Strimpler, A. M., Hori, H., Powers, J. C. (1987) *Biochemistry* 26, 1305–1314.

15 Froed, H. C., Wilson, I. B. (1984) *J. Biol. Chem.* 259, 11010–11013.

16 Malany, S., Sawai, M., Sikorski, R. S., Seravalli, J., Quinn, D. M., Radic, Z., Taylor, P., Kronman, C., Velan, B., Shafferman, A. (2000) *J. Am. Chem. Soc.* 122, 2981–2987.

17 Pryor, A. N., Selwood, T., Leu, L. S., Andracki, M. A., Lee, B. H., Rao, M., Rosenberry, T., Doctor, B. P., Silman, I., Quinn, D. M. (1992) *J. Am. Chem. Soc.* 114, 3896–900.

18 Stein, R. L. (1981) *J. Org. Chem.* 46, 3328–3330.

19 Rao, M., Barlow, P. N., Pryor, A. N., Paneth, P., O'Leary, M. H., Quinn, D. M., Huskey, W. P. (1993) *J. Am. Chem. Soc.* 115, 11676–11681.

20 Salih, E. (1992) *Biochem. J.* 285, 451–460.

21 Barlow, P. N., Acheson, S. A., Swanson, M. L., Quinn, D. M. (1987) *J. Am. Chem. Soc.* 109, 253–257.

22 Acheson, S. A., Dedopoulou, D., Quinn, D. M. (1987) *J. Am. Chem. Soc.* 109, 239–245.

23 Acheson, S. A., Barlow, P. N., Lee, G. C., Swanson, M. L., Quinn, D. M. (1987) *J. Am. Chem. Soc.* 109, 246–52.

24 Quinn, D. M., Swanson, M. L. (1984) *J. Am. Chem. Soc.* 106, 1883–1884.

25 Page, M. I., Laws, A. P. (1998) *Chem. Commun.*, 1609–1617.

26 Page, M. I., Vilanova, B., Layland, N. J. (1995) *J. Am. Chem. Soc.*, 117, 12092–12095.

27 Adediran, S. A., Deraniyagala, S. A., Xu, Y., Pratt, R. F. (1996) *Biochemistry* 35, 3604–3613.

28 Creighton, D. J., Schamp, D. J. (1980) *FEBS Lett.* 110, 313–318.

29 Creighton, D. J., Gessouroun, M. S., Heapes, J. M. (1980) *FEBS Lett.* 110, 319–322.

30 Wandinger, A., Creighton, D. J. (1980) *FEBS Lett.* 116, 116–122.

31 Case, A., Stein, R. L. (2003) *Biochemistry* 42, 9466–9481.

32 Zhang, Z. Y., Dixon, J. E. (1993) *Biochemistry* 32, 9340–9345.

33 Stuckey, J. A., Fauman, E. B., Schubert, H. L., Zhang, Z.-Y., Dixon, J. E. (1994) *Nature* 370, 571–575.

34 Zhang, Z. Y., Van Etten, R. L. (1991) *Biochemistry* 30, 8954–8959.

35 Bender, M. L., Hamilton, G. A. (1962) *J. Am. Chem. Soc.* 84, 2570–2576.

36 Pollock, E., Hogg, J. L., Schowen, R. L. (1973) *J. Am. Chem. Soc.* 95, 968–969.

37 Elrod, J. P., Hogg, J. L., Quinn, D. M., Venkatasubban, K. S., Schowen, R. L. (1980) *J. Am. Chem. Soc.* 102, 3917–3922.

38 Hogg, J. L., Elrod, J. P., Schowen, R. L. (1980) *J. Am. Chem. Soc.* 102, 2082–2086.

39 Quinn, D. M., Elrod, J. P., Ardis, R., Friesen, P., Schowen, R. L. (1980) *J. Am. Chem. Soc.* 102, 5358–5365.

40 Quinn, D. M., Venkatasubban, K. S., Kise, M., Schowen, R. L. (1980) *J. Am. Chem. Soc.* 102, 5365–5369.

41 Enyedy, E. J., Kovach, I. M. (2004) *J. Am. Chem. Soc.* 126, 6017–6024.

42 Stein, R. L., Strimpler, A. M., Hori,

H., POWERS, J. C. (1987) *Biochemistry* 26, 1301–1305.

43 MAGGIORA, G. M., and SCHOWEN, R. L. (1977) in *Bioorganic Chemistry*, VAN TAMELEN, E. E. (Ed.), Acedemic Press, New York.

44 KOVACH, I. M., LARSON, M., SCHOWEN, R. L. (1986) *J. Am. Chem. Soc. 108*, 3054–3056.

45 SZAWELSKI, R. J., WHARTON, C. W. (1981) *Biochem. J. 199*, 681–692.

46 STORER, A. C., CAREY, P. R. (1985) *Biochemistry 24*, 6808–6818.

47 ZHOU, X., MEDHEKAR, R., TONEY, M. D. (2003) *Anal. Chem. 75*, 3681–3687.

48 ZHANG, Z.-Y. (1995) *J.Biol. Chem. 270*, 11199–11204.

49 ZHANG, Z.-Y. (2003) *Acc. Chem. Res. 36*, 385–392.

50 RAINES, R. T. (1998) *Chem. Rev. 98*, 1045–1065.

51 EFTINK, M. R., BILTONEN, R. L. (1983) *Biochemistry 22*, 5123–5134.

52 MATTA, M. S., DIEP THI, V. (1986) *J. Am. Chem. Soc. 108*, 5316–5318.

53 ANSLYN, E., BRESLOW, R. (1989) *J. Am. Chem. Soc. 111*, 8931–8932.

54 SELWOOD, T., SINNOTT, M. L. (1990) *Biochem. J. 268*, 317–323.

55 SINNOTT, M. L. (1990) *Chem. Rev. 90*, 1171–1202.

56 ZECHEL, D. L., WITHER, S. G. (2000) *Acc. Chem. Res. 33*, 11–18.

57 RICHARD, J. P., HUBER, R. E., MCCALL, D. A. (2001) *Bioorg. Chem. 29*, 146–155.

58 SCHANSTRA, J. P., JANSSEN, D. B. (1996) *Biochemistry 35*, 5624–5632.

59 SCHANSTRA, J. P., KINGMA, J., JANSSEN, D. B. (1996) *J. Biol. Chem. 271*, 14747–14753.

60 PROKOP, Z., MONINCOVA, M., CHALOUPKOVA, R., KLVANA, M., NAGATA, Y., JANSSEN, D. B., DAMBORSKY, J. (2003) *J. Biol. Chem. 278*, 45094–45100.

61 VASELLA, A., DAVIES, G. J., BOHM, M. (2002) *Curr. Opin. Chem. Biol. 6*, 619–629.

62 NURIZZO, D., NAGY, T., GIBLERT, H. J., DAVIES, G. J. (2002) *Structure 10*, 547–556.

63 GOLAN, G., SHALLOM, D., TEPLITSKY, A., ZAIDE, G., SHULAMI, S., BAASOV, T., STOJANOFF, V., THOMPSON, A., SHOHAM, Y., SHOHAM, G. (2004) *J. Biol. Chem. 279*, 3014–3024.

64 JEFFCOAT, R., KIRKWOOD, S. (1987) *J. Biol. Chem. 262*, 1088–1091.

65 NORTHROP, D. B. (2001) *Acc. Chem. Res. 34*, 790–797.

66 COLL, M., KNOF, S. H., OHGA, Y., MESSERSCHMIDT, A., HUBER, R., MOELLERING, H., RUSSMANN, L., SCHUMACHER, G. (1990) *J. Mol. Biol. 214*, 597–610.

67 STEFANI, M., TADDEI, N., RAMPONI, G. (1997) *Cell. Mol. Life Sci. 53*, 141–151.

68 IZQUIERDO, M. C., STEIN, R. L. (1990) *J. Am. Chem. Soc. 112*, 6054–6062.

69 IZQUIERDO-MARTIN, M., STEIN, R. L. (1992) *J. Am. Chem. Soc. 114*, 325–31.

70 STEIN, R. L. (1988) *J. Am. Chem. Soc. 110*, 7907–7908.

71 HARRISON, R. K., CHANG, B., NIEDZWIECKI, L., STEIN, R. L. (1992) *Biochemistry 31*, 10757–1762.

72 IZQUIERDO-MARTIN, M., STEIN, R. L. (1992) *J. Am. Chem. Soc. 114*, 1527–1528.

73 IZQUIERDO-MARTIN, M., CHAPMAN, K. T., HAGMAN, W. K., STEIN, R. L. (1994) *Biochemistry 33*, 1356–1365.

74 HOLZ, R. C. (2002) *Coord. Chem. Rev. 232*, 5–26.

75 LEWIS, A. S., GLANTZ, M. D. (1974) *J. Biol. Chem., 249*, 3862–3866.

76 WEISS, P. M., COOK, P. F., HERMES, J. D., CLELAND, W. W. (1987) *Biochemistry 26*, 7378–7384.

77 MERKLER, D. J., KLINE, P. C., WEISS, P., SCHRAMM, V. L. (1993) *Biochemistry 32*, 12993–13001.

78 MERKLER, D. J., SCHRAMM, V. L. (1993) *Biochemistry 32*, 5792–5799.

79 WANG, Z.-X., QUIOCHO, F. A. (1998) *Biochemistry 37*, 8314–8324.

80 SNIDER, M. J., REINHARDT, L., WOLFENDEN, R., CLELAND, W. W. (2002) *Biochemistry 41*, 415–421.

81 HEDSTROM, L. (2002) *Chem. Rev. 102*, 4501–4524.

82 SUSSMAN, J. L., HAREL, M., FROLOW, F., OEFNER, C., GOLDMAN, A., TOKER, L., SILMAN, I. (1991) *Science 253*, 872–879.

83 DRENTH, J., KALK, K. H., SWEN, H. M. (1976) *Biochemistry 15*, 3731–3738.

84 O'Leary, M. H., Kluetz, M. D. (1970) J. Am. Chem. Soc. 92, 6089–6090.
85 Hengge, A. C., Stein, R. L. (2004) Biochemistry, 43, 742–747.
86 Case, A., Stein, R. L. (2003) Biochemistry 42, 3335–3348.
87 Zhang, Z., Harms, E., Van Etten, R. L. (1994) J. Biol. Chem. 269, 25947–25950.
88 Keng, Y. F., Wu, L., Zhang, Z.-Y. (1999) Eur. J. Biochem. 259, 809–814.
89 Bienvenue, D. L., Mathew, R. S., Ringe, D., Holz, R. C. (2002) J. Biol. Inorg. Chem. 7, 129–135.
90 Suh, J., Cho, W., Chung, S. (1985) J. Am. Chem. Soc. 107, 4530–4535.
91 Konsowitz, L. M., Cooperman, B. S. (1976) J. Am. Chem. Soc. 98, 1993–1995.
92 Welsh, K. M., Jacobyansky, A., Springs, B., Cooperman, B. S. (1983) Biochemistry 22, 2243–2248.

19
Hydrogen Atom Transfers in B_{12} Enzymes

Ruma Banerjee, Donald G. Truhlar, Agnieszka Dybala-Defratyka, and Piotr Paneth

19.1
Introduction to B_{12} Enzymes

B_{12} is a tetrapyrrolic-derived organometallic cofactor that supports three subfamilies of enzymes in microbes and in animals, the isomerases, the methyltransferases, and the dehalogenases [1]. The corrin ring system is more reduced than the porphyrin ring system, is heavily ornamented peripherally, and is distinguished by a central cobalt atom that is coordinated equatorially to four pyrrolic nitrogens. The cobalt can cycle between three oxidation states, +1 to +3, and the unique properties of each species are exploited in the chemistry of the reactions catalyzed by B_{12} enzymes. In both the methyltransferase and isomerase subfamilies of B_{12} enzymes, the upper axial ligand to cobalt is an alkyl group and the cobalt is formally in the +3 oxidation state. The alkyl group is methyl and deoxyadenosyl in methylcobalamin (MeCbl) and coenzyme B_{12} (AdoCbl), respectively. Their structures are shown in Scheme 19.1. Dichotomous pathways for cleaving the organometallic cobalt–carbon bond yield different products with different reactivities. In the methyltransferases, the cobalt–methyl bond ruptures heterolytically, and the products are cob(I)alamin (vitamin B_{12} with cobalt in the +1 oxidation state) and a carbocation equivalent that is transferred to a nucleophile. Cob(I)alamin is highly reactive and indeed, is regarded as nature's supernucleophile [2]. Its reactivity is exploited in biology for transferring methyl groups from unactivated methyl donors, viz. 5-methyltetrahydrofolate. In contrast, the cobalt–carbon bond is cleaved homolytically in the isomerases, where the upper axial ligand is a 5′-deoxyadenosyl group, and the coenzyme is called 5′-deoxyadenosyl cobalamin (AdoCbl). The radical products are cob(II)alamin (vitamin B_{12} with cobalt in the +2 oxidation state) and the reactive 5′-deoxyadenosyl radical, which is abbreviated dAdo•. The reactivity of the latter is harnessed to effect hydrogen atom abstractions in unusual and chemically challenging 1,2 rearrangement reactions involving the exchange of a variable group with a hydrogen atom on adjacent carbons.

Our understanding of the reaction mechanism of B_{12}-dependent reductive dehalogenations is quite limited [1]. However, the role of the cofactor appears to be sub-

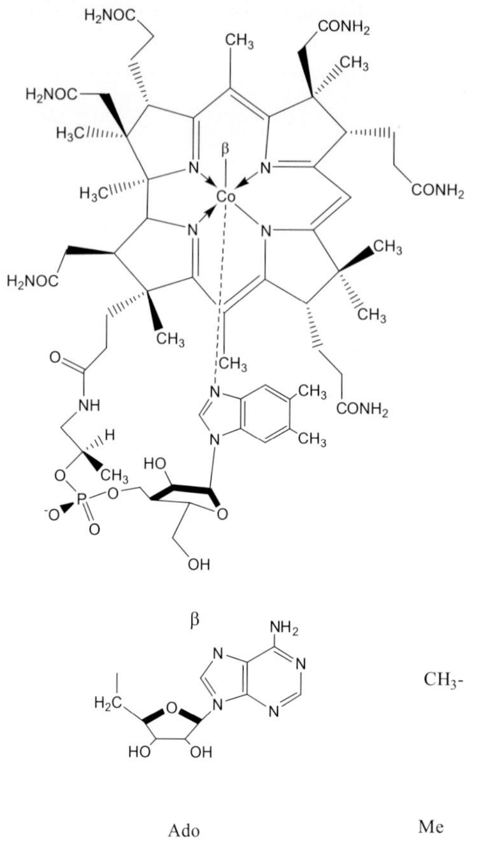

Scheme 19.1

stantially different from its role in the other two groups of B_{12}-dependent enzymes. It appears likely that the low redox potential of the Co(I) state of the cofactor is exploited to drive the reductive dehalogenation reactions.

The lower axial ligand in B_{12} derivatives is an extension of a peripheral propanolamine chain from ring D of the tetrapyrrolic structure. A variety of ligands are found in this position in nature, and the unusual base, 5,6-dimethylbenzimidazole, is the lower axial ligand in cobalamins. At acidic pH, the lower axial ligand is displaced via protonation. The cofactor can thus exist in two conformations, "base-on" and "base-off". However, the crystal structure of methionine synthase revealed yet another conformation, "base-off/His-on", in which the endogenous ligand is displaced and replaced by a histidine residue donated by the protein [3]. This ligand switch by an active site histidine embedded in a conserved DXHXXG motif [4] has since been observed in a number of other B_{12} enzymes including methylmalonyl-CoA mutase [5], glutamate mutase [6], and lysine amino mutase [7]. In contrast, a second subclass of AdoCbl-dependent enzymes, including diol dehydratase [8] and ribonucleotide reductase [9], binds the cofactor in the "base-on" conformation.

19.2
Overall Reaction Mechanisms of Isomerases

Isomerases that are dependent on coenzyme B_{12} constitute the largest subfamily of B_{12} enzymes and are components of a number of fermentative pathways in microbes [10, 11]. A single member of this group of enzymes, methylmalonyl-CoA mutase, is found in both bacteria and in mammals where it is a mitochondrial enzyme involved in the catabolism of odd-chain fatty acids, branched chain amino

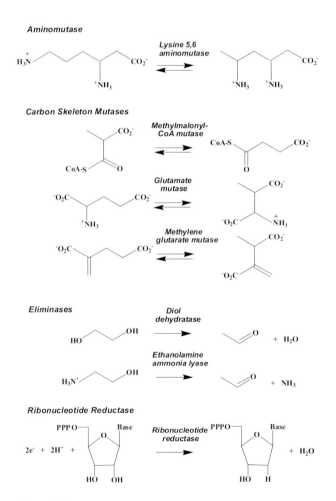

Scheme 19.2

acids, and cholesterol [12]. The general reaction catalyzed by the isomerases is a 1,2 interchange of a hydrogen atom and a variable group such as a group containing a heteroatom (hydroxyl or amino) or a carbon skeleton (see Scheme 19.2). One member, ribonucleotide reductase, uses the B_{12} cofactor to effect reductive elimination in the conversion of ribonucleotides to deoxyribonucleotides and represents a third class of this subfamily. As might be expected, individual enzymes differ somewhat in their radical generating strategies, as discussed below.

Whereas heterolytic cleavage in the methyltransferases is facilitated by methylcobalamin, homolytic cleavage is deployed by enzymes that resort to radical chemistry to effect difficult transformations. The preference for homolytic cleavage in AdoCbl may be related to the increased electron density on cobalt in the presence of the 5′-deoxyadenosyl group [13]. The chemical basis for the utility of the AdoCbl cofactor as a radical reservoir is the weak cobalt–carbon bond with a bond dissociation energy that is estimated to be $\sim$30 kcal mol^{-1} (in aqueous solution) in the "base-on" state [14]. Reversible cleavage and reformation of the cobalt–carbon bond during catalytic turnover results in the formation of transient radical intermediates.

The first common step in AdoCbl-dependent reactions is homolytic cleavage of the cobalt–carbon bond to generate a radical pair, cob(II)alamin and the carbon-centered dAdo• radical (Scheme 19.3). This reaction experiences a $\sim$10^{12}-fold rate enhancement in B_{12} enzymes [14, 15] in the presence of substrate, and the mechanism for this rate acceleration has been the subject of extensive scrutiny. Thus, in methylmalonyl-CoA mutase and in glutamate mutase, little if any destabilization of the cobalt–carbon bond is observed in the reactant state, as revealed by resonance Raman spectroscopy [16, 17], and the intrinsic substrate binding is utilized to labilize the bond. In contrast, approximately half of the destabilization of the cobalt–carbon bond in diol dehydratase is expressed in the reactant state. This re-

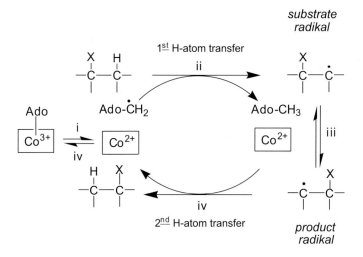

Scheme 19.3

actant destabilization may result in part from differences in the sizes of substrates that could translate into differences in binding energy. The destabilization renders the enzyme more prone to inactivation [10]; enzymes such as diol dehydratase can probably tolerate a higher inactivation rate due to the presence of repair chaperones that can catalyze the exchange of inactive cofactor for AdoCbl [18].

Rapid reaction studies on B_{12} enzymes reveal that homolysis is fast and not rate limiting [19–23]. Following homolysis, a series of controlled radical propagation steps result in migration of the organic radical (X in Scheme 19.3) to an adjacent carbon. The isomerization reaction is initiated by abstraction of a hydrogen atom from the substrate to generate a substrate-centered radical. This rearranges to a product-centered radical which reabstracts a 5′-hydrogen atom from 5′-deoxyadenosine. The dAdo• and cob(II)alamin radicals then recombine to complete a catalytic turnover cycle. In these 1,2 rearrangements, the hydrogen atom migrates intermolecularly, and a minimum of two hydrogen atom transfers, from substrate to dAdo• and back, are involved. A mechanistic complication may involve some competition of a 1,2-hydrogen shift along with the dominant 1,2 shift of the carbon-centered radical [24].

A key issue to understanding these reactions is ascertaining the role of the protein [25]. In bacterial methylmalonyl-CoA mutase, the substrate is bound inside an α/β barrel, which may be important for shielding the radical intermediates [5, 26, 27]. More significant catalytically may be the role of an active-site tyrosine [28, 29], which appears to sterically drive the adenosyl group off the Co, as a result of a conformational change upon substrate binding. EPR spectroscopy provides information about the distance of the dominant product radical from cob(II)alamin [30]. Finally, it is important to consider the role of entropy and the "solvating" power of the protein in promoting cobalt–carbon bond fission [31].

A critical issue for AdoCbl-dependent enzymes is controlling the timing of the homolysis step so that the radical pool is not dissipated. Homolysis of the cobalt–carbon bond takes place in the absence of substrate, as evidenced by the scrambling of label at the C5′ position in methylmalonyl-CoA mutase [32]. However, the equilibrium favors geminate recombination, and formation of the spectrally visible cob(II)alamin is not detected in the absence of substrate. Substrate binding triggers conformational adjustments, and the equilibrium shifts to favor the forward propagation of dAdo•. Thus, homolysis and hydrogen transfer from the substrate are kinetically coupled; evidence for this was first obtained with methylmalonyl-CoA mutase [20] and later with other enzymes, viz. glutamate mutase [33] and ethanolamine ammonia lyase [34]. In methylmalonyl-CoA mutase, substitution of the protons on the methyl group of methylmalonyl-CoA with deuterons decelerated the appearance of cob(II)alamin by ∼20-fold at 25 °C [20]. This unusual sensitivity of the homolysis reaction of the cofactor to isotopic substitution in the substrate was interpreted as evidence for kinetic coupling, whereby the detectable accumulation of cob(II)alamin was dependent on the extent of H-atom abstraction from the substrate. Kinetic coupling effectively shifts the equilibrium of the homolysis reaction, and it allows the substrate to gate mobilization of radicals from the AdoCbl reservoir.

Ribonucleotide reductase presents an exception to the above mechanism where the working radical is a thiyl derived from an active-site cysteine (C408 in the *Lactobacillus leichmannii* enzyme) rather than dAdo• [35]. Mutation of C408 leads to failure of the mutant enzyme to generate detectable levels of cob(II)alamin. However, the mutant catalyzes epimerization of AdoCbl that is stereoselectively deuterated at the 5' carbon bonded to cobalt [36]. This indicates that transient cleavage of the cobalt–carbon bond occurs, but when radical propagation to C408 is precluded, recombination of dAdo• and cob(II)alamin is favored.

19.3
Isotope Effects in B$_{12}$ Enzymes

Large primary kinetic isotope effects have been measured for the H-atom transfer steps from substrate to dAdo• and from dAdo• to the product radical in a number of AdoCbl-dependent enzymes as indicated in Table 19.1. In methylmalonyl-CoA mutase, the steady-state deuterium isotope effect is 5-6 in the forward direction, and the intrinsic isotope effect of step (i) in Scheme 19.3 is masked by the kinetically coupled but slower later steps [37–39]. The steady-state tritium kinetic isotope effect (k_H/k_T) in the forward direction has been reported to be 3.2 [38]. Note that the experiments with deuterium were performed with a fully deuterated methyl group, while those with tritium were carried out at the trace level and correspond to a single isotopic atom; therefore these two isotope effects should not be directly compared. For the reverse reaction, the deuterium kinetic isotope effect is also par-

Table 19.1. Summary of kinetic isotope effects reported for B$_{12}$-dependent isomerases.

Enzyme substrate	Overall kinetic isotope effect		k_H/k_D on Co(II) formation	Ref.
	k_H/k_T	k_H/k_D		
Diol dehydratase propanediol	83 (10 °C)	10 (37 °C)	3–4 (4 °C)	10, 45
Ethanoloamine lyase (EAL) ethanolamine	107 (23 °C)	7.4 (23 °C)	>10 (22 °C)	34, 44
Methylmalonyl-CoA mutase (MCM) methylmalonyl-CoA succinyl-CoA	3.2 (~30 °C)	5–6 (30 °C) 3.4 (30 °C)	43 (10 °C)	37–40 29
Glutamate mutase glutamate 3-methylaspartate	21 (10 °C) 19 (10 °C)	3.9 (10 °C) 6.3 (10 °C)	28 (10 °C) 35 (10 °C)	33, 43 33, 43

tially masked by kinetic complexity ($k_H/k_D = 3.4$) [38]. Under pre-steady-state conditions though, the measured kinetic isotope effects should not be affected by the product release step and, barring other complications, should be close to the intrinsic kinetic isotope effect. Under these conditions, a large deuterium isotope effect on cob(II)alamin formation has been reported for the conversion of methylmalonyl-CoA to succinyl-CoA [20, 40]. Since a protein-based hydrogen pool in methylmalonyl-CoA mutase, which could account for the anomalously large isotope effects, has been excluded [38], the involvement of tunneling was invoked [20]. An Arrhenius analysis of the temperature dependence of the pre-steady-state isotope effect has provided compelling evidence [40] that tunneling dominates the reaction in that the observed values of the ratio, A_H/A_D (0.078 ± 0.009), of pre-exponential factors and the difference, $E_{a,H} - E_{a,D}$ (3.41 ± 0.07 kcal mol^{-1}), of activation energies lie outside the ranges expected [41] ($A_H/A_D = 0.5$–1.4 and $E_{a,H} - E_{a,D}$ ca. <1.3 kcal mol^{-1}) in the absence of tunneling. The coupled homolysis/H-transfer steps catalyzed by methylmalonyl-CoA mutase are characterized by an equilibrium constant that is estimated to be close to unity and a phenomenological free energy of activation, $\Delta G^\ddagger$, of 13.1 ± 0.6 kcal mol^{-1} at 37 °C that corresponds to a $\sim 10^{12}$-fold [42] rate acceleration. In contrast, thermolysis of AdoCbl in solution is characterized by an unfavorable equilibrium, and a $\Delta G^\ddagger$ of 30 kcal mol^{-1} at 37 °C.

In glutamate mutase [43], the forward and reverse steady-state deuterium (k_H/k_D of 3.9 forward and 6.3 reverse) and tritium (k_H/k_T of 21 forward and 19 reverse) kinetic isotope effects are both suppressed. However large deuterium isotope effects of 28 and 35 in the forward and reverse directions respectively have been observed for cob(II)alamin formation under pre-steady-state conditions. These large kinetic isotope effects suggest that quantum mechanical tunneling also dominates this enzyme reaction.

Diol dehydratase and ethanolamine ammonia lyase exhibit the largest overall tritium isotope effects that have been measured in B$_{12}$-dependent enzymes [44, 45], the overall deuterium kinetic isotope effect is also substantial [10, 34, 45]. The observation of a deuterium isotope effect on the pre-steady-state formation of cob(II)alamin in diol dehyratase [10] and in ethanolamine ammonia lyase [25] is consistent with kinetic coupling between the homolysis and H-transfer steps.

Recently the secondary kinetic isotope effect has been measured for the Co–C homolysis step in the pre-steady-state reaction of glutamate mutase [46]. The result obtained was $k_H/k_T = 0.76 \pm 0.02$, which is a large inverse effect. The same study reported a secondary equilibrium isotope effect of $k_H/k_T = 0.72 \pm 0.04$. Thus the kinetic and equilibrium effects agree within the error bars, the most straightforward interpretation of which, in the absence of tunneling, would be that the dynamical bottleneck is close to the product, i.e., late. However in the light of the large role expected for tunneling, this conclusion is not justified. Tunneling would be expected to raise the secondary kinetic isotope effect, so the fact that the kinetic isotope effect is inverse seems very significant. Recall that the Co–C homolysis and the hydrogen transfer from substrate to dAdo•, though not likely to be concerted, are kinetically coupled. The homolysis step corresponds to a sp^3 → sp^2 hybridization change at C5′ and this direction of hybridization change usually makes a normal contribution to the secondary kinetic isotope effect [47], whereas the hydrogen transfer involves the opposite trend at C5′. Hence the net inverse kinetic isotope effect would seem to place the dynamical bottleneck of the kinetically coupled two-step process at the hydrogen transfer step.

19.4
Theoretical Approaches to Mechanisms of H-transfer in B_{12} Enzymes

It is not easy to infer the details of the hydrogen atom transfer steps from the experimental kinetics, and theoretical methods provide one possible way to increase understanding. Although computational approaches to the Co–C bond dissociation and radical rearrangement steps in B_{12}-dependent enzymes have been attempted [48–56], the hydrogen atom transfer has received less attention. The radical nature of the dAdo• reactant, the large size of the corrin moiety and the presence of a transition metal contribute to the difficulty of modeling this step in B_{12} enzymes. This difficulty is compounded by the paucity of reliable energetic data to calibrate the calculations. While good geometric information is sometimes sufficient for qualitative predictions of kinetic isotope effects for over the barrier processes, modeling the tunneling contribution [57–59] requires detailed knowledge of the ensemble of reaction paths, their barrier heights, the shapes (especially the widths) of the barriers, the curvature components of the reaction paths, and the potential energy in the tunneling swaths, which are the broad regions of configuration space through which tunneling from the reactant valleys to the product ones may proceed. Thus far only a few reports have been published on reactions that specifically aim at modeling the hydrogen atom transfer steps in B_{12}-dependent enzymes, and only one [60] addresses the tunneling contribution. Such calculations however are beginning to come within the realm of current computational technology, spurring new attempts at modeling H-atom transfer steps.

Several studies have addressed the energetics and geometry of H-atom transfer in reactions that serve as models for this step in B_{12}-dependent enzymes [61–65]. Although these studies do not include active site residues and do not attempt to address the non-classical behavior of this step, they do provide useful information. Since combined quantum mechanical/molecular mechanical (QM/MM) calculations [66] of the enzyme kinetics may require the inclusion of a large number of atoms in the QM part due to the size of the corrin moiety, it is advantageous to use an inexpensive quantum mechanical model such as semiempirical molecular orbital theory. Therefore, a study was carried out in which the performance of semiempirical methods was critically evaluated and compared to high-end theory levels for $C_nH_{2n+1} + C_nH_{2n+2}$ ($n = 1, 2, 3$) reactions [65]. Consensus values were evaluated from the high-level G3S//MP2(full)/6-31G(d), G3SX(MP3)//B3LYP/6-31G(2df,p), CBS-QB3//B3LYP/6-31G(d†), MCG3/3//MPW1K/6-31+G(d,p), MC-QCISD/3//MPW1K/6-31+G(d,p), MC-QCISD/3, MPW1K/MG3S, and MPW1K/MG3S//MPW1K/6-31+G(d,p) calculations. The energetics of the $n = 1$ species

Table 19.2. Calculated barrier heights (in kcal mol^{-1}) for $C_nH_{2n+1}{}^{\cdot} + C_nH_{2n+2}$ ($n = 2, 3$) reactions[a]

Method	$C_2H_5{}^{\cdot} + C_2H_6$	$C_3H_7{}^{\cdot} + C_3H_8$
Consensus barrier height	16.7	16.0
AM1	16.0	15.6
PM3	12.0	12.5
AM1-CHC-SRP	18.3	17.9
PM3-CHC-SRP	17.0	16.3
PM3(tm)	16.3	15.6
B3LYP/6-31+G(d,p)[b]	15.7	15.5
MP2/6-31+G(d,p)[b]	19.4	18.6

[a] average values for *gauche* and *trans* structures. [b] The basis set is given after the solidus, using conventional notation [67].

($CH_3{}^{\cdot}$ and CH_4) differ significantly from that obtained for the larger models, indicating the inadequacy of a methyl species as a model for the larger molecules. Some key results [65] for $n = 2$ and 3 are shown in Table 19.2. This table shows that the general AM1 semiempirical parametrization [67] is capable of reproducing the barrier heights for transfer of a hydrogen atom between two carbons centers (the "CHC" motif) within ~1 kcal mol^{-1}, which is quite encouraging. The equally inexpensive PM3 parametrization [67] is much less accurate, but the PM3(tm) method [67] is about as accurate as AM1. Use of specific reaction parameters (SRP) [68] for CHC systems [65] also improves PM3, but is unable to systematically improve AM1. Table 19.2 also shows a more expensive semiempirical method, B3LYP [67], which is a hybrid of Hartree–Fock theory and density functional theory, and it shows an *ab initio* post-Hartree–Fock level, MP2 [67]. Although B3LYP usually underestimates barriers for hydrogen atom transfers [69], for the CHC motif the magnitude of the underestimate is not large, only 0.5–1.0 kcal mol^{-1} in Table 19.2.

Toraya et al. [60–63] used B3LYP (with the 6-311G(d) basis set) for calculations on the H-atom transfer steps in diol dehydratase reaction. Both H-atom transfers, i.e., from the substrate and re-abstraction of a hydrogen atom from 5'-deoxyadenosine, were considered. The models used in these studies included the substrate, 1,2-propanediol, a potassium cation found in the active site, and an ethyl radical as a mimic of the dAdo• radical (Fig. 19.1). The activation barrier for the abstraction of the *pro-S* hydrogen atom of substrate by dAdo• was calculated to be 9.0 kcal mol^{-1}, while the activation barrier for the reverse reaction between product radical and 5'-deoxyadenosine was 15.7 kcal mol^{-1}. In the absence of the potassium cation the forward activation barrier is 9.6 kcal mol^{-1} indicating that coordination of the substrate by the potassium cation has a minimal energetic effect on the H-atom transfer step, but seems to hold the substrate and intermediates in

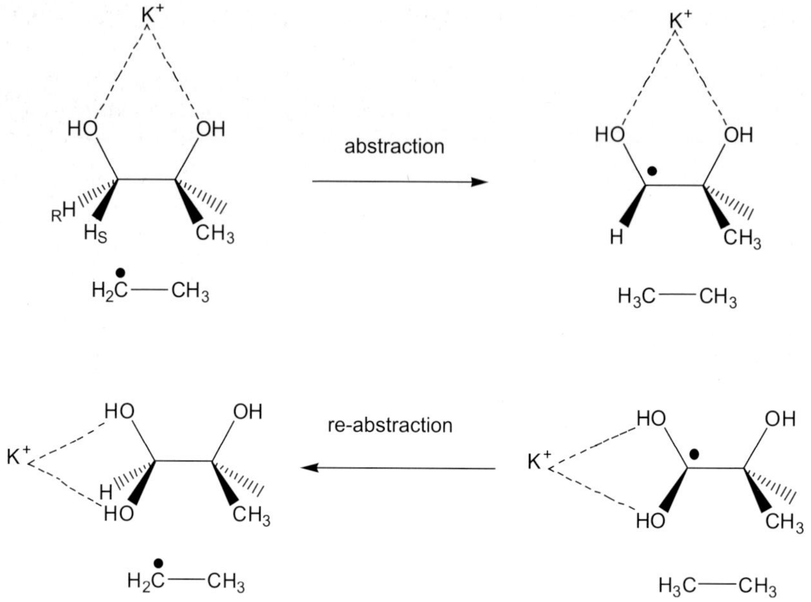

Figure 19.1. Model [61–63] of hydrogen atom transfer steps in diol dehydratase reaction.

position for the multistep sequence. These calculations disagree with the experimental [70] determination of the rate-determining step in that the barrier for hydrogen atom abstraction is lower than that for OH group migration, which is probably a consequence of omitting active-site residues, since calculations on other model systems show strong environmental effects on the OH migration [48, 49]. For the re-abstraction step, two pathways were considered that differ in the timing of dehydration (Fig. 19.1). When the dehydration step precedes the H-atom transfer step, an activation barrier of 19.8 kcal mol^{-1} is estimated. For the alternative pathway, in which hydrogen abstraction from deoxyadenosine by the product radical occurs prior to dehydration, the barrier is 15.1 kcal mol^{-1} and was proposed to be more likely. However, the lack of active site residues in this model precludes unequivocal exclusion of the first pathway.

The H-atom transfer steps in the reaction catalyzed by ethanolamine ammonia lyase reaction have also been examined computationally [64]. The simplest model employed a 1,5-dideoxyribose radical and 2-aminoethanol as the substrate (Fig. 19.2). The influence of full ($R^2 = H^+$) or partial protonation (R^2 = methyliminium = $NH_2CH_2^+$) of the nitrogen atom, as well as the synergetic presence of two ($R^1 = HCO_2^-$, R^2 = methyliminium = $NH_2CH_2^+$) hydrogen bonds, on the energetics of H-atom abstraction were evaluated. The hydrogen bonds mimic His and Asp residues in the active site, although the hydrogen bonds were arbitrarily placed since the 3D structure of the active site is not known. Two conformations of the ribose ring were considered. Calculations were carried out at B3LYP/6-

19.4 Theoretical Approaches to Mechanisms of H-transfer in B_{12} Enzymes

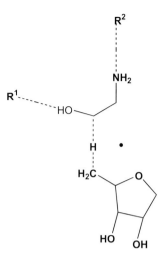

Figure 19.2. Model [64] of the hydrogen atom transfer between two CH_2 groups in ethanolamine ammonia lyase. See Section 19.4 of text for an explanation of R^1 and R^2.

311++G(d,p)//B3LYP/6-31G(d) and MP2/6-311++G(d,p)//B3LYP/6-31G(d) levels, and zero point vibrational energy was included. Activation enthalpies of 16.7 and 17.3 kcal mol^{-1} were found for the unprotonated substrate (R^1 and R^2 missing) and the ribose C3-*endo* and C2-*endo* conformers, respectively. The results indicate that the H-atom transfer step would be facilitated by protonation or by hydrogen bonding interactions to the substrate. In particular, activation enthalpies for models of fully protonated or singly hydrogen-bonded substrate were smaller than 15 kcal mol^{-1}, while the simultaneous presence of two hydrogen bonds to the nitrogen atom increased the activation barrier to over 25 kcal mol^{-1}.

Finally we consider some attempts to simulate the tunneling contributions in the hydrogen transfer step. This was first attempted using several models of differing complexity [60, 71]. PM3 calculations using conventional transition state theory (TST) [72] and a model comprising 37 atoms of the ribose radical and methylmalonyl with truncated CoA moiety (Fig. 19.3 with R^3 = H and R^4 missing) give a hydrogen kinetic isotope effect (for CH_3 vs. CD_3) of only about 10, indicating that TST without tunneling corrections is insufficient to account for the experimental results, which are summarized in column 2 of Table 19.4. To include tunneling, the barrier shape was estimated from the energies of three stationary points (the substrate, the transition state, and the product of the hydrogen atom reaction) by an algorithm called IVTST-0 that was developed earlier [73] for gas-phase reactions. This treatment allows the calculation of a multidimensional tunneling contribution. The resulting dynamical method [74, 75] is called TST/ZCT (where ZCT denotes zero-curvature tunneling, since this method ignores the curvature of the reaction path, which is discussed below). Calculated energetics (Table 19.3, column 3) and kinetic isotope effects (Table 19.4, column 4) compare reason-

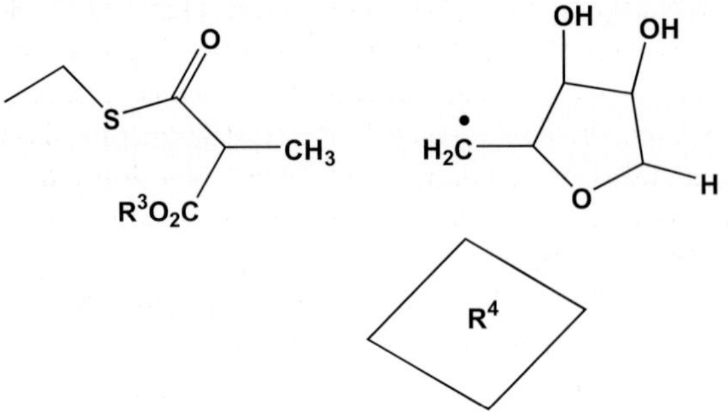

Figure 19.3. Models [60] used in calculations of hydrogen kinetic isotope effects with tunneling contributions.

Table 19.3. Classical barrier heights and energies of reaction (in kcal mol^{-1}) for models[a] of the hydrogen abstraction for methylmalonyl-CoA mutase.

	QM R^3 R^4	PM3 H ...	M[b] H ...	PM3 Arg ...	PM3(tm) H ...	PM3(tm) Arg Corrin+His	PM3 ... ...
Ref.		60	71	60	71	60	60
barrier height[c]		11.9	14.4	12.0	16.2	19.9	4.5
reaction energy[d]		−0.2	0.7	−0.6	−0.1	9.5	−2.6

[a] R^3 and R^4 are explained in Fig. 19.3. [b] M denotes MPW1K/6-31+G(d,p). [c] The change in potential energy (exclusive of zero-point energy) in proceeding from reactants to the saddle point. [d] The change in potential energy (exclusive of zero-point energy) in proceeding from reactants to products.

Table 19.4. Primary kinetic isotope effects for hydrogen abstraction from methylmalonyl-CoA.

T (°C)	Exp.	QM R^3 R^4	PM3 H ...	M[a] H ...	PM3 Arg ...	PM3(tm) H ...	PM3(tm) Arg Corrin+His	PM3 ... ...
Ref.	40		60	71	60	71	60	60
5	50		47	42	49	118	63	9.4
20	36		37	32	38	84	49	8.6

[a] M denotes MPW1K/6-31+G(d,p).

ably well with results carried out at the MPW1K/6-31+G(d,p) level (Table 19.3, column 4 and Table 19.4, column 5), which has been validated to give reliable results for hydrogen atom transfer reactions [69]. Inclusion of arginine, which is hydrogen-bonded to the carboxylate of the methylmalonyl moiety in the crystal structure, does not affect the results very much (Table 19.3 column 5 and Table 19.4, column 6).

However, the agreement between these results and the experimental values is only coincidental. When the PM3tm method, which gives a higher barrier height is used (Table 19.3, column 6), the TST/ZCT hydrogen kinetic isotope effect is much higher (Table 19.4, column 7). Further enlargement of the model to include the corrin moiety with an imidazole ring (Table 19.3, column 7 and Table 19.4, column 8) does not change the isotope effect very much, but deprotonation of the carboxyl group leads to substantial lowering of the barrier (Table 19.3, column 8) and consequently, a much smaller kinetic isotope effect (Table 19.4, column 9), illustrating the sensitivity of the calculated isotope effect to the barrier height.

As the next step in improving the dynamical description, it is important to include reaction path curvature (as well as zero point variation) in the description of the tunneling event. If the minimum energy path (MEP) from reactants to products were a straight line in the space of atomic Cartesian coordinates, there would be no internal centrifugal effect (bobsled effect) forcing the system's motion, on average, to deviate from the MEP. But the MEP for most reactions is curved, and there is a bobsled effect. For tunneling processes the bobsled effect is negative (because the semiclassical kinetic energy is negative) [76–78] and thus the dominant tunneling paths are on the concave side of the MEP. This phenomenon is called corner-cutting tunneling [78, 79].

To describe the tunneling process in more detail, we need to define some terminology. A transition state (or generalized transition state) is a dividing surface (technically a hypersurface) in phase space (the space of the nuclear coordinates and momenta); here we define transition states entirely in terms of their location in coordinate space, which, after separating translation and rotational motion, has $3N - 6$ dimensions, where N is the number of atoms; and we consider transition states that are orthogonal to the MEP. Distance along the MEP is the reaction coordinate. Because the reaction coordinate has a fixed value in a transition state, a transition state has $3N - 7$ vibrations, which are called its generalized normal modes (the word "generalized" is included because conventional normal modes are defined only at stationary points such as equilibrium geometries and saddle points). The conventional transition state passes through the saddle point, but generalized transition states intersect the MEP both earlier and later than the saddle point.

Reaction path curvature is actually a vector with $3N - 7$ components [80]. Each component is associated with a particular generalized normal mode, and it measures the extent to which the system curves into a particular direction as it progresses along the MEP. Corner-cutting tunneling involves a coupled motion involving the reaction coordinate and all the generalized normal modes that are associated with nonzero curvature components [81].

When reaction-path curvature is small, the tunneling is dominated by paths on the concave side of the MEP whose locations are determined by the reaction-path curvature components [81, 82]. These small-curvature tunneling paths are typically located close enough to the MEP that the harmonic approximation is valid, and thus the dominant tunneling paths may be calculated (approximately) from the $3N - 7$ curvature components and the $3N - 7$ harmonic force constants of the generalized normal modes. From this information and the shape of the potential energy along the MEP, one can then calculate the tunneling probabilities. This is called the small-curvature tunneling (SCT) approximation [82, 83].

Describing tunneling in terms of definite paths in coordinate space [84] is a classical-like approximation. This kind of approximation (like the well known WKB method [85]) is called "semiclassical" in chemical physics (and here) because it involves calculating a quantum mechanical quantity by classical-like methods; thus it is partly classical or semiclassical. Unfortunately the kinetic isotope effect community uses the word "semiclassical" to denote an entirely different approximation, namely including quantized vibrational energies in the treatment of the $3N - 7$ transition state vibrations, but treating the transmission coefficient entirely classically and thus completely neglecting tunneling. (This is sometimes called "quasiclassical" in chemical physics.) We hope this warning is sufficient to prevent confusion.

When reaction-path curvature is large, one requires a much more complicated semiclassical treatment to handle the tunneling. In the limit of large reaction-path curvature, tunneling tends to occur along straight-line tunneling paths because the shortest distance between two points is a straight line, and when reaction-path curvature is large, tunneling along short tunneling paths has a greatly enhanced probability [86]. In addition, tunneling tends to be much more delocalized, and systems may have appreciable probability of tunneling directly into vibrationally excited stretching modes [86, 87, 89]. A semiclassical theory that incorporates all of these features has been developed [82, 90, 91] and it is called the large-curvature tunneling (LCT) approximation.

In the general case one could obtain a good semiclassical tunneling approximation by optimizing the tunneling path somewhere between the small-curvature and large-curvature limits by a least-action approximation [92]. In practice, it has been found that simply choosing between the SCT and LCT approximations on the basis of whichever yields a larger tunneling probability (the tunneling is dominated by the most favorable tunneling paths at each tunneling energy) is enough optimization to yield accurate results [91, 93]. This is called microcanonically optimized multidimensional tunneling or μOMT [91, 93].

Tunneling can be included most consistently in transition state theory in the context of variational transition state theory, e.g., canonical variational theory (CVT) [75]. CVT calculations were performed [71] with zero-curvature tunneling (see Table 19.5, column 3), and they show that the IVTST-0 calculations overestimate ZCT contribution to the hydrogen kinetic isotope effect. Calculations were also performed including corner-cutting tunneling, and these are shown in the last five columns of Table 19.5. As can be seen from the comparison of the results in these columns, large-curvature tunneling (LCT) plays the dominant role. Col-

Table 19.5. Kinetic isotope effects for 37-atom model obtained using PM3 for electronic structure calculations.

T (°C)	CVT[a]	CVT/ZCT	CVT/SCT	CVT/LCT	CVT/µOMT		
QM model	PM3 M37[b]	PM3 M37	PM3 M37	PM3 M37	PM3 M37	PM3 M50[c]	AM1 M37
5	9.9	21	27	145	113	96	127
20	8.5	18	23	124	94	80	89

[a] No tunneling. [b] 37-atom model corresponding to R^3 = H, and R^4 not present. [c] 50-atom model: M37 + adenine.

umns 6 and 7 of Table 19.5 show that comparable results are obtained when the model is enlarged to include adenine, the last column contains results obtained for the AM1 Hamiltonian instead of PM3. Barriers for the AM1 parameterizations are 14.9 and 16.6 kcal mol^{-1} for the M37 and M50 models, respectively.

The last three columns of Table 19.5 report results obtained with the microcanonically optimized multidimensional tunneling approximation, which represents the currently most trusted method for including a tunneling contribution. These results predict that the hydrogen kinetic isotope effect for the hydrogen atom step in methylmalonyl-CoA reaction in the direction of succinyl formation is ~100.

The dynamical calculations in Tables 19.4 and 19.5 were performed with the POLYRATE [88] and MORATE [90] computer programs.

19.5
Free Energy Profile for Cobalt–Carbon Bond Cleavage and H-atom Transfer Steps

The interpretation of kinetic isotope effects observed in enzymes must take account of kinetic complexity. For example, the deuterium kinetic isotope effect on the methylmalonyl-CoA mutase-catalyzed reaction was measured under pre-steady-state conditions with UV–visible detection of cob(II)alamin formation [20, 42]. Thus, it reports on a combination of steps, including substrate binding and a concomitant conformational change in the enzyme, cobalt-carbon bond homolysis, and hydrogen-atom transfer to the dAdo• radical. The kinetics could be further complicated in other enzymes such as diol dehydratase [95] and glutamate mutase [96], where conformational changes in the dAdo• radical are expected to occur between the homolysis and hydrogen atom transfer steps, and in ribonucleoside triphosphate reductase, in which a protein-based cysteinyl radical functions as the working radical [97]. Because of these mechanistic complexities, observed kinetic isotope effects under pre-steady state conditions, although large, need not reflect the full intrinsic values for the kinetic isotope effects on the H-atom transfer steps in the respective enzymes. Thus it is unclear whether the difference between the large intrinsic kinetic isotope effect calculated for the hydrogen atom transfer step in methylmalonyl-CoA mutase and the isotope effect observed experimentally (which is also very large, but not as large as the calculated one) indicates that kinetic complexity causes about half of the isotope effect to be masked by isotope-insensitive steps in the observed pre-steady-state rate or whether it results from the quantitative uncertainty of the calculation. Since the calculation includes neither the full enzyme nor ensemble averaging, one should be very cautious about the former type of conclusion.

Further progress toward a quantitative resolution of the size of the intrinsic kinetic isotope effect requires a more complete mechanistic analysis, which is sometimes [98], but not always, possible. A step in this direction has been made for methylmalonyl-CoA mutase [42], for which a free energy profile extending across seven steps has been constructed on the basis of available kinetic and spectroscopic data. Similarly, a three-step profile has been presented for ethanolamine deaminase [99], and a six-step profile has been presented for glutamate mutase [22]. A feature seen in the profiles of both methylmalonyl-CoA mutase and glutamate mutase, which is also seen for many other enzymes, is that the energetic barriers to the interconversion of the various chemical intermediates are similar in height, which may be a general consequence of the tendency of enzymatic transformations to be partitioned into a series of discrete steps without the inefficiency of high energy release or high energy consumption in any one of them. In any event, it is precisely this feature that makes it hard to sort out elementary-step rate constants and kinetic isotope effects for the individual chemical steps.

19.6
Model Reactions

We discussed above some theoretical studies of model systems. There is an even larger literature devoted to experimental studies of model systems, dating back at least 20 years [100, 101]. Most recently, some model studies have appeared that are directly related to the present concerns. In particular, Finke and coworkers [102, 103] have studied the reaction of β-neopentyl-Cbl with ethylene glycol (a model of the diol dehydratase reaction) at temperatures up to 120 °C, and their results have engendered interesting discussions [104]. Their key finding is that the elevated-temperature kinetic isotope effects observed for the model reaction in solution (where the enzyme is not present to stabilize the dAdo• radical) are very similar to those obtained for methylmalonyl–CoA mutase at temperatures more typical of physiological action. The comparison is clouded by potential kinetic complexity, discussed above, that may suppress the intrinsic value of the kinetic isotope effect in the enzyme case. Nevertheless one clear conclusion of this work, which agrees with pure theoretical considerations as well as with studies of many nonenzymatic reactions, is that enzymes are not uniquely evolved to promote tunneling. Whether the fraction of a reaction that proceeds by tunneling in an enzyme reaction is greater than that in a similar nonenzymatic reaction will depend on the individual enzyme and on many detailed mechanistic factors. One scenario is that the environment can exert a greater leverage on a reaction by lowering the effective barrier height (which appears in an exponent) than by changing the transmission coefficient, and lower barriers are often (not always) associated with broader barrier tops; a lower, broader barrier makes it harder for tunneling to compete with over-barrier processes so that the fraction of reactive events that occurs by tunneling may be lower in the enzymatic system. If the intrinsic methylmalonyl–CoA mutase kinetic isotope effect is as large as the calculated values given above, though, then this reaction may prove to be an exception to that scenario.

19.7 Summary

Deconvoluting the contributions of the individual steps in the observed pre-steady-state rate constant for cob(II)alamin formation is challenging and awaits solution. In particular, details of the H-atom transfer steps elude direct determination. However, recent advances in the theoretical methodology hold promise for complementing the experimental analysis of a fundamental aspect of AdoCbl-dependent reactions, i.e., the H-atom transfer steps.

Acknowledgments

This work was supported by grants from the National Institutes of Health (Fogarty International Collaboration grant to P.P. and R.B. and DK45776 to R.B.) and the National Science Foundation (CHE-0349122 to D.T.).

References

1. BANERJEE, R., RAGSDALE, S. W. (2003) The many faces of vitamin B_{12}: Catalysis by cobalamin-dependent enzymes, *Annu. Rev. Biochem.* 72, 209–247.
2. SCHRAUZER, G. N., DEUTSCH, E. (1969) Reactions of cobalt(I) supernucleophiles. The alkylation of vitamin B_{12S}, cobaloximes(I), and related compounds, *J. Am. Chem. Soc.* 91, 3341–3350.
3. DRENNAN, C. L., HUANG, S., DRUMMOND, J. T., MATTHEWS, R., LUDWIG, M. L. (1994) How a protein binds B_{12}: A 3 Å X-ray structure of B_{12}-binding domains of methionine synthase, *Science* 266, 1669–1674.
4. MARSH, E. N. G., HOLLOWAY, D. E. (1992) Cloning and sequencing of glutamate mutase component S from *Clostridium tetanomorphum*, *FEBS Lett.* 310, 167–170.
5. MANCIA, F., KEEP, N. H., NAKAGAWA, A., LEADLAY, P. F., MCSWEENEY, S., RASMUSSEN, B., BÖSECKE, P., DIAT, O., EVANS, P. R. (1996) How coenzyme B_{12} radicals are generated: The crystal structure of methylmalonyl-coenzyme A mutase at 2 Å resolution, *Structure* 4, 339–350.

6 REITZER, R., GRUBER, K., JOGL, G., WAGNER, U. G., BOTHE, H., BUCKEL, W., KRATKY, C. (1999) Glutamate mutase from Clostridium cochlearium: the structure of a coenzyme B_{12}-dependent enzyme provides new mechanistic insights, *Structure Fold. Des.* 7, 891–902.

7 CHANG, C. H., FREY, P. A. (2000) Cloning, sequencing, heterologous expression, purification, and characterization of adenosylcobalamin-dependent D-lysine 5,6-aminomutase from Clostridium sticklandii, *J. Biol. Chem.* 275, 106–114.

8 SHIBATA, N., MASUDA, J., TOBIMATSU, T., TORAYA, T., SUTO, K., MORIMOTO, Y., YASUOKA, N. (1999) A new mode of B_{12} binding and the direct participation of a potassium ion in enzyme catalysis: X-ray structure of diol dehydratase, *Structure Fold Des.* 7, 997–1008.

9 SINTCHAK, M. D., ARJARA, G., KELLOGG, B. A., STUBBE, J., DRENNAN, C. L. (2002) The crystal structure of class II ribonucleotide reductase reveals how an allosterically regulated monomer mimics a dimer, *Nat. Struct. Biol.* 9, 293–300.

10 TORAYA, T. (2003) Radical catalysis in coenzyme B_{12}-dependent isomerization (eliminating) reactions, *Chem. Rev.* 103, 2095–2127.

11 BANERJEE, R. (2003) Radical carbon skeleton rearrangements: catalysis by coenzyme B_{12}-dependent mutases, *Chem. Rev.* 103, 2083–94.

12 BANERJEE, R., CHOWDHURY, S. (1999) *Methylmalonyl-CoA Mutase*, John Wiley and Sons, New York.

13 JENSEN, K. P., SAUER, S. P., LILJEFORS, T., NORRBY, P.-O. (2001) Theoretical investigation of steric and electronic effects in coenzyme B_{12} models, *Organometallics* 20, 550–556.

14 FINKE, R. G., HAY, B. P. (1984) Thermolysis of adenosylcobalamin: A product, kinetic and Co-C5′ bond dissociation energy study, *Inorg. Chem.* 23, 3041–3043.

15 HALPERN, J. (1985) Mechanisms of coenzyme B_{12}-dependent rearrangements, *Science* 227, 869–875.

16 DONG, S., PADMAKUMAR, R., MAITI, N., BANERJEE, R., SPIRO, T. G. (1998) Resonance Raman spectra of methylmalonyl-Coenzyme A mutase, a coenzyme B_{12}-dependent enzyme, reveal dramatic change in corrin ring conformation but little change in Co-C bond force constant in the cofactor upon its binding to the enzyme, *J. Am. Chem. Soc.* 120, 9947–9948.

17 HUHTA, M. S., CHEN, H. P., HEMANN, C., HILLE, C. R., MARSH, E. N. (2001) Protein-coenzyme interactions in adenosylcobalamin-dependent glutamate mutase, *Biochem. J.* 355, 131–137.

18 MORI, K., TORAYA, T. (1999) Mechanism of reactivation of coenzyme B_{12}-dependent diol dehydratase by a molecular chaperone-like reactivating factor, *Biochemistry* 38, 13170–13178.

19 HOLLAWAY, M. R., WHITE, H. A., JOBLIN, K. N., JOHNSON, A. W., LAPPERT, M. F., WALLIS, O. C. (1978) A spectrophotometric rapid kinetic study of reactions catalyzed by coenzyme B_{12}-dependent ethanolamine ammonia lyase, *Eur. J. Biochem.* 82, 143–154.

20 PADMAKUMAR, R., PADMAKUMAR, R., BANERJEE, R. (1997) Evidence that cobalt-carbon bond homolysis is coupled to hydrogen atom abstraction from substrate in methylmalonyl-CoA mutase, *Biochemistry* 36, 3713–3718.

21 BROWN, K. L., LI, J. (1998) Activation parameters for the carbon-cobalt bond homolysis of coenzyme B_{12} induced by the B_{12}-dependent ribonucleotide reductase from *Lactobacillus leichmannii*, *J. Am. Chem. Soc.* 120, 9466–9474.

22 CHIH, H. W., MARSH, E. N. (1999) Pre-steady-state kinetic investigation of intermediates in the reaction catalyzed by adenosylcobalamin-dependent glutamate mutase, *Biochemistry* 38, 13684–13691.

23 LICHT, S. S., LAWRENCE, C. C., STUBBE, J. (1999) Thermodynamic and kinetic studies on carbon-cobalt bond homolysis by ribonucleoside triphosphate reductase: the importance of

entropy in catalysis, *Biochemistry* 38, 1234–1242.

24 KUNZ, M., RÉTEY, J. (2000) Evidence for a 1,2 shift of a hydrogen atom in a radical intermediate of the methylmalonyl-CoA mutase reaction, *Bioorg. Chem.* 28, 134–139.

25 GARCIA-VILOCA, M., GAO, J., KARPLUS, M., TRUHLAR, D. G. (2004) How enzymes work: Analysis by modern rate theory and computer simulations, *Science* 303, 186–195.

26 BRANDEN, C., TOOZE, J. *Introduction to Protein Structure*, 2nd edn., Garland Publishing, New York, 1999, pp. 50–51.

27 THOMÄ, N. H., EVANS, P. R., LEADLAY, P. F. (2000) Protection of radical intermediates at the active site of adenosylcobalamin-dependent methylmalonyl-CoA mutase, *Biochemistry* 39, 9213–9221.

28 MANCIA, F., SMITH, G. A., EVANS, P. R. (1999) Crystal structure of substrate complexes of methylmalonyl-CoA mutase, *Biochemistry* 38, 7999–8005.

29 THOMÄ, N. H., MEIER, T. W., EVANS, P. R., LEADLAY, P. F. (1998) Stabilization of radical intermediates by an active site tyrosine residue in methylmalonyl-CoA mutase, *Biochemistry* 37, 14386–14393.

30 REED, G. H., MANSOORABADI, S. O. (2003) The positions of radical intermediates in the active sites of adenosylcobalamin-dependent enzymes, *Curr. Opin. Struct. Biol.* 13, 716–721.

31 PRATT, J. M. Cobalt in vitamin B_{12} and its enzymes, in *Handbook on Metalloproteins*, I. BERTINI, A. SIGEL, H. SIGEL (Eds.), Dekker, New York, 2001, pp. 603–668.

32 GAUDEMER, A., ZYBLER, J., ZYBLER, N., BARAN-MARSZAC, M., HULL, W. E., FOUNTOULAKIS, M., KONIG, A., WOLFE, K., RETEY, J. (1981) Reversible cleavage of the cobalt-carbon bond of coenzyme B_{12} catalyzed by methylmalonyl-CoA mutase from *Propionibacterium shermanii*: The use of coenzyme B_{12} stereospecifically deuterated in position 5′, *Eur. J. Biochem.* 119, 279–285.

33 MARSH, E. N. G., BALLOU, D. P. (1998) Coupling of cobalt-carbon bond homolysis and hydrogen atom abstraction in adenosylcobalamin-dependent glutamate mutase, *Biochemistry* 37, 11864–11872.

34 BANDARIAN, V., REED, G. H. (2000) Isotope effects in the transient phases of the reaction catalyzed by ethanol-amine ammonia-lyase: determination of the number of exchangeable hydrogens in the enzyme-cofactor complex, *Biochemistry* 39, 12069–12075.

35 LICHT, S. S., BOOKER, S., STUBBE, J. (1999) Studies on the catalysis of carbon-cobalt bond homolysis by ribonucleoside triphosphate reductase: Evidence for concerted carbon-cobalt bond homolysis and thiyl radical formation, *Biochemistry* 38, 1221–1233.

36 CHEN, D., ABEND, A., STUBBE, J., FREY, P. A. (2003) Epimerization at carbon-5′ of (5′R)-[5′-^{2}H]adenosylcobalamin by ribonucleoside triphosphate reductase: Cysteine 408-independent cleavage of the Co-C5′ bond, *Biochemistry* 42, 4578–4584.

37 MICHENFELDER, M., HULL, W. E., RETEY, J. (1987) Quantitative measurement of the error in the cryptic stereospecificity of methylmalonyl-CoA mutase, *Eur. J. Biochem.* 168, 659–667.

38 MEIER, T. W., THOMÄ, N. H., LEADLAY, P. F. (1996) Tritium isotope effects in adenosylcobalamin-dependent methylmalonyl-CoA mutase, *Biochemistry* 35, 11791–11796.

39 CHOWDHURY, S., THOMAS, M. G., ESCALANTE-SEMERENA, J. C., BANERJEE, R. (2001) The coenzyme B_{12} analog 5′-Deoxyadenosylcobinamide-GDP supports catalysis by methylmalonyl-CoA mutase in the absence of trans-ligand coordination, *J. Biol. Chem.* 276, 1015–1019.

40 CHOWDHURY, S., BANERJEE, R. (2000) Evidence for quantum mechanical tunneling in the coupled cobalt carbon bond homolysis-substrate radical generation reaction catalyzed by methylmalonyl-CoA mutase, *J. Am. Chem. Soc.* 122, 5417–5418.

41 SAUNDERS, W. H. JR. Kinetic isotope effects, in *Investigation of Rates and Mechanisms of Reactions, Part I*, 4th edn., C. F. BERNASCONI (Ed.), Wiley, New York, 1986, pp. 565–611.

42 CHOWDHURY, S., BANERJEE, R. (2000) Thermodynamic and kinetic characterization of Co-C bond homolysis catalyzed by coenzyme B_{12}-dependent methylmalonyl-CoA mutase, *Biochemistry* 39, 7998–8006.

43 CHIH, H. W., MARSH, E. N. (2001) Tritium partitioning and isotope effects in adenosylcobalamin-dependent glutamate mutase, *Biochemistry* 40, 13060–13067.

44 WEISBLAT, D. A., BABIOR, B. M. (1971) The mechanism of action of ethanolamine ammonia-lyase, a B_{12}-dependent enzyme. VIII. Further studies with compounds labeled with isotopes of hydrogen: Identification and some properties of the rate-limiting step, *J. Biol. Chem.* 246, 6064–6061.

45 ESSENBERG, M. K., FREY, P. A., ABELES, R. H. (1971) Studies on the mechanism of hydrogen transfer in the coenzyme B_{12} dependent dioldehydrase reaction II, *J. Am. Chem. Soc.* 93, 1242–1251.

46 CHENG, M. C., MARSH, E. N. G. (2004) Pre-steady-state measurement of intrinsic secondary tritium isotope effects associated with the homolysis of adenosylbcobalamin and the formation of 5′-deoxyadenosine in glutamate mutase, *Biochemistry* 43, 2155–2158.

47 SAUNDERS, W. H. JR., p. 598 of Ref 41.

48 SMITH, D. M., GOLDING, B. T., RADOM, L. (1999) Understanding the mechanism of B_{12}-dependent methylmalonyl-CoA mutase: Partial proton transfer in action, *J. Am. Chem. Soc.* 121, 9388–9399.

49 SMITH, D. M., GOLDING, B. T., RADOM, L. (2001) Understanding the mechanism of B_{12}-dependent diol dehydratase: a synergistic retro-push-pull proposal, *J. Am. Chem. Soc.* 123, 1664–1675.

50 ANDRUNIOW, T., ZGIERSKI, M. Z., KOZLOWSKI, P. M. (2001) Theoretical determination of the Co-C bond energy dissociation in cobalamins, *J. Am. Chem. Soc.* 123, 2679–2680.

51 JENSEN, K. P., RYDE, U. (2002) The axial N-base has minor influence on Co-C bond cleavage in cobalamins, *J. Mol. Struct. (Theochem)* 585, 239–255.

52 WETMORE, S. D., SMITH, D. M., BENNETT, J. T., RADOM, L. (2002) Understanding the mechanism of action of B_{12}-dependent ethanolamine ammonia-lyase: Synergistic interactions at play, *J. Am. Chem. Soc.* 124, 14054–14065.

53 FREINDORF, M., KOZLOWSKI, P. M. (2004) A combined density functional theory and molecular mechanics study of the relationship between the structure of coenzyme B_{12} and its binding to methylmalonyl-CoA mutase, *J. Am. Chem. Soc.* 126, 1928–1929.

54 LOFERER, M. J., WEBB, B. M., GRANT, G. H., LIEDL, K. R. (2003) Energetic and stereochemical effects of the protein environment on substrate: A theoretical study of methylmalonyl-CoA mutase, *J. Am. Chem. Soc.* 125, 1072–1078.

55 JENSEN, K. P., RYDE, U. (2003) Theoretical prediction of the Co-C bond strength in cobalamins, *J. Phys. Chem. A* 107, 7539–7545.

56 DOELKER, N., MASERAS, F., SIEGBAHN, P. E. M. (2004) Stabilization of the adenosyl radical in coenzyme B_{12} – a theoretical study, *Chem. Phys. Lett.* 386, 174–178.

57 TRUHLAR, D. G. Variational Transition State Theory and Multidimensional Tunneling for Simple and Complex Reactions in the Gas Phase, Solids, Liquids, and Enzymes, in *Isotope Effects in Chemistry and Biology*, H. LIMBACH, A. KOHEN (Eds.), Dekker, New York, pp. 579–619.

58 FERNÁNDEZ-RAMOS, A., MILLER, J. A., KLIPPENSTEIN, S. J., TRUHLAR, D. G. Modeling the Kinetics of Bimolecular Reactions, *Chem. Rev.* submitted for publication.

59 TRUHLAR, D. G., GAO, J., GARCIA-VILOCA, M., ALHAMBRA, C.,

Corchado, J., Sanchez, L., Poulsen, T. D. (2004) Ensemble-averaged variational transition state theory with optimized multidimensional tunneling for enzyme kinetics and other condensed-phase reactions, *Int. J. Quantum Chem.*, 100, 1136–1152.

60 Dybala-Defratyka, A., Paneth, P. (2001) Theoretical evaluation of the hydrogen kinetic isotope effect on the first step of the methylmalonyl-CoA mutase reaction, *J. Inorg. Biochem.* 86, 681–689.

61 Toraya, T., Yoshizawa, K., Eda, M., Yamabe, K. (1999) Direct participation of potassium ion in the catalysis of coenzyme B_{12}-dependent diol dehydratase, *J. Biochem. (Tokyo)* 126, 650–654.

62 Toraya, T., Eda, M., Kamachi, T., Yoshizawa, K. (2001) Energetic feasibility of hydrogen abstraction and recombination in coenzyme B_{12}-dependent diol dehydratase reaction, *J. Biochem. (Tokyo)* 130, 865–872.

63 Eda, M., Kamachi, T., Yoshizawa, K., Toraya, T. (2002) Theoretical study on the mechanism of catalysis of coenzyme B_{12}-dependent diol dehydratase, *Bull. Chem. Soc. Jpn.* 75, 1469–1481.

64 Semialjac, M., Schwarz, H. (2004) Computational investigation of hydrogen abstraction from 2-aminoethanol by the 1,5-dideoxyribose-5-yl radical: A model study of a reaction occurring in the active site of ethanolamine ammonia lyase, *Chem.: Eur. J.* 10, 2781–2788.

65 Dybala-Defratyka, A., Paneth, P., Pu, J., Truhlar, D. G. (2004) Benchmark results for hydrogen atom transfer between carbon centers and validation of electronic structure methods for bond energies and barrier heights, *J. Phys. Chem. A* 108, 2475–2486.

66 Gao, J., Truhlar, D. G. (2002) Quantum mechanical methods for enzyme kinetics, *Annu. Rev. Phys. Chem.* 53, 467–505.

67 Cramer, C. J. *Essentials of Computational Chemistry: Theories and Methods*, Wiley, Chichester, 2002.

68 Gonzàlez-Lafont, A., Truong, T. N., Truhlar, D. G. (1991) Direct dynamics calculations with neglect of diatomic differential overlap molecular orbital theory with specific reaction parameters, *J. Phys. Chem.* 95, 4618–4627.

69 Lynch, B. J., Fast, P. L., Harris, M., Truhlar, D. G. (2000) Adiabatic connection for kinetics, *J. Phys. Chem. A* 104, 4811–4815.

70 Eagar, R. G. Jr., Bachovchin, W. W., Richards, J. H. (1975) Mechanism of action of adenosylcobalamin: 3-Fluoro-1,2-propanediol as substrate for propanediol dehydratase-mechanistic implications, *Biochemistry* 14, 5523–5528.

71 Dybala-Defratyka, A. (2004) Ph. D. Thesis, Technical University of Lodz, Poland; Dybala-Defratyka, A., Paneth, P., Truhlar, D. G., unpublished results.

72 Eyring, H. (1938) The theory of absolute reaction rates, *Trans. Faraday Soc.* 34, 41–48.

73 Gonzàlez-Lafont, A., Truong, T. N., Truhlar, D. G. (1991) Interpolated variational transition state theory: Practical methods for estimating variational transition state properties and tunneling contributions to chemical reaction rates from electronic structure calculations, *J. Chem. Phys.* 95, 8875–8894.

74 Truhlar, D. G., Kuppermann, A. (1971) Exact tunneling calculations, *J. Am. Chem. Soc.* 93, 1840–1851.

75 Garrett, B. C., Truhlar, D. G., Grev, R. S., Magnuson, A. W. (1980) Improved treatment of threshold contributions in variational transition state theory, *J. Phys. Chem.* 84, 1730–1748.

76 Marcus, R. A. (1966) On the analytical mechanics of chemical reactions. Quantum mechanics of linear collisions, *J. Chem. Phys.* 45, 4493–4499.

77 Kuppermann, A., Adams, J. T., Truhlar, D. G. (1973) Streamlines of probability current density and

tunneling fractions for the collinear H + H$_2$ → H$_2$ + H reaction, in *Electronic and Atomic Collisions: Abstracts of Papers, Eighth International Conference on the Physics of Electronic and Atomic Collisions (VIII ICPEAC), Beograd, 1973*, B. C. COBIC, M. V. KUREPA (Eds.), Institute of Physics, Beograd, pp. 149–150.

78 SKODJE, R. T., TRUHLAR, D. G., GARRETT, B. C. (1982) Vibrationally adiabatic models for reactive tunneling, *J. Chem. Phys.* 77, 5955–5976.

79 KREEVOY, M. M., OSTOVIC, D., TRUHLAR, D. G., GARRETT, B. C. (1986) Phenomenological manifestations of large-curvature tunneling in hydride transfer reactions, *J. Phys. Chem.* 90, 3766–3774.

80 MILLER, W. H., HANDY, N. C., ADAMS, J. E. (1980) Reaction path Hamiltonian for polyatomic molecules, *J. Chem. Phys.* 72, 99–112.

81 TRUHLAR, D. G., ISAACSON, A. D., GARRETT, B. C. (1985) Generalized transition state theory, in *Theory of Chemical Reaction Dynamics*, M. BAER (Ed.), CRC Press, Boca Raton, FL, Vol. 4, pp. 65–137.

82 LU, D.-h., TRUONG, T. N., MELISSAS, V. S., LYNCH, G. C., LIU, Y.-P., GARRETT, B. C., STECKLER, R., ISAACSON, A. D., RAI, S. N., HANCOCK, G. C., LAUDERDALE, J. G., JOSEPH, T., TRUHLAR, D. G. (1992) POLYRATE 4: A new version of a computer program for the calculation of chemical reaction rates for polyatomics, *Comput. Phys. Commun.* 71, 235–262.

83 LIU, Y.-P., LYNCH, G. C., TRUONG, T. N., LU, D.-h., TRUHLAR, D. G., GARRETT, B. C. (1993) Molecular modeling of the kinetic isotope effect for the [1,5]-sigmatropic rearrangement of *cis*-1,3-pentadiene, *J. Am. Chem. Soc.* 115, 2408–2415.

84 MARCUS, R. A., COLTRIN, M. E. (1977) A new tunneling path for reactions such as H + H$_2$ → H$_2$ + H, *J. Chem. Phys.* 67, 2609–2613.

85 SCHATZ, G. C., RATNER, M. A. (1993) *Quantum Mechanics in Chemistry*, Prentice-Hall, Englewood Cliffs, NJ, pp. 167–181.

86 GARRETT, B. C., TRUHLAR, D. G., WAGNER, A. F., DUNNING, T. H. JR. (1983) Variational transition state theory and tunneling for a heavy-light-heavy reaction using an *ab initio* potential energy surface. ^{37}Cl + H(D)^{35}Cl → H(D)^{37}Cl + ^{35}Cl, *J. Chem. Phys.* 78, 4400–4413.

87 MARCUS, R. A. (1979) Similarities and differences between electron and proton transfers at electrodes and in solution. Theory of a hydrogen evolution reaction, in *Proceedings of the Third Symposium on Electrode Processes*, S. BRUCKENSTEIN, J. D. E. MCINTYRE, B. MILLER, and E. YEAGER (Eds.), Electrochemical Society, Princeton, pp. 1–12.

88 CHUANG, Y.-Y., CORCHADO, J. C., FAST, P. L., VILLÀ, J., HU, W.-P., LIU, Y.-P., LYNCH, G. C., JACKELS, C. F., NGUYEN, K. A., GU, M. Z., ROSSI, I., COITIÑO, E. L., CLAYTON, S., MELISSAS, V. S., LYNCH, B. J., STECKLER, R. B., GARRETT, C., ISAACSON, A. D., TRUHLAR, D. G. (2000) POLYRATE–Version 8.4.1.PL, University of Minnesota, Minneapolis.

89 GARRETT, B. C., ABUSALBI, N., KOURI, D. J., TRUHLAR, D. G. (1983) Test of variational transition state theory and the least-action approximation for multidimensional tunneling probabilities against accurate quantal rate constants for a collinear reaction involving tunneling in an excited state, *J. Chem. Phys.* 83, 2252–2258.

90 TRUONG, T. N., LU, D.-h., LYNCH, G. C., LIU, Y.-P., MELISSAS, V. S., STEWART, J. J. P., STECKLER, R., GARRETT, B. C., ISAACSON, A. D., GONZÀLEZ-LAFONT, A., RAI, S. N., HANCOCK, G. C., JOSEPH, T., TRUHLAR, D. G. (1993) MORATE: A program for direct dynamics calculations of chemical reaction rates by semiempirical molecular orbital theory, *Comput. Phys. Commun.* 75, 143–159.

91 LIU, Y.-P., LU, D.-h., GONZÀLEZ-LAFONT, A., TRUHLAR, D. G., GARRETT, B. C. (1993) Direct dynamics calculation of the kinetic isotope effect for an organic hydrogen-transfer

reaction, including corner-cutting tunneling in 21 dimensions, *J. Am. Chem. Soc.* 115, 7806–7817.
92 GARRETT, B. C., TRUHLAR, D. G. (1983) A least-action variational method for calculating multi-dimensional tunneling probabilities for chemical reactions, *J. Chem. Phys.* 79, 4931–4938.
93 ALLISON, T. C., TRUHLAR, D. G. Testing the accuracy of practical semiclassical methods: Variational transition state theory with optimized multidimensional tunneling, in *Modern Methods for Multidimensional Dynamics Computations in Chemistry*, D. L. THOMPSON (Ed.), World Scientific Singapore, 1998, pp. 618–712.
94 CHUANG, Y.-Y., FAST, P. L., HU, W.-P., LYNCH, G. C., LIU, Y.-P., TRUHLAR, D. G. (2001) MORATE-version 8.5, University of Minnesota, Minneapolis.
95 MASUDA, J., SHIBATA, N., MORIMOTO, Y., TORAYA, T., YASUOKA, N. (2000) How a protein generates a catalytic radical from coenzyme B_{12}: X-ray structure of a diol-dehydratase-adeninylpentylcobalamin complex, *Structure Fold. Des.* 8, 775–788.
96 GRUBER, K., REITZER, R., KRATKY, C. (2001) Radical shuttling in a protein: Ribose pseudorotation controls alkyl-radical transfer in the coenzyme B_{12} dependent enzyme glutamate mutase, *Angew. Chem. Int. Ed. Engl.* 40, 3377–3380.
97 LICHT, S., GERFEN, G. J., STUBBE, J. (1996) Thiyl radicals in ribonucleotide reductases, *Science* 271, 477–481.
98 BERTI, P. J., SCHRAMM, V. L. (1999) Enzymatic transition state structures constrained by experimental kinetic isotope effects: Experimental measurement of transition state variability, *ACS Symp. Ser.* 721, 473–488.
99 WARNCKE, K., SCHMIDT, J. C., KE, S.-C. (1999) Identification of a rearranged-substrate, product radical intermediate and the contribution of a product radical trap in vitamin B_{12} coenzyme-dependent ethanolamine deaminase catalysis, *J. Am. Chem. Soc.* 121, 10522–10528.
100 DOWD, P., TRIVEDI, B. K. (1985) On the mechanism of action of vitamin B_{12}. Model studies directed toward the hydrogen abstraction reaction, *J. Org. Chem.* 80, 206–217.
101 DOWD, P., WILK, B., WILK, B. K. (1992) First hydrogen abstraction-rearrangment modle for the coenyme B_{12}-dependent methylmalonyl-CoA to succinyl-CoA carbon skeleton rearrangment reaction, *J. Am. Chem. Soc.* 114, 7949–7951.
102 DOLL, K. M., FINKE, R. G. (2003) A compelling experimental test of the hypothesis that enzymes have evolved To enhance quantum mechanical tunneling in hydrogen transfer reactions: The β-neopentylcobalamin system combined with prior adocobalamin data, *Inorg. Chem.* 42, 4849–4856.
103 DOLL, K. M., BENDER, B. R., FINKE, R. G. (2003) The first experimental test of the hypothesis that enzymes have evolved to enhance hydrogen tunneling, *J. Am. Chem. Soc.* 125, 10877–10884.
104 KEMSLEY, J. (2003) Enzyme tunneling idea questioned, *Chem. Eng. News* 81(38), 29–30.

Part V
Proton Conduction in Biology

The motion of protons from one place to another is important in several areas of biochemistry and biology, particularly in bioenergetics where transmembrane proton-transport is a mechanism of energy storage and energy transduction as in the synthesis of high-energy molecules from a pre-established proton gradient. In enzyme action, there are other proton relay systems as well, for example in the transport of protons out of the active site of carbonic anhydrase. A number of these phenomena are explored elsewhere in these volumes. The sole paper in this section is Gutman and Nachliel's elegant examination of a paradigmatic system, the motion of protons at the surface of proteins in general and in such special circumstances as protein/membrane interfaces.

20
Proton Transfer at the Protein/Water Interface

Menachem Gutman and Esther Nachliel

20.1
Introduction

The bulk to surface proton transfer is the most voluminous of all reactions in the biosphere. Each O_2 molecule generated by the chloroplasts is coupled with 20 moles of H_3O^+ ions, which are sequentially taken up and released by membranal enzymes. Similarly, the mitochondrial oxidative phosphorylation is operated by a coupled cross-membranal proton flux, which is comparable in its number of protons to that of the photosynthesis process. This high proton flux has been, for a long time, a subject of intensive biochemical–biophysical research, and each level of refinement of the mechanism generated a new set of experimental/theoretical uncertainties. One line of research was aimed at elucidating the contribution of geometric features and the special physical properties of the reaction space on the mechanism of proton transfer near the surface. The other was focused on the proton transfer events at specific loci on the membrane or on proteins where the proton pumping activity is located. In the present chapter we shall discuss both approaches, with emphasis on the new possibilities opened up by the prevalence of structural information and the ability to reconstruct *in silico* complex chemical processes.

The free energy that mitochondria or chloroplasts generate is initially stored in the form of a proton gradient, which is utilized for the synthesis of ATP or other processes. Proton pumps, such as cytochrome c oxidase, are very efficient enzymes, having a turnover number of more than 1000 s^{-1}. Under physiological conditions there is, on average, less than one free proton in a mitochondrion; however, the measured rate constant of proton uptake at a certain step of the catalytic cycle of cytochrome c oxidase is still exceptionally fast, and as high as 2×10^{13} M^{-1} s^{-1} [1]. Such a rate is above the theoretical prediction by the Debye–Smoluchowski equation, implying that proton transfer within the cristae of the mitochondria utilizes an accessory mechanism that accelerates the reaction, either by imposing bias forces or high cross section for proton uptake events. Proton transfer reactions inside organelles are also mediated by the diffusion of mobile proton carriers, which facilitate the equilibration between the bulk and the surface. The density of proton

Hydrogen-Transfer Reactions. Edited by J. T. Hynes, J. P. Klinman, H.-H. Limbach, and R. L. Schowen
Copyright © 2007 WILEY-VCH Verlag GmbH & Co. KGaA, Weinheim
ISBN: 978-3-527-30777-7

binding sites on the surface of a protein can be high, and the binding residues can form clusters that interact with each other. Thus, the formulation of the dynamics should account for the local conditions, rather than the average distribution of binding sites.

The solvated proton is not a point charge; it exists in water in two solvated states of $H_3O^+_{aq}$ and $H_5O_2^+_{aq}$ and the location of the positive charge is smeared over the complex, as determined by quantum mechanical calculations. Any attempt to predict the location of a proton with 1 Å resolution is an 'overkill'. With the present advances in structural biology, ultra-fast kinetic measurements and computational capabilities, the understanding of the mechanism of proton transfer at the interface should be based on experimental measurements capable of observing molecular events at a real time [2], coupled with an analysis that matches the molecular details of the reaction space.

The mechanism of proton transfer at the membrane/bulk interface has been investigated since the acceptance of the Mitchell hypothesis as the paradigm of bioenergetics [3]. The early experiments were based on rather slow reactions, where the electron flux in the respiratory system was used to drive the proton translocation across the mitochondrial membrane [4]. Later, faster reactions were introduced where the triggering was initiated by photochemical reactions, using either the photosynthetic apparatus or the proton translocating reaction of the Bacteriorhodopsin. These systems suffered from a common disadvantage: the inherent complexity of the driving reaction. As a result, there is very detailed information about the mechanism of proton transfer in specific sites [5–11], but the general properties of proton transfer at the surface were lagging behind. To obtain direct experimental measurements of proton transfer at the interface, the protons have to be generated by an external source that is not part of the system under study. The laser induced proton pulse [12–15] is capable of generating a short pulse of a few μM of H_3O^+, within a nanosecond time frame, perturbing all acid–base equilibria in the reaction space. This technique enabled the monitoring of the reversible protonation of surface groups on the protein and membranes with high time resolution and the determination of the rate constants of these reactions [16–19].

The membrane/protein interface with the bulk is dominated by the discontinuity of the physical chemical properties of the reaction space. On one side of the borderline there is a low viscosity, high dielectric constant matrix where rapid proton diffusion can take place. On the other side of the boundary, there is a low dielectric matrix that is covered by a large number of rigidly fixed charged residues. The dielectric boundary amplifies the electrostatic potential of the fixed charges and, due to their organization on the surface of proteins, a complex pattern of electrostatic potentials is formed. These local fields determine the specific reactivity of the domain, either with free proton or with buffer molecules. In this chapter we shall discuss both the general properties of the interface and the manner in which they affect the kinetics of defined domains.

A special complication of proton transfer is the interruption of the proton diffusion by the formation of covalent bonds with the various proton-binding sites. Each site that interacts with a free proton will bind it for a time frame that is proportional to its pK value [13]. The combination of periods; where the proton is diffusing in the solution, is interrupted by the formation of covalent bonds, acquires the proton [2] a nature of "fly and perch" mode of propagation. The various proton-binding sites on the surface are unevenly dispersed, the distance between them being determined by the special structure of biomembranes [20–25] or a protein [2]. Clusters of nearby sites on a protein can serve as highly proton-reactive domains that fulfill special mechanistic roles, like the proton-collecting antenna of the bacteriorhodopsin [25]. For the most recent review see Ref. [26].

20.2
The Membrane/Protein Surface as a Special Environment

The motion of a proton near a protein or a membrane surface differs markedly from its random diffusion in the bulk, thus reflecting the inhomogeneity of the diffusion space. The various modes in which the proximity to the structure affects the ion are discussed below.

20.2.1
The Effect of Dielectric Boundary

The electric charging of a body requires investment of energy, called the self-energy, $E = q^2/2\varepsilon r$, where q is the charge, r is the radius of the body and ε is the dielectric constant of the surrounding medium. For a charged particle that is located in a high dielectric matrix at a distance of few nanometers from a low dielectric matrix, its self-energy will vary with the distance from the dielectric boundary. A convenient presentation of the charge's electrostatic potential and its decay as a function of distance is through the "image charge" model [27]. Consequently, the potential of an ion approaching the membrane is raised through interaction with its imaginary ion, with a charge almost identical located behind the membrane/water interface at the same distance from the interface as the real ion. This function decays very rapidly; for a monovalent ion dissolved in water, the repulsion will exceed the thermal energy only when the ion is less than ~4 Å from the surface. Thus, the first solvation layer of a neutral membrane can be regarded as "ion repellent" [28]. When the surface carries charges, the "image charge" formalism doubles their electrostatic potential. Thus, charged residues will serve as stronger attractors than the same charge dissolved in water.

20.2.2
The Ordering of the Water by the Surface

The charged residues on the surface of a membrane or proteins minimize their electrostatic potential through immobilization of the dipoles of the nearest water molecule. In parallel, stabilization can be gained through local interactions between adjacent residues with opposite charges. The competition between the two stabilization forces leads to a coupling between the solvation of a phospholipid membrane and their internal organization [29]. Accordingly, it is expected that local stress on the membrane, generated at the boundary between the membranal enzymes and the membrane matrix, would modulate the solvation of the phospholipid head-group with a subsequent effect on the diffusion of a proton in that domain.

The interaction of the water with the head-groups, was quantitated by the number of 'nonfreezing' water molecules per head-group [30], or through the dichroic ratio of the IR spectrum of the water at the interface [31–33]. These observations were also confirmed by molecular dynamics [34]. The extent of water immobilization by the phospho head-group is almost doubled in the presence of cholesterol [30], which acts as a spacer and allows more water molecules to get close enough to the head-group. Considering the high content of cholesterol in the mitochondrial membrane, one can imagine that the ordering of the water along the mitochondrial membrane will be nonhomogeneous. In some regions, the water will be more ordered than in others. This ordering of the water molecules has a direct effect on the dynamics of proton transfer, affecting two parameters, the rate of dissociation and the diffusion coefficient.

20.2.2.1 The Effect of Water on the Rate of Proton Dissociation

The dissociation of an acidic residue in water is essentially a downhill proton transfer between the donor and the solvent, and in the case of a large value for ΔpK_a, the reaction can be as fast as a barrier-less reaction [35]. To achieve the maximal rate of the reaction, the proton-donor atom must establish a hydrogen bond with the acceptor, and the reaction then takes place along this bond. As the rate of the reaction can be as fast as the hydroxyl vibration frequency, ~ 60–150 fs [36, 37], it is implicit that, for a fast reaction, the hydrogen bond must be established before the proton transfer event. In the case of proton transfer to water, pre-orientation of a few water molecules is required to stabilize the ejected proton and the conjugated base from which the proton was released. Accordingly, reduction in the availability of free water molecules will reduce the probability of the proton acceptor configurations, and this effect will appear as a reduction in the rate of dissociation.

The correlation between the availability of water and the rate constant of proton dissociation was measured in two systems. In one system, the ratio water: methanol of a mixed solution modulated the availability of water [38]. In the other system, made of concentrated electrolyte solutions, the activity of the water was modulated by the salt [39]. The dependence of the measured rate of dissociation [60, 67, 68], either from photoacid or ground state acids, on the activity of the solvent yielded a straight log–log correlation function with respect to the activity of the water

$$k_{dis} = k_{dis}^0 a_{water}^n$$

where k_{dis}^0 is the rate measured in pure water and the power n is specific for the dissociated acid.

The rate of proton dissociation from the excited pyranine was measured in different bio-environments, such as the inter-bilayer space of multilamellar vesicles [39a] and the heme-binding site of apomyoglobin and another protein [39b]. The results clearly indicated that the water molecules in such micro-spaces were highly immobilized, having a chemical potential well below that of bulk water. For example, in a small cavity such as the heme-binding site, the activity of the water was found to be as low as $a_{water} = 0.6$. The activity of the water in the inter-bilayer space of multilamellar phospholipid vesicles was determined to be of the order of $a_{water} = 0.8–0.9$, depending on the nature of the head-group and the width of the aqueous layer. Thus, the interaction between acidic residues and the water located on a surface, which is limited to the innermost solvation shells [2], can alter its tendency to dissociate.

20.2.2.2 The Effect of Water Immobilization on the Diffusion of a Proton

The mechanism of proton diffusion, usually referred as the Grotthuss mechanism, differs from that of other ions. Instead of a self-diffusion process, where the mass and the charge are inseparable, the proton diffusion is the movement of the protonic charge independently from the transport of the particle. This mode of propagation leads to the diffusion of the proton $D_{H^+} = 9.3 \times 10^{-5}$ cm^2 s^{-1} being faster than the diffusion of all other ions ($D \leq 2 \times 10^{-5}$ cm^2 s^{-1}), and of the self-diffusion of the water molecules in water as a solvent (for details see Ref. [40]).

The solvated proton assumes two basic structures in water: $H_3O^+_{aq}$ and $H_5O_2^+{}_{aq}$, which have almost the same potential [41, 42]. The diffusion of the proton in water is a sequence of transitions between these two states of the solvated proton, where the initiation of the transition is made by the random motion of water molecules in the second solvation shell of the proton [41, 43]. Naturally, the immobilization of the water molecules, which are in contact with the membrane or the protein's surface, will reduce their rate of orientation, leading to diminished diffusivity of the proton. Indeed, measurements of proton diffusion in immobilized water, ice, yielded a diffusion coefficient that is ~30% of the value in water at the same temperature [44].

It must be recalled that the ordering of the water next to the surface is limited to a few water molecules (4–7) per head-group, hardly enough to cover the surface with a continuous layer. Thus, the innermost solvation layer can exhibit lateral inhomogeneity, where ordered water forms patches over the surface of the membrane. Under such conditions, the most efficient trajectory for proton transfer between two sites on the surface will follow through the less ordered water molecules. This pathway may be longer, yet the overall passage time may be shorter. Indeed, direct measurements of proton dissociation in the ultra-thin water layers, only 8–11 Å thick, that are interspaced between the phospholipids layers in multilamellar vesicles, yielded values of $8–9 \times 10^{-5}$ cm^2 s^{-1} [45].

The diffusion coefficient of a particle, diffusing in a 3D space, is given by the expression $D = (\lambda^2 v)/6$ where λ is the length of the random step and v is the frequency (in s^{-1} units) of making a random step. For a particle that is diffusing in a nonhomogeneous space, made for example of water and fixed proton binding sites, the *apparent* diffusion coefficient becomes a function of the observation time. For a proton in bulk water, λ is the distance between the oxygen atoms in the water and v is the frequency of proton transfer between them [41, 45]. The diffusion coefficient of a proton in the inter-bilayer space, when measured with sub-nanosecond resolution, was comparable with that in water [45]. In such a short time frame, encounters of the proton with the surface sites depleted the free proton population. When the observation time is longer, or measured at steady state [46], more protons are removed from the system and another process may take place: the bound protons dissociate at the frequency of k_{diss} and replenish the diffusing population. In this case, the value λ will be the average distance between the proton binding sites. This mechanism makes a negligible contribution to proton diffusion. The calculated diffusion coefficient, using the experimental conditions of Zhang ($\lambda \sim 5$ Å, $k_{diss} \sim 10^6$ s^{-1}, compatible with p$K \sim 4$), is as small as $\sim 10^{-9}$ cm^2 s^{-1}, four orders of magnitude smaller than the measured value 1.5×10^{-5} cm^2 s^{-1} [46]. Accordingly, it must be concluded that the surface groups, when sufficiently close together, can very efficiently exchange a proton among themselves. This feature was noticed by Zundel and others, who termed it 'proton polarizability' [47, 48]. When the local configuration of proton binding sites can form a sequence of states where a proton is in more than one location, there is a rapid transition between them, at a frequency comparable to that of the IR spectrum. When the proximity between the sites and their pK values are suitable, the proton exchange reactions become intensive, leading to the appearance of a continuum IR absorption band.

20.3
The Electrostatic Potential Near the Surface

Biomembranes and proteins carry a large number of charges on their surface, generating a complex pattern of electrostatic interactions. In the case of proteins, where positive and negative residues can be packed together, the electrostatic potential is extremely nonhomogeneous, and local domains that differ in their charges from the surrounding structure are commonly found. Such local domains account for the kinetic peculiarities of the superoxide dismutase, where a local positive domain accelerates by funneling the encounter with the negative substrate [50].

Phospholipid membranes are more homogeneous in nature, due to the prevalence of the zwitterionic head-group of the PC, which contributes most of the membrane's mass. Langner and coworkers [51] used a variety of experimental techniques to study the effect of the monovalent lipid phosphtidylinositol (PI) and the trivalent lipid phosphtidylinositol 4,5-diphosphate (PIP2) on the electric potential adjacent to the bilayer membrane. When the membranes were formed from a mixture of PI and the zwitterionic lipid phosphatidyl choline (PC), the smeared charge theory of Gouy–Chapman–Stern adequately described the dependence of the potential on the distance from the membrane, the density of the negative charges and the screening electrolyte concentrations. However, with PC/PIP2 membranes, anionic probes reported less negative potentials than the counterions. The deviations of the experimental results from the theoretical predictions were greater for the counterions than for the anions. To improve the predictions, the authors formulated a consistent statistical mechanical theory that takes into account three effects which were ignored in the Gouy–Chapman–Stern theory: the finite size of the ions in the double layer, the electrical interaction between pairs of ions (correlation effects), and the mobile discrete nature of the surface charges. The improved model could predict the experimental observations. However, it must be stressed that deviations from the Gouy–Chapman–Stern theory are significant only for trivalent lipid headgroups, and are greater for anions than for counterions.

To test the limit of accuracy of the Gouy–Chapman–Stern theory, McLaughlin and coworkers [52] made a comparison between the predictions of the model with precise nonlinear Poisson–Boltzmann calculations of the electrostatic potentials in the aqueous phase adjacent to a molecular model of a phospholipid bilayers (phosphatidyl choline plus acidic lipids suspended in 0.1 M monovalent salt). When the bilayers contained less than 11% acidic lipid, the equipotential surfaces having a value of $-k_B T$ were sufficiently separated from each other to form discrete domes centered above the anionic residues. When the bilayers contained more than 25% acidic lipid, the -25 mV equipotential profiles had merged into an essentially flat surface and agreed well with the values calculated using the Gouy–Chapman theory. The membrane surface is not perfectly flat and ions, due to the Van-der-Waals repulsion, cannot get closer to the polar headgroups than 2 Å. However, these refined features hardly affect the calculated potential in the aqueous phase. Thus, the electrostatic potential at the interface can be calculated with high accuracy, using either the Gouy–Chapman–Stern method or the precise nonlinear Poisson–Boltzmann equations, depending on the level of structural accuracy of the domain under study.

20.4
The Effect of the Geometry on the Bulk-surface Proton Transfer Reaction

The protonation of a given site on a surface can proceed by two pathways. The proton can either react directly with the site, propagating to the target in a three dimensional (3D) diffusion, or it may first adsorb to the surface, which is huge with respect to the target, and then propagate in a 2D mechanism until it encounters its target. This mode of reaction was first introduced in the classical manuscript of Adam and Delbrück [53] and later developed by Berg and Purcell [54].

During a random three-dimensional walk in the bulk, a particle can encounter a surface many times. For a particle that is located at a given time at a distance of l_i from a surface, there is a certain probability that, during a random walk, it will re-encounter with the surface. This problem was treated by Berg and Purcell, who determined that the average number (n) of repeated encounters between the particle and the surface, before being "lost in the bulk", is given by by $n = R/l_i$ [54]. This expression implies that a large surface will be a better target than a small one. A convex surface, as we find in the folding of the mitochondrial inner membrane, will even increase the probability of the particle as the surface engulfs the space where the particle diffuses. This mechanism accelerates the reaction between a solute and a site on the surface by geometrical considerations, where the diffusion is all in 3D space.

The second mechanism by which a particle can encounter a target on a surface is based on the concept of "reduction of dimensionality" introduced by Adam and Delbrück [53]. The mechanism consists of two steps. At first, the particle encounters the surface by 3D diffusion and, due to attracting force, concentrates at the interface. The enhanced number of particles increases the probability of encounter with the target, which is further amplified by the higher efficiency of search on a 2D surface. The efficiency of the overall process is determined by two terms, the magnitude of the attractive force and the ratio between the diffusion coefficients in the 2D and 3D spaces. An intensive attraction will increase the density of particles at the interface, but may also reduce the frequency at which they execute a random stepping. As was argued by Berg and Purcell, for a ratio $D^{(2)}/D^{(3)} = 0.1$ the attraction to the surface should be more intensive than 3 kcal mol^{-1}, otherwise the surface will not facilitate the encounter between the particle and the target. For a proton near a phospholipid membrane, its tendency to concentrate near a negatively charged surface will lead to the formation of a covalent bond with the carboxylates on the surface, and in such state their diffusivity is zero. The macroscopic diffusion that Unwin and coworkers [46, 55] measured consisted of a binding of protons to the surface carboxylates providing the attractive force, and diffusion through the interspacing water molecules adjacent to the membrane's surface.

Stuchebrukchov and coworkers formulated a comprehensive reaction mechanism [56, 57]. According to this phenomenological model, the interaction of the low-pK protonable groups (carboxylates) with protons corresponds to the force detaining the particles near the surface, and their diffusion, either through the bulk or by the 'proton polarizability' mechanism of Zundel, corresponds to two-dimensional diffusion. The model enabled the formulation of time and length criteria defining the coupling between a source and a sink on the membrane surface, while protons generated by the source are fully equilibrated with the bulk. What is more, the model accounts well for the role of mobile buffers on the dynamics of bulk-surface proton transfer and for the delayed presence of proton near the surface of Purple Membrane [10, 58, 59]. Because of its empirical compatibility with the experimental observations, the model merits elaboration, defining its specific features.

The membrane of Stuchebrukchov's model is an infinite surface, where the multitude of proton binding sites (carboxylates with pK = 5) is represented by a density function (σ). The dwell time of a proton on any of the sites is determined by the pK ($\tau_{dwell} = K_{dis}k_{on}$), but during this time interval the proton can diffuse on the surface with a diffusion coefficient that is ~10% of the bulk value ($D_s \sim 0.1\ D_b$), screening an area with a radius L_s. On the surface, there is a proton-channel acting either as an absorbing sink, or a source which affects the immediate proton concentration, both at the surface and in the solution. The bulk phase in this model is an infinite reservoir, which is sufficiently far from the proton-consuming cluster to satisfy the demand $dC/dx = 0$, a definition that is based on a chemical function and not on physical properties. Encounter of a proton with the channel proceeds by two independent pathways; the direct collision of a free diffusing proton with the channel, or through surface diffusion. The former is a classical three-dimensional process, while the latter is described by an equation that incorporates the reduced dimensionality of the reaction space:

$$k_s = 2\pi D_s \sigma_{eq} / \ln(R_c/r_0)$$

The term R_c denotes the capture radius; the distance from the target at which the probability of capture decreases to zero and r_0 is the radius of the channel. Due to the logarithmic function, the radius of the channel r_0 has a minor effect on the rate. The quantitation of the contribution of the surface diffusion to the overall flux is given by the term R_0, which is the radius of channel that will support the same flux given that only the three-dimensional mechanism is operating. The typical values characterizing the proton flux in such a system are summarized in Table 20.1. The values in the table are characteristics for a negatively charged membrane suspended in unbuffered aqueous solution. The dwell time of a proton (~ 4 μs) was estimated from the pK of the proton binding sites [60]. During this time frame, the proton scans the surface sites that are within a radius of ~ 600 Å. Assuming that the surface carries a proton sink, which irreversibly consumes protons (or alternatively is a source of free protons), then the combination of repeated encounters plus the scanning of the surface during each encounter (τ_{dwell}) increases the effective capture range of the channel to 17 μm, a value that exceeds the dimensions of many sub-cellular organelles. The flux supported by the surface diffusion is so high that, in order to support the same rate just by protons coming directly from the bulk, the pore should be as large as 2 μm in radius. This value is $\sim 10^4$ times larger than the real size of a protonic channel (like gramicidin) which is ~ 2 Å. It is of interest to point out that the distance defined by Stuchebrukchov as the border with the bulk (~ 170 μm), is comparable to the width of the unstirred layer as measured by Pohl and coworkers (~ 200 μm) [61, 62].

In the presence of buffers, most of the proton flux is carried by the diffusion of

Table 20.1. The characteristics parameters for the bulk-surface proton transfer in the absence of soluble buffers.

Parameter	Value
σ (density of sites)	0.01 Å^{-2}
τ_{dwell} (average time of a proton on the surface)	4 μs
L_s (radius of search over the surface)	660 Å
R_c (effective capture radius)	17 μm
R_0 (real radius of channel)	2 μm
R_c/r_0 (efficiency factor)	$\sim 10^4$
L_0 (surface to bulk distance)	170 μm
τ_{eq} (equilibration time between surface and bulk)	~ 1 s

the buffer molecules. Indeed, the diffusion coefficient of buffer molecules is ~10–20% of that of the free proton yet the physiological concentration of the buffers (10–30 mM) exceeds the free proton concentration by 5 orders of magnitude. Consequently, in the presence of increasing buffer concentrations, the proton flux will shift from diffusion of free proton to a new regime, where the proton transfer is totally mediated by the diffusion of buffer molecules. This mechanism was first described by Junge and McLaughlin [63] and confirmed at the level of physiological systems. Vaughan-Jones and coworkers carried out measurements with single ventricular myocyte cells that were isolated from the rat, guinea pig, or rabbit. These measurements confirmed the accuracy of the Junge–McLaughlin equation [63] in predicting the rate of spreading of acidity on the surface and the role of the bicarbonate as a proton carrier [64].

It must be recalled that the model of Stuchebrukchov is phenomenological in nature and many discrete steps and events have been 'lumped together'. The understanding of all intricate relations between the rate constants of the chemical reaction of the proton binding sites with free proton and the macroscopic descriptors calls for molecular modeling and direct measurements, where the local properties of the reaction space must be included in the analysis. Accordingly, the Stuchebrukchov mechanism can be taken as a guideline and its predictions are approximations.

The basic feature derived by the model of Stuchebrukchov, that transient protonation of surface groups accelerated the migration of protons on the surface has been confirmed in real time measurements carried out by Nachliel and Gutman [17]. Figure 20.1 depicts time-resolved measurements of the protonation of the in-

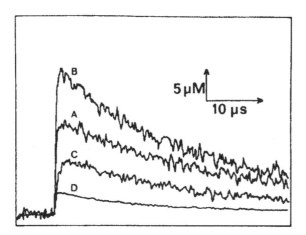

Figure 20.1. Transient protonation of Bromocresol Green adsorbed on mixed micelles. The reaction was carried out at pH 7.3 ± 0.1 in a solution containing 500 μM Bromocresol Green, 1 mM of the proton emitter 2-naphthol-3,6-disulfonate, 500 μM micellar concentration of Brij-58 (40 mg mL^{-1}), and 3 mM of phospholipids. The transient protonation was measured at 633 nm through an optical path of 2.0 mm: (A) control, no phospholipids were added; (B) phosphatidylcholine; (C) phosphatidylserine; (D) phosphatidic acid.

dicator Bromocresol Green, adsorbed on a neutral micelle, by a proton released in the bulk using a water-soluble photoacid. The kinetics were measured either in the absence of phospholipids (curve A) or in the presence of phosphatidylcholine (B), phosphatidylserine (C) and phosphatidic acid (D). As seen in the figure, the zwitterionic residues (PC), that can bind the proton for only a few nanoseconds, enhanced the protonation of the dye, confirming the mechanism of enhanced proton transfer between adjacent proton binding sites. On the other hand, residues with pK values that will retain the proton in a bound state for time frames that are comparable with the observation time appear as proton traps and reduce the probability of the proton reacting with the site under study.

20.5
Direct Measurements of Proton Transfer at an Interface

20.5.1
A Model System: Proton Transfer Between Adjacent Sites on Fluorescein

20.5.1.1 The Rate Constants of Proton Transfer Between Nearby Sites

The fluorescein molecule is a suitable model system for monitoring the mechanism of proton transfer between adjacent sites as it has two distinct proton binding sites ~6 Å apart, each having a distinct pK and spectral properties. The first is the oxyanion attached to the xanthene ring, whose protonation is associated with a spectral shift of the dye, while the second is the carboxylate on the benzene ring. The protonation of this site has no effect on the absorption spectrum of the dye (Scheme 20.1).

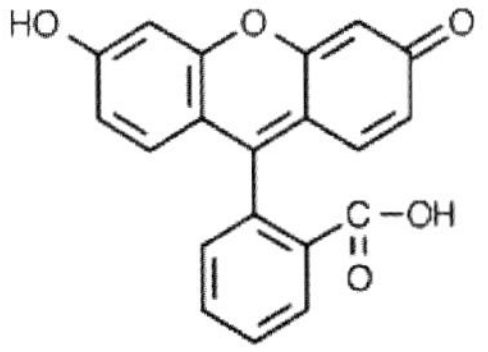

Scheme 20.1. The structure of fluorescein. The OH moiety of the xanthene ring controls the spectrum of the chromophore. In the ionized state, the dye has an intensive absorption at $\lambda_{max} = 496$ nm. The carboxylate of the benzene ring does not affect the spectrum of the dye.

The reversible protonation of the dye was measured in the time resolved domain using photoexcited pyranine as a proton emitter [65]. The rate constants of the various proton transfer reactions were determined by kinetic analysis [66, 66a] and the results are given in Table 20.2.

The rate constants determined by the analysis are subjected to quantitative evaluation by comparison with the predicted value according to the Debye–Smolu-

Table 20.2. The rate constants of proton transfer as measured for the pyranine fluorescein system in water, D_2O and in 100 mM NaCl at 25 °C. All rate constants are given in $M^{-1}\,s^{-1}$.

Reaction	H_2O ($M^{-1}\,s^{-1}$)	D_2O ($M^{-1}\,s^{-1}$)	KIE	100 mM NaCl ($M^{-1}\,s^{-1}$)	E_a (kcal mol^{-1}) in H_2O
$FLU^- + H^+$	2.5×10^{10}	1.7×10^{10}	1.47	1.6×10^{10}	3.3
$COO^- + H^+$	2.5×10^{10}	1.7×10^{10}	1.47	1.6×10^{10}	3.3
$COOH + FLU^-$	23.3×10^{10}	0.43×10^{10}	54.2	6.5×10^{10}	11.0

chowski equation, for the rate constant of the time independ diffusion controlled reaction an expression that correlates the rate constant with the radius of encounter ($\Sigma R_{j,i}$), the diffusion coefficients of the reactants $\Sigma D_{j,i}$ and the Avogadro number N_{av} [60, 67, 68].

$$k_{max} = (4\pi N_{av}/1000)\Sigma R_{j,i} \Sigma D_{j,i} \qquad (20.1)$$

The rate constants corresponding to the diffusion-controlled reaction between a free proton and acceptor and the rate constants were found to be compatible with the values predicted by the Debye–Smoluchowski equation for diffusion-controlled reactions (Table 20.2 rows 1 and 2), and so were the reactions that describe a collisional proton transfer between two free-diffusing reactants (not shown). In contrast with these reactions, the rate constant calculated for the intra-Coulomb cage proton transfer (Table 20.2, bottom row) had an extremely high value. The rate constant calculated by the Debye–Smoluchowski equation for a vanishing ionic strength $I \to 0$, is $k \sim 3 \times 10^9\,M^{-1}\,s^{-1}$. This value is smaller by two orders of magnitude than the measured one ($2.3 \times 10^{11}\,M^{-1}\,s^{-1}$). Such a fast reaction negates a mechanism based on encounter between two fluorescein molecules. Accordingly, the mechanism must be identified as an intra-Coulomb cage proton transfer, where the proton is first released from the benzene's carboxylate and then reacts preferentially with the nearest proton-binding site, the chromophore's oxyanion that is 6.7 Å apart. It should be stressed that the magnitude and the units of the reaction utilize second order rate constant units. For these reasons, the calculated rate constant of the intra-Coulomb cage proton transfer is suitable for comparison between measurements carried out under varying conditions (temperature, ionic strength, D_2O etc.), reflecting how the efficiency of the process is affected by the experimental conditions. A high rate constant implies that the pathway competes successfully with a parallel pathway, where the proton is dispersed into the bulk of the solution before re-encounter with the Coulomb cage of the fluorescein.

The kinetic features recorded for the intra-Coulomb cage proton transfer between the proton binding sites were reconfirmed with other dyes resembling the

Table 20.3. The solvent effect on the intra-Coulombic rate constant, determined by kinetic analysis of signals recorded at varying pH values. The second and third columns denote the rate constant for the proton transfer from the carboxylate to the chromophore as measured in water and in 100 mM NaCl. The column at the right denotes the KIE for the proton transfer reaction from the carboxylate to the chromophore.

Compound	Rate constant		KIE
	Water	100 mM NaCl	
Fluorescein	2.5×10^{11}	6.5×10^{10}	54
Rhodol green	$2. \times 10^{12}$	$1. \times 10^{11}$	105
5 carboxy fluorescein	9.3×10^{11}	$4. \times 10^{11}$	22.5
6 carboxy fluorescein	$1. \times 10^{12}$	2.5×10^{11}	38

fluorescein molecule. Careful measurements were carried out with Rhodol green, and the 6 and 5 di-carboxy-fluorescein. The rate constants for the intra-Coulomb cage proton transfer were calculated. As seen in Table 20.3, the rate constant for the intra proton binding site are high, the screening electrolyte shows the reaction and the kinetic isotope effect exceeds 20.

20.5.1.2 Proton Transfer Inside the Coulomb Cage

The intra-Coulomb-cage proton transfer differs from the diffusion controlled reactions by its kinetic and thermodynamic parameters. The kinetic isotope effect measured for the intra-Coulomb-cage proton transfer is extremely high ~ 50, significantly larger than the value of $\sqrt{2}$ measured for the reaction controlled by the diffusion of the proton. The same pattern is noticed when comparing the activation energies of the reactions. The protonation of the two proton binding sites of the fluorescein by free-diffusing proton ($E_a \sim 3$ kcal mol^{-1}) is compatible with the temperature dependence of the water's viscosity ($E_a = 4.2$ kcal mol^{-1}). In contrast with these values, the intra-Coulomb-cage proton transfer has significantly higher activation energy. This is a clear indication of a different reaction mechanism. Addition of a screening electrolyte reduced the diffusion-controlled reactions of free proton in accordance with the Debye–Smoluchowski equation. The intra-Coulomb-cage proton transfer was much more sensitive to the screening effect, which is another indication of a reaction mechanism that differs from the homogeneous diffusion-controlled reactions [66a, 66b].

The high activation energy and the large kinetic isotope effect values assigned to the intra-Coulomb-cage proton transfer are characteristic of proton transfer through a hydrogen bond that is slightly stretched beyond its equilibrium length [69, 70]. Accordingly, we investigated whether the solute molecule, the fluorescein, may affect the water molecules in its immediate vicinity, searching for the presence of hydrogen-bonded water molecules that may serve as pathways through which the proton can propagate from the carboxylate to the oxyanion. The distance between the two proton binding sites of the dye, 6.7 Å, is too large to support a direct proton transfer between them. Consequently, the proton transfer involves the water as a carrier, which makes it suitable to characterize the water in the inner solvation shell of solutes.

For a study of the fluorescein–water interactions, the fluorescein molecule with a single proton attached to the carboxylate moiety, in the presence of 639 water molecules and one Na^+, was simulated by molecular dynamics for 500 ps [66b]. The protonated carboxyl moiety of the fluorescein was hydrogen-bonded with water for 95% of the simulation time, and with an average of 5 hydrogen bonds. The acceptor oxygen atoms, attached to the xanthene rings, were permanently hydrogen-bonded with the water, averaging 4 hydrogen bonds. The hydrogen bonds between the dye and the water had a lifetime that was ∼2 times longer (5.80 ps and 5.13 ps for the carboxylate and the chromophore, respectively) than the average hydrogen bond between water molecules in the bulk (2.94 ps).

In the presence of so many water molecules, the proton can propagate from the donor to the acceptor by more than one trajectory. What is more, once the proton dissociates from the carboxylic moiety, it will affect the local structure of the water, by forming either $H_9O_4^+$ or $H_5O_2^+$ species. For these reasons, the search was limited to the water molecules in the vicinity of the dye that are at a suitable distance to form a hydrogen-bonded array, using the default value of the GROMACS program ($r \leq 3.5$ Å) for the length of the hydrogen bond. The search was carried out with respect to the two oxyanions of the chromophore, and 10 superpositioned trajectories are presented in Fig. 20.2.

The connection of the two sites by the water molecules is a flexible yet persistent feature. The length of the path varies with time, even within a 1 ps time-frame, and so does the shape of the path. Yet in all cases, the path remained close to the surface of the molecule, and did not extend out of the Coulomb cage of the molecule, which extends 14 Å into the bulk. The order imposed by the fluorescein on the inner solvation layer may account for the special features of the intra-Coulomb-cage proton transfer reactions. The proton transfer between two water molecules is regulated by the motion of the water molecules in the whole structure of the solvated proton; the hydrogen bonds in the second and third layer are rearranged, destabilizing bonds are broken and stabilizing ones are formed, and during the passage of the protonic charge, the hydrogen bonds are temporarily constricted [43]. These local rearrangements of structure are hampered by the structure imposed by the negative charges of the anion, thus forcing the proton transfer to take place in a more rigid environment where the O–O distances are larger than 2.4 Å. As a result, both the activation energy and the kinetic isotope effect (KIE) increase. Scheiner and coworkers [70, 71] have investigated the effect of the interatomic distance on the activation energy and the KIE of proton transfer. These quantum mechanical calculations show that the energy barrier for proton transfer increases with the distance between the oxygen atoms of the water, reaching a

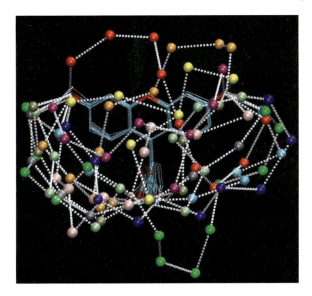

Figure 20.2. The array of water molecules that are at a hydrogen bond distance and form a possible proton-conducting pathway between the protonated state of the carboxylate (donor) on the benzene ring of fluorescein and the proton acceptor oxygen atoms on the xanthene ring. 10 trajectories are superpositioned. The figure also exhibiteds the structural fluctuations of the dye's structure.

value of 15–20 kcal mol^{-1} at a separation of ∼2.8 Å [70, 71]. Furthermore, under these conditions, the proton transfer operates both by a classical mechanism (transition state theory) and by tunneling, thus leading under certain conditions to a very high KIE [69]. In bulk water, the length of the hydrogen bond in the Eigen or the Zundel structures is of the order of 2.4 Å and the measured activation energy represents the reorganization of the solvent [43]. When the water molecules are held by the ordering forces of the charged scaffolding, the reaction operates under a different regime, where the concerted motion of the water molecule, as required to facilitate the proton transfer, is restricted. The rate-limiting step is shifted from the organization of the solvent to the passage of the proton, with subsequent enhancement of activation energy and KIE.

A question to be answered is why the proton remains in the vicinity of the acceptor water molecule (which is part of the interconnecting trajectory, as in Fig. 20.2) instead of escaping to the bulk. This can be explained by the presence of the intensive electrostatic potential that detains the proton within the space defined by the Coulomb cage, as depicted in Fig. 20.3. This figure depicts the 1 k_{Bol} T/e boundary around the Rhodol green molecule either at vanishing ionic strength or at 100 mM NaCl. At low ionic strength, the Coulomb cage is almost spheric and the acceptor moiety of the chromophore, the amine, is well under the "umbrella" of the negative charged Coulomb cage. At higher ionic strength, the cage had contracted and the amine moiety is protruding out of the cage. As a result, a proton

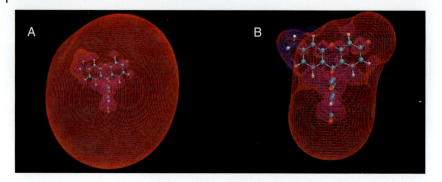

Figure 20.3. The electrostatic potential surrounding the Rhodol green molecule (at 1 k_{Bol} T/e) either in vanishing ionic strength (frame A) or in 100 mM NaCl (frame B). Please note that at high ionic strength the Coulomb cage had contracted, leaving the amine moiety of the dye, which serves as a proton acceptor, protruding out of the Coulomb cage, forcing the passage of a proton from the carboxylates on the benzene ring towards the acceptor site, to escape out of the attractive Coulomb cage.

transfer from the donor residues, the carboxylate on the benzene ring, to the acceptor must climb up in the potential field, before it can interact with the acceptor site. This energy-barrier accounts for the marked effect of the screening electrolyte on the rate constant of the intra-Coulomb cage reaction. In the case of the di-carboxylate fluorescein, where the molecule bears a charge of $Z = -3$, the Coulomb cage is large enough, both in low and high ionic strength, to allow proton transfer without breaching out of the cage. For this reason, the effect of the screening electrolyte on the intra-Coulomb cage reaction is much smaller. The diffusion out of the Coulomb cage is an uphill diffusion with escape time of the order of hundreds of picoseconds [67], while the intra-Coulomb-cage reaction hardly overcomes an electrostatic barrier. Thus, while the released proton attempts to diffuse out of the Coulomb cage, there are ample opportunities to propagate through the rigidified water molecules. The collapse of the local electrostatic potential brought about by high concentrations of electrolyte is consistent with our interpretation. In the presence of 100 mM NaCl, the ionic screening practically abolishes the wrapping Coulomb cage. In the absence of the retarding potential, the proton can readily disperse to the bulk and the intra-Coulomb-cage reaction is slowed to a level that reflects the proximity of the acceptor site to the site of release.

20.5.2
Direct Measurements of Proton Transfer Between Bulk and Surface Groups

Physically, the bulk phase of the solution is as close as a solute can get to a surface before it is affected by the image charges or the partial immobilization of the water molecules. We can surely assume that some 10–15 Å from the surface, equivalent to the third to fourth solvation layer, the solvent can be regarded as "bulk" [72, 73].

A system suitable for measuring the rate constants of proton transfer between bulk and surface should be made on a single bilayer, which is fully equilibrated with bulk water on both sides, and should contain a marker present at very low content, yet sensitive enough to report the interaction of the membrane with free proton. A phospholipid black lipid membrane impregnated with a small amount of monensin meets these experimental requirements. Monensin is an ionophore that supports an electroneutral proton/Na^+ exchange across a membrane. The kinetics of the exchange were measured by a system consisting of a black lipid membrane equilibrated with monensin and NaCl, with the photoacid pyranine added to the compartment facing one side of the membrane. When a laser pulse excited the pyranine, one face of the membrane was selectively acidified, and the released protons reacted with the membrane, thus initiating the exchange reaction. The first step is the formation of an unstable, ternary complex $MonHNa^+$ that dissociates into MonH and Na^+ ion. This enriches the MonH species concentration on one side of the membrane, and the perturbation is propagated to the other side by diffusion of MonH to one side of the back flux of MonNa. Each species, on arriving at the other side of the membrane, equilibrates with the bulk, releasing/binding either proton or Na^+ ions. Due to the short duration of the perturbation, some 1–2 μs during which the free proton concentration is higher than the pre-equilibrium state, the exchange reaction is highly synchronized and can be followed by monitoring the capacitance current of the system [74]. The selective protonation of one side of the membrane consists, physically, of the charging of one side of a capacitor and the propagation of the charge from one side to the other is a discharging event. By monitoring the current passing through a short-circuit which connects the two faces of the membrane, the charge dynamics can be observed and the chemical reactions associated with the charge translocation can be deduced.

The results of capacitance current, measured with increasing Na^+ concentrations, are presented in Fig. 20.4. As the reaction mechanism hinges on the protonation of the MonNa species, the shape of the signal varies with the Na^+ concentration. At low Na^+ (10 mM), the signal is small, and increases in magnitude and shape as more Na^+ ions are present. Above 100 mM, the signals are constant in shape and size. The perturbation has been expressed as a series of chemical reactions taking place simultaneously on the two sides of the membrane; it was converted into a set of chemical rate equations, and integrated over time to reconstruct the time derivative of the charge imbalance between the two faces of the membrane. The computed current and its variation with the Na^+ concentration are presented in the figure. The accuracy of the reconstruction is so high as to fall within the narrow band set by the electronic noise. The rate constants determined for the binding and release of the ions are given in Table 20.4.

The rate constants for the proton transfer between surface groups and the bulk, as well as the collisional proton transfer between the protonated surface sites and the soluble proton acceptors, are of the order of diffusion controlled reactions, with no indication as to an energy barrier that retains the free proton near the surface. The rate constants of the MonNa or the Mo^- imply that both species are located on the surface of the membrane, with the proton-binding site constantly exposed to

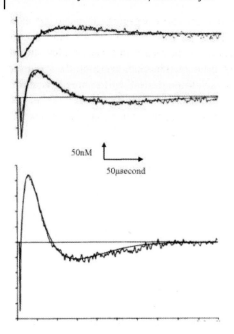

Figure 20.4. Capacitance current measured with monensin-impregnated, black lipid membrane subjected to pulse protonation of one side of the membrane. The traces represent the charge flux, already expressed in molar units as function of time. The signals were measured in 100 mM choline chloride as a conducting electrolyte, and in the presence of 10, 50 and 210 mM NaCl. Each experimental curve was reconstructed by integration of differential rate equations corresponding with the reaction mechanism. For details see [24].

Table 20.4. The rate constants of proton transfer bulk and surface. The rate constants reproduce the experimental capacitance currents presented in Fig. 20.3. Data taken from Ref. [74].

Reaction	Rate constant (M^{-1} s^{-1})
MonNa + H^+	1×10^{10}
Mon^- + H^+	1×10^{10}
Ps^- + H^+	1×10^{10}
PsH + ΦO^-	1×10^{10}
MonH + ΦO^-	1.2×10^{10}

Abbreviations: Monensin (Mon^-); Phosphatidyl serine (Ps^-); Pyranine (ΦO^-).

the bulk. To confirm this assumption, Ben-Tal and coworkers calculated the electrostatic potential of the various complexes of monensin, their solvation energy and the free energy changes as the complex is inserted into a slab of a low dielectric constant matter [75]. The results of these calculations, based on the structure of the MonNa crystals, confirmed the conclusions derived by the kinetic analysis, i.e. the stable location of the monensin in the membrane is at the water membrane interface and is oriented so that the proton-binding site is preferentially exposed to the bulk.

20.6
Proton Transfer at the Surface of a Protein

The scarcity of free proton in the cytoplasmic space of bacteria, eukaryotic cells or the mitochondrial matrix imposes a time limitation on the rate at which a free proton can diffuse towards the enzyme's active sites, so that the system appears to be rate-limited by the availability of free protons. However, the measured rates still seem to exceed the predicted values, for a review see Ref. [26], indicating that the protein's surface participates in channeling the proton to the orifice of the proton-conducting channels. This case was first demonstrated with bacteriorhodopsin, a membranal protein which utilizes the energy of a photon, absorbed by its chromophore, to drive protons from the cytoplasmic space of the bacteria to the external space.

The late phases of the bacteriorhodopsin's photocycle are kinetically limited by the intake of proton from the cytoplasmic side of the membrane, and by delivery of the proton to a carboxylate of residue D96, located below the surface of the protein. The carboxylate is connected to the surface through a shaft, which is too narrow to accommodate a water molecule [76]. To let the proton through, the shaft must expand to let the water in, an energy-consuming step estimated to be some 10–20 kcal mol^{-1}. To overcome this barrier, the protein retains, next to the shaft, a reservoir of available space, stored as micro-cavities [77]. The fusion of the micro-cavities with the shaft assists in its expansion, lowering the required investment of energy by $\sim$2.5 kcal mol^{-1}. Yet, even that makes the lifetime of the open state very short. To assure a rapid proton transfer to the D96, when the "moment of grace" is coming, the protein must retain an available proton in the immediate vicinity of the shaft's opening. The local proton storage is provided by the carboxylate of D38, which is partially exposed to the surface and, due to its semi-hydrophobic environment, has a rather high pK value (pK = 5.1) [25]. The partial exposure of D38 renders its reaction with free proton somewhat slow. To ensure that whenever the shaft is expanded, the proton will be available, an accessory mechanism is required. This enhancement of protonation of D38 is attained by a cluster of three carboxylates that assist in the protonation of D38 (the proton-collecting antenna) [78].

The proton-collecting antenna was first identified in studies carried out with the ground state form of the protein, which is a resting configuration. To establish the physiological role of the antenna, evidence must be supported on the monitoring of the process on the surface of the M state, which is a short-time intermediate where the Schiff-base has lost its proton and is re-protonated from the cytoplasmic side. To monitor the dynamics at this state of the photocycle Nachliel and co-workers [79] used mutated BR preparation D96N and a triple mutant (D96G/F171C/F219L). Both mutations permit, under steady background illumination, the accumulation of the late M state. The protein, while being kept in the M state, was subjected to reversible pulse protonation caused by repeated excitation of pyranine present in the reaction mixture, and the re-protonation dynamics of the pyranine anion were recorded and subjected to kinetic analysis. The calculated rate constants indicated that, in the late M state of bacteriorhodopsin, there is an efficient mechanism of proton delivery to the unoccupied and most basic residue on its cytoplasmic surface (D38), see Fig. 20.5. This machinery was even more efficient in the M state than in the ground state (BR) configuration of the protein.

The presence of a proton-collecting antenna has been demonstrated for the matrix surface of bovine cytochrome c oxidase [26] and the intracellular surface of the *Rhodobacter sphaeroides* cytochrome oxidase [65]. These proteins operate under temporal restriction, where the protein must pump protons at a rate compatible with the physiological requirements of the cell. On the other hand, membranal proteins that utilize the proton-motive force for driving a slow reaction, whose rate-limiting step is the binding and release of the substrate, such as the Lac-permease or the transhydrogenase, are devoid of a proton-collecting antenna [80, 81].

20.7
The Dynamics of Ions at an Interface

The kinetic analysis of proton transfer at the bulk/surface interface is based on standard chemical kinetic formalism, a procedure that cannot account for the multitude of forces operating in the reaction space, or for the dynamics of the protein's surface groups. A full account of the reaction mechanism can be compiled by molecular dynamics, with one drawback: the calculations carried out for a solvated proton are extremely laborious, due to the need to account for the breaking and formation of covalent bonds, which are the essence of the proton's dissociation mechanism [82, 83]. An alternative procedure, which can shed light on the proton's propagation at the interface, is to make a minimal generalization, and to substitute the proton, with its special chemistry, by charged ions which propagate by self-diffusion and with the assumption that the forces operating on a proton will affect other ions in the same manner.

A molecular dynamics (MD) simulation was carried out, simulating the propagation of Na^+ and Cl^- ions in the immediate vicinity of a small globular protein [85]. The protein selected as a model for this study is the S6, which forms part of the bacterial 30S ribosome central domain [84], and does not have a function associ-

20.7 The Dynamics of Ions at an Interface | 1519

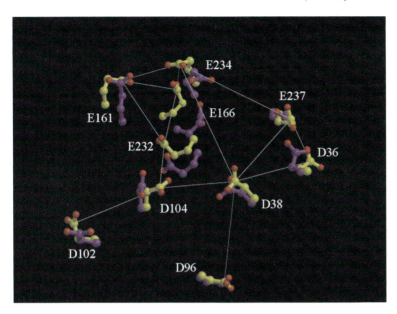

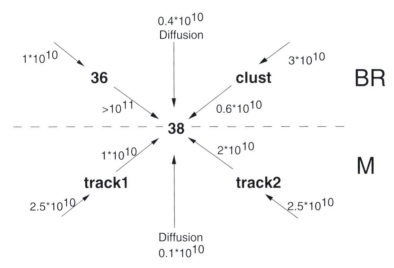

Figure 20.5. Surface proton-transfer pathways on the cytoplasmic surface of bacteriorhodopsin in the BR and the M-states. (A) Atomic model of the carboxylic side chains at the cytoplasmic surface of bacteriorhodopsin. Initial state is colored purple and the M state yellow. (B) Schematic representation of the proton conduction pathways in the Br and the M states. The numbers on the arrows represent the rate constants and are given in $M^{-1}\ s^{-1}$.

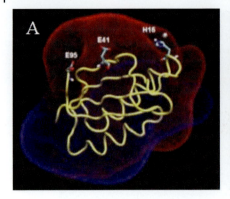 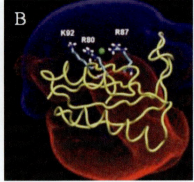

Figure 20.6. The electrostatic potential surface around the small protein S6. (A): Residue His16 (which is transiently located in the vicinity of the ion) and the two attractor sites Glu41 and Glu95. Right: Residues Arg80 and Arg87, which are the strongest ion attractors and Lys92, which is located in their vicinity and forms a weak ion attractor, are presented as the ion is detained by Arg80 and Arg87. The Coulomb cages for the positive (blue) and negative (red) domains are drawn at the distance where the electrostatic potential equals 1 $k_B T/e$.

ated with ion transport on its surface. The S6 is a globular protein of 101 amino acids, 32 of which are charged at a physiological pH. Moreover, due to its high charge density and globular structure, no amino acid is totally buried in the protein matrix and all the amino acids are at least partially exposed to the bulk. The uneven distribution of the charged residues on the protein generates two potential lobes, one having a positive potential and the other a negative (Fig. 20.6).

To follow the dynamics of the ions near the protein, molecular dynamics simulations were carried out in the presence of 6639 water molecules, and Na^+ and Cl^- ions were added to a formal concentration of 30 and 120 mM. The simulations were carried out for a period of 10 ns, and the diffusion coefficients of the ions were calculated to be comparable with the values determined by experimental methods [85]. A great advantage of molecular dynamics calculations is the possibility to visualize the motion of each ion. On inspecting the various ions, it became evident that their spatial distribution was not random. There was a clear tendency of some anions and cations to remain in the vicinity of the protein, as if the local forces detained them next to the protein's surface.

Molecular dynamics simulation of the protein in the presence of a small number of ions (4 Na^+ and 4 Cl^- ions, comparable with 30 mM NaCl solution) revealed that the ions were scanning the whole reaction space. Yet, the ions were not evenly distributed in the solution and had a tendency to linger in the immediate proximity (6 Å) of the protein for a relatively long time, ~1 ns (see Fig. 20.7). The simulation revealed fast exchange of the ions between the protein's surface and the bulk, reflecting competition between two forces: the electrostatic attraction that favors the

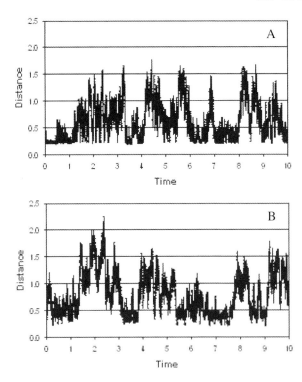

Figure 20.7. The minimal distance, in nm, between any of the Cl⁻ (A) and Na⁺ (B) ions and the protein as a function of simulation time. The distances are given in nm and the time in ns. The minimal distance is dictated by the steric interferences between the Van-der-Waals radii of the ions.

detainment and the entropic drive that prefers the free state of the ion. Therefore, throughout most of the simulation time, the ions diffuse in a Brownian motion in the bulk, but once an ion is trapped by the protein's Coulomb cage, it is drawn to the nearest attractor site. Sooner or later, depending on the strength of the attractor site, the ion will escape its detainment and will diffuse either within the Coulomb cage to the next attractor or out of the Coulomb cage. Thus, an ion that is already located inside the Coulomb cage has a higher probability to encounter with other attractor sites [85].

There is a strong resemblance between the mechanism of ion motion next to the protein and the proton-collecting antenna reported for bacteriorhodopsin [78, 79] or cytochrome c oxidase [2]. These domains consist of a cluster of carboxylates that function as proton binding sites. The protonation on any carboxylate of the cluster leads to rapid proton exchange reactions that finally deliver the proton to the immediate vicinity of the proton-conducting channel of the protein.

20.8
Concluding Remarks

Proton transfer at the surface of a protein or biomembrane is a cardinal reaction in the biosphere, yet its mechanism is far from clarification. The reaction, in principle, should be considered as a quantum chemistry event, and the reaction space as a narrow layer, 3–5 water molecules deep. What is more, local forces are intensive and vary rapidly with the precise molecular features of the domain. For this reason, approximate models that are based on pure chemical models or on continuum physical approximations are somewhat short of being satisfactory models with quantitative prediction power.

At the present time, when the structure of the proton pumping protein and the membrane's surface can be gained at atomic resolution, when the dissociation of a proton can be recorded with sub-nanosecond resolution and molecular dynamics can be extended to tens of nanoseconds, it seems that a combination of these methods will be required to elucidate the mechanism of the reaction. Thus, combination of specific labeling of sites of interest by a photoacid or indicator, coupled with time-resolved measurements and molecular dynamics of the reaction, will be the next step in the research. Once these combined experiments are available, the generalization of the process, like the role of local electrostatic potential, orientation of water and the relative motion of side-chains, will be quantitated, with a subsequent improvement in the theoretical predicting power.

Acknowledgments

The authors are grateful to Ran Friedman, from Tel-Aviv University, Israel for his help with the theoretical calculations. The research in the Laser Laboratory for Fast Reactions in Biology is supported by the Israeli Science Foundation (472/01-2) and the American Israel Binational Science Foundation (2002-129). The molecular dynamics calculations were carried out in the High Performance Computing Unit of the Inter University Computing Center.

References

1 NAMSLAUER, A., et al., Intramolecular proton-transfer reactions in a membrane-bound proton pump: the effect of pH on the peroxy to ferryl transition in cytochrome c oxidase, *Biochemistry*, 2003, **42**, 1488–98.

2 MARANTZ, Y., et al., Proton-collecting properties of bovine heart cytochrome C oxidase: kinetic and electrostatic analysis, *Biochemistry*, 2001, **40**, 15086–15097.

3 MITCHELL, P., Chemiosmotic coupling in oxidative and photosynthetic phosphorylation, *Biol. Rev. Camb. Philos. Soc.*, 1966, **41**, 445–502.

4 WIKSTROM, M.K., Proton pump coupled to cytochrome c oxidase in mitochondria, *Nature*, 1977, **266**, 271–273.

5 PADDOCK, M.L., et al., Pathway of proton transfer in bacterial reaction centers: replacement of serine-L22 by

alanine inhibits electron and proton transfers associated with reduction of quinone to dihydroquinone, *Proc. Natl. Acad. Sci. USA*, 1990, **87**, 6803–6807.
6. McPherson, P.H., et al., Protonation and free energy changes associated with formation of QBH2 in native and Glu-L212 → Gln mutant reaction centers from Rhodobacter sphaeroides, *Biochemistry*, 1994, **33**, 1181–1193.
7. Lancaster, R.C., D., Michel, H., The coupling of light-induced electron transfer and proton uptake as derived from crystal structures of reaction centers from Rhodopseudomonas viridis modified at the binding site of the secondary quinone, QB, *Structure*, 1997, **5**, 1339–1359.
8. Axelrod, H.L., et al., Determination of the binding sites of the proton transfer inhibitors Cd^{2+} and Zn^{2+} in bacterial reaction centers, *Proc. Natl. Acad. Sci. USA*, 2000, 97, 1542–1547.
9. Dencher, N.A., et al., Structure-function relationship of the light-driven proton pump bacteriorhodopsin, *J. Protein Chem.*, 1989, **8**, 340–343.
10. Heberle, J., et al., Proton migration along the membrane surface and retarded surface to bulk transfer, *Nature*, 1994, **370**, 379–382.
11. Riesle, J., et al., D38 is an essential part of the proton translocation pathway *in* bacteriorhodopsin, *Biochemistry*, 1996, **35**, 6635–6643.
12. Gutman, M., Huppert, D., Rapid pH and deltamuH+ jump by short laser pulse, *J. Biochem. Biophys. Methods*, 1979, **1**, 9–19.
13. Gutman, M., The pH jump: probing of macromolecules and solutions by a laser-induced, ultrashort proton pulse-theory and applications in biochemistry, *Methods Biochem. Anal.*, 1984, **30**, 1–103.
14. Gutman, M., Application of the laser-induced proton pulse for measuring the protonation rate constants of specific sites on proteins and membranes, *Methods Enzymol.*, 1986, **127**, 522–538.
15. Viappiani, C., Bonetti, G., Carelli, M., Ferrati, F., Sternieni A., Study of proton transfer processes in solutions using the laser induced pH jump: A new experimental setup and an improved data analysis based on genetic algorithm, *Rev. Sci. Instrum.*, 1998, **69**, 270–276.
16. Gutman, M., et al., Time-resolved protonation dynamics of a black lipid membrane monitored by capacitative currents, *Biochim. Biophys. Acta*, 1987, **905**, 390–398.
17. Nachliel, E., Gutman, M., Time resolved proton-phospholipid interaction. Methodology and kinetic analysis, *J. Am. Chem. Soc.*, 1988, **110**, 2635.
18. Yam, R., Nachliel, E., Gutman, M., Time-resolved proton-protein interaction. Methodology and kinetic analysis, *J. Am. Chem. Soc.*, 1988, **110**, 2636–2640.
19. Kotlyar, A.B., et al., The dynamics of proton transfer at the C side of the mitochondrial membrane: picosecond and microsecond measurements, *Biochemistry*, 1994, **33**, 873–879.
20. Haines, T.H., Dencher, N. A., Cardiolipin: a proton trap for oxidative phosphorylation, *FEBS Lett.*, 2002, **528**, 35–39.
21. Brzustowicz, M.R., et al., Molecular organization of cholesterol in polyunsaturated membranes: microdomain formation, Biophys. J., 2002, **82**, 285–298.
22. Samsonov, A.V., Mihalyov, I., Cohen, F.S., Characterization of cholesterol-sphingomyelin domains and their dynamics in bilayer membranes, *Biophys. J.*, 2001, **81**, 1486–1500.
23. Shaikh, S.R., et al., Lipid phase separation in phospholipid bilayers and monolayers modeling the plasma membrane, *Biochim. Biophys. Acta*, 2001, **1512**(2), 317–328.
24. Bransburg, Z.S., Nachliel, E., Gutman, M., Utilization of monensin for detection of microdomains in cholesterol containing membrane, *Biochim. Biophys. Acta*, 1996, **1285**, 146–154.

25 Checover, S., et al., Mechanism of proton entry into the cytoplasmic section of the proton-conducting channel of bacteriorhodopsin, Biochemistry, 1997, **36**, 13919–13928.
26 Adelroth, P., Brzezinski, P., Surface-mediated proton-transfer reactions in membrane-bound proteins, Biochim. Biophys. Acta, 2004, **1655**, 102–115.
27 Jaskson, J.D., Classical Electrodynamics, John Wiley & Sons, New York, 1975.
28 Cohen, B., et al., Excited state proton transfer in reverse micelles, J. Am. Chem. Soc., 2002, **124**(25), 7539–7547.
29 Binder, H., et al., Hydration of the dionic lipid dioctadecadienoylphosphatidylcholine in the lamellar phase. An infrared linear dichroism and X-ray study on headgroup orientation, water ordering, and bilayer dimensions, Biophys. J., 1998, **74**, 1908–1923.
30 Milhaud, J., New insights into water-phospholipid model membrane interactions, Biochim. Biophys. Acta, 2004, **1663**, 19–51.
31 Bach, D., Miller, I.R., Attenuated total reflection (ATR) Fourier transform infrared spectroscopy of dimyristoyl phosphatidylserine-cholesterol mixtures, Biochim. Biophys. Acta, 2001, **1514**, 318–326.
32 Miller, I.R., Bach, D., Organization of water molecules by adhering to oriented layers of dipalmitoylphosphatidyl serine in the presence of varying concentrations of cholesterol, Biochim. Biophys. Acta, 2000, **1468**, 199–202.
33 Bach, D., Miller, I.R., Hydration of phospholipid bilayers in the presence and absence of cholesterol, Biochim. Biophys. Acta, 1998, **1368**, 216–224.
34 Nicklas, K., et al., Molecular dynamics studies of the interface between a model membrane and an aqueous solution, Biophys. J., 1991, **60**(1), 261–272.
35 Kiefer, P.M., Hynes, J.T., Nonlinear free energy relations for adiabatic proton transfer reactions in a polar environment. II. Inclusion of the hydrogen bond vibration, J. Phys. Chem., 2002, **106**, 1850–1861.
36 Douhal, A., Lahmani, F., Zewail, A.H., Proton-transfer reaction dynamics, Chem. Phys., 1996, **207**, 477–498.
37 Rini, M., et al., Real-time observation of bimodal proton transfer in acid-base pairs in water, Science, 2003, **301**, 349–352.
38 Lee, J., Griffin, R.D., Robinson, G.W., 2-naphthol: a simple example of proton transfer effected by water structure, J. Chem. Phys., 1985, **82**, 4920–4925.
39 Huppert, D.K., Gutman, M., Nachliel, E., The effect of water activity on the rate constant of proton dissociation, J. Am. Chem. Soc., 1982, **104**, 6944–6953.
39a Gutman, M., Nachliel, E., Kiryati, S., Dynamic studies of proton diffusion in mesoscopic heterogeneous matrix. The interbilayer space between phospholipids membranes. Biophys. J., 1992, **63**, 281–290.
39b Shimoni, E., Tsfadia, Y., Nachliel, E., Gutman, M., Gaugement of the inner space of the apomyoglobin's heme binding site by a single free diffusing proton. I. Proton in the cavity. Biophys. J., 1993b, **64**(2), 472–479.
39c Yam, R., Nachliel, E., Kiryati, S., Gutman, M., Huppert, D., Proton transfer dynamics in the non-homogeneous electric field of a protein. Biophys. J., 1991, **59**(1), 4–11.
40 Agmon, N., The Grotthuss mechanism, Chem. Phys. Lett., 1995, **244**, 456–462.
41 Agmon, N., Hydrogen bonds, water rotation and proton mobility, J. Chem. Phys., 1996, **93**, 1714–1736.
42 Agmon, N., Estimation of hydrogen-bond lengths to H_3O^+ and $H_5O_2^+$ in liquid water, J. Mol. Liq., 1997, **73/74**, 513–520.
43 Lapid, H., et al., A bond-order analysis of the mechanism for hydrated proton mobility in liquid water, J. Chem. Phys., 2005, **122**, 14506.

44 PINES, E. HUPPERT, D., Kinetics of proton transfer in ice via the pH-jump method: evaluation of the proton diffusion rate in polycrystalline doped ice, *Chem. Phys. Lett.*, 1985, **116**, 295–301.

45 BRANSBURG, Z.S., et al., Biophysical aspects of intra-protein proton transfer, *Biochim. Biophys. Acta*, 2000, **1458**, 120–134.

46 ZHANG, J., UNWIN, P.R., Proton diffusion at phospholipid assemblies, *J. Am. Chem. Soc.*, 2002, **124**, 2379–2383.

47 ZUNDEL, G., MERZ, H., On the role of hydrogen bonds and hydrogen-bonded systems with large proton polarizability for mechanisms of proton activation and conduction in bacteriorhodopsin, *Progr. Clin. Biol. Res.*, 1984, **164**, 153–164.

48 ZUNDEL, G., Proton polarizability of hydrogen bonds: infrared methods, relevance to electrochemical and biological systems, *Methods Enzymol.*, 1986, **127**, 439–455.

49 WANG, J., EL-SAYED, M.A., Time-resolved Fourier transform infrared spectroscopy of the polarizable proton continua and the proton pump mechanism of bacteriorhodopsin, *Biophys. J*, 2001, **80**, 961–971.

50 KLAPPER, I., et al., Focusing of electric fields in the active site of Cu-Zn superoxide dismutase: effects of ionic strength and amino-acid modification, *Proteins*, 1986, **1**, 47–59.

51 LANGNER, M., et al., Electrostatics of phosphoinositide bilayer membranes. Theoretical and experimental results, *Biophys. J.*, 1990, **57**, 335–349.

52 PEITZSCH, R.M., et al., Calculations of the electrostatic potential adjacent to model phospholipid bilayers, *Biophys. J.*, 1995, **68**, 729–738.

53 ADAM, G.A., DELBRUCK., M., Reduction of dimensionality in biological diffusion processes, *Struct. Chem. Mol. Biol.*, 1968, 198–215.

54 BERG, H.C., PURCELL, E.M., Physics of chemoreception, *Biophys. J.*, 1977, **20**, 193–219.

55 SELEVIA, C.J., UNWIN, P.R., Lateral proton diffusion rates along stearic monolayer, *J. Am. Chem. Soc.*, 2000, **122**, 2597–2602.

56 GEORGIEVSKII, Y., MEDVEDEV, E.S., STUCHEBRUKHOV, A.A., Proton transport via coupled surface and bulk diffusion, *J. Chem. Phys.*, 2002, **116**, 1692–1699.

57 GEORGIEVSKII, Y., MEDVEDEV, E.S., STUCHEBRUKHOV, A.A., Proton transport via the membrane surface, *Biophys. J.*, 2002, **82**, 2833–2846.

58 HEBERLE, J., Proton transfer reactions across bacteriorhodopsin and along the membrane, *Biochim. Biophys. Acta*, 2000, **1458**, 135–147.

59 HEBERLE, J., et al., Bacteriorhodopsin: the functional details of a molecular machine are being resolved, *Biophys. Chem.*, 2000, **85**, 229–248.

60 GUTMAN, M., NACHLIEL, E., The dynamic aspects of proton transfer processes, *Biochim. Biophys. Acta*, 1990, **1015**, 391–414.

61 SEROWY, S., et al., Structural Proton Diffusion along Lipid Bilayers, *Biophys. J.*, 2003, **84**, 1031–1037.

62 POHL, P., SAPAROV, S.M., ANTONENKO, Y.N., The size of the unstirred layer as a function of the solute diffusion coefficient, *Biophys. J.*, 1998, **75**, 1403–1409.

63 JUNGE, W., MCLAUGHLIN, S., The role of fixed and mobile buffers in the kinetics of proton movement, *Biochim. Biophys. Acta*, 1987, **890**, 1–5.

64 ZANIBONI, M., et al., Proton permeation through the myocardial gap junction, *Circ. Res.*, 2003, **93**, 726–735.

65 MARANTZ, Y., et al., The proton collecting function of the inner surface of cytochrome c oxidase from Rhodobacter sphaeroides, *Proc. Natl. Acad. Sci. U.S.A.*, 1998, **95**, 8590–8595.

66 MOSCOVITCH, D., et al., Determination of a unique solution to parallel proton transfer reactions using the genetic algorithm, *Biophys. J.*, 2004, **87**, 47–57.

66a MEZER, A., FRIEDMAN, R., NOIVIRT, O., NACHLIEL, E., GUTMAN, M., The mechanism of proton transfer between adjacent sites exposed to water. *J. Phys. Chem. B. Condens Matter Mater. Surf. Interfaces Biophys.*, 2005, **109**(22), 11379–11388.

66b GUTMAN, M., NACHLIEL, E., FRIEDMAN, R., The mechanism of proton transfer between adjacent sites on the molecular surface. *Biochim. Biophys. Acta.*, 2006b, in press.

67 GUTMAN, M., NACHLIEL, E., The dynamics of proton exchange between bulk and surface groups, *Biochim. Biophys. Acta*, 1995, **1231**, 123–138.

68 GUTMAN, M., NACHLIEL, E., Time resolved dynamics of proton transfer in proteinous systems, *Annu. Rev. Phys. Chem.*, 1997, **48**, 329–356.

69 SCHEINER, S., LATAJKA, Z., Kinetics of Proton-Transfer in (H3ch..Ch3)-, *J. Phys. Chem.*, 1987, **91**, 724–730.

70 HILLENBRAND, E.A., SCHEINER, S., Effects of Molecular Charge and Methyl Substitution on Proton-Transfer between Oxygen-Atoms, *J. Am. Chem. Soc.*, 1984, **106**, 6266–6273.

71 SCHEINER, S., Proton transfers in hydrogen-bonded systems. Cationic oligomers of water, *J. Am. Chem. Soc.*, 1981, **103**, 315–320.

72 PARSEGIAN, V.A., RAU, D.C., Water near intracellular surfaces, *J. Cell Biol.*, 1984, **99**, 196s–200s.

73 RAND, R.P., et al., Variation in hydration forces between neutral phospholipid bilayers: evidence for hydration attraction, *Biochemistry*, 1988, **27**, 7711–7722.

74 NACHLIEL, E., FINKELSTEIN, Y., GUTMAN, M., The mechanism of monensin-mediated cation exchange based on real time measurements, *Biochim. Biophys. Acta*, 1996, **1285**, 131–145.

75 BEN-TAL, N., et al., Theoretical calculations of the permeability of monensin-cation complexes in model bio-membranes, *Biochim. Biophys. Acta*, 2000, **1466**, 221–233.

76 SCHATZLER, B., et al., Sub-second proton-hole propagation in bacteriorhodopsin, *Biophys. J.*, 2003, **84**, 671–686.

77 FRIEDMAN, R., NACHLIEL, E., GUTMAN, M., The role of small intraprotein cavities in the catalytic cycle of bacteriorhodopsin, *Biophys. J.*, 2003, **85**, 886–896.

78 CHECOVER, S., et al., Dynamics of the proton transfer reaction on the cytoplasmic surface of bacteriorhodopsin, *Biochemistry*, 2001, **40**, 4281–4292.

79 NACHLIEL, E., et al., Proton Transfer Dynamics on the Surface of the Late M State of Bacteriorhodopsin, *Biophys. J.*, 2002, **83**, 416–426.

80 NACHLIEL, E., et al., Time-resolved study of the inner space of lactose permease, *Biophys. J.*, 2001, **80**, 1498–1506.

81 NACHLIEL, E., GUTMAN, M., Probing the substrate binding domain of lactose permease by a proton pulse, *Biochim. Biophys. Acta*, 2001, **1514**, 33–50.

82 LAASONEN, K., et al., Structures of small water clusters using gradient-corrected density functional theory, *Chem. Phys. Lett.*, 1993, **207**, 208–213.

83 TUCKERMAN, M.E., MARX, D., PARRINELLO, M., The nature and transport mechanism of hydrated hydroxide ions in aqueous solution, *Nature*, 2002, **417**, 925–929.

84 AGALAROV, S.C., et al., Structure of the S15,S6,S18-rRNA Complex: Assembly of the 30S Ribosome Central Domain, *Science*, 2000, **288**, 107–112.

85 FRIEDMAN, R., NACHLIEL, E., GUTMAN, M., Molecular dynamics of a protein surface: ion-residues interactions. *Biophys. J.*, 2005, **89**(2), 768–781.

86 GUTMAN, M., NACHLIEL, E., FRIEDMAN, R., The dynamics of proton transfer between adjacent sites. *Photochem. Photobiol. Sci.*, 2006a, **5**(6), 531–537.

Index

a
absorbance 1410
absorbance, excited-state 450, 453, 471
absorbance, transient 465
absorption, excited state (ESA) 354, 356f.
absorption spectroscopy 245, 256f., 386f., 388, 435, 463, 466, 530, 1408, 1432
absorption, transient (TA) 509, 532, 540, 544, 547, 555
– time-resolved 352ff., 468, 1434
absorption of H_2 753
acceleration 1063, 1072, 1109
acceleration factor 1045ff.
acetaldehyde 519, 957, 1037, 1430f.
– hemiacetal 980
acetate anion 957, 980
acetic acid 173, 210f., 450, 519, 529, 967f., 1001
– methanol complex 932f.
acetic anhydride 980f.
acetone 952f., 955, 967
acetonitrile 172, 199, 400, 516ff., 529, 543, 887, 962, 1278
acetophenone 1004
acetyl group 225ff.
acetylene 519
N-acetylmethionineethyl ester 1030
acids 1502
acids, mineral 387, 392, 394, 398
acid-base
– catalysis 975ff., 1079ff., 1162, 1241, 1361
  – mechanism 977, 980
– equilibrium 370, 410
– neutralization 445, 448f.
– pairs 108f., 377f., 434, 1069, 1178
– reactions 110
– systems 303, 309, 400, 406f., 444
Acinetobacter calcoaceticus 1070
Aquifex pyrophilus 1161

acridinium 1051, 1066
activation energy 748
– of diffusion 753, 763
additives, electronegative and electropositive 773
adenine 934
adenosine triphosphate (ATP) 552, 1152, 1154ff., 1378, 1499
S-adenosylhomocysteine 1062
adiabatic longitudinal transport after dissociation engenders net alignment (ALTADENA) 657, 662ff.
adiabatic regime 303, 307, 309ff., 315, 320ff., 327
– excited state 597
– ground state 588, 597
– ground state potential 837, 840f., 846f., 851
– PES 368
– surface 1185
– theory of reactions 834, 840, 845
adsorption 688ff., 698, 773ff.
– activated H 756f.
– activation barrier 757
– energetic heterogeinity 759
– energy 758, 771, 773
– kinetics 773
advanced materials 245
Aequorea victoria 435
Aeromonas caviae 1151
agostic interaction 611
alanine (Ala) 1026, 1081, 1092, 1102, 1129, 1139ff., 1147, 1161, 1161, 1218, 1397, 1430, 1447
Albery-Kreevoy-Lee approach 1054
alcohols 228f., 427ff., 775, 992f., 1037, 1048
– aromatic 400ff.
– tertiary 993
aldehyde 1048, 1107

aldimine 1141, 1143f., 1150
aldol-keto isomerization 1120
Alexander-Binsch formalism 622, 670
alkali metals 773
alkaline fuel cells (AFCs) 709ff., 718
alkane 957
– aggregates, physical state 116ff.
– as proton acceptors 105ff., 115ff., 126ff.
– σ-basicity 110f.
– crystallisation 126, 128ff.
– halogenated (Cl, Br) 112ff.
– systems 991
alkene 1001
alkyne 1001
allostery 1381ff., 1422, 1452
alloys 771ff., 787, 804ff.
AlO_4 tetrahedron 685, 700
aluminium 779, 782, 809, 812
aluminophosphates (AlPOs) 686
aluminosilicates *see* zeolites
Alzheimer's disease 1029, 1379
amides 1017
amidines 196
amidines, diaryl- 203
amidines, N-N′-di-(p-F-phenyl) 193ff.
amidines, fluoro 172
amines 205, 591, 1006, 1017ff.
amines, chloro 1017
amines, tertiary 399
amino acid 232, 729, 1013, 1017, 1110, 1439, 1475
– activation 1459ff.
– ester 967
– radical generation 548ff.
aminopyrenes 398, 400, 445
1-aminopyrene (1-AP) 228ff., 382
8-aminopyrene-1,3,6-trisulfonate (APTS) 382, 400
2-amino-pyridine 446
aminopyrimidine 1427
1-amino-8-trifluoroacetylaminonaphthalene 979
ammonia 431f., 446, 593, 641, 688ff., 697
ammonium ions 688ff., 954, 1001
amphiphilic helices 1081
4-androstene-3,17-dione 1109, 1125
5-androstene-3,17-dione 1109, 1125
angular momentum 59
anion propagation 214
anisotropy, elastic 789
anisotropy experiments 226, 234, 238, 256ff., 262ff.
– time-resolved 262

ansa-bridging 628
anthracene 517ff.
antibiotics, petide-based 1154
antioxidants 1014
anvil cell, diamond 740, 742ff.
anvil cell, sapphire 819
apomyoglobin 1503
approximate instanton method (AIM) 896, 904f., 914, 923, 927, 938f.
aqueous solution 377f., 410, 428ff., 443f., 542, 714ff., 724, 867, 949, 992, 1014, 1052, 1098, 1116, 1186, 1507
– of KOH 711, 714
– NaOH 714
arginine (Arg) 528, 1084, 1099, 1145, 1407, 1485, 1519
Arrhenius behavior 333ff., 489, 577, 823
Arrhenius curves 135ff., 174, 176, 178ff., 188ff., 197ff., 205ff., 212ff., 264, 326, 337ff., 429f., 536f., 677, 765, 798, 877ff., 884ff., 891f., 917f., 923f., 927ff., 1255, 1289f., 1327, 1384f.
– for single and multiple H transfer 146, 150f., 157, 160f., 163f., 167f., 181, 185ff.
Arrhenius expression 1245
Arrhenius factor
– pre-exponential factor 9, 135, 150, 167f., 185, 206ff., 213, 217, 719, 762, 798f., 877, 879, 884f., 891f., 1245
Arrhenius law 148, 162, 164, 167, 655, 797f., 812
Arrhenius prefactor 1254ff., 1268, 1274, 1276ff., 1326f., 1342
Arrhenius relation 762, 1060, 1172
Arrhenius rate 655f.
arsenic 761
artificial light-harvesting 245
aryl groups 203f.
asparagine (Asn) 1108, 1397
aspartate (Asp) 528, 1108, 1113, 1126f., 1131, 1145, 1154, 1161, 1398, 1440
atmospheric reactions 834
attractive forces 759
autoionisation 443f.
autoprotolysis 216
– constant 214
– enthalpy 214
– mechanism 214
7-azaindole 925, 935f.
azenes, diaryltri- 203
azenes, tri- 205f.
azimuthal polarization mode 266
azophenine 155, 172, 176, 197ff.

b

Bacillus stearothermophilus   212f., 1143f., 1151, 1217, 1268, 1325, 1329, 1384, 1425
backbone, aromatic   445
backbone structure   377, 435
backdonation (BD)   604, 629
bacterial cell walls   1161
bacteriochlorin   175, 184
bacteriorhodopsin   1500f., 1506, 1516ff.
Baker Campbell Hausdorf Theorem   1212
band mode   787
band model   767f.
band structure, atomic   766f.
barrier   141, 143, 147, 150ff., 157, 160f., 189, 193ff., 199, 203, 251ff., 265, 275, 278, 282f., 291f., 584ff., 596, 645, 652, 687, 834, 840ff., 855, 928ff., 1007, 1110f., 1190f., 1196, 1232, 1481ff.
– above/over-the-barrier   84, 100, 199, 304, 309, 632, 810, 858, 869, 875, 904, 915, 936
– fluctuations   824, 1260ff., 1363
– oscillations   880
– quadruple-barrier   161, 163f.
– through-the-barrier   84f., 100, 136, 307, 364, 904, 914
– two- /double-   162, 164
barrier, Eckart   879
barrier, Gaussian   879
barrier, reaction   315ff., 322ff., 328ff., 338f., 510, 513, 522, 1179
barrier, single   164, 211, 218
barrier, transition   34ff., 39, 44
barrier, triple   159ff.
barrier, trunctuated parabolic   879
basal plane of crystals   809
base, scavenging   448ff.
BaZrO$_3$, Y-doped   732
barrel domain   1134
BBO   352
Becke-Lee-Yang-Parr (BLYP)   288ff., 697ff.
– B3LYP   249, 265, 571, 919f., 927f., 939, 1480ff.
BDE   511, 516ff.
– Bell tunneling model   136f., 144ff., 146ff., 251, 584, 623, 641, 653ff., 656, 674, 676, 844, 847, 1250ff., 1257
– dependence on hydrogen isotope   1250
– tunnel correction   1258, 1270, 1293, 1341f., 1350
BDE, one-dimensional   1293
BDE, truncated   1288, 1294, 1303
Bell-Evans-Polanyi (BEP) principle   585, 590

Bell-Limbach model   135, 137f., 146ff., 153, 155, 174, 177, 189, 191f., 198f., 208, 216
– parameters   169ff.
Bema Hapothle   584
benzaldehyde   1247, 1247
benzaldehyde acetals   983, 992, 998
benzene   517, 519, 689, 877
benzenesulfonate   239
benzimidazole   721, 727
benzisoxazole   989, 1006
benzoic acid   135, 171, 188, 296, 529
– crystalline   923, 1214ff., 1234
– dimer   278ff., 284, 919, 922
– methyl ester   568
p-benzoquinone   525f.
S-benzoylglutathione   1099f.
1-benzoyl-3-phenyl-1,2,4-triazole   983
benzyl alcohol   1266
benzylamine   1162
1-benzyl-1,4-dihydronicotinamide   1049
benzylsulfonium cation   984f.
bifurcation point   910, 920
Bigeleisen limits   1291ff.
Bigeleisen theory   140f., 148
Bigeleisen-Wolfsberg formalism   1287, 1291
bilayer   737ff., 747, 749, 780 (of water molecules)
bimetallic systems   771, 773
biological reactions   436
biological systems   232, 433ff., 537, 548ff., 729, 733, 778, 932, 1241
biomedical application   498
biomolecular processes   834
biomolecule   213, 224, 445
[2,2 bipyridyl]-3,3'-diol (BP(OH)$_2$)   370f.
2,5-bis(2'-benzaoxazolyl)-hydroquinone (BBXHQ)   353, 358, 361, 365
1,3-bis(4-fluorophenyl)[1,3]-N$_2$]triazene   203, 205
bleach   450, 464
Bloch orbitals   768f.
Bloch states   812
body centered cubic cell (bcc)   737, 740, 789, 797ff., 814, 817, 825
Bloembergen, Purcell and Pound model (BPP)   791
Boltzmann constant   148
Boltzmann distribution   149, 655f.
Boltzmann law   147
Boltzmann probability   1196f.
Bombyx mori   1152
bond   *see also* hydrogen bond
– angle H–O–H   718, 737

bond, C–H 515f.
– activation 1405
– order 1055
bond, C=N 1395ff., 1400
– activation 1398
bond, C=O 1395ff.
– C–S 1432
– Co–C 1476ff., 1487
– dissociation energy 109f., 1018, 1023, 1027
– formation/breaking 1083, 1097, 1163, 1229, 1351, 1487
bond, C–C 1430
bond, σ- 107f., 111
– rotation, C–C 1097
– rotation, C–N 1108
– stretch 1245f., 1250
bond, O–H 515
– three-centre three-electron bond 111
bond, ribosidic 1233
bond, S–S 1432
– valences 141ff.
bond energy – bond order model (BEBO) 325, 394ff.
Born-Oppenheimer 276, 461, 480, 837f., 1175
bottleneck 838ff., 1303, 1479f.
brain-derived neurotrophic factor (BDNF) 1370f.
– receptor (trkB) 1370f.
Bravais sublattice 801
bridges, carbon 525
bridges, salt 494, 527ff., 554, 1081, 1386
Brillouin zone 766f.
brome mosaic virus (BMV) particle 1378
bromocresol green 1508f.
4-bromopyrazole (4BrP) 193ff.
Brønsted acidity 397, 962
Brønsted acids 377, 409f., 443, 585, 719, 954, 965f., 969f.
Brønsted bases 377, 443, 585, 965f., 970, 1049
Brønsted coefficient 313, 317ff., 334f., 583ff., 599, 954, 1049, 1054ff., 1095, 1116f.
Brønsted correlation 589f., 591ff., 1000f.
Brønsted equation 977, 1001, 1093
Brønsted relation 585, 997
Brønsted site 685ff., 693ff., 700ff.
Brønsted-type plot 427, 510
Brownian dynamics (BD) simulation 1193
buffers 954, 969, 976
Bruice's proximity effect 1044
n-butylamine 691

c

Ca^{2+} 597, 1152, 1374, 1378f.
cage 223, 229
Calcium 808
calix[4]arene 173, 213ff., 223, 939f.
– p-tert-butyl- 215, 939
CAMEL-SPIN 174
canonical ensemble 833
canonical mean shape (CMS) 862f.
canonical variational theory (CVT) 836, 842, 867, 1486
carbanions 949ff., 958ff.
– aromatic 566
– α-carbonyl carbon 957, 963ff., 968f.
carbanions, α-cyanomethyl 955
– enamine 1419, 1429, 1434
– intermediates 565, 573ff., 581f., 1001, 1158
carbanions, vinyl 1001
carbenes 429
– /ylide 963
carbenium ions 106, 110, 116, 123f., 129f., 686ff., 703ff.
– tert-butyl cation 704
– of xylene 704
2-carbethoxycyclopentanone 1293
carbon acid 949ff., 960, 964, 966f., 1107, 1110
carbocations 106, 1300
carbocations, oxo- 994
carbocations, 1-phenylethyl 983
carbohydrates 1023f.
carbonium ions 106, 111, 116, 123f., 127
carbonyl groups 225, 230ff., 957, 963, 1002ff.
carbonyl/carboxylate acceptor 1114ff.
carboxylamide orientation 1403
carboxylate anions 1107
carboxylic acids 462, 466, 523, 1107
– dimeric 896
carboxylic bases 395, 426
Car-Parrinello
– molecular dynamics (CPMD) 286ff., 293, 692ff., 696ff., 721f.
– path 1185
catalytic cleavage 1375
catalytic cracking 686
catalytic mechanism
– of cages 223
– of enzyme reaction (KSI) 1126, 1133, 1435
  – one-base mechanism 1141, 1157
  – two-base mechanism 1141, 1158
catalytic proficiency 1046f.
catalytic triad 1460
cation propagation 214

cavity 1100
– CD 223ff., 230
– porphyrins 245, 248ff., 254, 267f.
– zeolites 686, 691
CCl_4 172, 208f.
CCl_3F 208f.
CCSD(T) 705
$CDCl_3$ 185
CD_2Cl_2 172, 198
$C_2D_2Cl_4$ 198
$CDFCl_2$ 185
CDF_2Cl 185
cephalosporin 1156
3-center 2-electron bonding 604
centroid path integral approach 1183
cesium perfluorooctane 233
m/p-$CF_3C_6H_4CHClCH_2Cl$ 578
$CF_2=CCl_6H_5$ 575
$CF_2=Cl_2$ 573
CF_3COOH 613
charge carriers 714ff.
charge interactions 1079f., 1096, 1098
charge relay system 1460
charge separation 506
charge transfer 417
– intramolecular 420, 445
– metal-to-ligand 530f.
CHARMM program 1219, 1291
– /MOPAC 1227
Chaudret 630
$C_6H_5CHBrCH_2Br$ 580
$C_6H_5CHBrCH_2Cl$ 580
$C_6H_5CHBrCH_2F$ 580
$C_6H_5CH(CF_3)_2$ 571
$C_6H_5C(CF_3)=CF_2$ 573
$C_6H_5CH(CF_3)CF_2OCH_3$ 573
$C_6H_5CH(CH_3)CH_2Br$ 577
$C_6H_5CHClCF_3$ 571
$C_6H_5CHClCF_2Cl$ 577
$C_6H_5CHClCH_2Cl$ 579
CH_2Cl_2 523, 533
$CHCl_2CF_3$ 573, 580
chelate interaction 958f.
chemical step 1316
chemical reactions 833f., 869, 875ff., 1175, 1227, 1447
– addition 1001, 1107
– bimolecular 834, 836, 843ff., 887ff.
 – Claisen-type 952
– carbocation-nucleophile addition 958
– catalysis 965ff., 980ff., 988ff.
– catalysis, acidic 967, 982f., 993ff., 998ff., 1089, 1094
– catalysis, base 966, 983ff., 999ff., 1004ff., 1089
– cyclization 985, 1001, 1007
– dehydration 1107, 1188
– elimination 952, 1107
 – E2 1006
– fragmentation 1018
 – α-β- 1014
– 1,2-H-shift 1014ff., 1031, 1107, 1477
– 1,5-H-shift 1017f.
– 1,1-proton transfer 1108, 1139, 1157
– hydration 1107
– hydride transfer 1107, 1187, 1230
– hydrolysis 976, 882ff., 976ff., 988ff., 993ff., 1002, 1005ff., 1086ff., 1101ff.
– inversion 883ff.
– intramolecular 987ff., 993ff., 998ff., 1002ff.
 – cyclisation 987
– mechanisms 835, 981ff., 1006
 – acetal cleavage mechanism 991
– nucleophilic 958, 984ff., 993ff., 998ff., 1089
 – S_N1 994
 – S_N2 984f., 994, 1008, 1187
 – S_Ni 985
– oxidation 1013, 1328, 1353
– sterically hindered 876ff.
– transesterification 1086
– unimolecular 1242f.
chemisorption 754ff., 766
chirality 433, 448, 939
chlorine 175, 184, 849ff., 853ff.
chloroacetate 1293
p-chlorobenzyl alcohol 1267
4-chlorobutanol 985
chlorodecanes 126, 130
1-chloro-2,4-dinitrobenzene 1383
chloroethene 860
1-chlorohexane 126
1-chloropentane 126, 129
chloroplasts 1499
2-chloropyridine 546
cholesterol 1476
$CH_3OCF_2CHCl_2$ 573
$CH_3OH_2^+$ 692
chromophore 260, 266, 367, 378, 435, 448, 455, 528, 991, 1512
chromium 771, 798ff., 805ff.
chromous acid 278
Chudley-Elliott model 801ff.
circular dichroism (CD) 1085, 1142f., 1149ff., 1163
– magnetic 245
classical mechanics 639

m-ClC$_6$H$_4$CHBrCH$_2$Br 578
m-ClC$_6$H$_4$CHClCH$_2$Cl 578
m/p-ClC$_6$H$_4$CHClCH$_2$F 578
CLIO 69f.
Clostridium stricklandii 1157
clusters 689
– of metals 774
cluster and slab approaches 758
CO 773, 777f.
CO$_2$ 431, 777, 936, 1188
Co^{1+}, Co^{2+}, Co^{3+} 1473, 1476
Co^{2+} 1140, 1152
co-adsorbed molecules 771
coalescence 623
cobalt 771, 810
coenzymes 1064ff.
– B$_{12}$ (AdoCbl) 1473ff.
– 4-chlorobenzoyl-CoA 1131
– 4-(N,N-dimethylamino)cinnamoyl-CoA 1128
– enoyl CoA 1128f.
– flavin adenine dinucleotide (FAD/FADH$_2$) 1039f., 1065, 1107, 1426, 1432ff.
– flavin adenosine mononucleotide (FMN/FMMH$_2$) 1039f., 1064, 1113, 1350, 1354
– 4-fluorobenzoyl-CoA 1131
– N^5,N^{10}-methylene-5,6,7,8-tetrahydrofolate 1322
– methylcobalamin (MeCbl) 1473
– nicotinamide 1048ff., 1060ff., 1071
  – adenosine dinucleotide (NAD$^+$/NADH) 212, 1038f., 1055, 1065, 1157f., 1165, 1217ff., 1265, 1325, 1393ff., 1425
  – adenosine dinucleotide phosphate (NAPD$^+$/NADPH) 1038f., 1297, 1301, 1322, 1348ff., 1393ff., 1439ff.
– 4-nitrobenzoyl-CoA 1131
– pyridoxal 5' phosphate (PLP) 1139ff., 1151f.
– quinone 1039, 1041, 1068
  – cysteine tryptophyl (CTQ) 1041
  – lysine tyrosyl-(LTQ) 1041
  – pyrolloquinoline (PQQ) 1041, 1069
  – 2,4,5-trihydroxyphenylalanine (TPQ) 1041, 1273
  – tryptophan tryptophyl (TTQ) 1041
– thiamine diphosphate (ThDP) 1419ff.
  – Mg^{2+}-, Apo- 1426
complex 432, 444, 451ff.
– amidinium-carboxylate interfaces 492, 494, 527ff., 554
– DNA-acrylamide 492, 496
complex, encounter 960

complex, electron and proton donor 1075
– guanidinium-carboxylate 528
– dihydrogen 603ff., 640
  – elongated 409
complex, macromolecular 1378f.
complex, metal 480, 1064
complex, Michaelis 1154, 1354f., 1399ff., 1406, 1429, 1448f., 1463
complex, oxygen evolving 551
– phenol:py 544
complex, precursor 509
– stoichiometry 225ff.
complex, thymine-acrylamide 496
– of cyclodextrin 225ff., 230ff.
complex, transition 513ff., 519, 603ff.
– coordination sphere 603ff., 633
– (μ-η^2: η^2-peroxo)dicopper(II) 514
– bis(μ-oxo)-dicopper(III) 514
– Cr 615
– Fe 613, 621, 627
– Fe^{IV}=O 515ff., 549f.
– Fe^{III}–OH 550f.
– Fe^{III}=O 550
– iron bi-imidazoline (Fe^{III}Hbim) 492f., 513f.
– [Fe(H$_2$)H(dppe)$_2$]$^+$ 623, 632
– [Fe(H)$_2$(H$_2$)(PEtPh$_2$)$_3$] 625
– [FeH$_2$(H$_2$)(PEtPh$_2$)$_2$] 632
– cis-[FeH(H$_2$)(L)$_4$]$^+$ 624
– PFe^{III}=O, P = porphyrin 550
– [(N$_4$Py)Fe^{IV}=O]$^{2+}$ 514ff.
– Ir 613, 619
– IrClH$_2$(H$_2$)(P^{i}PrR$_3$)$_2$ 622, 632
– [Ir(H$_2$)H(bq)(PPh$_3$)$_2$]$^+$ 623
– IrH$_2$X(P^{t}Bu$_2$Ph)$_2$ 626
– IrXH$_2$(H$_2$)(PR$_3$)$_2$ 627ff.
– [TpIrH(H$_2$)(PR$_3$)]$^+$ 624
– [Mn(CO)(dppe)$_2$(H$_2$)]$^+$ 620
– Mo 606, 615, 617, 624f.
– MoH$_2$(CO)(Et$_2$OCH$_4$PEt$_2$)$_2$ 619
– [Cp*MoH$_4$(H$_2$)(PR$_3$)]$^+$ 624
– [Cp$_2$MoH$_3$]$^+$ 628
– Me$_2$X(C$_5$R$_4$)$_2$Mo(H$_2$)]$^+$ 628ff.
– Nb 615, 619, 629
– Os 627
– [Os(H$_2$)(ethylenediamine)$_2$(acetate)]$^+$ 605
– trans-[OsCl(H$_2$)dppe)$_2$]$^+$ 609
– OsH$_3$X(P^{i}PR$_3$)$_2$ 626
– polyhydrides 619
– [ReH$_9$]$^{2-}$
– [ReH$_8$(PR$_3$)]$^-$ 606
– [ReH$_4$(CO)(H$_2$)(PR$_3$)$_3$]$^+$ 624

- $[Re(H_2)(H)_2(PMe_2Ph)_3(CO)]^+$ 625
- $[ReH_4(CO)L_3]^+$ 628f.
- $[Rh(cod)(dppb)]^+$ 664
- $[(bpy)_2(py)Ru^{IV}=O]^{2+}$ 514, 517
- MeC(O)O-[Ru]-Y   540ff.
- CpRuH(PP), PP = diphosphine   613
- $[Cp^*Ru(\eta^4\text{-}CH_3CH=CHCH=CHCOOH)]$ $[CF_3SO_3]$ 662f.
- $[CpRu(H_2)(dppm)]^+$ 619
- $Tp^*RuH(H_2)_2$ 631
- $Cp^*RuH_3(PCy_3)$, $Cp^* = C_5Me_5$ 622f.
- $RuH_2(H_2)_2(PCy_3)_2$ 632f.
- $(L_2)(H_2)Ru(\mu\text{-}H)(\mu\text{-}Cl)_2RuH(PPh_3)_2$ 671f.
- trans-$[Ru(D_2)Cl(dpp)_2]PF_6$ 671ff.
- Ru   619ff., 627
- $[RuH(H_2)(dppe)_2]^+$ 621
- $[RuH(H_2)(CO)_2(PR_3)_2]^+$ 624f.
- $[RuH(H_2)(PR_3)]^+$ 628
- $[RuH_2(H_2)_2(PCy_3)_2]$ 630
- ruthenium polypyridyl   492ff., 524, 530f.
- $Cp^*_2Zr(H_2)$, deuterated   625
- SISHA interactions   632
- Ta   615, 629
- tris-bipyridine   497, 538 $[Ru(bpy)_3]^{2+}$ 529ff.
- W   606, 615, 617, 625
- $W(CO)_3(PR_3)2(H_2)$ 618, R=Cy, iPr   603ff., 609f., 621ff., 674ff.
- Zn^{II} porphyrin   523ff., 528ff.
- Zn^{II} tetraphenylporphyrin   533f.
- compression variable   1058ff.
- comproportionation reaction   495
- computer simulations   1171ff., 1193
- condensed phase   597ff., 864, 867ff., 1241ff.
- Condon field emission   653
- configuration   72, 372, 858
- boat-shaped   15 (tropolon)
- configuration interaction singles methodology (CIS)   21
- confined geometry   223ff.
- conformation   228, 1371ff., 1394, 1447
- base-on, base-off   1474
- β-barrel   435f.
- in alkane radical cations, alkanes   107f., 119f., 122, 128ff.
- bowl-shape   1462
- helical   1027ff., 1080ff., 1097, 1102, 1112, 1370f.
  - wheel   1082
- helix-loop-helix   1080, 1086, 1089, 1096, 1378ff.
- conformation, β-sheet   1027, 1029, 1100, 1349, 1386, 1444

conformational
- changes   1361, 1367, 1378
  - dependence   213
  - during H transfer reactions   197ff., 203ff.
- E- and Z-isomers of vinyl ethers   575
- conformational flexibility of glycopyranoside units   231
- conformers   1362f.
- continuum theory   499, 539
- copper   757, 759, 764, 772ff., 707, 800, 812, 1017, 1028f.
- corrin   1473, 1480
- corner cutting   1209, 1215ff., 1485
- corrphycene   245ff.
- Coulomb cage   1511ff., 1519
- Coulomb cage, intra-   1510ff.
- Coulomb energy   163
- Coulomb field   124f.
- Coulomb interaction   445
- coumarins   230ff., 448
- CPMD   see Car-Parrinello molecular dynamics
- Cram boxes   223
- criss-cross mechanism   1123
- cryogenic matrices
  - chlorofluorocarbons   107f.
    - CCl_3F 105, 115ff., 129
    - CCl_2FCF_2Cl 122
  - electron acceptors
    - chloroalkane   123ff.
    - CO_2 123ff.
    - SF_6 122
  - zeolites   122f.
- crystal structure
  - of enzyme complex   1153
  - orthorhombic   289
- crystal structure, single   794
- CRYSTAL   286f.
- $CsHSO_4$ 732, 749
- Cu^{2+} 1279
- phenoxyl radical   1278
- cubic force constants   470
- Cumene   517f.
- cyanoalkanes   962
- 3-cyanomethyl-4-methylthiazolium cation 961ff.
- cyclic dimers   204, 462, 466ff., 523ff.
- cyclodextrin (CD)   223ff., 228ff.
- structure   224
- 1,3-cyclohexadiene   359, 519
- cyclohexane   350, 354f., 370, 519
- cyclohexene   518
- 2-cyclohexenone   1348ff.
- N-cyclohexylformamide (CXF)   1397

cyclopentene 689
cyclopentadiene 519
cyclopropane 876f., 882
cysteamine 1025f.
cysteine (Cys) 1023, 1027, 1086, 1145, 1159, 1161f.
cytochrome 1342f.
cytochrome c 1372
cytochrome oxidase 528, 1499, 1516ff.
cytochrome P450 549f.
cytoplasma 1515
cytosine 525, 542, 548, 934

d

de Broglie 1256
– wavelength 765
Debye-length 727
Debye-Smoluchovski equation (DSE) 391, 422f., 448ff., 1499, 1510f.
cis-decalin-d_{18} 112ff.
decane 129
decarboxylation 1018f., 1048, 1085f., 1130, 1301, 1419, 1425ff.
decarboxylation, oxidative 1432
DeDp-ApAe systems 513, 524, 527, 531ff.
De-[H^+]-Ae systems 528
deformation 932
degenerate four-wave mixing (DFWM) 5, 14
degree of freedom (DOF) 80f., 84, 88, 223f., 275, 643, 837, 841, 864, 1008, 1079, 1213f., 1232, 1245f.
dehydrohalogenation 576ff.
– alkoxide promoted 579
π-delocalization 566, 580
denaturing 1373, 1421
dendrimers 240
density functional theory (DFT) 8, 13, 41ff., 205, 245, 262, 286f., 296, 360, 493ff., 571, 627ff., 689, 692, 696ff., 704f., 758, 777, 930, 1174, 1291
density matrix 648, 661, 666f., 668
– evolution 1196
– theory 649ff.
density operator 648, 659, 666
deoxyadenosine 542, 1477
5′-deoxy-5′-adenosylcobalamin 1023
deoxyuridine 542, 1020f.
2′-deoxyuridine-5′-monophosphate, cyclic (cUMP) 1322
depolarization 259ff.
deprotonation 229, 496, 949ff., 960, 965ff., 1032, 1158, 1421, 1427
– energy 687
– rate 1422ff.

designed catalyst 1085ff., 1100
desorption 782
– kinetics 773
– spectroscopy (TDS) 755, 776f., 780
desorption, thermal 775
deuterium 764f., 767
Dewar-Chatt-Duncanson model 604
DFT see density functional theory
DFWM see degenerate four-wave mixing
DHAQ 365
diabatic
– curves 588
– electronically 314, 327, 330f.
– states 1173ff.
diamond structure 737
diarylformamidines 204
dibenzo-tetraaza[14]annulene (DTAA) 171, 185f.
Dieckmann condensation 1130
dielectric boundary 1501
dielectric constant 232, 377, 511, 529, 723
– electronic 480
– inertial 480
dielectric continuum 480
– of the environment 486ff., 936
dielectric environment 505f.
dielectric relaxation 283
1,1-difluoroalkenes 573ff.
β-β-difluorostyrenes 575
diffusion 717, 787ff.
– barrier 764f.
– control 976f., 1014
– coefficients 720, 739, 745, 748f., 765, 790, 795ff., 810, 1503
– equation 741
– energy 758, 764, 773
– long range 789, 817
– mechanism 770, 788, 801ff.
– path 792
– tracer 792
diffusion, anomolous 811
diffusion, self- 727, 793
dihydroanthracene (DHA) 513ff.
7,8-dihydrofolate 1322, 1398, 1439f., 1440
dihydrogen (H_2) 615ff.
– bridging (η) 607, 609, 615ff., 623ff., 632, 640
– cis-interaction 612
– cleavage 608, 610ff.
– deuterated 639
– dynamic behavior 603ff., 633
– exchange 623ff., 627ff., 675
– pK_a 612
– physisorbed 604

dihydroxyacetone phosphate (DHAP) 955, 967, 1088, 1101, 1109, 1118, 1157
diketopiperazines 1024f.
diiron cofactor 549
diiron metalloprotein 1087
dimethylamine 205
p-(dimethylamino)benzaldehyde (DABA) 1397
8-dimethylamino-1-naphthol derivatives 992, 995f., 1004
N,N-dimethyl-1-aminopyrene (DMAP) 382
5,6-dimethylbenzimidazole 1474
dimethylbutane 517
1,3-dimethylcyclopentadiene 689
3,6-dimethylene-cyclohexa-1,4-diene 689
N-N′-di-(p-F-phenyl)amidine (DFFA) 194, 201
1,3-dimethyl-2-phenylbenzimidazoline derivative 1061
3,5-dimethylpyrazole (DMP) 172, 190f., 193ff.
3,4-dinitrobenzoic acid 523f., 529ff.
2,4-dinitrophenyl triester 984
dioxane 173, 208f., 237f., 992
dipeptides 1014
diphenylamine 888
3,5-diphenyl-4-bromopyrazole (DPBrP) 190, 193ff.
3,5-diphenyl-4-pyrazole (DPP) 172, 191, 193f.
dipicolylamine ligands 545
dipole 266f., 527, 898, 1178
– moment 59, 63, 287, 1082
– relaxation 620, see relaxation
dipole-dipole interactions 462, 791
Dirac exchange interaction 644f.
direct methanol fuel cells (DMFCs) 709, 777f.
discrete molten globule states 1373f.
dispersed polaron 1181
dissolution 753
distance
– d_{HH} 630
– continuum 605f.
– donor-acceptor 366ff., 484, 493, 499, 548, 1198, 1225, 1230, 1326, 1332f.
– off-pathway 548
– in crystals 810
– equilibrium 277f., 289, 291
– N–N in porphyrins 245, 249ff., 255
– Om-H in zeolites 692
– O–H in water clusters in zeol 696ff.
– O–O 919f., 940, 1512
– shrinkage 231
– reduction 362ff.
– transfer 511, 754

distant residues 1452
– pattern 787f.
1,4-dithiothreitol 1020ff., 1025
di-tertbutyl-2-hydroxyphenoxyl radical 173
distortion 816
DMSO 386f., 399, 407ff., 431f., 542, 544, 950f., 1116
DNA 232, 240, 541, 752, 1374, 1377ff.
– damage 1028
– repair 934
dividing surface 833, 843, 869, 905
docking motif 1375f.
docking program (DOT) 1377
DOIT program 908, 923, 927, 938
doublet separation (DS) 4ff., 15, 17, 25, 27
DPBrP crystal 171
drugs 236
– metabolism 1020
dyads 536, 1131, 1430
dyes 225, 228ff., 1509ff.
– aromatic 410
– Dansyl 234
– organic 378

e

Escherichia coli 553, 1023, 1161, 1398, 1425ff., 1443f., 1450
echo attenuation 731
– spin- 790
Eckart barriers, one-dimensional 1293
effect of excitation 1315
effective molarity 987ff., 1001ff., 1008
Ehrenfest's theorem 358
Eigen cation see $H_9O_4^+$ ion
eigenfunction 61, 64, 639ff.
– nodes through C–C bonds 385
– symmetric (gerade) and anti-symmetric (ungerade) linear combinations 642
Eigen's scheme of H-transfer 165, 174
Eigen solvation core 454
Eigen-type mechanism 962
eigenstates 24, 89, 98, 359, 644f., 768f.
eigenvalues 61f., 768f., 802
eigenvectors 802, 1174
Eigen-Weller model 444, 451f.
elastic after-effect see Gorsky effect
elastic incoherent structure factor (EISF) 806
elbow plot 757
electrocatalysts 711
electrode 710ff., 771
electrolytes 709ff., 718, 732, 749, 771
electron
– lone-pairs 110
– π-systems 110

electron attachment 124
electron
– acceptor (Ae) 503, 507, 523
– bonding 480
– donor (De) 503, 507, 523
– relay 548, 552
– transfer (ET) 480, 503ff.
  – stepwise 511
electron capture 887
electron diffraction 42f.
electron-energy loss spectroscopy 776
electron paramagnetic resonance (EPR)
  spectroscopy 105, 107, 115ff., 124, 127f.,
  876, 883ff., 1030, 1384, 1477
– powder spectra 112ff.
– spin trapping 1017
– time resolved 265, 527
electron spin echo spectroscopy 265
electron-stimulated desorption ion angular
  distribution 780
electron transfer 378, 1048ff., 1178
electronic dipole transitions 371
electronic polarization 1233
electronic states
– ground state 4, 19, 38, 48f., 100, 225ff.,
  398ff., 417, 463, 922ff.
  – of porphycenes 247ff., 253ff., 268
– excited state 4, 15, 38ff., 48f., 225ff., 369,
  377ff., 396, 401, 420, 446, 544, 922ff.
  – of porphycenes 253ff., 258ff., 268
– models 445ff.
– transitions 385ff., 399, 404ff., 447ff.
electronic switching 363
electroosmosis 725
electroosmotic drag 711, 726f.
electrostatic interaction 1395
electrostatic potential 1504
electrostatic stabilization 966, 997
emission decay 229
emission spectroscopy 225, 228, 230, 357,
  425, 780, 1408
– time-resolved 534ff.
enamine 1000
energy
– binding 688
– deprotonation 687, 690
– electrical 709
energy hopping 224
energy of activation 156, 214, 394, 836, 866,
  869, 1255
enol 237f., 1000
– ethers 1002ff.
enolate 953f., 957, 966, 970, 1000, 1004,
  1110ff., 1134

enolization 1002ff.
Enterobacter cloacae 1461
enthalpy 1316
entrance channel 757
entropic activation 1404
entropy 213, 428, 978, 987, 1007, 1079, 1316,
  1382
– contribution in water, protein 234
enzymatic system 135, 835, 1296
enzyme 212f., 224, 232, 523, 548ff., 598,
  975ff., 1006, 1009, 1175, 1186f., 1196, 1209,
  1393, 1473ff.
– acetohydroxyacid synthase 1430
– acetylcholinesterase 1456, 1460ff.
– aconitase 1107
– actinomycin synthetase II (ACMSII) 1140,
  1156
– acylaminoacid racemase 1140, 1145
– acyl-CoA dehydrogenase 1107, 1114
– acyl CoA desaturase 1255, 1329
– acylphosphatase 1456
– adenosine deaminase 1456
– S-adenosylhomocysteine hydrolase 1062
– alanine racemase 1113, 1139ff., 1166
– alcohol dehydrogenase (ADH) 135, 212ff.,
  1037, 1069, 1209f., 1217ff., 1230ff., 1247f.,
  1251, 1255, 1291, 1304f., 1325f., 1334,
  1370, 1384f.
  – horse liver (HLADH) 1218ff., 1265ff., 1341
  – thermophile (ht-ADH) 1266ff.
  – yeast (YADH) 1244f., 1264ff., 1341
– amino acid racemases 1107, 1140, 1151
– amino acid transaminases 1107
– amino dehydrogenase 1231f.
– δ-L-(α-aminoadipoyl)-L-cysteinyl-D-valine
  (ACV) 1140, 1156
– aspartate racemase 1140, 1145, 1159
– bovine serum amine oxidase (BSAO)
  1262ff., 1273ff., 1329, 1341
– carbohydrate epimerases 1165
– carbonic anhydrase 597, 1334
  – II 936ff.
  – III, human 1188ff.
– carboxypeptidase A 1456
– catalytic efficiency 1458
– catalytic power 1045, 1071ff., 1312f., 1341
– chymotrypsin 1456, 1460ff.
– citrate synthase 1107
– creatinase 1456
– diaminopimelate epimerase (DAP) 1140,
  1145, 1159, 1162
– dihydrofolate reductase (DHFR) 528, 868,
  1244, 1255, 1322, 1329f., 1334, 1398, 1403,
  1405, 1439ff.

- diol dehydratase   1475, 1478ff.
- dopamine β-monooxygenase (DβM)   1251, 1279f.
- dTDP-L-rhamnose synthase   1140, 1165
- enolase   1107ff., 1131ff.
- enoyl-CoA hydratase (ECH)   1109, 1116f., 1127ff., 1134
- ethanolamine ammonia lyase   1475, 1478f.
- fumarase   1107
- galactose oxidase   549, 1329
- β-galactosidase   1456, 1464
- α-glucoronidase   1456
- glucose dehydrogenase   1070
- glucose 6-phosphate isomerase   967, 1107
- glucose oxidase   1270ff.
- glutamate dehydrogenase   1297f., 1303
- glutamate mutase   1474ff., 1487
- glutamate racemase   1140, 1145, 1159ff.
- glutathione transferase (GSTM1)   1383
- glycerol-3-phosphate dehydrogenase (G3PDH)   1403ff.
- glycosidases   993
- haloalkane dehalogenase   1456, 1464
- hydrogenase   679
- hydrolases   1455ff.
  - classification   1455f.
- inorganic pyrophosphatase   1456
- 3-isopropylmalate dehydrogenase   1217
- α-ketoacid decarboxylase   1048, 1432
- keto-L-gulonate 6-phosphate decarboxylase   969
- ketosteroid enolase   956
- ketosteroid isomerase (KSI)   956, 1109ff., 1125ff., 1134
- kinetic control   1223ff.
- Lac-permease   1518
- β-lactamase   1456, 1461
- lactate dehydrogenase (LDH)   1062, 1209, 1223ff., 1394ff., 1405
- leucyl aminopeptidase   1456
- leukocyte elastase   1460, 1462
- lipoxygenase   492, 498f., 549, 1198, 1231, 1251
  - soybean (SLO)   1244f., 1255, 1263, 1271f., 1276ff., 1329ff., 1334, 1345
- lysine 5,6 aminomutase   1474f.
- lysozyme   988, 993, 998, 1179
- malate dehydrogenase
  - cytoplasmatic   1403
  - mitochondrial   1403
- malate synthase   1367
- mandelate dehydrogenase   1113
- mandelate racemase   956, 1004, 1108ff., 1116, 1131ff., 1140, 1145, 1152ff.
- mannose 6-phosphate isomerase   967
- methane monooxygenase   1278, 1329
- methanol dehydrogenase   1070
- methylamine dehydrogenase   1251, 1255, 1329, 1334
- methylene glutarate mutase   1475
- methylmalonyl-CoA epimerase (MMCE)   1140, 1152, 1156f.
- methylmalonyl-CoA mutase   1329, 1474ff., 1488f.
- methylmalonyl-CoA racemase   1145
- microperoxidases   549
- mitogen activated protein kinases (MAPKs)   1375f.
  - kinase (MKK1)   1383
- monoamine oxidase B   1275f., 1329, 1352
- monooxygenases   1080
- morphinone reductase (MR)   1343, 1347ff.
- nitric oxide synthase   528
- pancreatic ribonuclease   1456
- papain   1456, 1461ff.
- pentaerythritol tetranitrate (PETN) reductase   1343, 1347ff., 1349
- pepsin   1456
- peptidylglycine-α-hydroxylating monooxygenase (PHM)   1244, 1251, 1255, 1279f., 1322, 1329
- phenylalanine racemase   1140, 1154f.
- proline oxidase   1157
- proline racemase   1109, 1112, 1139, 1145, 1157ff., 1302
- prostaglandin H synthase   1016
- protein-tyrosine-phosphatase   1456, 1461ff.
- purine nucleoside phosphorylase   1233
- pyruvate decarboxylase   1419ff.
  - multienzyme complex   1425
- pyruvate oxidase   1419, 1425f., 1432ff.
  - holo   1426
- D-ribulose 5-phosphate 3-epimerase   1140, 1145, 1157
- ribonuclease   985ff., 1458, 1463f.
- ribonucleotide reductase   549, 553, 1028, 1475, 1478
  - benzylsuccinate synthase   1020, 1022
  - class I (RNR1)   1016, 1024
  - class II (RNR1)   1023
  - class III (RNR), pyruvate formate lyase   1020, 1022
- ribunucleoside triphosphate reductase   1487
- sarcosine dehydrogenase   1255, 1329
  - heterotetrameric (TSOX)   1343, 1348, 1350f.
- serine proteases   1452, 1459, 1463

- serine racemase 1140, 1152
- serum albumens 1006
- succinate dehydrogenase 1107
- sugar epimerases 1157f., 1165
- superoxide dismutase 1504
- thermolysin 1456
- thymidylate synthase 1322, 1329f.
- transhydrogenase 1518
- transketolase 1424ff.
- trimethylamine dehydrogenase (TMADH) 1255, 1329, 1343, 1348, 1350ff.
- triosephosphate-isomerase (TIM) 956, 967, 1080, 1101, 1107ff., 1115ff., 1134, 1291, 1334
- tyrosyl-tRNA synthetase 1114
- UDP-N-acetylglucosamine-2-epimerase (UDP-GlcNAc) 1140, 1163
- UDP-galactose-4-epimerase 1140, 1157, 1406
- vicinal oxygen chelate (VOC) superfamily 1152
- xylose isomerase 868

enzyme catalysts 34, 53, 860, 949ff., 955ff., 970, 988, 1079ff., 1195, 1316, 1341ff., 1428, 1439ff.
enzyme motion 1447ff.
enzyme-substrate complex 3, 28, 987, 1114ff., 1242f., 1265, 1352ff., 1393ff., 1439
enzyme reactions 137, 498ff., 834f., 1110, 1139ff., 1172, 1241ff., 1311ff., 1440
- mechanism 1045ff., 1052ff., 1351, 1421, 1426ff., 1456ff., 1475ff.
  - double displacement 1457ff., 1462ff.
  - merge 1054
  - ping-pong 1271, 1458
  - single displacement 1457
equations of motion 910
equilibrium constants 378, 388, 964, 1056f.
- pK_a 565, 949ff., 957ff., 967, 977, 981, 1002, 1085ff., 1090ff., 1101, 1107ff., 1127f., 1190, 1194, 1352ff., 1440
equilibrium overshoot 1142f., 1159, 1162ff.
equilibrium perturbation 1146ff.
- washout 1148f., 1160
ESI (electron spray ionization) 1369, 1378, 1387
ESIPT see excited-state intramolecular proton-transfer
esters 976, 1086, 1089ff., 1097ff.
- acetate 980
- cationic 963
ethanol 235, 385, 427, 519, 983, 1217ff., 1305
ethanolic sodium ethoxide 573ff.

ethyl acetate 954f.
ethylbenzene 516ff.
ethylene glycol 1024
ethylene oxide 729f.
Euclidian action 904
eukaryotic cells 1515
Evans window 462
Evans-Polanyi relation 318, 510, 515, 517, 519, 588
excimer/exciplex formation 224
exit channel 757
excitation-emission cycle scenarios 387
excited-state intramolecular proton-transfer (ESIPT) 225, 228, 349ff., 357ff., 362, 366ff., 372f.
- multidimensional model 363ff.
Eyring equation 676ff.
Eyring's transition state theory 140, 217, 1245

f

face centered cubic center (fcc) 755, 797ff., 810
facile intramolecular site exchange 623
fatty acids 1475
- hydroperoxides 1276
- polyunsaturated fatty acids 1014, 1025, 1028
  - arachidonate 1015
  - linoleate, linolenate 1015
  - linoleic acid 498, 1276f., 1328f.
FDMR see fluorescence detected magnetic resonance
Fe^{2+} 1140, 1152, 1381
Fe^{3+}-OH 1276, 1328
Feit and Fleck approximate propagator 1212
FEL see free electron laser
FELIX (free electron laser for infrared experiments) 56ff., 66ff., 75
femtochemistry 223ff.
femtosecond studies 230ff., 471
Fe–N bonds 493
Fe(III)–OH cofactor/Fe(III) center 498
fermentative pathways 1475
Fermi resonance 64f., 460, 462f., 469
Fermi symmetrization rules 639, 678
ferricinium ion 1049
ferricyanide 1049ff.
ferrierite (FER) 704
Feynman's path integral formulation 1184
fibrin 1382
fibrinogen 1382
Fick's diffusion coefficient 793
Fick's first law 762

Fick's second law   741, 793, 796
field emission fluctuation method   763
flash spectroscopy   443
fluctuation dissipation theorem   1210
fluctuations   791, 803, 1333, 1342, 1447
– of the environment   1177ff.
fluorescein   1509ff.
fluorene   516, 518
fluoroethanols   591
fluorescence   226f., 239, 258f., 378, 383, 408, 434ff., 449, 523ff., 534
– decay   256, 262, 391, 935
– detected magnetic resonance (FDMR)   107
– detection   266
– dip IR spectra (FDIRS)   20, 41
– enhancement   224f.
– excitation spectroscopy   14ff., 253ff., 259, 924
– lifetime   230, 236, 259, 264, 389ff., 431
– polarization   259
– quantum yields   392
– spectroscopy   256f., 350, 387f., 397, 400
– time resolved   1411
– titration   387ff., 422
fluorine bond   426
flux   870
– correlation function formalism   1212
– equilibrium one-way   833, 862
– net reactive   833
fluxional behavior   606, 610, 631
fly and perch propagation   1501
Fokker-Planck theory   86
folding   1363
– unfolding   1363, 1372
  – foldons   1373
formaldehyde   519
formamide   386
– dimer   921
formamidine dimer   921
formamidine-formamide dimer   921
formic acid
– dimer   897ff., 913, 918ff.
– formic acid derivatives   983
– formic acid-formamidine dimer   921
Förster acidities   421
Förster calculation   422
Förster cycle   378f., 383ff., 389ff., 397f., 405f., 410
– energy gaps (FEG)   379, 383f., 406
Förster equation   417
Fourier transforms/transformation   355, 359, 370, 464, 467f., 471, 793f.
fractal network   810
fractional negative charge   965

Franck-Condon   527, 907
Franck-Condon factors   461, 473f., 481, 1179
Franck-Condon overlap   485, 1332, 1345f.
Franck-Condon region   349, 360, 363, 366f., 372
free electron laser (FEL)   56f., 66, 69, 75
free energy   310ff., 328, 949ff., 1072f., 1110, 1114, 1149, 1175, 1343ff., 1365, 1372, 1398
– activation   303, 310ff., 329, 339, 394, 512f., 1054, 1190
– bare   310f., 315
– contour plot   311
– correlation   393ff., 510f.
– equilibrium   1008
– excited state   383f.
– ground state   303, 307, 383
– perturbation (FEP)   1175, 1181f., 1185f., 1195
  – perturbation-umbrella sampling   1176, 1185f.
– relation (FER)   313, 315ff., 320ff., 332, 335, 597, 1187, 1193
  – models   584ff., 590ff.
  – relation, linear (LFER)   583ff., 594, 1171ff., 1185ff., 1199
– reaction   303, 307, 342, 496, 503, 1054, 1186
– relaxed   383ff.
– surface
  – ET diabatic   481ff.
  – reactant and product   486f., 510
– total   310, 394
Frenkel defects   804
freon   891
– mixture   185, 633
FTIR   see infrared spectroscopy
fuel cells   709ff., 732, 752, 771
fullerene (C_{60})   544ff.
FWHM (full-width half-maximum)   632

g
gallium   761
Gamow model   653
gas phase   383, 594, 679, 835
– clusters   431f., 446
– ionization   115
– reactions   834, 843, 863f., 870
  – bimolecular   843ff.
  – unimolecular   857ff.
– studies   580
gas phase spectroscopy   35, 41, 45, 53ff., 61ff., 256, 431f.
gas phase ionization   115
gas voltaic battery   710

Gaussian model 810
Gaussian cross correlation 357
Gaussian98 program package 360
geminate recombination 445, 448f.
generalized gradient approximations (GGA) 758
Gerlt-Gassman/Cleland-Kreevoy proposal 1114ff., 1126
giant planets 740
glass transition 188
glasses 256ff., 265f.
– organic 887ff.
– sol-gel 235, 240
D-glucanpyranose 224
glucosamines 1018f.
α-glucosyl fluoride 983
glutamate (Glu) 1102, 1117ff., 1127ff., 1131, 1145, 1156, 1161, 1179, 1423f., 1430f., 1474ff., 1519
glutamine (Gln) 1092, 1354f., 1397
glutathione (GSH) 1014, 1028
glyceraldehyde 3-phosphate (G3P) 1088, 1102, 1109, 1118
glycine (Gly) 1017ff., 1026f., 1029f., 1349
glycolytic pathway 1107
glycosides 993f.
glyoxylate 1367
germanium 757, 761
gold 757, 772f.
Golden Rule calculations 931
Gorsky effect 789f., 795, 798ff., 802
gramacidin S 1154
green solvent 419
GROMACS program 1511
Grote-Hynes theory 313
Grotthus-type hopping proton 452f.
Grotthuss mechanism 399, 418f., 443f., 454, 717, 1192, 1503
ground state destabilization 1316
group theory 639
Grove 710
guanine 525, 541f., 548, 932ff.
guanidinium chloride (GdmCl) 1372f.
guanidinium group 998
guest-host 224ff., 230, 236, 240
Guoy-Chapman theory 723
Guoy-Chapman-Stern theory 1504f.

h

hafnium 800, 805ff.
Hagen-Poisseuille 726
Hamiltonian 12f., 45, 59f., 81f., 84, 87ff., 489, 645ff., 657ff., 670, 813, 904ff., 1173, 1176ff., 1210ff., 1231, 1234
– model, 2D 275ff.
– pure spin 644
– Zwanzig 1210
Hammes-Schiffer and coworkers' model 541
Hammett
– equation 401ff.
– plot 568
– value 1095
Hammond see Bema Hapothle
Hammond postulate 318, 590, 1054, 1269
Hangman porphyrin models 550
harmonic approximation 834, 844, 1287
harmonic frequency 854
– calculation 287, 288ff.
harmonic twofold potential 643
Hartree
– time-dependent Hartree approximation 82
Hartree-Fock theory 689, 1481
– methodology, restricted (RHF) 21ff.
Haven ratios 722, 731, 790
heart 1223ff.
Heaviside step function 839
Hessian matrix 838f.
hexamethylphosphoramide 1062
H-bonded crystals 273ff.
– N–H–N fragments 277
– containing quasi-symmetric O H O fragments 277ff., 288ff.
H-chelate ring 361, 363, 366, 371
HCl 309ff., 394
$HClO_4$ 392, 425
$HClO_4 \cdot 2H_2O$ 288ff.
heat of solution 753
helix, coaxial 435
hemiporphycene 245ff.
hemoglobin (Hb) 1381
heptane 126ff., 173
hexagonal close-packed (hcp) 770
hexafluoro-2-propanol (HFIPA) 546
hexamethylbenzene 689
hexane 386
hexan-6-ol-1-al 1314
3-hexyne-1-ol 662f.
HF 308ff., 394, 593ff., 983
HgCdTe 749
high-resolution electron-energy loss spectroscopy (HREELS) 768, 776, 779ff.
Hilbert space 643ff., 659, 670
histidine (His) 936, 1026, 1029, 1083f., 1087, 1090ff., 1097ff., 1108, 1113ff., 1120, 1131, 1145, 1155, 1188, 1193, 1228, 1352ff., 1394, 1430, 1474
$H_3O_2^+(H_2O)_2$ 699
H_3O^+ 1500

$H_5O_2^+$ ion   286, 288ff., 454, 696ff., 714ff., 1500, 1504
– structures in crystal and in gas phase   289
$H_7O_3^+$ ion   697
$H_9O_4^+$ ion   454, 694, 715ff.
Hoffmann-Löffler-Freytag reaction   1018
hole refilling   763
hole-burning experiments   256
homogeneous media   714ff.
HOMO   *see* molecular orbital, highest occupied
homocysteine   1063
host   226, 230, 239
Hove correlation function   793
Hoz, Yang and Wolfe (HYW)   592, 595
HPLC   1369, 1378, 1381
Hückel aromaticity   184
human genome   1013
humidification   711
Hund   641f.
Huppert-Agmon model   428
Hwang Aqvist Warshel (HAW)   1186ff.
hydration   231ff., 655ff., 723ff.
– shell   717
hydride   771, 804ff.
– dynamic behavior   603ff.
– transfer   136, 835, 1393ff., 1450
  – 1,1-   1157
  – alternate routes   1055, 1301, 1391
  – formal   1037, 1048ff., 1052, 1071
  – mechanism   1050, 1069f.
  – model studies   1037ff., 1045, 1048ff., 1061ff.
– rate   1445
hydrocarbons   685, 704ff., 771
– ligands   516ff.
– saturated   106
– unsaturated   658
hydrodynamic theory under stick conditions   227
hydrogen
– dissociation   753, 775
  – homolytic   747
  – probability   771, 782
  – spontaneous   761
– freezing point   755
– quantum delocalization   770
hydrogen atom abstraction   512ff., 522, 553, 887ff., 1023, 1322, 1351
hydrogen atom transfer (HAT)   136, 223f., 503ff., 509ff., 555, 843, 1013ff., 1473ff.
– geometry   504ff.
– mechanism   504ff.
– synchronous   505f., 513, 522

hydrogen bond   426, 463ff.
– bihalide anions   60ff., 91ff.
– bihalide neutrals   95
– breaking   90ff., 511, 539, 565ff., 588, 878, 975, 981, 1000, 1362
– CHC   207
– compression   141, 144, 159, 196f., 200
– crystalline environment   274
– energy   715
– formation   775, 960f., 975, 981, 1007
  – C–H   957
– geometric hydrogen bond correlation   142f.
– geometry   141, 143, 193ff.
– length   193ff., 248
– low barrier (LBHB)   1195, 1459
– OHN   207
– OHO   207
– pre-equilibria   203ff.
– quasi-linear   274
– strength   53, 75f., 185, 188, 245, 274f., 280ff., 288ff.
– halides (Br, I)   59
hydrogen bonded chain   776, 783
hydrogen bonding   459ff., 717, 730, 997
– polarized   527ff.
– symmetric   523ff.
hydrogen-bonding wire   423
hydrogen bromide   849ff.
hydrogen cyanide   960
hydrogen diffusion   754
hydrogen halides   454
hydrogen jump   787ff., 801ff., 812ff., 823ff., 881ff.
– in binary-hydrogen systems   802
– long-range   793
– low-temperature hopping   821ff.
hydrogen migration   753
hydrogen subway   84
hydrogen motion   787ff., 807, 875
– quantum   812
hydrogen storage   825
hydrogen transfer (HT)
– between carbon and oxygen   565ff.
– in complex systems   212ff., 1318
– dynamics
  – ground state   368
  – fast   224ff.
  – ultrafast   79ff., 224, 230ff., 236ff.
– enzymatic   1209ff., 1241ff., 1316ff., 1419ff.
– exchange   641, 1361ff.
  – intramolecular   670
  – rates   1361ff., 1371
– excited state (ESHT)   410, 446
– from carbon   1311ff.

- four-state diagram 503, 510, 537
- heavy atom motions 174ff., 197ff.
  - environmental influence 187f.
- HH 153ff., 160, 162, 170, 177f., 196ff., 210f.
  - free energy diagram, free energy correlation 155ff.
- HHH 159ff., 210f.
- HHHH 161ff., 196, 215
- in condensed phases 135ff.
  - application 168ff.
  - in liquids 136, 188
  - in solids 136, 835
- intermolecular 137, 188ff.
- intramolecular 83ff., 166, 185ff.
  - excited state 443f.
- multiple 136, 1139ff.
  - concerted 151f.
  - stepwise 152ff., 159f., 161ff., 182, 185ff.
- nonclassical 1245
- on metals 751, 754, 756, 761ff., 775ff.
- photoinduced 460, 523f.
- probability 1262
- single 136ff.
- reactions 833
- with pre-equilibria 165ff.
hydrogenation 658, 664, 771
- single/two-step 666ff.
hydrogenolysis 771
hydrolytic cleavage 1455
hydronium ion 953ff., 967
hydroperoxides 1014
hydrophilic domain 723
hydrophilic pore 235
hydrophobic
- core 1098
- interaction 239
- pocket 1376
- pore 235f.
- residues 1082, 1091
hydrophobic/hydrophilic nano-separation 711
hydroxide 961f., 964, 967, 1188ff.
hydroxy dimers 467ff.
- deuterated 469ff.
hydroxy groups 754
hydroxyacetone 968f.
1-hydroxy-2-acetonaphthone 350, 353ff., 361, 366
1'-hydroxy-2'-acetophenone (HAN) 225ff., 364
hydroxyarenes 420f.
o-hydroxybenzaldehyde (OHBA) 350f., 353, 356f., 369

10-hydroxybenzo[h]quinoline (10-HBQ) 353, 360ff., 369
p-hydroxybenzylidenediazolone 435
3-hydroxybutyryl CoA 1109
3-hydroxybutyrylpanthetheine 1128
10-hydroxycamptothecin (10HCT) 434, 448
2-(2'-hydroxy-4'-methylphenyl)benzoxazole (MeBO) 206f.
2-(2'-hydroxy-5'-methylphenyl)benzotriazole (TINUVIN P) 353, 362, 366, 369, 473
9-hydroxyphenalenone derivatives 279
2-hydroxyphenoyxlradical 173
2-(2'-hydroxyphenyl)benzoxazole (HBO) 350f., 353ff., 360ff., 366
2-(2'-hydroxyphenyl)benzothiazole (HBT) 350f., 353ff., 360ff., 366ff., 463ff., 471ff.
1-(2'-hydroxyphenyl)-4-methyloxazole (HMPO) 236ff.
2-(2'-hydroxyphenyl)-5-phenyloxazole 362
2-(2'-hydroxyphenyl)-triazole 369
2-hydroxypropyl phosphate 997
1-hydroxypyrene (1HP) 382, 394ff., 396, 403, 405
8-hydroxypyrene-1,3,6-dimethylsulfamide (HPTA) 382, 401ff.
8-hydroxypyrene-1,3,6-trisdimethylsulfonamide 426
8-hydroxypyrene-1,3,6-trisulfonate (HPTS) 382, 289ff., 396, 401ff., 419, 446ff., 1502, 1509, 1515
- fingerprint spectrum 447
- photoexcited 1509
hydroxyquinoline 434
6-hydroxyquinoline (6HQ) 423, 434
6-hydroxyquinoline-N-oxide (6HQNO) 424f., 434
7-hydroxyquinoline (7HQ) 423
3-hydroxystilbene 420
4-hydroxystilbene 420
Hynes model 429
hyper-coordination 718

i

Ibach's and Lehwald's structure model 779f.
ice 737ff., 747, 749, 781
- crystal structure 737ff.
- deuterated 745
- high pressure phases 742ff.
  - VII 737f., 740, 746ff.
  - VIII 737f., 740
- hot 740
ICR see photodissiociation, – ion cyclotron resonance
image charge model 1501

imidazole   714, 720ff., 729f., 987, 1002, 1115
imidazolium cation   963
improved canonical variational theory (ICVT)   837, 847, 852f.
impurities   1066
IN6 spectrometer   820
Indene   516, 517
indigodiimin   170f.
inelastic electron tunneling spectroscopy (IETS)   770
inelastic neutron scattering (INS)   273, 283, 285ff., 291, 607, 616, 622, 627, 632f., 640f., 671, 675ff., 771f., 813ff., 817ff.
inertial polarization potential   481
inflammatory conditions   1017
infrared array detectors   749
infrared reflection   742ff.
– absorption spectroscopy (IRAS)   776f., 779
infrared spectroscopy (IR)   6ff., 16f., 20, 23, 25, 37, 63ff., 72ff., 79f., 248ff., 273, 278, 285ff., 609, 679, 688f., 694ff., 740ff., 1395ff.
– 2D   555
– FTIR (fourier transform IR)   19, 26, 45ff., 1369ff.
– attenuated total reflection   1371
– mid-IR   389, 471
– ultrafast   554
Ingold radical   173
INS   *see* inelastic neutron scattering
instanton   910f.
– action   908, 913ff.
interconversion   258, 266, 1088, 1162, 1223, 1365
– cis-trans   260, 263, 267
– trans-trans   259f., 263, 266f.
intermediates   982ff., 1069, 1294, 1462
– quinoid   1145
– zwitterionic   159, 161f., 184, 199, 206, 921f.
intermetallic compounds   788, 804ff.
internal conversion   368f., 410
internal friction   789
internal return   1142
– mechanism   566, 572, 575
interstitial sites   787, 797f., 804ff., 818
intersystem crossing   535
intramolecular hypervalent interactions (IHI)   632
inverse electrolysis   710
ionization potential   517ff.
ionizing radiation   *see* radiation chemistry
IR-PD   *see* photodissociation, – infrared
iridium   776, 779, 782
isobacteriochlorin   175, 184
isobestic point   535f.

isobutene   687ff., 704
isoinertial coordinates   836ff.
isoleucine (Ile)   1029, 1278, 1444, 1451
isomerization   876ff.
isopentane   877, 882
isoporphycene   245f., 932f.
isooctane   882
isotope effects   45, 797, 1093, 1115, *see also* kinetic isotope effects
– H and D   3, 7, 20f., 25f., 34, 65ff., 196f., 253ff., 277ff., 645, 678, 745f., 822, 896, 983, 1003, 1009, 1049, 1183, 1266ff.
isotope effects, equilibrium (EIE)   618, 1315ff., 1321
isotope effects, heavy atom   576ff., 746, 1165
– carbon   1301
– oxygen   19, 25f.
– metal   678
isotope effects, inverse   764
– on isotope effects   1297f., 1301
isotope effects, solvent   1349, 1463
isotopic exchange   1299
isotopic fractionation   154f., 157f., 161ff., 176
isotopic labeling   1253, 1365f., 1371, 1427
– ^{14}C   1248, 1305
– ^{1}H   1248
– ^{18}O   1396
– spin, site-directed   1384
isotopic labeling, mixed   1323f.
isotopic labeling, stereospecific   1297
isotopic scrambling   755, 779, 783
isotopologs   193, 530
isotopomer   530, 639, 679, 920, 923
iron   759, 810, 1017

j

Jacobian   848
jellyfish   435
Jencks   584
– anchor principle   1042
– libido rule   981, 989, 1002
JN   1092
JNII   1092, 1094
JNIIOR   1096
Johnston and Parr   585

k

Kamlet-Taft analysis   409, 426, 445
Karplus calculations   1124
Kemp elimination   1006, 1007
dTDP-4-keto-6-deoxy-D-Glucose
ketone (keto-type, keto-bonds)   225ff., 357, 363, 958, 962, 967, 1107
– mono-keto isomer, di-keto-form   370

$K_3H(SO_4)_2$ 291
kinematic coupling 275
kinetic acidity 566
kinetic analysis 1443
kinetic energy 838
kinetic experiments 509ff., 521f., 675
kinetic isotope effects 135f., 493, 513, 517, 522, 536, 539f., 835, 867
– H/D (hydrogen/deuterium) 140, 144, 148f., 151ff., 157f., 160ff., 169ff., 174, 176, 179, 189ff., 198ff., 205ff., 210ff., 251, 303ff., 315ff., 323ff., 333ff., 429, 497, 511f., 544f., 573ff., 622, 843f., 878, 884ff., 914ff., 931, 938ff., 1142ff., 1162, 1241, 1271, 1285, 1313ff., 1342ff., 1478ff., 1512f.
– competitive 1248f., 1273, 1305
– intrinsic 1320ff., 1442
– magnitude 1249
– masking 1215f., 1317
– measurement 1247
– model 1245
– kinetic isotope effects, multiple 1146ff., 1286ff., 1304ff.
– non-competitive 1247f.
– kinetic isotope effects, primary 154, 157, 162, 565ff., 574, 577, 858, 868, 1158ff., 1218f., 1246, 1251f., 1260, 1268ff., 1273, 1287, 1291, 1298, 1315
– kinetic isotope effects, secondary 154f., 157, 1052, 1060, 1247, 1252f., 1268ff., 1287, 1315, 1480
– kinetic isotope effects, temperature dependence of 1326
kinetics 975ff., 1045ff., 1108ff., 1164, 1242
kinetics, exchange 1421
– equivalence 979
kinetics, first order 876, 881f., 1046
kinetics, saturation 1087
kinetics, second order 571, 876, 1046
Kondo parameter 86, 816
KO-42 1087ff.
Kreevoy model 1054

l

α-lactalbumin 1374
lactate 1223
Lactobacillus fermenti 1161
Lactobacillus leichmannii 1478
Lactobacillus plantarum 1425f., 1432f.
Langevin equation 1210
Langevin dynamics (LD) simulations 1193
large-curvature tunneling (LCT) 842f., 850ff., 862f., 1486
Larmor frequencies 665
laser methods 509 (PCET)

laser-pulse 83ff., 362
– control 92
– infrared 83, 85, 91ff.
– UV 90ff.
laser-pulse, Ti:sapphire 352
laser spectroscopy 33, 39, 46, 75, 79ff., 423
lattice 816ff.
– cubic (C15-type) 798ff., 805ff.
– cubic (A15-type) 809
– hexagonal (C14-type) 805ff.
– hexagonal (C36-type) 805
lattice expansion 787
lattice-hydrogen interactions 764
Laves phases 798, 805, 823
least-action tunneling (LAT) 852f.
least motion 1009
Leffler 590 *see* Bema Hapothle
Lennard-Jones potential, H on metals 756
leucine (Leu) 1278, 1349, 1444ff.
leuco crystal violet 1292
level inversion 404
Lewis studies 420
ligand exchange 606
linear dichroism 248, 262, 1371
linear response approximation 1181
Liouville-von-Neumann equation 670
Liouville space 648ff., 659f., 670
Liouville super operator 648ff., 670
Liu-Siegbahn-Truhlar-Horowitz (LSTH) potential energy surface 843
lipid breakdown 1156
lipid vesicles 240
liquid chromatography (LC) 1369, 1378, 1387
liquids 459, 463, 732, 835, 860
liquid crystals 240
liquid films 752
London-Eyring-Polanyi-Sato 585
Lorentzian term 793ff., 801ff., 821
low-energy electron diffraction (LEED) 761, 765, 768, 772, 780
Löwenstein rule 685
luminescence 509
LUMO *see* molecular orbital, lowest unoccupied
lutetium 821
2,6-lutidine 984
lyonium/lyate species 975
lyoxide ion 955
lysine (Lys) 1006, 1083, 1085, 1092, 1099, 1120, 1133, 1143, 1154, 1188f., 1474f., 1519

m

magic-angle
– configuration 354
– emission decay 227

magnesium 810
– in enzymes 969
maleamic acids 988
mammalian brain tissue 1152
malonaldehyde (MA) 3ff., 14, 369
mandelamide 1001
mandelate 1109, 1131, 1152f.
mandelic acid 956
manganese 788, 805f., 817ff.
Mannich derivative 434
mapping potential 1175f.
Marcus
– barrier 489, 583f., 594f., 598
– intrinsic 958ff., 969, 1110
Marcus charge-transfer theory (MCT) 394ff.
Marcus-Coltrin method 834, 842, 848
Marcus equation 394ff., 592, 600, 958ff., 1188ff.
Marcus formalism 1110
Marcus-like model 1329ff.
Marcus parabolas 1176, 1198
Marcus relation 305, 310, 597f.
Marcus theory 417f., 493, 510, 522, 538ff., 587, 714, 962, 1049, 1054ff., 1185, 1214, 1261, 1332
MARI neutron spectrometer 819
Markov approximation 86
Markov process 793
mass-skew skew 849
mass spectrometry (MS) 432, 1277, 1367ff., 1375, 1378
– accurate mass tag analysis 1367
– C-terminal carboxypeptidase Y digestion 1367
– Fourier transform ion cyclotron resonance (FT-ICR) 1369
– MALDI- 1375ff.
– MALDI-SUPREX (stability of unpurified proteins from rates of hydrogen exchange) 1380, 1387
– MALDI-TOF 1367f.
– /MS 1367, 1386f.
– post-source decay 1367
– radio frequency (RF) 57
– tandem MS 57
master equation 801f.
matrix element 282, 642
Matthieu type differential equation 643
MCSCF calculations 918
mean-square-displacement of H 287
Me-BO 173
medical applications (porphyrins) 245
Meisenheimer complex 1131
membrane 240, 443, 552, 711, 724, 733, 752, 1330

– bilayers 1241, 1514
– transporter proteins 1371
membrane, biological 1499ff., 1504ff.
2-mercaptoethanol 1020
metal-hydrogen binding energy 758f.
metal-hydrogen potential 765
metals 787ff.
– amorphous 795, 810ff.
metal surfaces 751ff., 766ff.
– heterogenous 754
metallic fluid 740
metallic glass 812
metamorphosis 1152
methane 853ff., 1278
methanol 173, 210ff., 232, 350, 399, 402, 406, 425, 427ff., 565ff., 689, 691f., 711, 726, 775ff., 867, 887ff.
– deuterated 567
– dimer 689, 691f.
methanolic sodium methoxide 565
methionin (Met) 1026, 1029ff., 1448
methotrexate 1439
methoxide 214, 776ff.
methoxonium ion 214, 693f.
p-methoxybenzylamine 1275
methoxymethyl acetal 993, 993f.
1-methyl-1,10-dihydroacridan 1049
N-methylacridinium ion $(MA(H)^+$ 1053
N-methyl-1-aminopyrene (MAP) 382
N-methyl-hydroxyquinolinium species 423
4-methylimidazole 1087, 1093, 1096, 1101
methylchlorocarbene 860, 864
methylcyclohexane 172, 199
methylethylether 203, 205
2-methyl-6-hydroxyquinoline-N-oxide (MeHQNO) 424f.
3-methylindene 689
methylisocyanide 887
methylmalonyl-CoA 1156
3-methylpentane 237
3-methylphenyl-(2,4-dimethylphenyl)-methane 689
methyl salicylate 351
2-methyltetrahydrofuran 537
methyl viologen acceptor 497
Mg^{2+} 597, 1112, 1131ff., 1140, 1152, 1426
Mg-Mg breathing 868
micelle 229, 232ff., 240, 448, 752, 1509
Michael addition 1102
Michaelis mimics 1403f.
microcanonical emsemble 833
microcanonical optimized curvature tunneling (μOMT) 856, 862, 867
microcanonical variational theory 837ff.
microscopic reversibility 1049

microwave spectroscopy 13, 37
Miller-index 759
minimum energy for tunneling 135, 145, 150, 188f., 202, 213
minimum energy path (MEP) 835ff., 846ff., 854f., 863ff., 910ff., 1485f.
Mitchell hypothesis 1500
mitochondria 1499f., 1505, 1515
mixed isotopic exponents see Saunders exponents
MN 1092
Mn^{2+} 1140, 1152, 1426
MNDO study 591, 1401
MO see molecular orbital
molecular diffusion 746ff.
molecular dynamics (MD) 230f., 287, 490ff., 692, 696, 1449, 1518
– classical 287f.
– with quantum transition (MDQT) 1196
molecular hydrodynamic theory 231
mobile loop 1407
molecular mechanics (MM) 1171ff., 1193, 1228, 1233, 1291, 1334, 1342, 1349, 1440, 1480
molecular mechanics calculation 230
molecular memories 245
molecular orbital (MO) 583, 594f., 757, 1172
– computations 16, 20, 22f., 590ff.
– highest occupied (HOMO) 107, 513
  – of 2-naphtholate 420
– lowest unoccupied (LUMO) 245, 1022
  – of 2-naphtholate 420f.
– semi-occupied (SOMO) 107, 115, 1022
molecular pocket 223ff.
molecular structures, derivatives of 2- and 4- membered aromatic rings 380ff.
Møller-Plesset perturbation theory (MP2, MP3, MP4) 21ff., 42ff., 71, 249, 596, 689, 695ff., 705, 1481
molybdenum 807
moment of inertia 40, 1288, 1315
– (a-type, b-type) 40, 47
monensin 1514f.
monolayer 779, 774, 779
mono-p-nitrophenyl fumarate 1089, 1091ff.
Monte Carlo calculations 761, 791, 939
– diffusion (DMC) 60, 71ff.
More O'Ferrall 584
More O'Ferrall-Jencks diagram 592
Morse potential 1175
motion 1341
motion, coherent low-frequency 459ff.
motion, coupled 1313, 1449ff.
motion, gating 1345, 1350f.
motion, low-frequency 471ff.
motion, primary-secondary coupled 1323
mouse brain enzyme 1152
Mulliken charge transfer picture 308f.
multilamellar vesicles 1502f.
multimode (MM) methodology 59f., 69, 71, 75
multiphoton process 55f., 66ff., 76
multishell continuum model 231
muonium 843ff.
Murdoch 584
muscle 1223ff.
mutagenesis 1217ff., 1375f., 1446ff.
mutants 1029, 1099, 1102, 1108, 1132, 1145, 1154, 1161, 1189ff., 1263, 1267f., 1278, 1325, 1332, 1352ff., 1516
mutation 1383, 1430
– multiple 1445, 1452
– point 934

n

Na^+ 597, 1514f.
Nafion® 711, 724ff., 732
nanoscopic pools 232
nanostructure 228, 713
nanotubes 225, 240
NaOH 425
naphthalene 386, 420, 446
naphthalenediimide (NI) carboxylate 534ff.
naphthazarin 896, 899, 913, 926, 942
naphthols 445, 448
– 1-naphthol (1-NP) 228ff., 380, 386f., 398f., 401ff., 420ff., 428, 431f., 446
– 2-naphthol (2-NP) 380f., 385f., 389, 398f., 401ff., 420ff., 428
– 1-naphthol-3,6-disulfonate (1N3,6diS) 380, 401ff.
– 1-naphthol-4-chlorate (1N2Cl) 380, 403
– 1-naphthol-2-sulfonate (1N2S) 380, 403
– 1-naphthol-3-sulfonate (1N3S) 380, 403
– 1-naphthol-4-sulfonate (1N4S) 380
– 1-naphthol-5-sulfonate (1N5S) 380, 402
– 2-naphthol-6-sulfonate 449
– 5-cyano-1-naphthol (1N5CN/DCN1) 380, 401, 422, 449
– 5-tButyl-1-naphthol (1NtBu) 380, 402ff., 421
– 5-cyano-2-naphthol (2N5CN/5CN2) 380f., 401, 421ff., 428, 431f.
– 6-cyano-2-naphthol (6CN2) 422f.
– 7-cyano-2-naphthol (7CN2) 422f.
– 8-cyano-2-naphthol (2N8CN) 380f., 401, 421
– 5,8-dicyano-2-naphthol (DCN2) 401, 421ff., 428ff.

- 2-naphthol-6,8-disulfonate (2N6,8diS) 380f., 401ff.
- 5-methanesulfonyl-1-naphthol 448
- 5-methanesulfonyl-2-naphthol (5MSN2) 422
- 6-methyl-2-naphthol (2N6Me) 380f.
- 5,8-dicyano-2-naphthol (2N5,8diCN) 380f.
- 2-naphthol-3,6-disulfonate (2N3,6diS) 380f., 1508
- 3,5,8-tricyano-2-naphthol 421
naphthoquinone 547
narcisstic type of reaction 260
native chemical ligation 1086
near attack conformations (NAC) 1044, 1404
neopentane 877
Nernst-Einstein relation 719
Neurospora crassa 1380
neurotoxic effects 1029
neutralization 214
neutron diffraction crystallography 142f., 273, 277, 605, 695
Newtonian dynamics 1210
Newtonial principle of parsimony 1042
Ni^{2+} 1140, 1152
nickel 755, 758f., 764ff., 771ff., 776, 779, 781, 800, 809ff.
Nile Blue a 235
niobium 753, 762, 787f., 798f., 801f., 809, 814ff., 821ff.
NIR 352
nitrogen monoxide 1020
nitrogen dioxide 1020
2-nitropropane 1285, 1290
p-nitrophenyl acetate 1087
4-nitropyrazole 192
p-nitrophenyl esters 987
NMR lineshapes 655ff., 677, 1293
NMR, natural abundance techniques 1248
NMR pulse sequence 174
NMR relaxation 188
NMR relaxometry 923, 939
NMR spectroscopy 15, 38, 136, 142f., 187, 197, 201ff., 238, 248ff., 265, 534, 610f., 615ff., 629f., 639ff., 666f., 787, 791, 927, 1080, 1084, 1098, 1364, 1375, 1451
- ^{13}C 615, 628, 1300, 1406, 1421
- CLEANEX (clean chemical exchange) 1366
  - phase modulated (CLEANEX-PM) 1366f., 1386
- coupling constant 619, 662, 671
- downfield shift 693
- ^{1}H 671, 700ff., 951, 1092, 1115, 1428f.
  - pulsed field gradient (PFG) 716, 719ff., 731, 792, 798
- ^{2}H 623, 671ff., 677
- HSQC (heteronuclear single quantum coherence) 1366, 1374
- FHSQC (2D-fast) 1266f.
- liquid 615, 628, 251, 640, 646, 671
- NOE (heteronuclear Overhauser effects) 1365, 1427, 1448
- ^{17}O-PFG 716
- ^{31}P 628
- PFG 719f.
- solid state 251f., 604f., 621ff., 627, 671ff.
- TROSY 1367, 1386
NMR studies 283, 630ff.
- ^{15}N CPMAS NMR 251ff.
- magnetic field-cycling 283
m/p-$NO_2C_6H_4CH=CF_2$ 575
nonadiabatic coupling 483ff.
nonaqueous solution 428ff.
non-enzymic systems 1037
nonlinear vibrational spectroscopy 463
nonxollinearly phase matched optical parametric amplifiers (NOPAs) 352f.
norbornene 547
NovoTim 1.2.4 1088
nuclear motion 349ff.
nuclear spin 140
nucleoside
- diphosphates 553
nucleoside, deoxy- 553
nucleoside, pyrene-modified 542

o

octane 126ff.
2,3,6,7,11,12,17,18-octaethylcorrphycene 247
2,3,6,7,11,12,16,17-octaethylhemiporphyrene 247
2,3,6,7,11,12,16,17-octaethylporphyrene 247
olefin coordination 604
oligosaccharides 224
one-frequency models 1287, 1291
optical density 532
optical spectroscopy 38, 389, 927, 1147, 1219ff., 1407
optical trigger
- ultrafast 445ff.
- trigger pulse 443
optimized curvature tunneling (OCT) 842f.
optimized multidimensional tunneling (OMT) 856f.
Orange II 239f.
organelles 1499
organic ethers 614
organic solution 528ff.
organic solvent 399
organic solvent, aprotic 542

organometallic chemistry 640
orthoesters 982
ortho hydrogen 656
oscillator 461ff., 469ff., 642
oscillator, harmonic 81f., 84, 87f., 1211, 1258
– approximation 254, 258
– approximation, quantum mechanical 654ff.
– undamped 491
oscillator, one-dimensional symmetric double 138f.
overlayer-underlayer transitions 762
oxalamidines 155, 172, 177, 197ff.
– OA5 200f.
– OA6 200f.
– OA7 198f., 200f.
oxidative damage 541
oxidative cleavage 1322
oxidative stress 1014, 1020
oxidic support materials 771
oxo-acids 727, 732
oxolacetate 1085f.
oxygen
– carboxylic 1179, 1214
– glycosidic 1179

p
palladium 753, 755, 759, 762, 771ff., 776f., 779ff., 787, 796ff., 801, 810ff.
– film 796
PAM 1140
$(PAMA^+H_3PO_4^-)nH_3PO_4$ 727f.
para-hydrogen 655ff., 665, 669
para-hydrogen induced polarization (PHIP) 640, 657ff., 662ff., 668f., 679
– ^{1}H PHIP 662ff.
– ^{13}C PHIP NMR 664f.
– lineshapes 665ff.
para hydrogen and synthesis allow dramatically enhanced nuclear alignment (PASADENA) 656ff., 662, 667ff.
partition functions 140, 834, 844, 914f.
partition functions, reactant 836f.
path integral quantum TST (PI-QTST) 869
pathological processes 1013
Pauli exclusion principle 640, 670
Pauli matrices 660
Pauli repulsion 757
P branch 7, 14
PDLD/S-LRA approach 1189
PEF see potential energy function
penicillin 1156
1,3-pentadiene 858
pentafluorobenzene (PFB) 565ff., 576, 580

– tritiated 565
2,3,4,5,6-pentafluorobenzyl alcohol 1219
pentane 129
pepsin 1367
peptide 435, 1013, 1017, 1027f., 1079ff., 1134, 1322, 1363
– backbone 1081, 1361ff., 1370, 1446ff.
– EF hand 1380
– scaffolds 1079f., 1085, 1095ff., 1345
peptide, β-amyloid-(β-AP) 1029f.
peptide, unstructured 1361
Perdew-Burke-Ernzerhofer (PBE) 697f.
Perdew-Wang91 (PW91) 697f.
perfluorocarbons
– as cryogenic matrix 107
perfluorohexanesulfonyl-2-naphthol (6pFSN2) 425
permanganate, tetrabutylammonium (Bu_4NMnO_4) 514ff.
permeation methods 795f.
permutation operator 645ff.
peroxynitrite 1020
perturbation treatment 1174
PES see potential energy surface
pH 975, 990f., 999, 1062, 1087, 1090ff., 1101f., 1128, 1150, 1361ff., 1371f., 1400
– dependence 496, 538ff., 1352, 1362, 1423
– of a solution 397
– jump 378, 419, 445
phase diagram
– H-on-Ru(0001) 760
– of ice 737f.
– solid-solution 787
phase transition, 2D–3D 754
phenols 446, 519ff., 548
– fluorinated 426
– oxidation potential 520ff.
– oxidation rate constants 522
phenolate 981
phenol-ammonia clusters 410
phenylalanine (Phe) 1026, 1154f., 1218ff., 1376, 1435, 1444
1-phenylcloproanolate 1000
phenylethylbromide 1305
9-phenylfluorene 419, 570, 572, 580
– deuterated 572
– tritiated [9-PhFl-9-t] 565ff.
phenyl glycidyl ether 426
phonon-vibron coupling 764
phosphate 1001
– monoesters 995
phosphatidyl choline 1504, 1508f.
phosphatidyl inositol 1504
phosphatidyl inositol 4,5-diphosphate 1594

phosphatidyl serine 1508f., 1515
phosphatidylic acid 1508f.
phosphazene 169
2-phosphoglycerate 956, 1109, 1131ff.
phosphoglycolohydroxamate 1123
phospholipid 1501ff.
– bilayer (membrane) 1503
– black lipid membrane 1514
– head group 1502, 1504
phosphonic acid 714, 720, 729ff., 732
4′-phosphopantethein (PAN) 1155
phosphorane dianion 986
phosphorescence enhancement 224
phosphoric acid 711f., 714, 719ff., 732
phosphoric acid fuel cells (PACFs) 709ff.
phosphorolysis 1223
photoacidity 379ff., 389ff., 410
– determination 387
– effects of substituents 400ff.
– solvent assisted 377ff.
– ^1L_a ^1L_b paradigm 404ff.
– solvent effects 398ff.
photoacids 377ff., 383ff., 389ff., 397ff., 410, 417ff., 422ff., 433ff., 445ff., 1502
– biological 436
– in the gas phase 431
– pK_a values 401f., 408, 419, 421ff., 445
– reactivity 393
photoacids, cationic 398
photoacids, neutral 398
photoacids, super 417, 422ff., 426ff.
photoactivation 1370
photobases 377ff., 383ff., 390ff., 411, 429
photochemistry 224ff.
photochemistry, radical organic 507
photocleavage 224
photodeoxygenation 424
photodetachment
– of electrons 91ff.
photodissociation 427ff., 782
– infrared (IR-PD) 55ff., 61ff., 65ff., 75, 91, 100
– ion cyclotron resonance (ICR) 106
– rate, protolytic 434
photodynamics 228
photoexcitation 57, 225, 266, 432, 435, 496
photohydrolysis 426
photoinitiation 509
photoisomerization 224f.
photolysis 876, 882
photolysis, laser flash 1016
photon echo techniques 460
photons 55f.
– absorption 387
– emission 387
photophysics 224ff., 436
photosynthetic systems 245, 418
– bacterial 552
photosynthesis 1499
Photosystem II 538
phototherapy 236, 245
photovoltaic systems 245
phthalocyanine 169, 176
physisorption 753ff.
pigments of life 245
platinum 710ff., 763f., 771ff., 776ff., 779ff., 800
Platt's notation 385
PMP 194
Poisson-Boltzmann equation 723, 1505
Polanyi see Bema Hapothle, Evans-Polanyi
polar headgroup 229
polarity 232
polarity, solvent 325, 529
polarity reversal catalysis (PRC) 1021
polyacrylamide hydrogel 235
polybenzimidazole (PBI) 713, 727ff.
polymer electrolyte membrane or proton exchange membrane fuel cells (PEM) 709ff., 725, 731f.
polymeric fibrils 1378
polymers 240, 256, 685, 713, 727, 749
– films/sheets 256ff., 262ff.
polymers, hydrated acidic 723ff.
polymethylmethacrylate 267
polypeptide chains 523
polyphosphate 728
polystyrene 187
polyvinyl butyral 256, 258, 263f.
population dynamics 85ff., 390f.
porphine 928f.
porphycenes 171, 185f., 193ff., 245ff., 896, 929f., 842
– crystalline 252f., 255, 265
– nonsymmetric 256
porphyrinoids 267
porphyrins 135, 155, 169f., 184f., 193ff., 206, 245ff., 266, 523ff., 529ff., 896, 913, 1293
– analogs of p. 174ff.
– constitutional isomers (reshuffled) 245f.
– hydroporphyrins 184ff.
– inverted (N-confused) 245
– metal complexes 245f., 528ff.
– symmetric 174ff.
– unsymmetrically substituted 181ff.
porphyrins, free base 547
positive hole density 105, 107, 109

positive hole transfer   125
potassium acetate   395
potassium bicarbonate
potassium bromide   743
potassium formate   395
potential energy   760
potential energy curves   238
– enthalpy changes   591
potential energy diagram   753, 856
potential energy function (PEF)   3ff., 15f., 24, 180, 233, 1174
– double-well   4, 27, 40, 54, 79, 138, 143ff., 174, 275, 812
  – inter-well transition   282
  – intra-well transition   282
– single-well   53f.
potential energy surface (PES)   3, 5, 8ff., 14, 21, 23, 27f., 36, 39f., 58, 61f., 68, 75, 79ff., 87, 90ff., 223, 358, 350, 365, 372, 461, 583, 589ff., 694, 835, 849, 870, 880, 885, 901ff., 910f., 961, 1174ff., 1190ff., 1227, 1245
– LEPS   1293
– multidimensional   843
– SPES   854
– topology   366ff.
– two-dimensional   273ff., 280ff., 291ff.
potential energy well   756ff.
potential function   see potential energy function
potential of mean force (PMF)   863, 863
potential surface   see potential energy surface
power law potential   676
PQR rotational envelopes   6f.
praseodymium   810
primary metabolic steps   548
probability density   88, 768
probability distribution function, phase-space   833
probability flux correlation function   490ff.
proline   1157ff.
propane   882
propanol   430
propene   689
propinquity catalysis   see Bruice's proximity effect
Propionibacterium shermanii   1156
propionyl-CoA   1156
N-propylisonicotinamidine   533f.
protein   232, 236, 240, 435, 490, 523, 598, 733, 1013ff., 1028, 1171ff., 1195ff., 1343, 1361ff.
– design   1080ff., 1100
– dynamics   1209ff., 1313f., 1333, 1356, 1371ff., 1382ff., 1406, 1439ff.

– environment   1261
protein, flavo   1275, 1341ff.
protein, globular   1518
protein, green fluorescent (GFP)   435f., 455
– human serum albumin (HSA)   236ff.
– interactions   1374ff., 1379ff., 1452
– interactions, interloop   1446f.
protein-ligand recognition   236
– mobility   1361, 1382, 1385
– Monellin   234
protein, native   1363
protein, photosensor   443
– Sublitsin Carlsberg (SC)   234
– structure   1112, 1373, 1443ff.
  – tertiary   549
– water interface   1499
proteolysis   1367
proteome   1013
protic solvent   448
protodetriation rate   565ff.
proton
– abstraction   1108f., 1118ff., 1351, 1455, 1459ff.
– acceptor (Ap)   377, 444, 503, 507, 527, 1052
– donation   1455
– donor (Dp)   503, 507, 527
– flux   1507
proton affinity   689, 697ff., 704, 1110
proton, bridging   273, 278, 1458, 1462
– energy diagram for neutral alkyl radicals   109
proton-collecting antenna   1501, 1516ff.
proton-conducting channel   1519
proton conduction   709ff., 715ff., 728, 731f.
– confinement and interfacial effects   723
– mechanism   718
proton-coupled electron transfer (PCET)   479ff., 483ff., 503ff., 526ff., 552, 1015, 1017
– across interfaces   523ff.
– asynchronous   505f., 510f., 513, 519, 522
– bidirectional   508, 537ff., 549, 553
– bimolecular   531ff.
– concerted   503ff.
– in protein   498f.
  – temperature dependence   499, 537
– in solution   492ff.
– non-specific 3-point   538
– redox state   508, 523, 525, 537
– site differentiated   523ff.
– solvent dependence   511
– study methods   509ff., 514
– temperature dependence   512
– theoretical formulation   480ff.
  – dynamical effects   485

– types   507, 512, 523, 537f., 541, 543, 549ff.
  – non-specific 3-point   538ff.
  – site-specified 3-point   543ff.
  – symmetric   see H bond
  – polarized   see H bond
– unidirectional   508, 512ff., 553
– unimolecular   533
proton, defect   718
proton diffusion   737ff., 953, 960
– at high pressure   740
  – spectral analysis   745ff.
proton-dissociation   388, 392, 395, 398ff., 404, 1502f.
– lifetimes   400
proton donor   377, 444
proton donor-acceptor   714, 932, 960
– motion   483, 492, 498
proton dynamics   273ff.
proton, free   110
proton, geminate   377, 389
proton jump   700ff., 739f., 749, 1004
proton mobility   724, 727
proton polarizability   1503ff.
proton-proton coupling/correlation   895ff., 908ff., 918ff., 927, 941
proton pump   771, 1499
proton recombination   388ff., 448f.
proton repellent   1501
proton solvents
– covalently bound   728
proton sponge   245
proton transfer   105ff., 112ff., 122ff., 136, 213ff., 223ff., 236, 370, 424, 481, 686ff., 1048, 1114, 1171ff., 1230, 1455, 1499
– asymmetric   110, 123ff.
  – proton-donor site selectivity   124f.
  – proton-acceptor site selectivity   125f.
– at carbon   958ff., 965ff., 1107ff.
– bimolecular   443ff.
– by low-frequency mode excitation   279
– carbon acid to methoxide   565ff.
– concerted   418, 895ff., 910, 940ff.
– conduits   932ff.
– coordinate   274, 279, 283
– exited state (ESPT)   420ff., 427ff., 433ff., 443f.
  – in a box   435ff.
  – intramolecular (ESIPT)   443f.
– geometry   1505f.
– in vivo   433
– intramolecular   926ff.
– mechanism   1069
– methanol to carbanion   573
– methoxide promoted   576

– multiple   136, 370f., 895ff., 908ff., 919ff., 932f., 939ff.
– photoinduced   226, 430
– rate   398
– rate constant   229, 394, 909, 918, 932
– reaction   228, 377ff., 1185ff., 1428ff., 1499
  – activation energy   394
– stepwise   418, 511, 895ff., 910, 921, 928, 941
– symmetric   110, 115ff.
  – proton-acceptor site selectivity   120f.
  – proton-donor site selectivity   119f.
– thermoneutral   960ff.
– theoretical aspects in polar environment   303ff.
– theoretical simulations   583ff., 1171ff.
– to and from carbon   949ff., 960, 970, 1000
– ultrafast   349ff., 424, 453
proton translocation (PTR)   1171, 1193, 1199
proton transport, long range   714ff., 732
Pseudomonas putida   1151
Pseudomonas qetrolens   1151
pseudorotation   986
pulse radiolysis   1014, 1018, 1025
pump-control scheme   91
pump-dump scheme   83, 90ff.
pump-probe experiments   352, 360f., 447ff., 459, 463f., 466ff., 1407
purine   1439
purpurin   528ff.
pyranine   see 8-hydroxypyrene-1,3,6-trisulfonate
pyrazoles   135, 189ff., 721
pyrenols   445
pyridine (py)   545, 593, 689, 984
– substituted   1285
pyridinium   1062
pyridone group   434
2-pyridone-2-hydroxypyridine dimer   924f.
pyrimidine
– biosynthesis   1439
– nucleobases   542
pyrrole rings   245, 267
pyruvate   955, 1223ff., 1367, 1395ff., 1407f.
PZD2   1087, 1101

q

Q bands   245, 261f., 528
Q branch   14f.
– spikes   7, 19
Q frequencies   312
quadrupolar Pake pattern   672
quantum average   317, 324f.
quantum classical path (QCP)   1184f., 1196f.

quantum coherent vibrational dynamics 459
quantum correction factor 840
quantum dynamics 1209
quantum Kramers model 1209ff., 1231ff., 1342
quantum Kramers flux autocorrelation 1213
quantum mechanical activation barrier 1184
quantum mechanical exchange coupling (QEC) 619f., 629
quantum mechanical integrated reaction probability 148
quantum mechanical rotor, one-dimensional free 652
quantum mechanical studies 135, 143, 175, 217
quantum mechanics (QM) 639ff., 665, 1171ff., 1193, 1228, 1233, 1291, 1316, 1334, 1342, 1349, 1440, 1480
quantum mechanics, nuclear (NQM) 1177ff., 1195ff.
quasiclassical hybrid 837
quasi-elastic neutron scattering (QNS/QENS) 283, 632f., 787, 792ff., 801ff., 807ff., 821f.
quasiparticles 1184
quenching 225ff., 390ff., 523ff., 534, 544, 818, 1421
quinone 552
quinuclidines 943, 967
QSID (quadratic configuration interaction including single and double substitutions) 211, 596

r
racemization 1142ff.
radiation chemistry 106, 107, 1020
– γ-irradiation 106, 112ff., 119ff., 126f.
  – radiolytic process, mechanism 115ff., 123ff.
radicals 512f., 519, 542f., 544, 546, 553, 876ff., 1013ff.
– acridine 1052
– alkoxyl 1014
– alkyl
  – cations 105ff., 115ff., 123ff.
  – electronic absorption 106
  – neutrals 112ff., 123
  – paramagnetic properties 107f., 117
– α-(alkylthio)alkyl 1029f.
– bi- 542
– benzyloxyl 1015
– bis(trifluoromethyl)nitroxide 880ff.
– bromine 513
– carbon centered 1019f.
– chlorine 512f.
– cumylperoxide 521f.
– cyclohexydienyl 1025
– cyclopropyl 883f.
– cysteinyl 1016, 1023f.
– deoxyuridin-1-yl 1020
– 2,3-dimethyloxiranyl 885
– N-(6,6-diphenyl-5-hexenyl)acetamidyl 1018
– dioxolanyl 886
– 3,5-di-tert-butylneophyl 207ff.
– free 867ff.
– 2-hydroxyphenoxyl 208f.
  – 3,6-di-tert-butyl-2-hydroxyphenoxyl 208f.
– hydroxyl 1013, 1052
– iodine 512
– isobutyl 883
– methyl 887ff.
– 1-methylcyclopropyl 883f.
– nitrogen centered 1017
  – amidyl 1017ff.
  – aminyl 1017ff.
  – in biological samples 1017
– octamethyloctahydroanthracen-9-yl 881
– oxiranyl 884f.
– oxygen centered 1013
– phenoxyl 1020
– sulfur centered 1019ff.
  – cations, sulfide 1019, 1029ff.
  – sulfonyl 1027
  – thiyl 553f., 1019ff., 1027
– tetramethylgermacene 882
– trapping 888f.
– 2,4,6-tri(1′-adamantyl)phenyl 881
– tri-n-butyltin 882
– trifluoromethyl 883
– trimethyltin 876
– trimethylsilicon 876
– 2,4,6-tri-tert-butylphenyl 207ff., 876ff.
– 2,4,6-tri-tert-butylneophyl 876
– tyrosyl 538, 553, 1015f., 1023
radiolytic processes 106
Raman spectroscopy 45, 285, 288, 361, 464f., 468ff., 1131, 1395ff., 1398, 1476
– third-order (TOR) 555
Raman studies 473f.
rate constants 184, 188, 191, 197f., 210, 303, 337, 431, 837, 840, 867, 951ff., 958ff., 961, 964, 967, 1052, 1056, 1068, 1092, 1171, 1196, 1430
– first order 888ff., 987, 1046, 1463
– intrinsic 1361
– predictions 833
– pre-steady-state 1489
– pseudo-first order 166, 201, 544

– second order 949ff., 957, 967, 987, 1046, 1057, 1071, 1095
– third order 967
rate equilibrium
– correlations 961
– studies 1058
rate law
– second order 211
rate limiting step 1316
$Rb_3H(SO_4)_2$ 291
R branch 7, 14
reaction asymmetry 333, 341
reaction coordinate 505, 585, 587ff., 840ff., 908ff., 935, 841, 1316
– intrinsic 589
reaction dynamics 449ff.
reaction mechanism
– biochemical 419
– OA/RE type 633
reaction path 366, 373, 505
reaction progress
– coordinate 1058
– variable 1058ff.
reaction rates 388, 512, 869
reaction scheme 371, 484ff.
rearrangement 204, 225, 411, 1027, 1031, 1300, 1473
– 1,2- 1477
– electronic 385
– heavy atom 1457f.
– intramolecular, of H_2 623ff.
– of water molecules 452
– sigmatropic 858f.
recoverin 1379f.
redox reaction 515ff., 1057, 1062, 1067
reduced mass 274f., 838, 1325
reductive methylation 1322
refractive index 287f.
regio-selectivity 668
regio-specificity 666
Rehm-Weller equation 417, 585
relaxation 235, 648
– anelastic 787ff., 825
– dynamics 223
– molecular 223
– rate 82f., 651, 790
– reorientation 789
– spin-lattice 282ff., 623, 673ff., 790, 821ff., 1448
– spin-spin 1448
– solvent 138, 140, 427ff., 446
– time 79, 186, 253, 618
– vibrational 138
reorganization 499

– inner-sphere 493
– outer-sphere 482f.
repair reaction 1020
repulsive forces 758ff.
resonance effects (of substitutents) 961ff.
resonance frequency 791
resonance structure 1175
respiration 1223ff.
– anaerobic 1224
reversible oxidation 1325
dTDP-L-rhamnose 1165
rhodium 759, 764, 767, 776, 779ff.
Rhodobacter sphaeroides 1516
rhodopsin 1370, 1379
riboflavin tetraacetate 1067
ribonucleotide 1476
– in TTUTT 987
ribosome 1518
ribozyme 1241
ring opening 1006
RMS fitting error 72f.
RNA 985ff., 1028, 1087
ROESY 174
rotamer 1091
– keto 225ff.
rotation 225ff., 604, 609, 739
– dihydrogen 615ff., 627, 657
rotational constants 42f.
rotational diffusion 262
rotational states 7, 13f.
rotational transitions 19, 69
rotational-vibrational
– state 39, 43, 46f.
– lines 66
RRKM
– system 1314
– theory 857
rubredoxin, mesophilic 1384
rule of geometric mean (RGM) 157, 160, 190, 210, 1252f., 1266, 1285f., 1297ff., 1304ff., 1324
– breakdown 1298ff.
– RS (RGM and Swain-Schaad) exponents 1252, 1267
ruthenium 759f., 764, 770ff., 776f., 779ff.
– $Ru^{IV}O^{2+}$ 1051

S
S-824 1087
SA-42 1090
Saccharomyces cerevisiae 1421ff., 1427
saddle-point (SP) 4, 12, 21, 589, 687, 757, 833f., 844, 850ff., 858
– first order 895ff., 909, 913, 926
– second order 249, 899ff., 909, 913, 930

salicylic acid derivatives 988ff., 995, 998, 1002
salicylaldimine 88ff.
sapphyrin 267
sarcosine 1348
Saunders exponents 1304, 1306
scandium 788ff.
scanning tunneling microscope (STM) 752, 770, 772, 781
scattering law 287
SCF *see* self-consistent field methodology
Schiff base 554, 1069, 1401
Schönbein 710
Schrödinger equation 5, 21, 276f., 280, 327, 643f., 670
SCSAG method 849
SD *see* spectral doublets
self-consistent field methodology (SCF) 21, 926
semi classical method 8f., 21, 23, 36, 58, 136f., 938, 1179f., 1219, 1245, 1256, 1318, 1324, 1442, 1486
semi empirical method 837, 841, 1227, 1401, 1481
semiconductors 757, 761, 771ff.
semiquinone 546
serine (Ser) 1026f., 1152, 1376
syn-sesquinorbornene disulfone 931
SHAKE algorithm 1220
shift
– blue 227f., 230, 472
– red 232, 432, 460, 463
Sievert's law 810
Silicon 757, 761, 810f.
silicon-aluminiumphosphates (SAPOs)
– (H-SAPO-34) 686, 695ff.
siloxane backbone 732
silver 757, 772ff., 781
single crystal 759, 764, 802
single molecule spectroscopy 266
single photon counting detection 449
SiO_4 tetrahedron 685, 700
skew angle 849, 857
small-curvature tunneling (SCT) 842f., 848ff., 855f., 862, 867, 1486
Snoek effect 788f.
sodium bifluoride 285
sodium hydrogen bis (4-nitrophenoxide)dihydrage 278
solid acids 749
solute-solvent
– interaction 1174
– system 491
– terms 1192

solution 1171, 1176
solvation 305, 398, 417, 490, 1501
– dynamics 230f., 447
 – ultrafast 231
– layer 1503
– shells 451ff.
solvatochromism 426, 445
solvent chaperoned 418
solvent configuration 310
solvent fluctuation 1196
solvent mixtures 427ff.
solvent reaction coordinate 310, 330
solvent reorganization 955, 1197, 1513
– energy 1181ff., 1186, 1198
solvent-solvent terms 1192
SOMO *see* molecular orbital, – semi-occupied
Soret bands 245, 261f., 528
sorption energy 753
SP *see* saddle-point
spacer 532f., 713, 732
spectral doublets (SD) 13, 27
spectroscopic signatures 1404
spill-over effect 754, 773f.
spin 1/2 645f., 670, 790
spin 1 646f., 670
spin >1 790
spin −1/2 660
spin boson *see* dispersed polaron
spin-lattice correlation rates 923
spin system, 2 state 1232
staphylococcal nuclease 1367
Stark effect 13
static-secondary-zone result 866
steady-state 1243, 1294
– studies 230, 389ff., 432
– velocity 1242
Stern-Volmer quenching analysis 449, 531
stereochemistry 433, 940
stereoinversion 1152ff., 1158f., 1165f.
stereoselective reactions 662f.
stereoselectivity 433, 1020
stereospecific reduction 1322
stereospecificity 433
Stokes-Einstein relation 719
Stokes radius 1378
Stokes shift 236, 378, 408
stopped-flow methods 1351, 1355, 1434
– FRET 1322
strain 821
– tensor 788
stress 1224
structure activity relationships 996, 999
structure diffusion 716f.

structure reactivity studies   1059
Stuchebrukchov's model   1506ff.
styrene   1305
substituent effects   957, 958ff.
substrate activation   1395
substrate binding   1242
substrate transport   1241ff.
subsurface-site population   754
succinyl-CoA   1156
sugars   992, 1158, 1165, 1271, 1419
– ketol transfer   1424
surface coverage   768ff., 778
surface diffusion   761ff., 762, 773ff.
sulfonic acid groups   723
sulfuric acid   710
superacids   110, 686
– $SFO_3H\text{-}SbF_5\text{-}SO_2$ 692
supercritical fluids   419, 431
superionic fluid   740
supersonic jet studies   253
support materials (metals)   774
supramolecular chemistry   224
supramolecule   523, 528f.
surface hopping method   1196
surface orientation   759, 764
surface reconstruction   753f.
Swain-Schaad exponent   1259, 1275, 1280, 1285ff., 1290ff., 1304ff., 1318ff.
– kinetic complexity   1319
– mechanistic complexity   1294
– primary   1320
Swain-Schaad relation   320, 322ff., 333, 340ff., 565ff., 577, 914ff., 920, 928f., 1252f., 1266, 1318ff.
– primary   1320ff.
– secondary   1323ff.
swinging door   1142, 1157
symmetry
– effects (on NMR lineshapes)   655, 670
– in dihydrogen transfer   639ff.
symmetry groups
– C_2 19, 65f., 462
– C_{2h} 36, 46, 49, 260, 897, 921
– C_{2v} 5, 21ff., 260, 627, 897, 921f.
– C_4 939
– C_{4v} 767
– $C_{\infty h}$ 61
– C_s 262
– D_{2h} 36, 46, 248, 261, 920f.
– $D_{\infty h}$ 61
– E-type   767
– inversion operations   4
– G_4 4f.
symmetry related quantum effects   640

π-System   420, 1000, 1007
Szabo-Collins-Kimball model (SCK)   451
Szymanski and Scheurer   649

t

tantalum   753, 762, 788, 794, 798f., 801, 805ff., 814, 821ff.
tautomer   4, 23ff., 184, 188, 202, 610, 618, 926, 996
– amidine-carboxylic acid   529
– phototautomer   225ff., 266, 425
tautomerism   135, 154f., 174ff., 181, 189f., 198ff., 207ff.
tautomerization   15f., 19, 23, 27, 623ff., 934ff., 1124, 1427, 1461
– keto-enol   226ff., 237f., 366f., 372, 463, 471ff., 926, 976, 1439ff.
– kinetics   259f.
– in metal complexes   619
– in porphycenes   245ff., 251ff., 258ff.
– in quinolines   424
– rates   262ff.
temperature jump relaxation spectroscopy   1407ff.
tensor   647ff.
tetrachloroethylene   351
1,2,4,5-tetracyanobenzene (TCB)   545f.
tetrahedral jump mechanism   606
tetrahydrofuran   428
5,6,7,8-tetrahydrofolate   1439, 1447
1,5,6,6-tetramethyl-3-methylene-cyclohexa-1,4-diene   689
tetraphenylchlorin (TPC)   170, 183ff.
tetraphenylbacteriochlorin (TPBC)   184f.
tetraphenylisobacteriochlorin (TpiBC)   170, 183ff.
tetraphenoloxalamidine (TPOA)   198ff.
2,7,12,17-tetra-n-propyl-9-acetoxyporphyrene   256ff.
tetratoloylporphyrin (TPP)   174, 176ff., 185
TFXA spectrometer   818
thermal annealing   744ff.
thermal equilibrium   *see* canonical ensemble
thermodynamic acidity   580
thermodynamic cycle   379
thermodynamic miscibility   773
thermodynamic parameters   569f., 1056
thermodynamically symmetric reaction   317
thermodynamics (of PCET)   503f., 513ff., 521ff., 1231ff.
– pK_a value/E_{red} value   504, 508, 523, 527, 537, 554
– coupling   548
Thermotoga maritima   1374

Thermus thermophilus 1217
THF 172f., 203, 210, 225ff., 529
– d_8 672, 933
thiamine 958
thioacetylone 85ff.
thioester 1107, 1156
thiol 984, 1020
thiophenol 1017
thioredoxin 1087, 1101
thorium 810
Thornton *see* Bema Hapothle
threonine (Thr) 1026f., 1222
thrombin 1382
thrombomodulin 1382
through-atom axis (^1L_a state) 445f.
through-bond axis (^1L_b state) 445f.
thymine (T) 542, 934, 1028
tight-binding 758
time correlation function 490
TIP3P potential 1219
titanium 753, 761, 805, 810
titration 528 *see* fluorescence titration
– Mataga's titration treatment 544
α-tocopherol 1017
toluene 463, 516ff., 689, 877, 891
topoisomerase I inhibitors 434
topotecan 434
transition metals 772ff.
– reductive cleavage 1014
transition moment 258ff.
– directions 260ff.
transition path sampling 1209
transition state 513, 590ff., 909, 924, 938, 958, 1008, 1098, 1110, 1245f., 1394, 1442, 1483
– descriptors 1056
– inhibitors 1209ff.
– inhibitors, multiple 1073
– stabilization 1047, 1052, 1072, 1087, 1108, 1115, 1194, 1316, 1458
– structure 589ff., 919ff., 981ff., 994ff., 1251, 1268, 1280
– theory (TST) 309ff., 904, 918, 1056, 1074, 1186, 1195, 1241, 1287ff., 1299ff., 1326ff., 1513
　– zero-curvature tunneling (TST/ZCT) 1483ff.
transmission change 355f., 370
transmission coefficient 840f., 870
transmission factor 1196f.
trapping of hydrogen 817
– rate 795, 803
trifluoroacetic acid (TFA) 546
trifluoroethanol (TFE) 546
trifluoroethanolysis 983f.

6-tri-fluoromethanesulfonyl-2-naphthol 425
trihydrogen 617
– in complexes 624
trimethoprim 1439
trimethylamine 205, 691, 1348, 1353f.
N,N,N-trimethylammonium glycine methyl ester 954
triphenylmethane 516f.
tripodal phosphine 628
TRIS 1001
tritium 764f., 767
TRN *see* tropolone
trolox C 1014
tropolone (TRN) 3ff., 8, 13ff., 79
– diethyl acetale 982
– derivatives 3, 26f., 82
Trouillier-Martins pseudo-potentials 288ff.
tryptophan (W, Trp) 232, 540ff., 554, 1218ff., 1352ff., 1448
TSTH model 25
TTAA 171, 193ff.
tungsten 763f.
tunneling 136, 258, 303ff., 326ff., 333, 336, 340f., 364, 483, 499, 641, 783, 788, 812ff., 1007, 1187, 1195, 1217ff., 1241ff., 1255ff., 1279, 1285ff., 1475f.
– coenzymatic reactions 1060ff., 1068
– coherent 3ff., 13ff., 35ff., 138ff., 183, 642ff., 649ff., 675, 817ff., 1314
　– at higher temperatures 647
– hydrogen 817ff., 1285ff., 1341ff., 1347ff., 1439ff.
– DNA 33f.
– Eigenstates 21
– electron-proton 505
– fast 675
– frequency 147, 642, 645
– ground-state nuclear 1313
– hot bands 19
– phonon assisted 714, 797ff., 816
– intermediate (speed) 671ff.
– masses 145f., 149ff., 189, 205
– modes 906f.
– models 135ff., 251, 897ff.
　– one-dimensional 143ff.
　– two-dimensional 143ff., 505
– nonadiabatic 303, 307, 333ff., 429
– pathway 84f., 145, 156
– quantum mechanical (QMT) considerations 33f., 39, 48, 623, 678, 875ff., 884ff., 1341
– parameters 182, 188
– properties for carboxylic acids 35ff.
　– aromatic dimers 37
　– benzoic acid dimer 35ff., 48f.

- cyclic carboxylic dimers 37
- N,N-dimethyl carbamic acid dimer 36f.
- formic acid dimers (FAD) 36f., 42ff.
- properties of malonaldehyde 5ff., 27f., 138, 279
- properties of tropolon 13ff., 138
tunneling, correction 148
tunneling, enhanced 1314
tunneling, environmentally coupled 1314
tunneling, incoherent 138ff., 649ff., 675
tunneling, multidimensional (MT) 834f., 842, 869, 1293f., 1342ff.
tunneling, nonthermally activated quantum 770
tunneling, proton 3ff.
  - dynamics 904ff., 908ff.
  - fundamental, overtone, combination, hot band vibrational transitions 5
  - rotation-tunneling transitions 14, 617, 634
  - slow 671f.
  - splitting 9f., 15ff., 21, 23ff., 27f., 35, 37, 44, 48f., 254, 264, 267, 279, 283, 327, 652, 679, 813ff., 909, 914, 920, 926f.
  - damping 15ff., 21
  - multidimensional tunneling splitting 9
  - zero-point (ZP) 5, 12ff., 19, 21, 24, 27f., 48, 79
tunneling, vibrationally enhanced (VET) 1195
TURBOMOLE 288ff.
turkey ovalbumin third domain protein 1364
turn-over rates 1243
two-mode linear model 274
two-dimension lattice gas behavior 761
two-state model 795
tyrosine (Tyr, Y) 538ff., 553f., 1015, 1118, 1126f., 1143ff., 1151, 1352ff.
- oxidation 492, 496, 543
TZ2P calculation 249

u

Ubbelohde effect 277f.
ultrafast optical Kerr effects spectroscopy 235
ultrafast transient lens measurements (UTL) 239f.
ultrasonic loss 789
unpaired electrons 107, 109
α,β-, β,γ-unsaturated 3-oxo-steroids 1125
uridine 3′-2,2,2-trichloroethylphosphte 1087
uridyl esters 986
UV photoemission (UPS) 776ff.
UV pulse 354
UV spectroscopy 14f., 38f., (80)
UV/vis 443, 447ff., 689, 1112

v

valence bond (VB) 308f., 343, 583, 587, 595
- empirical (MS-EVB) 724, 1177, 1264
- model, four-state 480ff.
valence bond, empirical (EVB) 1171ff., 1185ff., 1195
valence bond, multistate 1264
valine (Val) 1026, 1029, 1156, 1218, 1222
vanadium 761, 788, 794, 798f., 800ff., 805ff., 814, 821ff.
van-der-Waals forces 751, 778
van-der-Waals potential function 1175
van-der-Waals radii 1520
VASP see Vienna ab initio simulation program
variational transition state theory (VTST) 833ff., 858, 869, 1303, 1342ff.
- ensemble-averaged VTST with static secondary zone (EA-VTST-SSZ) 861, 865ff.
- ensemble-averaged VTST with equilibrium static secondary zone (EA-VTST-ESZ) 861
- equilibrium solvation path (ESP-VTST) 861, 864f., 867f.
- nonequilibrium solvation path (NESP-VTST) 861, 864, 867f.
VCI see virtual configuration interaction
vehicle mechanism 717, 726
vehicular diffusion 718
ventricular myocyte cells 1508
vibration 8, 137, 554
- aromatic ring 41
- bending 21, 64, 71, 140f., 289ff., 371f., 688, 845, 878, 1370
  - SiOH 695
- cyclic 281
- C=O 1370, 1407ff.
- coherently excited 359, 371
- coupling 1216, 1232
  - stretch-stretch 1289
vibration, H atom 787
- in-plane 8, 364, 372, 609
- librational 233, 514, 616, 620
- out-of-plane 8, 19, 21, 24, 27, 469, 609
vibration, promoting 1209, 1213ff., 1217ff., 1230ff., 1347, 1209, 1213ff., 1217ff., 1230ff., 1314
- rotation-contortion-vibration electronic states 4
- skeletal 17, 24ff., 362, 472
- stretching 61ff., 66ff., 94, 98, 140f., 144ff., 156, 460, 610, 849f., 878, 1246
  - antisymmetric 1409
    - COO^- 1409

- asymmetric   277, 284, 289f., 609, 917
- 3-atom N-H O   536
- C–C   367
- C=C   1403
- CD   565, 1257
- CH   12f., 19f., 565, 854f., 859, 888, 918, 1180
- C–N   1370
- C=O   45ff., 284, 363f., 450, 453, 462, 471ff., 555, 918, 1395, 1403
- H–H   755, 845f.
- H–Cl   854
- M–H   608
- NH   248, 254, 258, 267, 471, 555, 720, 1355, 1370
- OD   453, 463ff., 466ff., 475
- OH   8, 12f., 17, 19f., 23ff., 27f., 41, 45f., 56, 71f., 74, 79, 88, 280ff., 289f., 364, 367, 459ff., 466ff., 474f., 539, 555, 688, 695, 743ff., 777f., 919f., 926
- O H O   284ff., 289f.
- O O   12, 21, 23, 27, 70, 74, 280ff.
- strong coupling   275ff., 296
- torsion   19
- twisting   16, 226
- wagging   16, 469
vibrational calculation   59f.
vibrational frequencies   13, 16, 19, 42, 45f., 61ff., 98, 329
vibrational loss spectroscopy   768
vibrational marker   449
vibrational modes   55f., 74, 360, 447ff., 460, 476, 812, 837, 844, 848, 854ff., 1195
vibrational polarization   464
vibrational relaxation   39, 223, 459, 475
vibrational-rotational energy   839
vibrational-rotational transitions   35, 37
vibrational self-consistent field (VSCF)   7f., 71f.
vibrational spectroscopy   12, 41, 45, 53ff., 61ff., 75, 287, 812, 1397, 1407
vibrational splitting   8
vibrational states   768
- anharmonicity   5, 7f., 10f., 20, 55, 68, 70, 460
- harmonicity   10f., 70f., 75
- quasiharmonicity   24f.
- specific doublets   14, 17
vibrational structures   3
vibrational transitions   447
vibrationally adiabatic zero-curvature approximation   841

vibronic
- absorption spectrum   14
- state   1179f.
- surface   490
- transitions   253ff.
Vienna ab initio simulation program   286f.
viral capsids   1378
virtual configuration interaction (VCI)   69, 71ff.
vitamin B_1 1419
vitamin B_{12} 1473
vitamin E   1014
volatile hydrides   761
VSCF   see vibrational self-consistent field
VTST, potential-of-mean-force (PMF-VTST)   861f., 864f.
- based on a single-reaction-coordinate (SRC-PMF-VTST)   861
VTST, quantum mechanical effects   835ff., 840ff.
VTST, separable equilibrium solvation (SES-VTST)   861f., 864f., 867f.
VTZ   72, 288ff.

w

Wasserstoff-Brücken-Bindung   751
water   226ff., 230ff., 237, 311, 314ff., 377, 386, 395, 399f., 406f., 427ff., 443f., 446, 689ff., 723ff., 955ff., 967, 1116ff., 1192, 1502ff.
- absorption   711
- activation   1459ff.
- and DNA bases   932ff.
- autodissociation   597
- biological   233, 240, 435, 1219
- bridges   453ff., 598
- channel, hard wired   550
- clusters   694ff., 718
- coordination   718
- cyclic hexamer   780
- diffusion   724
- dimer   689
- immobilization   1503
- interactions with proteins   232
  - dynamic equilibrium   233
- magnetization transfer experiments   1365f.
- migration   739
- Mn-bound   551
- nucleophile   977f., 993f.
- on metal   775, 778ff.
- ordering   1501
- ortho-/para   679
- solvent   962

- tetramer 689
- trimer 689, 697ff., 939f.
Watson-Crick base bairs 525ff.
wave function 36, 83ff., 92ff., 307, 309, 317, 329, 481ff., 639ff., 656, 679, 812f., 1174, 1257ff., 1313
- harmonic 275f.
- overlap 1261, 1263
- vibrational 493, 499, 1442
wave packet 223, 238, 240, 349
- dynamics 370
  - low-frequency 463, 474
- motion
  - ballistic 357ff., 363
  - coherent 371, 468f.
- oscillating 359
wave particle duality of matter 1256
well 139, 408, 813
- frequency 316
Wenzel, Kramers, Brillouin (WKB) approximation 34, 146, 653, 837, 845ff., 1486
Westheimer-effect 146
Westheimer and Melander (W-M) 304, 323f., 326, 565
Wigner tunneling correction 833, 844ff., 853

x

Xanthene 516ff.
Xanthobacter autotrophicus 1464
X-ray crystallography 202, 969, 1325, 1387, 1398, 1424, 1435, 1439, 1443, 1447
X-ray diffraction 65, 142, 273, 280, 288f., 435, 605
X-ray photoelectron spectroscopy (XPS) 776, 782
X-ray structure 236, 239, 248, 252, 598, 628, 730, 937, 1120, 1127, 1192, 1279, 1374, 1381ff.
m-xylene 689
D-xylulose 5-phosphate 1165

y

$YC_6H_4CH=CF_2$ 571
yttrium 805ff., 821

z

Zassenhaus expansion 1212
zeolites 223f., 240, 685ff., 691ff., 700ff.
- acidic catalysts 685ff.
  - chabasite H-CHA, CHA 686, 690ff., 696ff.
  - FER 692ff.
  - H-FAU, FAU 686, 690f., 700ff.
  - H-MFI (H-ZSM-5), MFI 685f., 692, 701ff.
  - H-SSZ-13 697ff.
  - MOR 690f., 704
  - quantum mechanical studies 690
  - sodalite (SOD) 691ff.
  - TON 692
  - Z-ZSM-5 692, 695, 599
- unit cell 692
zero-curvature tunneling (ZCT) 842f.
zero-point 60, 71
- energies 136, 143ff., 258, 305f., 309ff., 315ff., 321ff., 340, 511, 536f., 565, 599, 842, 845f., 878, 896, 907, 918, 1060, 1187, 1245ff., 1274, 1291, 1296, 1315, 1325 (ZPE)
- excited state 1318
- ground state 1318
- splitting 918, 939
- vibration 275
- vibrational (EXC) 1288
zinc 810, 936ff., 1067
- in enzymes 968
- Zn^{2+} 597, 968, 1188, 1217, 1341, 1397
ZP see zero-point
zirconium 761, 798ff., 805ff., 810f.
Zundel ion see $H_5O_2^+$ ion
Zundel mechanism 1506
Zwanzig approach 1210, 1213
Zymomonas mobilis 1422, 1428ff.

Further Titles of Interest

S.M. Roberts

**Catalysts for Fine Chemical Synthesis V 5 –
Regio and Stereo-controlled Oxidations and Reductions**

2007
ISBN 0-470-09022-7

M. Beller, C. Bolm (Eds)

**Transition Metals for Organic Synthesis.
Building Blocks and Fine Chemicals**

Building Blocks and Fine Chemicals

2004
ISBN 3-527-306137

G. Dyker (Ed.)

Handbook of C-H Transformations

Applications in Organic Synthesis

2005
ISBN 3-527-310746

H. Yamamoto, K. Oshima (Eds.)

Main Group Metals in Organic Synthesis

2004
ISBN 3-527-305084

G. A. Olah, Á. Molnár

Hydrocarbon Chemistry

2003
ISBN 0-471-417823